Quantum Mechanics for Mathematicians

Quantum Mechanics for Mathematicians

Leon A. Takhtajan

Graduate Studies
in Mathematics

Volume 95

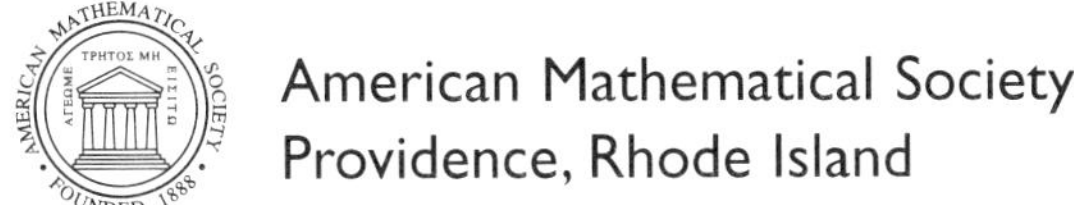

American Mathematical Society
Providence, Rhode Island

2000 *Mathematics Subject Classification.* Primary 81–01.

For additional information and updates on this book, visit
www.ams.org/bookpages/gsm-95

Library of Congress Cataloging-in-Publication Data

Takhtadzhian, L. A. (Leon Armenovich)
  Quantum mechanics for mathematicians / Leon A. Takhtajan.
    p. cm. — (Graduate studies in mathematics ; v. 95)
  Includes bibliographical references and index.
  ISBN 978-0-8218-4630-8 (alk. paper)
  1. Quantum theory.   2. Mathematical physics.   I. Title.

QC174.12.T343   2008
530.12—dc22   2008013072

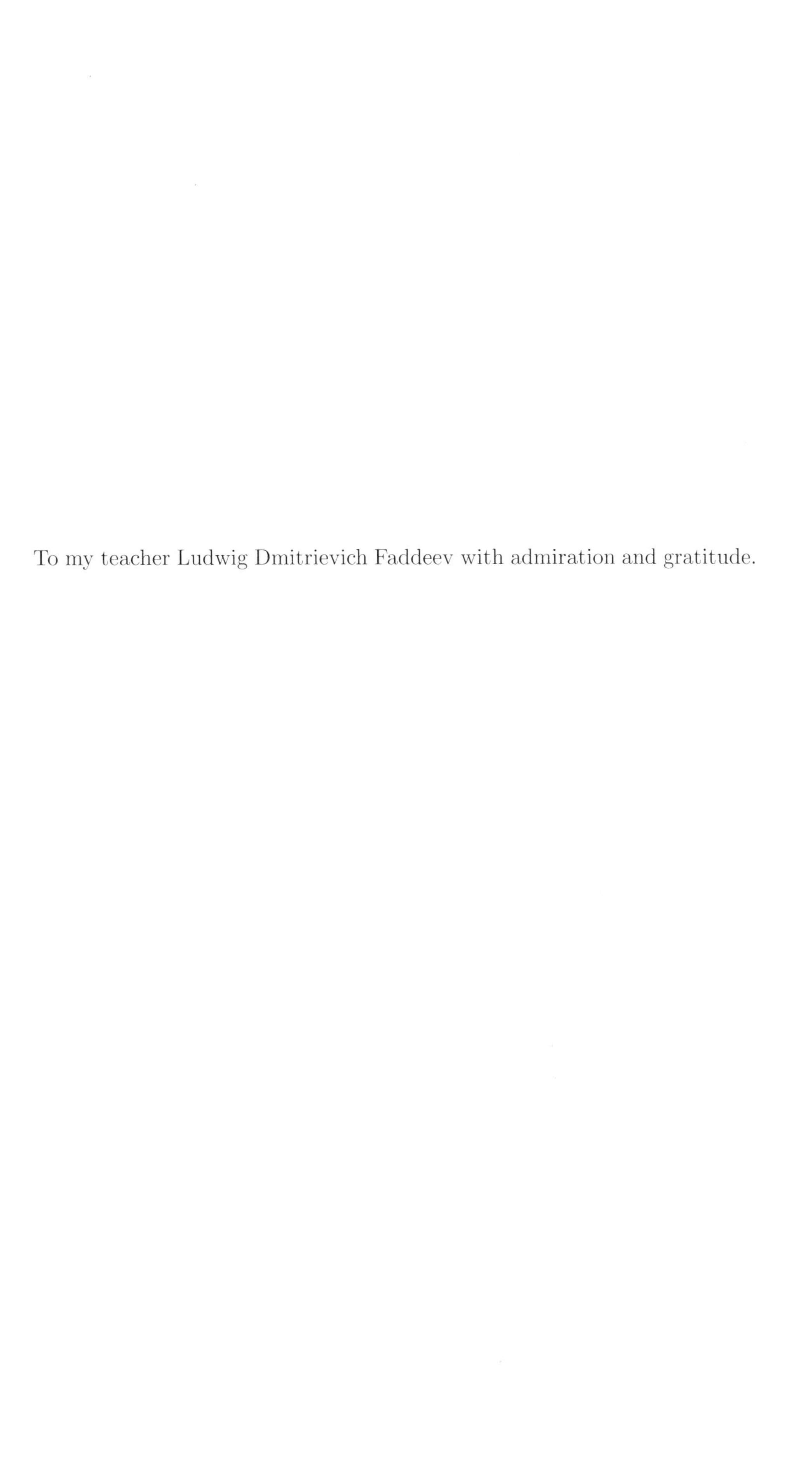

To my teacher Ludwig Dmitrievich Faddeev with admiration and gratitude.

Contents

Preface

This book is based on graduate courses taught by the author over the last fourteen years in the mathematics department of Stony Brook University. The goal of these courses was to introduce second year graduate students with no prior knowledge of physics to the basic concepts and methods of quantum mechanics. For the last 50 years quantum physics has been a driving force behind many dramatic achievements in mathematics, similar to the role played by classical physics in the seventeenth to nineteenth centuries. Classical physics, especially classical mechanics, was an integral part of mathematical education up to the early twentieth century, with lecture courses given by Hilbert and Poincaré. Surprisingly, quantum physics, especially quantum mechanics, with its intrinsic beauty and connections with many branches of mathematics, has never been a part of a graduate mathematics curriculum. This course was developed to partially fill this gap and to make quantum mechanics accessible to graduate students and research mathematicians.

L.D. Faddeev was the first to develop a course in quantum mechanics for undergraduate students specializing in mathematics. From 1968 to 1973 he regularly lectured in the mathematics department of St. Petersburg State University in St. Petersburg, Russia[1], and the author enjoyed the opportunity to take his course. The notes for this book emerged from an attempt to create a similar course for graduate students, which uses more sophisticated mathematics and covers a larger variety of topics, including the Feynman path integral approach to quantum mechanics.

---

[1] At that time in Leningrad, Soviet Union.

There are many excellent physics textbooks on quantum mechanics, starting with the classic texts by P.A.M. Dirac [**Dir47**], L.D. Landau and E.M. Lifshitz [**LL58**], and V.A. Fock [**Foc78**], to the encyclopedic treatise by A. Messiah [**Mes99**], the recent popular textbook by J.J. Sakurai [**Sak94**], and many others. From a mathematics perspective, there are classic monographs by J. von Neumann [**vN96**] and by H. Weyl [**Wey50**], as well as a more recent book by G.W. Mackey [**Mac04**], which deal with the basic mathematical formalism and logical foundations of the theory. There is also a monumental project [**DEF$^+$99**], created with the purpose of introducing graduate students and research mathematicians to the realm of quantum fields and strings, both from a mathematics and a physics perspective. Though it contains a very detailed exposition of classical mechanics, classical field theory, and supersymmetry, oriented at the mathematical audience, quantum mechanics is discussed only briefly (with the exception of L.D. Faddeev's elegant introduction to quantum mechanics in [**Fad99**]). Excellent lecture notes for undergraduate students by L.D. Faddeev and O.A. Yakubovskiĭ [**FY80**] seems to be the only book on quantum mechanics completely accessible to mathematicians[2]. Recent books by S.J. Gustafson and I.M. Sigal [**GS03**] and by F. Strocchi [**Str05**] are also oriented at mathematicians. The latter is a short introductory course, while the former is more an intermediate level monograph on quantum theory rather than a textbook on quantum mechanics. There are also many specialized books on different parts of quantum mechanics, like scattering theory, the Schrödinger operator, $\mathbb{C}^*$-algebras and foundations, etc.

The present book gives a comprehensive treatment of quantum mechanics from a mathematics perspective and covers such topics as mathematical foundations, quantization, the Schrödinger equation, the Feynman path integral and functional methods, and supersymmetry. It can be used as a one-year graduate course, or as two one-semester courses: the introductory course based on the material in Part 1, and a more advanced course based on Part 2. Part 1 of the book, which consists of Chapters 1-4, can be considered as an expanded version of [**FY80**]. It uses more advanced mathematics than [**FY80**], and contains rigorous proofs of all main results, including the celebrated Stone-von Neumann theorem. It should be accessible to a second-year graduate student. As in [**FY80**], we adopt the approach, which goes back to Dirac and was further developed by Faddeev, that classical mechanics and quantum mechanics are just two different realizations of the fundamental mathematical structure of a physical theory that uses the notions of observables, states, measurement, and the time evolution — dynamics. Part 2, which consists of Chapters 5-8, deals with functional methods in quantum

---

[2]The English translation will appear in the AMS "The Student Mathematical Library" series.

mechanics, and goes beyond the material in [**FY80**]. Exposition there is less detailed and requires certain mathematical sophistication.

Though our presentation freely uses all the necessary tools of modern mathematics, it follows the spirit and tradition of the classical texts and monographs mentioned above. In this sense it can be considered "neoclassical" (as compared with a more abstract approach in [**DF99a**]). Each chapter in the book concludes with a special *Notes and references* section, which provides references to the necessary mathematics background and physics sources. A courageous reader can actually learn the relevant mathematics by studying the main text and consulting these references, and with enough sophistication, could "translate" corresponding parts in physics textbooks into the mathematics language. For the physics students, the book presents an opportunity to become familiar with the mathematical foundations and methods of quantum mechanics on a "case by case" basis. It is worth mentioning that development of many mathematics disciplines has been stimulated by quantum mechanics.

There are several ways to study the material in this book. A casual reader can study the main text in a cursory manner, and ignore numerous remarks and problems, located at the end of the sections. This would be sufficient to obtain basic minimal knowledge of quantum mechanics. A determined reader is supposed to fill in the details of the computations in the main text (a pencil and paper are required), which is the only way to master the material, and to attempt to solve the basic problems[3]. Finally, a truly devoted reader should try to solve all the problems (probably consulting the corresponding references at the end of each section) and to follow up on the remarks, which may often be linked to other topics not covered in the main text.

The author would like to thank the students in his courses for their comments on the draft of the lecture notes. He is especially grateful to his colleagues Peter Kulish and Lee-Peng Teo for the careful reading of the manuscript. The work on the book was partially supported by the NSF grants DMS-0204628 and DMS-0705263. Any opinions, findings, and conclusions or recommendations expressed in this book are those of the author and do not necessarily reflect the views of the National Science Foundation.

---

[3]We leave it to the reader to decide which problems are basic and which are advanced.

Part 1

Foundations

Chapter 1

Classical Mechanics

We assume that the reader is familiar with the basic notions from the theory of smooth (that is, C^∞) manifolds and recall here the standard notation. Unless it is stated explicitly otherwise, all maps are assumed to be smooth and all functions are assumed to be smooth and real-valued. Local coordinates $\boldsymbol{q} = (q^1, \ldots, q^n)$ on a smooth n-dimensional manifold M at a point $q \in M$ are Cartesian coordinates on $\varphi(U) \subset \mathbb{R}^n$, where (U, φ) is a coordinate chart on M centered at $q \in U$. For $f : U \to \mathbb{R}^n$ we denote $(f \circ \varphi^{-1})(q^1, \ldots, q^n)$ by $f(\boldsymbol{q})$, and we let

$$\frac{\partial f}{\partial \boldsymbol{q}} = \left(\frac{\partial f}{\partial q^1}, \ldots, \frac{\partial f}{\partial q^n} \right)$$

stand for the gradient of a function f at a point $\boldsymbol{q} \in \mathbb{R}^n$ with Cartesian coordinates $(q^1, \ldots, q^n)$. We denote by

$$\mathcal{A}^\bullet(M) = \bigoplus_{k=0}^{n} \mathcal{A}^k(M)$$

the graded algebra of smooth differential forms on M with respect to the wedge product, and by d the de Rham differential — a graded derivation of $\mathcal{A}^\bullet(M)$ of degree 1 such that df is a differential of a function $f \in \mathcal{A}^0(M) = C^\infty(M)$. Let $\mathrm{Vect}(M)$ be the Lie algebra of smooth vector fields on M with the Lie bracket $[\,,\,]$, given by a commutator of vector fields. For $X \in \mathrm{Vect}(M)$ we denote by $\mathcal{L}_X$ and i_X, respectively, the Lie derivative along X and the inner product with X. The Lie derivative is a degree 0 derivation of $\mathcal{A}^\bullet(M)$ which commutes with d and satisfies $\mathcal{L}_X(f) = X(f)$ for $f \in \mathcal{A}^0(M)$, and the inner product is a degree -1 derivation of $\mathcal{A}^\bullet(M)$ satisfying $i_X(f) = 0$

and $i_X(df) = X(f)$ for $f \in \mathcal{A}^0(M)$. They satisfy Cartan formulas

$$\mathcal{L}_X = i_X \circ d + d \circ i_X = (d + i_X)^2,$$

$$i_{[X,Y]} = \mathcal{L}_X \circ i_Y - i_Y \circ \mathcal{L}_X.$$

For a smooth mapping of manifolds $f : M \to N$ we denote by $f_* : TM \to TN$ and $f^* : T^*N \to T^*M$, respectively, the induced mappings on tangent and cotangent bundles. Other notations, including those traditional for classical mechanics, will be introduced in the main text.

1. Lagrangian Mechanics

1.1. Generalized coordinates. Classical mechanics describes systems of finitely many interacting *particles*[1]. A system is called *closed* if its particles do not interact with the outside material bodies. The position of a system in space is specified by positions of its particles and defines a point in a smooth, finite-dimensional manifold M, the *configuration space* of a system. Coordinates on M are called *generalized coordinates* of a system, and the dimension $n = \dim M$ is called the number of *degrees of freedom*[2].

The *state* of a system at any instant of time is described by a point $q \in M$ and by a tangent vector $v \in T_q M$ at this point. The basic principle of classical mechanics is the *Newton-Laplace determinacy principle* which asserts that a state of a system at a given instant completely determines its motion at all times t (in the future and in the past). The motion is described by the *classical trajectory* — a path $\gamma(t)$ in the configuration space M. In generalized coordinates $\gamma(t) = (q^1(t), \ldots, q^n(t))$, and corresponding derivatives $\dot{q}^i = \dfrac{dq^i}{dt}$ are called *generalized velocities*. The Newton-Laplace principle is a fundamental experimental fact confirmed by our perception of everyday experiences. It implies that *generalized accelerations* $\ddot{q}^i = \dfrac{d^2 q^i}{dt^2}$ are uniquely defined by generalized coordinates q^i and generalized velocities $\dot{q}^i$, so that classical trajectories satisfy a system of second order ordinary differential equations, called *equations of motion*. In the next section we formulate the most general principle governing the motion of mechanical systems.

1.2. The principle of the least action. A *Lagrangian system* on a configuration space M is defined by a smooth, real-valued function L on $TM \times \mathbb{R}$ — the direct product of a tangent bundle TM of M and the time axis[3] — called the *Lagrangian function* (or simply, *Lagrangian*). The motion of a

---

[1] A particle is a material body whose dimensions may be neglected in describing its motion.

[2] Systems with infinitely many degrees of freedom are described by classical field theory.

[3] It follows from the Newton-Laplace principle that L could depend only on generalized coordinates and velocities, and on time.

Lagrangian system (M, L) is described by the *principle of the least action in the configuration space* (or *Hamilton's principle*), formulated as follows.

Let

$$P(M)_{q_0,t_0}^{q_1,t_1} = \{\gamma : [t_0, t_1] \to M; \ \gamma(t_0) = q_0, \ \gamma(t_1) = q_1\}$$

be the space of smooth parametrized paths in M connecting points q_0 and q_1. The path space $P(M) = P(M)_{q_0,t_0}^{q_1,t_1}$ is an infinite-dimensional real Fréchet manifold, and the tangent space $T_\gamma P(M)$ to $P(M)$ at $\gamma \in P(M)$ consists of all smooth vector fields along the path γ in M which vanish at the endpoints q_0 and q_1. A smooth path Γ in $P(M)$, passing through $\gamma \in P(M)$, is called a *variation with fixed ends* of the path $\gamma(t)$ in M. A variation Γ is a family $\gamma_\varepsilon(t) = \Gamma(t, \varepsilon)$ of paths in M given by a smooth map

$$\Gamma : [t_0, t_1] \times [-\varepsilon_0, \varepsilon_0] \to M$$

such that $\Gamma(t, 0) = \gamma(t)$ for $t_0 \leq t \leq t_1$ and $\Gamma(t_0, \varepsilon) = q_0, \Gamma(t_1, \varepsilon) = q_1$ for $-\varepsilon_0 \leq \varepsilon \leq \varepsilon_0$. The tangent vector

$$\delta\gamma = \left.\frac{\partial\Gamma}{\partial\varepsilon}\right|_{\varepsilon=0} \in T_\gamma P(M)$$

corresponding to a variation $\gamma_\varepsilon(t)$ is traditionally called an *infinitesimal variation*. Explicitly,

$$\delta\gamma(t) = \Gamma_*(\tfrac{\partial}{\partial\varepsilon})(t, 0) \in T_{\gamma(t)}M, \quad t_0 \leq t \leq t_1,$$

where $\frac{\partial}{\partial\varepsilon}$ is a tangent vector to the interval $[-\varepsilon_0, \varepsilon_0]$ at 0. Finally, a tangential lift of a path $\gamma : [t_0, t_1] \to M$ is the path $\gamma' : [t_0, t_1] \to TM$ defined by $\gamma'(t) = \gamma_*(\frac{\partial}{\partial t}) \in T_{\gamma(t)}M$, $t_0 \leq t \leq t_1$, where $\frac{\partial}{\partial t}$ is a tangent vector to $[t_0, t_1]$ at t. In other words, $\gamma'(t)$ is the velocity vector of a path $\gamma(t)$ at time t.

Definition. The *action functional* $S : P(M) \to \mathbb{R}$ of a Lagrangian system (M, L) is defined by

$$S(\gamma) = \int_{t_0}^{t_1} L(\gamma'(t), t)dt.$$

Principle of the Least Action (Hamilton's principle). A path $\gamma \in PM$ describes the motion of a Lagrangian system (M, L) between the position $q_0 \in M$ at time t_0 and the position $q_1 \in M$ at time t_1 if and only if it is a critical point of the action functional S,

$$\left.\frac{d}{d\varepsilon}\right|_{\varepsilon=0} S(\gamma_\varepsilon) = 0$$

for all variations $\gamma_\varepsilon(t)$ of $\gamma(t)$ with fixed ends.

The critical points of the action functional are called *extremals* and the principle of the least action states that a Lagrangian system (M, L)

moves along the extremals[4]. The extremals are characterized by equations of motion — a system of second order differential equations in local coordinates on TM. The equations of motion have the most elegant form for the following choice of local coordinates on TM.

Definition. Let (U, φ) be a coordinate chart on M with local coordinates $\boldsymbol{q} = (q^1, \ldots, q^n)$. Coordinates

$$(\boldsymbol{q}, \boldsymbol{v}) = (q^1, \ldots, q^n, v^1, \ldots, v^n)$$

on a chart TU on TM, where $\boldsymbol{v} = (v^1, \ldots, v^n)$ are coordinates in the fiber corresponding to the basis $\dfrac{\partial}{\partial q^1}, \ldots, \dfrac{\partial}{\partial q^n}$ for $T_q M$, are called *standard coordinates*.

Standard coordinates are Cartesian coordinates on $\varphi_*(TU) \subset T\mathbb{R}^n \simeq \mathbb{R}^n \times \mathbb{R}^n$ and have the property that for $(q, v) \in TU$ and $f \in C^\infty(U)$,

$$v(f) = \sum_{i=1}^n v^i \frac{\partial f}{\partial q^i} = \boldsymbol{v} \frac{\partial f}{\partial \boldsymbol{q}}.$$

Let (U, φ) and (U', φ') be coordinate charts on M with the transition functions $F = (F^1, \ldots, F^n) = \varphi' \circ \varphi^{-1} : \varphi(U \cap U') \to \varphi'(U \cap U')$, and let $(\boldsymbol{q}, \boldsymbol{v})$ and $(\boldsymbol{q}', \boldsymbol{v}')$, respectively, be the standard coordinates on TU and TU'. We have $\boldsymbol{q}' = F(\boldsymbol{q})$ and $\boldsymbol{v}' = F_*(\boldsymbol{q})\boldsymbol{v}$, where $F_*(\boldsymbol{q}) = \left\{ \dfrac{\partial F^i}{\partial q^j}(\boldsymbol{q}) \right\}_{i,j=1}^n$ is a matrix-valued function on $\varphi(U \cap U')$. Thus "vertical" coordinates $\boldsymbol{v} = (v^1, \ldots, v^n)$ in the fibers of $TM \to M$ transform like components of a tangent vector on M under the change of coordinates on M.

The tangential lift $\gamma'(t)$ of a path $\gamma(t)$ in M in standard coordinates on TU is $(\boldsymbol{q}(t), \dot{\boldsymbol{q}}(t)) = (q^1(t), \ldots, q^n(t), \dot{q}^1(t), \ldots, \dot{q}^n(t))$, where the dot stands for the time derivative, so that

$$L(\gamma'(t), t) = L(\boldsymbol{q}(t), \dot{\boldsymbol{q}}(t), t).$$

Following a centuries long tradition[5], we will usually denote standard coordinates by

$$(\boldsymbol{q}, \dot{\boldsymbol{q}}) = (q^1, \ldots, q^n, \dot{q}^1, \ldots, \dot{q}^n),$$

where the dot *does not* stand for the time derivative. Since we only consider paths in TM that are tangential lifts of paths in M, there will be no confusion[6].

---

[4]The principle of the least action does not state that an extremal connecting points q_0 and q_1 is a minimum of S, nor that such an extremal is unique. It also does not state that any two points can be connected by an extremal.

[5]Used in all texts on classical mechanics and theoretical physics.

[6]We reserve the notation $(\boldsymbol{q}(t), \boldsymbol{v}(t))$ for general paths in TM.

Theorem 1.1. *The equations of motion of a Lagrangian system (M, L) in standard coordinates on TM are given by the Euler-Lagrange equations*

$$\frac{\partial L}{\partial \boldsymbol{q}}(\boldsymbol{q}(t), \dot{\boldsymbol{q}}(t), t) - \frac{d}{dt}\left(\frac{\partial L}{\partial \dot{\boldsymbol{q}}}(\boldsymbol{q}(t), \dot{\boldsymbol{q}}(t), t)\right) = 0.$$

Proof. Suppose first that an extremal $\gamma(t)$ lies in a coordinate chart U of M. Then a simple computation in standard coordinates, using integration by parts, gives

$$\begin{aligned}
0 &= \left.\frac{d}{d\varepsilon}\right|_{\varepsilon=0} S(\gamma_\varepsilon) \\
&= \left.\frac{d}{d\varepsilon}\right|_{\varepsilon=0} \int_{t_0}^{t_1} L\left(\boldsymbol{q}(t,\varepsilon), \dot{\boldsymbol{q}}(t,\varepsilon), t\right) dt \\
&= \sum_{i=1}^{n} \int_{t_0}^{t_1} \left(\frac{\partial L}{\partial q^i}\delta q^i + \frac{\partial L}{\partial \dot{q}^i}\delta \dot{q}^i\right) dt \\
&= \sum_{i=1}^{n} \int_{t_0}^{t_1} \left(\frac{\partial L}{\partial q^i} - \frac{d}{dt}\frac{\partial L}{\partial \dot{q}^i}\right) \delta q^i dt + \sum_{i=1}^{n} \left.\frac{\partial L}{\partial \dot{q}^i}\delta q^i\right|_{t_0}^{t_1}.
\end{aligned}$$

The second sum in the last line vanishes due to the property $\delta q^i(t_0) = \delta q^i(t_1) = 0$, $i = 1, \ldots, n$. The first sum is zero for arbitrary smooth functions δq^i on the interval $[t_0, t_1]$ which vanish at the endpoints. This implies that for each term in the sum the integrand is identically zero,

$$\frac{\partial L}{\partial q^i}(\boldsymbol{q}(t), \dot{\boldsymbol{q}}(t), t) - \frac{d}{dt}\left(\frac{\partial L}{\partial \dot{q}^i}(\boldsymbol{q}(t), \dot{\boldsymbol{q}}(t), t)\right) = 0, \quad i = 1, \ldots, n.$$

Since the restriction of an extremal of the action functional S to a coordinate chart on M is again an extremal, each extremal in standard coordinates on TM satisfies Euler-Lagrange equations. $\qquad\square$

Remark. In calculus of variations, the directional derivative of a functional S with respect to a tangent vector $V \in T_\gamma P(M)$ — the *Gato derivative* — is defined by

$$\delta_V S = \left.\frac{d}{d\varepsilon}\right|_{\varepsilon=0} S(\gamma_\varepsilon),$$

where γ_ε is a path in $P(M)$ with a tangent vector V at $\gamma_0 = \gamma$. The result of the above computation (when γ lies in a coordinate chart $U \subset M$) can be written as

$$\begin{aligned}
\delta_V S &= \int_{t_0}^{t_1} \sum_{i=1}^{n} \left(\frac{\partial L}{\partial q^i} - \frac{d}{dt}\frac{\partial L}{\partial \dot{q}^i}\right)(\boldsymbol{q}(t), \dot{\boldsymbol{q}}(t), t) v^i(t) dt \\
&= \int_{t_0}^{t_1} \left(\frac{\partial L}{\partial \boldsymbol{q}} - \frac{d}{dt}\frac{\partial L}{\partial \dot{\boldsymbol{q}}}\right)(\boldsymbol{q}(t), \dot{\boldsymbol{q}}(t), t) \boldsymbol{v}(t) dt.
\end{aligned} \tag{1.1}$$

Here $V(t) = \sum_{i=1}^{n} v^i(t) \dfrac{\partial}{\partial q^i}$ is a vector field along the path γ in M. Formula (1.1) is called the formula for the *first variation of the action with fixed ends*. The principle of the least action is a statement that $\delta_V S(\gamma) = 0$ for all $V \in T_\gamma P(M)$.

Remark. It is also convenient to consider a space $\widehat{P(M)} = \{\gamma : [t_0, t_1] \to M\}$ of all smooth parametrized paths in M. The tangent space $T_\gamma \widehat{P(M)}$ to $\widehat{P(M)}$ at $\gamma \in \widehat{P(M)}$ is the space of all smooth vector fields along the path γ in M (no condition at the endpoints). The computation in the proof of Theorem 1.1 yields the following formula for the *first variation of the action with free ends*:

$$(1.2) \qquad \delta_V S = \int_{t_0}^{t_1} \left(\frac{\partial L}{\partial \boldsymbol{q}} - \frac{d}{dt} \frac{\partial L}{\partial \dot{\boldsymbol{q}}} \right) \boldsymbol{v} \, dt + \frac{\partial L}{\partial \dot{\boldsymbol{q}}} \boldsymbol{v} \Big|_{t_0}^{t_1}.$$

Problem 1.1. Show that the action functional is given by the evaluation of the 1-form Ldt on $TM \times \mathbb{R}$ over the 1-chain $\tilde{\gamma}$ on $TM \times \mathbb{R}$,

$$S(\gamma) = \int_{\tilde{\gamma}} Ldt,$$

where $\tilde{\gamma} = \{(\gamma'(t), t); t_0 \leq t \leq t_1\}$ and $Ldt \left(w, c\frac{\partial}{\partial t} \right) = cL(q, v)$, $w \in T_{(q,v)}TM$, $c \in \mathbb{R}$.

Problem 1.2. Let $f \in C^\infty(M)$. Show that Lagrangian systems (M, L) and $(M, L+df)$ (where df is a fibre-wise linear function on TM) have the same equations of motion.

Problem 1.3. Give examples of Lagrangian systems such that an extremal connecting two given points (i) is not a local minimum; (ii) is not unique; (iii) does not exist.

Problem 1.4. For γ an extremal of the action functional S, the *second variation* of S is defined by

$$\delta^2_{V_1 V_2} S = \frac{\partial^2}{\partial \varepsilon_1 \partial \varepsilon_2} \Big|_{\varepsilon_1 = \varepsilon_2 = 0} S(\gamma_{\varepsilon_1, \varepsilon_2}),$$

where $\gamma_{\varepsilon_1, \varepsilon_2}$ is a smooth two-parameter family of paths in M such that the paths $\gamma_{\varepsilon_1, 0}$ and $\gamma_{0, \varepsilon_2}$ in $P(M)$ at the point $\gamma_{0,0} = \gamma \in P(M)$ have tangent vectors V_1 and V_2, respectively. For a Lagrangian system (M, L) find the second variation of S and verify that for given V_1 and V_2 it does not depend on the choice of $\gamma_{\varepsilon_1, \varepsilon_2}$.

1.3. Examples of Lagrangian systems. To describe a mechanical phenomena it is necessary to choose a *frame of reference*. The properties of the *space-time* where the motion takes place depend on this choice. The space-time is characterized by the following postulates[7].

---

[7]Strictly speaking, these postulates are valid only in the non-relativistic limit of special relativity, when the speed of light in the vacuum is assumed to be infinite.

Newtonian Space-Time. The space is a three-dimensional affine Euclidean space E^3. A choice of the *origin* $0 \in E^3$ — a *reference point* — establishes the isomorphism $E^3 \simeq \mathbb{R}^3$, where the vector space $\mathbb{R}^3$ carries the Euclidean inner product and has a fixed orientation. The time is one-dimensional — a time axis $\mathbb{R}$ — and the space-time is a direct product $E^3 \times \mathbb{R}$. An *inertial* reference frame is a coordinate system with respect to the origin $0 \in E^3$, initial time t_0, and an orthonormal basis in $\mathbb{R}^3$. In an inertial frame the space is *homogeneous* and *isotropic* and the time is *homogeneous*. The laws of motion are invariant with respect to the transformations

$$\boldsymbol{r} \mapsto g \cdot \boldsymbol{r} + \boldsymbol{r}_0, \quad t \mapsto t + t_0,$$

where $\boldsymbol{r}, \boldsymbol{r}_0 \in \mathbb{R}^3$ and g is an orthogonal linear transformation in $\mathbb{R}^3$. The time in classical mechanics is *absolute*.

The Galilean group is the group of all affine transformations of $E^3 \times \mathbb{R}$ which preserve time intervals and which for every $t \in \mathbb{R}$ are isometries in E^3. Every Galilean transformation is a composition of rotation, space-time translation, and a transformation

$$\boldsymbol{r} \mapsto \boldsymbol{r} + \boldsymbol{v}t, \quad t \mapsto t,$$

where $\boldsymbol{v} \in \mathbb{R}^3$. Any two inertial frames are related by a Galilean transformation.

Galileo's Relativity Principle. The laws of motion are invariant with respect to the Galilean group.

These postulates impose restrictions on Lagrangians of mechanical systems. Thus it follows from the first postulate that the Lagrangian L of a closed system does not explicitly depend on time. Physical systems are described by special Lagrangians, in agreement with the experimental facts about the motion of material bodies.

Example 1.1 (Free particle). The configuration space for a free particle is $M = \mathbb{R}^3$, and it can be deduced from Galileo's relativity principle that the Lagrangian for a free particle is

$$L = \tfrac{1}{2}m\dot{\boldsymbol{r}}^2.$$

Here $m > 0$[8] is the mass of a particle and $\dot{\boldsymbol{r}}^2 = |\dot{\boldsymbol{r}}|^2$ is the length square of the velocity vector $\dot{\boldsymbol{r}} \in T_{\boldsymbol{r}}\mathbb{R}^3 \simeq \mathbb{R}^3$. Euler-Lagrange equations give *Newton's law of inertia,*

$$\ddot{\boldsymbol{r}} = 0.$$

---

[8]Otherwise the action functional is not bounded from below.

Example 1.2 (Interacting particles). A closed system of N interacting particles in $\mathbb{R}^3$ with masses $m_1, \ldots, m_N$ is described by a configuration space

$$M = \mathbb{R}^{3N} = \underbrace{\mathbb{R}^3 \times \cdots \times \mathbb{R}^3}_{N}$$

with a position vector $\boldsymbol{r} = (\boldsymbol{r}_1, \ldots, \boldsymbol{r}_N)$, where $\boldsymbol{r}_a \in \mathbb{R}^3$ is the position vector of the a-th particle, $a = 1, \ldots, N$. It is found that the Lagrangian is given by

$$L = \sum_{a=1}^{N} \tfrac{1}{2} m_a \dot{\boldsymbol{r}}_a^2 - V(\boldsymbol{r}) = T - V,$$

where

$$T = \sum_{a=1}^{N} \tfrac{1}{2} m_a \dot{\boldsymbol{r}}_a^2$$

is called *kinetic energy* of a system and $V(\boldsymbol{r})$ is *potential energy*. The Euler-Lagrange equations give *Newton's equations*

$$m_a \ddot{\boldsymbol{r}}_a = \boldsymbol{F}_a,$$

where

$$\boldsymbol{F}_a = -\frac{\partial V}{\partial \boldsymbol{r}_a}$$

is the *force* on the a-th particle, $a = 1, \ldots, N$. Forces of this form are called *conservative*. It follows from homogeneity of space that potential energy $V(\boldsymbol{r})$ of a closed system of N interacting particles with conservative forces depends only on relative positions of the particles, which leads to the equation

$$\sum_{a=1}^{N} \boldsymbol{F}_a = 0.$$

In particular, for a closed system of two particles $\boldsymbol{F}_1 + \boldsymbol{F}_2 = 0$, which is the equality of action and reaction forces, also called *Newton's third law*.

The potential energy of a closed system with only pair-wise interaction between the particles has the form

$$V(\boldsymbol{r}) = \sum_{1 \leq a < b \leq N} V_{ab}(\boldsymbol{r}_a - \boldsymbol{r}_b).$$

It follows from the isotropy of space that $V(\boldsymbol{r})$ depends only on relative distances between the particles, so that the Lagrangian of a closed system of N particles with pair-wise interaction has the form

$$L = \sum_{a=1}^{N} \tfrac{1}{2} m_a \dot{\boldsymbol{r}}_a^2 - \sum_{1 \leq a < b \leq N} V_{ab}(|\boldsymbol{r}_a - \boldsymbol{r}_b|).$$

If the potential energy $V(\mathbf{r})$ is a homogeneous function of degree ρ, $V(\lambda\mathbf{r}) = \lambda^{\rho}V(\mathbf{r})$, then the average values $\overline{T}$ and $\overline{V}$ of kinetic energy and potential energy over a closed trajectory are related by the *virial theorem*

$$(1.3) \qquad\qquad 2\overline{T} = \rho\overline{V}.$$

Indeed, let $\mathbf{r}(t)$ be a periodic trajectory with period $\tau > 0$, i.e., $\mathbf{r}(0) = \mathbf{r}(\tau)$, $\dot{\mathbf{r}}(0) = \dot{\mathbf{r}}(\tau)$. Using integration by parts, Newton's equations, and Euler's homogeneous function theorem, we get

$$2\overline{T} = \frac{1}{\tau}\int_0^\tau \sum_{a=1}^N m_a \dot{\mathbf{r}}_a^2 dt = -\frac{1}{\tau}\int_0^\tau \sum_{a=1}^N m_a \mathbf{r}_a \ddot{\mathbf{r}}_a dt$$

$$= \frac{1}{\tau}\int_0^\tau \sum_{a=1}^N \mathbf{r}_a \frac{\partial V}{\partial \mathbf{r}_a} dt = \rho\overline{V}.$$

Example 1.3 (Universal gravitation). According to *Newton's law of gravitation*, the potential energy of the gravitational force between two particles with masses m_a and m_b is

$$V(\mathbf{r}_a - \mathbf{r}_b) = -G\frac{m_a m_b}{|\mathbf{r}_a - \mathbf{r}_b|},$$

where G is the gravitational constant. The configuration space of N particles with gravitational interaction is

$$M = \{(\mathbf{r}_1, \dots, \mathbf{r}_N) \in \mathbb{R}^{3N} : \mathbf{r}_a \neq \mathbf{r}_b \text{ for } a \neq b,\, a, b = 1, \dots, N\}.$$

Example 1.4 (Particle in an external potential field). Here $M = \mathbb{R}^3$ and

$$L = \tfrac{1}{2}m\dot{\mathbf{r}}^2 - V(\mathbf{r}, t),$$

where potential energy can explicitly depend on time. Equations of motion are Newton's equations

$$m\ddot{\mathbf{r}} = \mathbf{F} = -\frac{\partial V}{\partial \mathbf{r}}.$$

If $V = V(|\mathbf{r}|)$ is a function only of the distance $|\mathbf{r}|$, the potential field is called *central*.

Example 1.5 (Charged particle in electromagnetic field[9]). Consider a particle of charge e and mass m in $\mathbb{R}^3$ moving in a time-independent electromagnetic field with scalar and vector potentials $\varphi(\mathbf{r})$ and $\mathbf{A}(\mathbf{r}) = (A_1(\mathbf{r}), A_2(\mathbf{r}), A_3(\mathbf{r}))$. The Lagrangian has the form

$$L = \frac{m\dot{\mathbf{r}}^2}{2} + e\left(\frac{\dot{\mathbf{r}}\,\mathbf{A}}{c} - \varphi\right),$$

---

[9]This is a non-relativistic limit of an example in classical electrodynamics.

where c is the speed of light. The Euler-Lagrange equations give Newton's equations with the *Lorentz force*,

$$m\ddot{\boldsymbol{r}} = e\left(\boldsymbol{E} + \frac{\dot{\boldsymbol{r}}}{c} \times \boldsymbol{B}\right),$$

where $\times$ is the cross-product of vectors in $\mathbb{R}^3$, and

$$\boldsymbol{E} = -\frac{\partial\varphi}{\partial\boldsymbol{r}} \quad \text{and} \quad \boldsymbol{B} = \operatorname{curl}\boldsymbol{A}$$

are electric and magnetic[10] fields, respectively.

Example 1.6 (Small oscillations). Consider a particle of mass m with n degrees of freedom moving in a potential field $V(\boldsymbol{q})$, and suppose that potential energy U has a minimum at $\boldsymbol{q} = 0$. Expanding $V(\boldsymbol{q})$ in Taylor series around 0 and keeping only quadratic terms, one obtains a Lagrangian system which describes small oscillations from equilibrium. Explicitly,

$$L = \tfrac{1}{2}m\dot{\boldsymbol{q}}^2 - V_0(\boldsymbol{q}),$$

where V_0 is a positive-definite quadratic form on $\mathbb{R}^n$ given by

$$V_0(\boldsymbol{q}) = \tfrac{1}{2}\sum_{i,j=1}^{n}\frac{\partial^2 V}{\partial q^i \partial q^j}(0)q^i q^j.$$

Since every quadratic form can be diagonalized by an orthogonal transformation, we can assume from the very beginning that coordinates $\boldsymbol{q} = (q^1, \ldots, q^n)$ are chosen so that $V_0(\boldsymbol{q})$ is diagonal and

$$(1.4) \qquad L = \tfrac{1}{2}m\big(\dot{\boldsymbol{q}}^2 - \sum_{i=1}^{n}\omega_i^2(q^i)^2\big),$$

where $\omega_1, \ldots, \omega_n > 0$. Such coordinates $\boldsymbol{q}$ are called *normal coordinates*. In normal coordinates Euler-Lagrange equations take the form

$$\ddot{q}^i + \omega_i^2 q^i = 0, \quad i = 1, \ldots, n,$$

and describe n decoupled (i.e., non-interacting) *harmonic oscillators* with *frequencies* $\omega_1, \ldots, \omega_n$.

Example 1.7 (Free particle on a Riemannian manifold). Let (M, ds^2) be a Riemannian manifold with the Riemannian metric ds^2. In local coordinates $x^1, \ldots, x^n$ on M,

$$ds^2 = g_{\mu\nu}(x)dx^\mu dx^\nu,$$

where following tradition we assume the summation over repeated indices. The Lagrangian of a free particle on M is

$$L(v) = \tfrac{1}{2}\langle v, v\rangle = \tfrac{1}{2}\|v\|^2, \ v \in TM,$$

---

[10]Notation $\boldsymbol{B} = \operatorname{rot}\boldsymbol{A}$ is also used.

where $\langle\,,\,\rangle$ stands for the inner product in fibers of TM given by the Riemannian metric. The corresponding functional

$$S(\gamma) = \tfrac{1}{2}\int_{t_0}^{t_1}\|\gamma'(t)\|^2 dt = \tfrac{1}{2}\int_{t_0}^{t_1}g_{\mu\nu}(x)\dot{x}^\mu\dot{x}^\nu dt$$

is called the action functional in Riemannian geometry. The Euler-Lagrange equations are

$$g_{\mu\nu}\ddot{x}^\mu + \frac{\partial g_{\mu\nu}}{\partial x^\lambda}\dot{x}^\mu\dot{x}^\lambda = \frac{1}{2}\frac{\partial g_{\mu\lambda}}{\partial x^\nu}\dot{x}^\mu\dot{x}^\lambda,$$

and after multiplying by the inverse metric tensor $g^{\sigma\nu}$ and summation over ν they take the form

$$\ddot{x}^\sigma + \Gamma^\sigma_{\mu\nu}\dot{x}^\mu\dot{x}^\nu = 0, \quad \sigma = 1,\dots,n,$$

where

$$\Gamma^\sigma_{\mu\nu} = \frac{1}{2}g^{\sigma\lambda}\left(\frac{\partial g_{\mu\lambda}}{\partial x^\nu} + \frac{\partial g_{\nu\lambda}}{\partial x^\mu} - \frac{\partial g_{\mu\nu}}{\partial x^\lambda}\right)$$

are Christoffel's symbols. The Euler-Lagrange equations of a free particle moving on a Riemannian manifold are geodesic equations.

Let ∇ be the Levi-Civita connection — the metric connection in the tangent bundle TM — and let ∇_ξ be a covariant derivative with respect to the vector field $\xi \in \text{Vect}(M)$. Explicitly,

$$(\nabla_\xi\eta)^\mu = \left(\frac{\partial\eta^\mu}{\partial x^\nu} + \Gamma^\mu_{\nu\lambda}\eta^\lambda\right)\xi^\nu, \quad \text{where} \quad \xi = \xi^\mu(x)\frac{\partial}{\partial x^\mu}, \ \eta = \eta^\mu(x)\frac{\partial}{\partial x^\mu}.$$

For a path $\gamma(t) = (x^\mu(t))$ denote by $\nabla_{\dot\gamma}$ a covariant derivative along γ,

$$(\nabla_{\dot\gamma}\eta)^\mu(t) = \frac{d\eta^\mu(t)}{dt} + \Gamma^\mu_{\nu\lambda}(\gamma(t))\dot{x}^\nu(t)\eta^\lambda(t), \quad \text{where} \quad \eta = \eta^\mu(t)\frac{\partial}{\partial x^\mu}$$

is a vector field along γ. Formula (1.1) can now be written in an invariant form

$$\delta S = -\int_{t_0}^{t_1}\langle\nabla_{\dot\gamma}\dot\gamma, \delta\gamma\rangle dt,$$

which is known as the formula for the first variation of the action in Riemannian geometry.

Example 1.8 (The rigid body). The configuration space of a rigid body in $\mathbb{R}^3$ with a fixed point is a Lie group $G = \text{SO}(3)$ of orientation preserving orthogonal linear transformations in $\mathbb{R}^3$. Every left-invariant Riemannian metric $\langle\,,\,\rangle$ on G defines a Lagrangian $L : TG \to \mathbb{R}$ by

$$L(v) = \tfrac{1}{2}\langle v, v\rangle, \quad v \in TG.$$

According to the previous example, equations of motion of a rigid body are geodesic equations on G with respect to the Riemannian metric $\langle\,,\,\rangle$. Let $\mathfrak{g} = \mathfrak{so}(3)$ be the Lie algebra of G. A velocity vector $\dot{g} \in T_gG$ defines the *angular velocity of the body* by $\Omega = (L_{g^{-1}})_*\dot{g} \in \mathfrak{g}$, where $L_g : G \to G$ are left translations on G. In terms of angular velocity, the Lagrangian takes the form

$$L = \tfrac{1}{2}\langle \Omega, \Omega \rangle_e,$$

where $\langle\ ,\ \rangle_e$ is an inner product on $\mathfrak{g} = T_eG$ given by the Riemannian metric $\langle\ ,\ \rangle$. The Lie algebra $\mathfrak{g}$ — the Lie algebra of 3×3 skew-symmetric matrices — has the invariant inner product $\langle u, v\rangle_0 = -\tfrac{1}{2}\operatorname{Tr} uv$ (the Killing form), so that $\langle \Omega, \Omega \rangle_e = \langle \boldsymbol{A}\cdot\Omega, \Omega\rangle_0$ for some symmetric linear operator $\boldsymbol{A} : \mathfrak{g} \to \mathfrak{g}$ which is positive-definite with respect to the Killing form. Such a linear operator $\boldsymbol{A}$ is called the *inertia tensor* of the body. The *principal axes of inertia* of the body are orthonormal eigenvectors e_1, e_2, e_3 of $\boldsymbol{A}$; corresponding eigenvalues I_1, I_2, I_3 are called the *principal moments of inertia*. Setting $\Omega = \Omega_1 e_1 + \Omega_2 e_2 + \Omega_3 e_3{}^{11}$, we get

$$L = \tfrac{1}{2}(I_1\Omega_1^2 + I_2\Omega_2^2 + I_3\Omega_3^2).$$

In this parametrization, the Euler-Lagrange equations become *Euler's equations*

$$I_1\dot{\Omega}_1 = (I_2 - I_3)\Omega_2\Omega_3,$$

$$I_2\dot{\Omega}_2 = (I_3 - I_1)\Omega_1\Omega_3,$$

$$I_3\dot{\Omega}_3 = (I_1 - I_2)\Omega_1\Omega_2.$$

Euler's equations describe the rotation of a free rigid body around a fixed point. In the system of coordinates with axes which are the principal axes of inertia, principal moments of inertia of the body are I_1, I_2, I_3.

Problem 1.5. Determine the motion of a charged particle in a constant uniform magnetic field. Show that if the initial velocity $v_3 = 0$ in the z-axis (taken in the direction of the field, $\boldsymbol{B} = (0, 0, B)$), the trajectories are circles of radii $r = \dfrac{cmv_t}{eB}$ in a plane perpendicular to the field (the xy-plane), where $v_t = \sqrt{v_1^2 + v_2^2}$ is the initial velocity in the xy-plane. The centers (x_0, y_0) of circles are given by

$$x_0 = \frac{cmv_1}{eB} + x, \quad y_0 = -\frac{cmv_2}{eB} + y,$$

where (x, y) are points on a circle of radius r.

Problem 1.6. Show that the Euler-Lagrange equations for the Lagrangian $L(v) = \|v\|$, $v \in TM$, coincide with the geodesic equations written with respect to a constant multiple of the natural parameter.

Problem 1.7. Prove that for a particle in a potential field, discussed in Example 1.4, the second variation of the action functional, defined in Problem 1.4, is given by

$$\delta^2 S = \int_{t_0}^{t_1} \mathcal{J}(\delta_1\boldsymbol{r})\delta_2\boldsymbol{r}dt,$$

---

[11]This establishes the Lie algebra isomorphism $\mathfrak{g} \simeq \mathbb{R}^3$, where the Lie bracket in $\mathbb{R}^3$ is given by the cross-product.

where $\delta_1 \boldsymbol{r}, \delta_2 \boldsymbol{r} \in T_\gamma P\mathbb{R}^3$, $\gamma = \boldsymbol{r}(t)$ is the classical trajectory, $\mathcal{J} = -m\dfrac{d^2}{dt^2}I - \dfrac{\partial^2 V}{\partial \boldsymbol{r}^2}(t)$,

I is the 3×3 identity matrix, and $\dfrac{\partial^2 V}{\partial \boldsymbol{r}^2}(t) = \left\{ \dfrac{\partial^2 V}{\partial \boldsymbol{r}_a \partial \boldsymbol{r}_b}(\boldsymbol{r}(t)) \right\}_{a,b=1}^{3}$. A second-order linear differential operator $\mathcal{J}$, acting on vector fields along γ, is called the *Jacobi operator*.

Problem 1.8. Find normal coordinates and frequencies for the Lagrangian system considered in Example 1.6 with $V_0(\boldsymbol{q}) = \frac{1}{2}a^2 \sum_{i=1}^{n}(q^{i+1}-q^i)^2$, where $q^{n+1} = q^1$.

Problem 1.9. Prove that the second variation of the action functional in Riemannian geometry is given by

$$\delta^2 S = \int_{t_0}^{t_1} \langle \mathcal{J}(\delta_1\gamma), \delta_2\gamma \rangle dt.$$

Here $\delta_1\gamma, \delta_2\gamma \in T_\gamma PM$, $\mathcal{J} = -\nabla_{\dot\gamma}^2 - R(\dot\gamma, \cdot)\dot\gamma$ is the Jacobi operator, and R is a curvature operator — a fibre-wise linear mapping $R : TM \otimes TM \to \mathrm{End}(TM)$ of vector bundles, defined by $R(\xi, \eta) = \nabla_\eta \nabla_\xi - \nabla_\xi \nabla_\eta + \nabla_{[\xi,\eta]} : TM \to TM$, where $\xi, \eta \in \mathrm{Vect}(M)$.

Problem 1.10. Choosing the principal axes of inertia as a basis in $\mathbb{R}^3$, show that the Lie algebra isomorphism $\mathfrak{g} \simeq \mathbb{R}^3$ is given by $\mathfrak{g} \ni \begin{pmatrix} 0 & -\Omega_3 & \Omega_2 \\ \Omega_3 & 0 & -\Omega_1 \\ -\Omega_2 & \Omega_1 & 0 \end{pmatrix} \mapsto (\Omega_1, \Omega_2, \Omega_3) \in \mathbb{R}^3$.

Problem 1.11. Show that for every symmetric $\boldsymbol{A} \in \mathrm{End}\,\mathfrak{g}$ there exists a symmetric 3×3 matrix A such that $\boldsymbol{A} \cdot \Omega = A\Omega + \Omega A$, and find A for diagonal $\boldsymbol{A}$.

Problem 1.12. Derive Euler's equations for a rigid body. (*Hint:* Use that $L = -\frac{1}{2}\mathrm{Tr}\, A\Omega^2$, where $\Omega = g^{-1}\dot{g}$ and $\delta\Omega = -g^{-1}\delta g\,\Omega + g^{-1}\delta\dot{g}$, and obtain the Euler-Lagrange equations in the matrix form $A\dot\Omega + \dot\Omega A = A\Omega^2 - \Omega^2 A$.)

1.4. Symmetries and Noether's theorem. To describe the motion of a mechanical system one needs to solve the corresponding Euler-Lagrange equations — a system of second order ordinary differential equations for the generalized coordinates. This could be a very difficult problem. Therefore of particular interest are those functions of generalized coordinates and velocities which remain constant during the motion.

Definition. A smooth function $I : TM \to \mathbb{R}$ is called the *integral of motion* (*first integral*, or *conservation law*) for a Lagrangian system (M, L) if

$$\frac{d}{dt}I(\gamma'(t)) = 0$$

for all extremals γ of the action functional.

Definition. The *energy* of a Lagrangian system (M, L) is a function E on $TM \times \mathbb{R}$ defined in standard coordinates on TM by

$$E(\boldsymbol{q}, \dot{\boldsymbol{q}}, t) = \sum_{i=1}^{n} \dot{q}^i \frac{\partial L}{\partial \dot{q}^i}(\boldsymbol{q}, \dot{\boldsymbol{q}}, t) - L(\boldsymbol{q}, \dot{\boldsymbol{q}}, t).$$

Lemma 1.1. *The energy* $E = \dot{\boldsymbol{q}}\,\dfrac{\partial L}{\partial \dot{\boldsymbol{q}}} - L$ *is a well-defined function on* $TM \times \mathbb{R}$.

Proof. Let (U, φ) and (U', φ') be coordinate charts on M with the transition functions $F = (F^1, \ldots, F^n) = \varphi' \circ \varphi^{-1} : \varphi(U \cap U') \to \varphi'(U \cap U')$. Corresponding standard coordinates $(\boldsymbol{q}, \dot{\boldsymbol{q}})$ and $(\boldsymbol{q}', \dot{\boldsymbol{q}}')$ are related by $\boldsymbol{q}' = F(\boldsymbol{q})$ and $\dot{\boldsymbol{q}}' = F_*(\boldsymbol{q})\dot{\boldsymbol{q}}$ (see Section 1.2). We have $d\boldsymbol{q}' = F_*(\boldsymbol{q})d\boldsymbol{q}$ and $d\dot{\boldsymbol{q}}' = G(\boldsymbol{q}, \dot{\boldsymbol{q}})d\boldsymbol{q} + F_*(\boldsymbol{q})d\dot{\boldsymbol{q}}$ (for some matrix-valued function $G(\boldsymbol{q}, \dot{\boldsymbol{q}})$), so that

$$
\begin{aligned}
dL &= \frac{\partial L}{\partial \boldsymbol{q}'}d\boldsymbol{q}' + \frac{\partial L}{\partial \dot{\boldsymbol{q}}'}d\dot{\boldsymbol{q}}' + \frac{\partial L}{\partial t}dt \\
&= \left(\frac{\partial L}{\partial \boldsymbol{q}'}F_*(\boldsymbol{q}) + \frac{\partial L}{\partial \dot{\boldsymbol{q}}'}G(\boldsymbol{q}, \dot{\boldsymbol{q}}) \right) d\boldsymbol{q} + \frac{\partial L}{\partial \dot{\boldsymbol{q}}'}F_*(\boldsymbol{q})d\dot{\boldsymbol{q}} + \frac{\partial L}{\partial t}dt \\
&= \frac{\partial L}{\partial \boldsymbol{q}}d\boldsymbol{q} + \frac{\partial L}{\partial \dot{\boldsymbol{q}}}d\dot{\boldsymbol{q}} + \frac{\partial L}{\partial t}dt.
\end{aligned}
$$

Thus under a change of coordinates

$$
\frac{\partial L}{\partial \dot{\boldsymbol{q}}'}\,F_*(\boldsymbol{q}) = \frac{\partial L}{\partial \dot{\boldsymbol{q}}} \quad \text{and} \quad \dot{\boldsymbol{q}}'\frac{\partial L}{\partial \dot{\boldsymbol{q}}'} = \dot{\boldsymbol{q}}\frac{\partial L}{\partial \dot{\boldsymbol{q}}},
$$

so that E is a well-defined function on TM. $\qquad\square$

Corollary 1.2. *Under a change of local coordinates on* M, *components of* $\dfrac{\partial L}{\partial \dot{\boldsymbol{q}}}(\boldsymbol{q}, \dot{\boldsymbol{q}}, t) = \left(\dfrac{\partial L}{\partial \dot{q}^1}, \ldots, \dfrac{\partial L}{\partial \dot{q}^n} \right)$ *transform like components of a 1-form on* M.

Proposition 1.1 (Conservation of energy). *The energy of a closed system is an integral of motion.*

Proof. For an extremal γ set $E(t) = E(\gamma(t))$. We have, according to the Euler-Lagrange equations,

$$
\begin{aligned}
\frac{dE}{dt} &= \frac{d}{dt}\left(\frac{\partial L}{\partial \dot{\boldsymbol{q}}} \right)\dot{\boldsymbol{q}} + \frac{\partial L}{\partial \dot{\boldsymbol{q}}}\ddot{\boldsymbol{q}} - \frac{\partial L}{\partial \boldsymbol{q}}\dot{\boldsymbol{q}} - \frac{\partial L}{\partial \dot{\boldsymbol{q}}}\ddot{\boldsymbol{q}} - \frac{\partial L}{\partial t} \\
&= \left(\frac{d}{dt}\left(\frac{\partial L}{\partial \dot{\boldsymbol{q}}} \right) - \frac{\partial L}{\partial \boldsymbol{q}} \right)\dot{\boldsymbol{q}} - \frac{\partial L}{\partial t} = -\frac{\partial L}{\partial t}.
\end{aligned}
$$

Since for a closed system $\dfrac{\partial L}{\partial t} = 0$, the energy is conserved. $\qquad\square$

Conservation of energy for a closed mechanical system is a fundamental law of physics which follows from the homogeneity of time. For a general closed system of N interacting particles considered in Example 1.2,

$$
E = \sum_{a=1}^{N} m_a \dot{\boldsymbol{r}}_a^2 - L = \sum_{a=1}^{N} \tfrac{1}{2}m_a \dot{\boldsymbol{r}}_a^2 + V(\boldsymbol{r}).
$$

In other words, the total energy $E = T + V$ is a sum of the kinetic energy and the potential energy.

Definition. A Lagrangian $L : TM \to \mathbb{R}$ is invariant with respect to the diffeomorphism $g : M \to M$ if $L(g_*(v)) = L(v)$ for all $v \in TM$. The diffeomorphism g is called the *symmetry* of a closed Lagrangian system (M, L). A Lie group G is the *symmetry group* of (M, L) (group of *continuous symmetries*) if there is a left G-action on M such that for every $g \in G$ the mapping $M \ni x \mapsto g \cdot x \in M$ is a symmetry.

Continuous symmetries give rise to conservation laws.

Theorem 1.3 (Noether). *Suppose that a Lagrangian $L : TM \to \mathbb{R}$ is invariant under a one-parameter group $\{g_s\}_{s \in \mathbb{R}}$ of diffeomorphisms of M. Then the Lagrangian system (M, L) admits an integral of motion I, given in standard coordinates on TM by*

$$I(\boldsymbol{q}, \dot{\boldsymbol{q}}) = \sum_{i=1}^{n} \frac{\partial L}{\partial \dot{q}^i}(\boldsymbol{q}, \dot{\boldsymbol{q}}) \left(\frac{dg_s^i(\boldsymbol{q})}{ds} \bigg|_{s=0} \right) = \frac{\partial L}{\partial \dot{\boldsymbol{q}}} \boldsymbol{a},$$

where $X = \displaystyle\sum_{i=1}^{n} a^i(\boldsymbol{q}) \dfrac{\partial}{\partial q^i}$ *is the vector field on M associated with the flow g_s. The integral of motion I is called the Noether integral.*

Proof. It follows from Corollary 1.2 that I is a well-defined function on TM. Now differentiating $L((g_s)_*(\gamma'(t))) = L(\gamma'(t))$ with respect to s at $s = 0$ and using the Euler-Lagrange equations we get

$$0 = \frac{\partial L}{\partial \boldsymbol{q}} \boldsymbol{a} + \frac{\partial L}{\partial \dot{\boldsymbol{q}}} \dot{\boldsymbol{a}} = \frac{d}{dt} \left(\frac{\partial L}{\partial \dot{\boldsymbol{q}}} \right) \boldsymbol{a} + \frac{\partial L}{\partial \dot{\boldsymbol{q}}} \frac{d\boldsymbol{a}}{dt} = \frac{d}{dt} \left(\frac{\partial L}{\partial \dot{\boldsymbol{q}}} \boldsymbol{a} \right),$$

where $\boldsymbol{a}(t) = \big(a^1(\gamma(t)), \ldots, a^n(\gamma(t)) \big)$. $\qquad\square$

Remark. A vector field X on M is called an *infinitesimal symmetry* if the corresponding local flow g_s of X (defined for each $s \in \mathbb{R}$ on some $U_s \subseteq M$) is a symmetry: $L \circ (g_s)_* = L$ on U_s. Every vector field X on M lifts to a vector field X' on TM, defined by a local flow on TM induced from the corresponding local flow on M. In standard coordinates on TM,

$$(1.5) \quad X = \sum_{i=1}^{n} a^i(\boldsymbol{q}) \frac{\partial}{\partial q^i} \quad \text{and} \quad X' = \sum_{i=1}^{n} a^i(\boldsymbol{q}) \frac{\partial}{\partial q^i} + \sum_{i,j=1}^{n} \dot{q}^j \frac{\partial a^i}{\partial q^j}(\boldsymbol{q}) \frac{\partial}{\partial \dot{q}^i}.$$

It is easy to verify that X is an infinitesimal symmetry if and only if $dL(X') = 0$ on TM, which in standard coordinates has the form

$$(1.6) \qquad \sum_{i=1}^{n} a^i(\boldsymbol{q}) \frac{\partial L}{\partial q^i} + \sum_{i,j=1}^{n} \dot{q}^j \frac{\partial a^i}{\partial q^j}(\boldsymbol{q}) \frac{\partial L}{\partial \dot{q}^i} = 0.$$

Remark. Noether's theorem generalizes to time-dependent Lagrangians $L : TM \times \mathbb{R} \to \mathbb{R}$. Namely, on the *extended configuration space* $M_1 = M \times \mathbb{R}$ define a time-independent Lagrangian L_1 by

$$L_1(\boldsymbol{q}, \tau, \dot{\boldsymbol{q}}, \dot{\tau}) = L\left(\boldsymbol{q}, \frac{\dot{\boldsymbol{q}}}{\dot{\tau}}, \tau\right)\dot{\tau},$$

where $(\boldsymbol{q}, \tau)$ are local coordinates on M_1 and $(\boldsymbol{q}, \tau, \dot{\boldsymbol{q}}, \dot{\tau})$ are standard coordinates on TM_1. The Noether integral I_1 for a closed system (M_1, L_1) defines an integral of motion I for a system (M, L) by the formula

$$I(\boldsymbol{q}, \dot{\boldsymbol{q}}, t) = I_1(\boldsymbol{q}, t, \dot{\boldsymbol{q}}, 1).$$

When the Lagrangian L does not depend on time, L_1 is invariant with respect to the one-parameter group of translations $\tau \mapsto \tau + s$, and the Noether integral $I_1 = \dfrac{\partial L_1}{\partial \dot{\tau}}$ gives $I = -E$.

Noether's theorem can be generalized as follows.

Proposition 1.2. *Suppose that for the Lagrangian $L : TM \to \mathbb{R}$ there exist a vector field X on M and a function K on TM such that for every path γ in M,*

$$dL(X')(\gamma(t)) = \frac{d}{dt}K(\gamma'(t)).$$

Then

$$I = \sum_{i=1}^{n} a^i(\boldsymbol{q})\frac{\partial L}{\partial \dot{q}^i}(\boldsymbol{q}, \dot{\boldsymbol{q}}) - K(\boldsymbol{q}, \dot{\boldsymbol{q}})$$

is an integral of motion for the Lagrangian system (M, L).

Proof. Using Euler-Lagrange equations, we have along the extremal γ,

$$\frac{d}{dt}\left(\frac{\partial L}{\partial \dot{\boldsymbol{q}}}\boldsymbol{a}\right) = \frac{\partial L}{\partial \boldsymbol{q}}\boldsymbol{a} + \frac{\partial L}{\partial \dot{\boldsymbol{q}}}\dot{\boldsymbol{a}} = \frac{dK}{dt}. \qquad \square$$

Example 1.9 (Conservation of momentum). Let $M = V$ be a vector space, and suppose that a Lagrangian L is invariant with respect to a one-parameter group $g_s(q) = q + sv$, $v \in V$. According to Noether's theorem,

$$I = \sum_{i=1}^{n} v^i \frac{\partial L}{\partial \dot{q}^i}$$

is an integral of motion. Now let (M, L) be a closed Lagrangian system of N interacting particles considered in Example 1.2. We have $M = V = \mathbb{R}^{3N}$, and the Lagrangian L is invariant under simultaneous translation of coordinates $\boldsymbol{r}_a = (r_a^1, r_a^2, r_a^3)$ of all particles by the same vector $\boldsymbol{c} \in \mathbb{R}^3$. Thus $v = (\boldsymbol{c}, \dots, \boldsymbol{c}) \in \mathbb{R}^{3N}$ and for every $\boldsymbol{c} = (c^1, c^2, c^3) \in \mathbb{R}^3$,

$$I = \sum_{a=1}^{N} \left(c^1 \frac{\partial L}{\partial \dot{r}_a^1} + c^2 \frac{\partial L}{\partial \dot{r}_a^2} + c^3 \frac{\partial L}{\partial \dot{r}_a^3} \right) = c^1 P_1 + c^2 P_2 + c^3 P_3$$

is an integral of motion. The integrals of motion P_1, P_2, P_3 define the vector

$$\boldsymbol{P} = \sum_{a=1}^{N} \frac{\partial L}{\partial \dot{\boldsymbol{r}}_a} \in \mathbb{R}^3$$

(or rather a vector in the dual space to $\mathbb{R}^3$), called the *momentum* of the system. Explicitly,

$$\boldsymbol{P} = \sum_{a=1}^{N} m_a \dot{\boldsymbol{r}}_a,$$

so that the total momentum of a closed system is the sum of momenta of individual particles. Conservation of momentum is a fundamental physical law which reflects the homogeneity of space.

Traditionally, $p_i = \dfrac{\partial L}{\partial \dot{q}^i}$ are called *generalized momenta* corresponding to generalized coordinates q^i, and $F_i = \dfrac{\partial L}{\partial q^i}$ are called *generalized forces*. In these notations, the Euler-Lagrange equations have the same form

$$\dot{\boldsymbol{p}} = \boldsymbol{F}$$

as Newton's equations in Cartesian coordinates. Conservation of momentum implies Newton's third law.

Example 1.10 (Conservation of angular momentum). Let $M = V$ be a vector space with Euclidean inner product. Let $G = \mathrm{SO}(V)$ be the connected Lie group of automorphisms of V preserving the inner product, and let $\mathfrak{g} = \mathfrak{so}(V)$ be the Lie algebra of G. Suppose that a Lagrangian L is invariant with respect to the action of a one-parameter subgroup $g_s(q) = e^{sx} \cdot q$ of G on V, where $x \in \mathfrak{g}$ and e^x is the exponential map. According to Noether's theorem,

$$I = \sum_{i=1}^{n} (x \cdot q)^i \frac{\partial L}{\partial \dot{q}^i}$$

is an integral of motion. Now let (M, L) be a closed Lagrangian system of N interacting particles considered in Example 1.2. We have $M = V = \mathbb{R}^{3N}$, and the Lagrangian L is invariant under a simultaneous rotation of coordinates $\boldsymbol{r}_a$ of all particles by the same orthogonal transformation in $\mathbb{R}^3$.

Thus $x = (u, \dots, u) \in \underbrace{\mathfrak{so}(3) \oplus \cdots \oplus \mathfrak{so}(3)}_{N}$, and for every $u \in \mathfrak{so}(3)$,

$$I = \sum_{a=1}^{N} \left((u \cdot \boldsymbol{r}_a)^1 \frac{\partial L}{\partial \dot{r}_a^1} + (u \cdot \boldsymbol{r}_a)^2 \frac{\partial L}{\partial \dot{r}_a^2} + (u \cdot \boldsymbol{r}_a)^3 \frac{\partial L}{\partial \dot{r}_a^3} \right)$$

is an integral of motion. Let $u = u^1 X_1 + u^2 X_2 + u^3 X_3$, where $X_1 = \left(\begin{smallmatrix} 0 & 0 & 0 \\ 0 & 0 & -1 \\ 0 & 1 & 0 \end{smallmatrix} \right), X_2 = \left(\begin{smallmatrix} 0 & 0 & 1 \\ 0 & 0 & 0 \\ -1 & 0 & 0 \end{smallmatrix} \right), X_3 = \left(\begin{smallmatrix} 0 & -1 & 0 \\ 1 & 0 & 0 \\ 0 & 0 & 0 \end{smallmatrix} \right)$ is the basis in $\mathfrak{so}(3) \simeq \mathbb{R}^3$ corresponding to the rotations about the vectors e_1, e_2, e_3 of the standard orthonormal basis in $\mathbb{R}^3$ (see Problem 1.10). We get

$$I = u^1 M_1 + u^2 M_2 + u^3 M_3,$$

where $\boldsymbol{M} = (M_1, M_2, M_3) \in \mathbb{R}^3$ (or rather a vector in the dual space to $\mathfrak{so}(3)$) is given by

$$\boldsymbol{M} = \sum_{a=1}^{N} \boldsymbol{r}_a \times \frac{\partial L}{\partial \dot{\boldsymbol{r}}_a}.$$

The vector $\boldsymbol{M}$ is called the *angular momentum* of the system. Explicitly,

$$\boldsymbol{M} = \sum_{a=1}^{N} \boldsymbol{r}_a \times m_a \dot{\boldsymbol{r}}_a,$$

so that the total angular momentum of a closed system is the sum of angular momenta of individual particles. Conservation of angular momentum is a fundamental physical law which reflects the isotropy of space.

Problem 1.13. Find how total momentum and total angular momentum transform under the Galilean transformations.

1.5. One-dimensional motion. The motion of systems with one degree of freedom is called one-dimensional. In terms of a Cartesian coordinate x on $M = \mathbb{R}$ the Lagrangian takes the form

$$L = \tfrac{1}{2} m \dot{x}^2 - V(x).$$

The conservation of energy

$$E = \frac{1}{2} m \dot{x}^2 + V(x)$$

allows us to solve the equation of motion in a closed form by separation of variables. We have

$$\frac{dx}{dt} = \sqrt{\frac{2}{m}(E - V(x))},$$

so that

$$t = \sqrt{\frac{m}{2}} \int \frac{dx}{\sqrt{E - V(x)}}.$$

The inverse function $x(t)$ is a general solution of Newton's equation

$$m\ddot{x} = -\frac{dV}{dx},$$

with two arbitrary constants, the energy E and the constant of integration.

Since kinetic energy is non-negative, for a given value of E the actual motion takes place in the region of $\mathbb{R}$ where $V(x) \le E$. The points where $V(x) = E$ are called *turning points*. The motion which is confined between two turning points is called *finite*. The finite motion is periodic — the particle oscillates between the turning points x_1 and x_2 with the period

$$T(E) = \sqrt{2m} \int_{x_1}^{x_2} \frac{dx}{\sqrt{E - V(x)}}.$$

If the region $V(x) \le E$ is unbounded, then the motion is called *infinite* and the particle eventually goes to infinity. The regions where $V(x) > E$ are forbidden.

On the phase plane with coordinates (x, y) Newton's equation reduces to the first order system

$$m\dot{x} = y, \quad \dot{y} = -\frac{dV}{dx}.$$

Trajectories correspond to the phase curves $(x(t), y(t))$, which lie on the level sets

$$\frac{y^2}{2m} + V(x) = E$$

of the energy function. The points $(x_0, 0)$, where x_0 is a critical point of the potential energy $V(x)$, correspond to the equilibrium solutions. The local minima correspond to the stable solutions and local maxima correspond to the unstable solutions. For the values of E which do not correspond to the equilibrium solutions the level sets are smooth curves. These curves are closed if the motion is finite.

The simplest non-trivial one-dimensional system, besides the free particle, is the harmonic oscillator with $V(x) = \frac{1}{2}kx^2$ ($k > 0$), considered in Example 1.6. The general solution of the equation of motion is

$$x(t) = A\cos(\omega t + \alpha),$$

where A is the *amplitude*, $\omega = \sqrt{\dfrac{k}{m}}$ is the *frequency*, and α is the *phase* of a simple harmonic motion with the period $T = \dfrac{2\pi}{\omega}$. The energy is $E = \frac{1}{2}m\omega^2 A^2$ and the motion is finite with the same period T for $E > 0$.

Problem 1.14. Show that for $V(x) = -x^4$ there are phase curves which do not exist for all times. Prove that if $V(x) \ge 0$ for all x, then all phase curves exist for all times.

Problem 1.15. The simple pendulum is a Lagrangian system with $M = S^1 = \mathbb{R}/2\pi\mathbb{Z}$ and $L = \frac{1}{2}\dot\theta^2 + \cos\theta$. Find the period T of the pendulum as a function of the amplitude of the oscillations.

Problem 1.16. Suppose that the potential energy $V(x)$ is even, $V(0) = 0$, and $V(x)$ is a one-to-one monotonically increasing function for $x \geq 0$. Prove that the inverse function $x(V)$ and the period $T(E)$ are related by the Abel transform

$$T(E) = 2\sqrt{2m} \int_0^E \frac{dx}{dV} \frac{dV}{\sqrt{E-V}} \quad \text{and} \quad x(V) = \frac{1}{2\pi\sqrt{2m}} \int_0^V \frac{T(E)dE}{\sqrt{V-E}}.$$

1.6. The motion in a central field and the Kepler problem. The motion of a system of two interacting particles — the *two-body problem* — can also be solved completely. Namely, in this case (see Example 1.2) $M = \mathbb{R}^6$ and

$$L = \frac{m_1\dot{\boldsymbol{r}}_1^2}{2} + \frac{m_2\dot{\boldsymbol{r}}_2^2}{2} - V(|\boldsymbol{r}_1 - \boldsymbol{r}_2|).$$

Introducing on $\mathbb{R}^6$ new coordinates

$$\boldsymbol{r} = \boldsymbol{r}_1 - \boldsymbol{r}_2 \quad \text{and} \quad \boldsymbol{R} = \frac{m_1\boldsymbol{r}_1 + m_2\boldsymbol{r}_2}{m_1 + m_2},$$

we get

$$L = \tfrac{1}{2}m\dot{\boldsymbol{R}}^2 + \tfrac{1}{2}\mu\dot{\boldsymbol{r}}^2 - V(|\boldsymbol{r}|),$$

where $m = m_1 + m_2$ is the *total mass* and $\mu = \dfrac{m_1 m_2}{m_1 + m_2}$ is the *reduced mass* of a two-body system. The Lagrangian L depends only on the velocity $\dot{\boldsymbol{R}}$ of the center of mass and not on its position $\boldsymbol{R}$. A generalized coordinate with this property is called *cyclic*. It follows from the Euler-Lagrange equations that generalized momentum corresponding to the cyclic coordinate is conserved. In our case it is a total momentum of the system,

$$\boldsymbol{P} = \frac{\partial L}{\partial \dot{\boldsymbol{R}}} = m\dot{\boldsymbol{R}},$$

so that the center of mass $\boldsymbol{R}$ moves uniformly. Thus in the frame of reference where $\boldsymbol{R} = 0$, the two-body problem reduces to the problem of a single particle of mass μ in the external central field $V(|\boldsymbol{r}|)$. In spherical coordinates in $\mathbb{R}^3$,

$$x = r\sin\vartheta\cos\varphi, \ y = r\sin\vartheta\sin\varphi, \ z = r\cos\vartheta,$$

where $0 \leq \vartheta < \pi$, $0 \leq \varphi < 2\pi$, its Lagrangian takes the form

$$L = \tfrac{1}{2}\mu(\dot{r}^2 + r^2\dot\vartheta^2 + r^2\sin^2\vartheta\,\dot\varphi^2) - V(r).$$

It follows from the conservation of angular momentum $\boldsymbol{M} = \mu\boldsymbol{r}\times\dot{\boldsymbol{r}}$ that during motion the position vector $\boldsymbol{r}$ lies in the plane P orthogonal to $\boldsymbol{M}$ in $\mathbb{R}^3$. Introducing polar coordinates (r,χ) in the plane P we get $\dot\chi^2 = \dot\vartheta^2 + \sin^2\vartheta\,\dot\varphi^2$, so that

$$L = \tfrac{1}{2}\mu(\dot{r}^2 + r^2\dot\chi^2) - V(r).$$

The coordinate χ is cyclic and its generalized momentum $\mu r^2 \dot{\chi}$ coincides with $|\boldsymbol{M}|$ if $\dot{\chi} > 0$ and with $-|\boldsymbol{M}|$ if $\dot{\chi} < 0$. Denoting this quantity by M, we get the equation

$$(1.7) \qquad \mu r^2 \dot{\chi} = M,$$

which is equivalent to *Kepler's second law*[12]. Using (1.7) we get for the total energy

$$(1.8) \qquad E = \tfrac{1}{2}\mu(\dot{r}^2 + r^2\dot{\chi}^2) + V(r) = \tfrac{1}{2}\mu\dot{r}^2 + V(r) + \frac{M^2}{2\mu r^2}.$$

Thus the radial motion reduces to a one-dimensional motion on the half-line $r > 0$ with the effective potential energy

$$V_{eff}(r) = V(r) + \frac{M^2}{2\mu r^2},$$

where the second term is called the *centrifugal energy*. As in the previous section, the solution is given by

$$(1.9) \qquad t = \sqrt{\frac{\mu}{2}} \int \frac{dr}{\sqrt{E - V_{eff}(r)}}.$$

It follows from (1.7) that the angle χ is a monotonic function of t, given by another quadrature

$$(1.10) \qquad \chi = \frac{M}{\sqrt{2\mu}} \int \frac{dr}{r^2\sqrt{E - V_{eff}(r)}},$$

yielding an equation of the trajectory in polar coordinates.

The set $V_{eff}(r) \leq E$ is a union of annuli $0 \leq r_{min} \leq r \leq r_{max} \leq \infty$, and the motion is finite if $0 < r_{min} \leq r \leq r_{max} < \infty$. Though for a finite motion $r(t)$ oscillates between r_{min} and r_{max}, corresponding trajectories are not necessarily closed. The necessary and sufficient condition for a finite motion to have a closed trajectory is that the angle

$$\Delta\chi = \frac{M}{\sqrt{2\mu}} \int_{r_{min}}^{r_{max}} \frac{dr}{r^2\sqrt{E - V_{eff}(r)}}$$

is commensurable with 2π, i.e., $\Delta\chi = 2\pi\dfrac{m}{n}$ for some $m, n \in \mathbb{Z}$. If the angle $\Delta\chi$ is not commensurable with 2π, the orbit is everywhere dense in the annulus $r_{min} \leq r \leq r_{max}$. If

$$\lim_{r \to \infty} V_{eff}(r) = \lim_{r \to \infty} V(r) = V < \infty,$$

the motion is infinite for $E > V$ — the particle goes to ∞ with finite velocity $\sqrt{\frac{2}{\mu}(E - V)}$.

---

[12]It is the statement that *sectorial velocity* of a particle in a central field is constant.

A very important special case is when

$$V(r) = -\frac{\alpha}{r}.$$

It describes Newton's gravitational attraction ($\alpha > 0$) and Coulomb electrostatic interaction (either attractive or repulsive). First consider the case when $\alpha > 0$ — Kepler's problem. The effective potential energy is

$$V_{eff}(r) = -\frac{\alpha}{r} + \frac{M^2}{2\mu r^2}$$

and has the global minimum

$$V_0 = -\frac{\alpha^2 \mu}{2M^2}$$

at $r_0 = \dfrac{M^2}{\alpha\mu}$. The motion is infinite for $E \geq 0$ and is finite for $V_0 \leq E < 0$. The explicit form of trajectories can be determined by an elementary integration in (1.10), which gives

$$\chi = \cos^{-1} \frac{\dfrac{M}{r} - \dfrac{M}{r_0}}{\sqrt{2\mu(E - V_0)}} + C.$$

Choosing a constant of integration $C = 0$ and introducing notation

$$p = r_0 \quad \text{and} \quad e = \sqrt{1 - \frac{E}{V_0}},$$

we get the equation of the orbit (trajectory)

$$(1.11) \qquad\qquad \frac{p}{r} = 1 + e \cos\chi.$$

This is the equation of a conic section with one focus at the origin. Quantity $2p$ is called the *latus rectum* of the orbit, and e is called the *eccentricity*. The choice $C = 0$ is such that the point with $\chi = 0$ is the point nearest to the origin (called the *perihelion*). When $V_0 \leq E < 0$, the eccentricity $e < 1$ so that the orbit is the ellipse[13] with the major and minor semi-axes

$$(1.12) \qquad a = \frac{p}{1 - e^2} = \frac{\alpha}{2|E|}, \quad b = \frac{p}{\sqrt{1 - e^2}} = \frac{|M|}{\sqrt{2\mu|E|}}.$$

Correspondingly, $r_{min} = \dfrac{p}{1 + e}$, $r_{max} = \dfrac{p}{1 - e}$, and the period T of elliptic orbit is given by

$$T = \pi\alpha\sqrt{\frac{\mu}{2|E|^3}}.$$

The last formula is *Kepler's third law*. When $E > 0$, the eccentricity $e > 1$ and the motion is infinite — the orbit is a hyperbola with the origin as

---

[13]The statement that planets have elliptic orbits with a focus at the Sun is *Kepler's first law*.

internal focus. When $E = 0$, the eccentricity $e = 1$ — the particle starts from rest at ∞ and the orbit is a parabola.

For the repulsive case $\alpha < 0$ the effective potential energy $V_{eff}(r)$ is always positive and decreases monotonically from ∞ to 0. The motion is always infinite and the trajectories are hyperbolas (parabola if $E = 0$)

$$\frac{p}{r} = -1 + e \cos \chi$$

with

$$p = \frac{M^2}{\alpha \mu} \quad \text{and} \quad e = \sqrt{1 + \frac{2EM^2}{\mu \alpha^2}}.$$

Kepler's problem is very special: for every $\alpha \in \mathbb{R}$ the Lagrangian system on $\mathbb{R}^3$ with

$$(1.13) \qquad L = \tfrac{1}{2}\mu \dot{\boldsymbol{r}}^2 + \frac{\alpha}{r}$$

has three extra integrals of motion W_1, W_2, W_3 in addition to the components of the angular momentum $\boldsymbol{M}$. The corresponding vector $\boldsymbol{W} = (W_1, W_2, W_3)$, called the *Laplace-Runge-Lenz vector*, is given by

$$(1.14) \qquad \boldsymbol{W} = \dot{\boldsymbol{r}} \times \boldsymbol{M} - \frac{\alpha \boldsymbol{r}}{r}.$$

Indeed, using equations of motion $\mu \ddot{\boldsymbol{r}} = -\dfrac{\alpha \boldsymbol{r}}{r^3}$ and conservation of the angular momentum $\boldsymbol{M} = \mu \boldsymbol{r} \times \dot{\boldsymbol{r}}$, we get

$$\begin{aligned}
\dot{\boldsymbol{W}} &= \mu \ddot{\boldsymbol{r}} \times (\boldsymbol{r} \times \dot{\boldsymbol{r}}) - \frac{\alpha \dot{\boldsymbol{r}}}{r} + \frac{\alpha (\dot{\boldsymbol{r}} \cdot \boldsymbol{r})\boldsymbol{r}}{r^3} \\
&= (\mu \ddot{\boldsymbol{r}} \cdot \dot{\boldsymbol{r}})\boldsymbol{r} - (\mu \ddot{\boldsymbol{r}} \cdot \boldsymbol{r})\dot{\boldsymbol{r}} - \frac{\alpha \dot{\boldsymbol{r}}}{r} + \frac{\alpha (\dot{\boldsymbol{r}} \cdot \boldsymbol{r})\boldsymbol{r}}{r^3} \\
&= 0.
\end{aligned}$$

Using $\mu(\dot{\boldsymbol{r}} \times \boldsymbol{M}) \cdot \boldsymbol{r} = \boldsymbol{M}^2$ and the identity $(\boldsymbol{a} \times \boldsymbol{b})^2 = \boldsymbol{a}^2 \boldsymbol{b}^2 - (\boldsymbol{a} \cdot \boldsymbol{b})^2$, we get

$$(1.15) \qquad \boldsymbol{W}^2 = \alpha^2 + \frac{2\boldsymbol{M}^2 E}{\mu}$$

where

$$E = \frac{\boldsymbol{p}^2}{2\mu} - \frac{\alpha}{r}$$

is the energy corresponding to the Lagrangian (1.13). The fact that all orbits are conic sections follows from this extra symmetry of Kepler's problem.

Problem 1.17. Prove all the statements made in this section.

Problem 1.18. Show that if

$$\lim_{r \to 0} V_{eff}(r) = -\infty,$$

then there are orbits with $r_{min} = 0$ — "fall" of the particle to the center.

Problem 1.19. Prove that all finite trajectories in the central field are closed only when

$$V(r) = kr^2, \ k > 0, \quad \text{and} \quad V(r) = -\frac{\alpha}{r}, \ \alpha > 0.$$

Problem 1.20. Find parametric equations for orbits in Kepler's problem.

Problem 1.21. Prove that the Laplace-Runge-Lenz vector $\boldsymbol{W}$ points in the direction of the major axis of the orbit and that $|\boldsymbol{W}| = \alpha e$, where e is the eccentricity of the orbit.

Problem 1.22. Using the conservation of the Laplace-Runge-Lenz vector, prove that trajectories in Kepler's problem with $E < 0$ are ellipses. (*Hint:* Evaluate $\boldsymbol{W} \cdot \boldsymbol{r}$ and use the result of the previous problem.)

1.7. Legendre transform. The equations of motion of a Lagrangian system (M, L) in standard coordinates associated with a coordinate chart $U \subset M$ are the Euler-Lagrange equations. In expanded form, they are given by the following system of second order ordinary differential equations:

$$\frac{\partial L}{\partial q^i}(\boldsymbol{q}, \dot{\boldsymbol{q}}) = \frac{d}{dt}\left(\frac{\partial L}{\partial \dot{q}^i}(\boldsymbol{q}, \dot{\boldsymbol{q}})\right)$$

$$= \sum_{j=1}^{n}\left(\frac{\partial^2 L}{\partial \dot{q}^i \partial \dot{q}^j}(\boldsymbol{q}, \dot{\boldsymbol{q}})\,\ddot{q}^j + \frac{\partial^2 L}{\partial \dot{q}^i \partial q^j}(\boldsymbol{q}, \dot{\boldsymbol{q}})\,\dot{q}^j\right), \quad i = 1, \ldots, n.$$

In order for this system to be solvable for the highest derivatives for all initial conditions in TU, the symmetric $n \times n$ matrix

$$H_L(\boldsymbol{q}, \dot{\boldsymbol{q}}) = \left\{\frac{\partial^2 L}{\partial \dot{q}^i \partial \dot{q}^j}(\boldsymbol{q}, \dot{\boldsymbol{q}})\right\}_{i,j=1}^{n}$$

should be invertible on TU.

Definition. A Lagrangian system (M, L) is called *non-degenerate* if for every coordinate chart U on M the matrix $H_L(\boldsymbol{q}, \dot{\boldsymbol{q}})$ is invertible on TU.

Remark. Note that the $n \times n$ matrix H_L is a Hessian of the Lagrangian function L for vertical directions on TM. Under the change of standard coordinates $\boldsymbol{q}' = F(\boldsymbol{q})$ and $\boldsymbol{q}' = F_*(\boldsymbol{q})\boldsymbol{v}$ (see Section 1.2) it has the transformation law

$$H_L(\boldsymbol{q}, \dot{\boldsymbol{q}}) = F_*(\boldsymbol{q})^T H_L(\boldsymbol{q}', \dot{\boldsymbol{q}}')F_*(\boldsymbol{q}),$$

where $F_*(\boldsymbol{q})^T$ is the transposed matrix, so that the condition $\det H_L \neq 0$ does not depend on the choice of standard coordinates.

For an invariant formulation, consider the 1-form θ_L, defined in standard coordinates associated with a coordinate chart $U \subset M$ by

$$\theta_L = \sum_{i=1}^{n} \frac{\partial L}{\partial \dot{q}^i} dq^i = \frac{\partial L}{\partial \dot{\boldsymbol{q}}} d\boldsymbol{q}.$$

It follows from Corollary 1.2 that θ_L is a well-defined 1-form on TM.

Lemma 1.2. *A Lagrangian system (M, L) is non-degenerate if and only if the 2-form $d\theta_L$ on TM is non-degenerate.*

Proof. In standard coordinates,

$$d\theta_L = \sum_{i,j=1}^{n} \left(\frac{\partial^2 L}{\partial \dot{q}^i \partial \dot{q}^j} d\dot{q}^j \wedge dq^i + \frac{\partial^2 L}{\partial \dot{q}^i \partial q^j} dq^j \wedge dq^i \right),$$

and it is easy to see, by considering the $2n$-form $d\theta_L^n = \underbrace{d\theta_L \wedge \cdots \wedge d\theta_L}_{n}$, that the 2-form $d\theta_L$ is non-degenerate if and only if the matrix H_L is non-degenerate. $\qquad\square$

Remark. Using the 1-form θ_L, the Noether integral I in Theorem 1.3 can be written as

$$(1.16) \qquad\qquad I = i_{X'}(\theta_L),$$

where X' is a lift to TM of a vector field X on M given by (1.5). It also immediately follows from (1.6) that if X is an infinitesimal symmetry, then

$$(1.17) \qquad\qquad \mathcal{L}_{X'}(\theta_L) = 0.$$

Definition. Let (U, φ) be a coordinate chart on M. Coordinates

$$(\boldsymbol{p}, \boldsymbol{q}) = (p_1, \ldots, p_n, q^1, \ldots, q^n)$$

on the chart $T^*U \simeq \mathbb{R}^n \times U$ on the cotangent bundle T^*M are called *standard coordinates*[14] if for $(p, q) \in T^*U$ and $f \in C^\infty(U)$

$$p_i(df) = \frac{\partial f}{\partial q^i}, \quad i = 1, \ldots, n.$$

Equivalently, standard coordinates on T^*U are uniquely characterized by the condition that $\boldsymbol{p} = (p_1, \ldots, p_n)$ are coordinates in the fiber corresponding to the basis $dq^1, \ldots, dq^n$ for T_q^*M, dual to the basis $\dfrac{\partial}{\partial q^1}, \ldots, \dfrac{\partial}{\partial q^n}$ for T_qM.

---

[14]Following tradition, the first n coordinates parametrize the fiber of T^*U and the last n coordinates parametrize the base.

Definition. The 1-form θ on T^*M, defined in standard coordinates by

$$\theta = \sum_{i=1}^{n} p_i dq^i = \boldsymbol{p}d\boldsymbol{q},$$

is called *Liouville's canonical 1-form.*

Corollary 1.2 shows that θ is a well-defined 1-form on T^*M. Clearly, the 1-form θ also admits an invariant definition

$$\theta(u) = p(\pi_*(u)), \quad \text{where} \quad u \in T_{(p,q)}T^*M,$$

and $\pi : T^*M \to M$ is the canonical projection.

Definition. A fibre-wise mapping $\tau_L : TM \to T^*M$ is called a *Legendre transform* associated with the Lagrangian L if

$$\theta_L = \tau_L^*(\theta).$$

In standard coordinates the Legendre transform is given by

$$\tau_L(\boldsymbol{q},\dot{\boldsymbol{q}}) = (\boldsymbol{p},\boldsymbol{q}), \quad \text{where} \quad \boldsymbol{p} = \frac{\partial L}{\partial \dot{\boldsymbol{q}}}(\boldsymbol{q},\dot{\boldsymbol{q}}).$$

The mapping τ_L is a local diffeomorphism if and only if the Lagrangian L is non-degenerate.

Definition. Suppose that the Legendre transform $\tau_L : TM \to T^*M$ is a diffeomorphism. The *Hamiltonian* function $H : T^*M \to \mathbb{R}$, associated with the Lagrangian $L : TM \to \mathbb{R}$, is defined by

$$H \circ \tau_L = E_L = \dot{\boldsymbol{q}}\,\frac{\partial L}{\partial \dot{\boldsymbol{q}}} - L.$$

In standard coordinates,

$$H(\boldsymbol{p},\boldsymbol{q}) = \left.(\boldsymbol{p}\dot{\boldsymbol{q}} - L(\boldsymbol{q},\dot{\boldsymbol{q}}))\right|_{\boldsymbol{p}=\frac{\partial L}{\partial \dot{\boldsymbol{q}}}},$$

where $\dot{\boldsymbol{q}}$ is a function of $\boldsymbol{p}$ and $\boldsymbol{q}$ defined by the equation $\boldsymbol{p} = \dfrac{\partial L}{\partial \dot{\boldsymbol{q}}}(\boldsymbol{q},\dot{\boldsymbol{q}})$ through the implicit function theorem. The cotangent bundle T^*M is called the *phase space* of the Lagrangian system (M,L). It turns out that on the phase space the equations of motion take a very simple and symmetric form.

Theorem 1.4. *Suppose that the Legendre transform $\tau_L : TM \to T^*M$ is a diffeomorphism. Then the Euler-Lagrange equations in standard coordinates on TM,*

$$\frac{d}{dt}\frac{\partial L}{\partial \dot{q}^i} - \frac{\partial L}{\partial q^i} = 0, \quad i = 1,\dots,n,$$

are equivalent to the following system of first order differential equations in standard coordinates on T^*M:

$$\dot{p}_i = -\frac{\partial H}{\partial q^i}, \quad \dot{q}^i = \frac{\partial H}{\partial p_i}, \quad i = 1, \dots, n.$$

Proof. We have

$$dH = \frac{\partial H}{\partial \boldsymbol{p}} d\boldsymbol{p} + \frac{\partial H}{\partial \boldsymbol{q}} d\boldsymbol{q}$$

$$= \left(\boldsymbol{p} d\dot{\boldsymbol{q}} + \dot{\boldsymbol{q}} d\boldsymbol{p} - \frac{\partial L}{\partial \boldsymbol{q}} d\boldsymbol{q} - \frac{\partial L}{\partial \dot{\boldsymbol{q}}} d\dot{\boldsymbol{q}} \right) \Big|_{\boldsymbol{p} = \frac{\partial L}{\partial \dot{\boldsymbol{q}}}}$$

$$= \left(\dot{\boldsymbol{q}} d\boldsymbol{p} - \frac{\partial L}{\partial \boldsymbol{q}} d\boldsymbol{q} \right) \Big|_{\boldsymbol{p} = \frac{\partial L}{\partial \dot{\boldsymbol{q}}}}.$$

Thus under the Legendre transform,

$$\dot{\boldsymbol{q}} = \frac{\partial H}{\partial \boldsymbol{p}} \quad \text{and} \quad \dot{\boldsymbol{p}} = \frac{d}{dt} \frac{\partial L}{\partial \dot{\boldsymbol{q}}} = \frac{\partial L}{\partial \boldsymbol{q}} = -\frac{\partial H}{\partial \boldsymbol{q}}. \qquad \square$$

Corresponding first order differential equations on T^*M are called *Hamilton's equations* (*canonical equations*).

Corollary 1.5. *The Hamiltonian H is constant on the solutions of Hamilton's equations.*

Proof. For $H(t) = H(\boldsymbol{p}(t), \boldsymbol{q}(t))$ we have

$$\frac{dH}{dt} = \frac{\partial H}{\partial \boldsymbol{q}} \dot{\boldsymbol{q}} + \frac{\partial H}{\partial \boldsymbol{p}} \dot{\boldsymbol{p}} = \frac{\partial H}{\partial \boldsymbol{q}} \frac{\partial H}{\partial \boldsymbol{p}} - \frac{\partial H}{\partial \boldsymbol{p}} \frac{\partial H}{\partial \boldsymbol{q}} = 0. \qquad \square$$

For the Lagrangian

$$L = \frac{m\dot{\boldsymbol{r}}^2}{2} - V(\boldsymbol{r}) = T - V, \quad \boldsymbol{r} \in \mathbb{R}^3,$$

of a particle of mass m in a potential field $V(\boldsymbol{r})$, considered in Example 1.4, we have

$$\boldsymbol{p} = \frac{\partial L}{\partial \dot{\boldsymbol{r}}} = m\dot{\boldsymbol{r}}.$$

Thus the Legendre transform $\tau_L : T\mathbb{R}^3 \to T^*\mathbb{R}^3$ is a global diffeomorphism, linear on the fibers, and

$$H(\boldsymbol{p}, \boldsymbol{r}) = (\boldsymbol{p}\dot{\boldsymbol{r}} - L)|_{\dot{\boldsymbol{r}} = \frac{\boldsymbol{p}}{m}} = \frac{\boldsymbol{p}^2}{2m} + V(\boldsymbol{r}) = T + V.$$

Hamilton's equations

$$\dot{r} = \frac{\partial H}{\partial p} = \frac{p}{m},$$

$$\dot{p} = -\frac{\partial H}{\partial r} = -\frac{\partial V}{\partial r}$$

are equivalent to Newton's equations with the force $\boldsymbol{F} = -\dfrac{\partial V}{\partial \boldsymbol{r}}$.

For the Lagrangian system describing small oscillators, considered in Example 1.6, we have $\boldsymbol{p} = m\dot{\boldsymbol{q}}$, and using normal coordinates we get

$$H(\boldsymbol{p}, \boldsymbol{q}) = (\boldsymbol{p}\dot{\boldsymbol{q}} - L(\boldsymbol{q}, \dot{\boldsymbol{q}}))\big|_{\dot{\boldsymbol{q}} = \frac{\boldsymbol{p}}{m}} = \frac{\boldsymbol{p}^2}{2m} + V_0(\boldsymbol{q}) = \frac{1}{2m}\Big(\boldsymbol{p}^2 + m^2 \sum_{i=1}^{n} \omega_i^2 (q^i)^2\Big).$$

Similarly, for the system of N interacting particles, considered in Example 1.2, we have $\boldsymbol{p} = (\boldsymbol{p}_1, \ldots, \boldsymbol{p}_N)$, where

$$\boldsymbol{p}_a = \frac{\partial L}{\partial \dot{\boldsymbol{r}}_a} = m_a \dot{\boldsymbol{r}}_a, \quad a = 1, \ldots, N.$$

The Legendre transform $\tau_L : T\mathbb{R}^{3N} \to T^*\mathbb{R}^{3N}$ is a global diffeomorphism, linear on the fibers, and

$$H(\boldsymbol{p}, \boldsymbol{r}) = (\boldsymbol{p}\dot{\boldsymbol{r}} - L)\big|_{\dot{\boldsymbol{r}} = \frac{\boldsymbol{p}}{m}} = \sum_{a=1}^{N} \frac{\boldsymbol{p}_a^2}{2m_a} + V(\boldsymbol{r}) = T + V.$$

In particular, for a closed system with pair-wise interaction,

$$H(\boldsymbol{p}, \boldsymbol{r}) = \sum_{a=1}^{N} \frac{\boldsymbol{p}_a^2}{2m_a} + \sum_{1 \le a < b \le N} V_{ab}(\boldsymbol{r}_a - \boldsymbol{r}_b).$$

In general, consider the Lagrangian

$$L = \sum_{i,j=1}^{n} \tfrac{1}{2} a_{ij}(\boldsymbol{q}) \dot{q}^i \dot{q}^j - V(\boldsymbol{q}), \ \boldsymbol{q} \in \mathbb{R}^n,$$

where $A(\boldsymbol{q}) = \{a_{ij}(\boldsymbol{q})\}_{i,j=1}^{n}$ is a symmetric $n \times n$ matrix. We have

$$p_i = \frac{\partial L}{\partial \dot{q}^i} = \sum_{j=1}^{n} a_{ij}(\boldsymbol{q}) \dot{q}^j, \quad i = 1, \ldots, n,$$

and the Legendre transform is a global diffeomorphism, linear on the fibers, if and only if the matrix $A(\boldsymbol{q})$ is non-degenerate for all $\boldsymbol{q} \in \mathbb{R}^n$. In this case,

$$H(\boldsymbol{p}, \boldsymbol{q}) = (\boldsymbol{p}\dot{\boldsymbol{q}} - L(\boldsymbol{q}, \dot{\boldsymbol{q}}))\big|_{\boldsymbol{p} = \frac{\partial L}{\partial \dot{\boldsymbol{q}}}} = \sum_{i,j=1}^{n} \tfrac{1}{2} a^{ij}(\boldsymbol{q}) p_i p_j + V(\boldsymbol{q}),$$

where $\{a^{ij}(\boldsymbol{q})\}_{i,j=1}^{n} = A^{-1}(\boldsymbol{q})$ is the inverse matrix.

Problem 1.23 (Second tangent bundle). Let $\pi : TM \to M$ be the canonical projection and let $T_{\mathrm{V}}(TM)$ be the *vertical tangent bundle* of TM along the fibers of π — the kernel of the bundle mapping $\pi_* : T(TM) \to TM$. Prove that there is a natural bundle isomorphism $i : TM \simeq T_{\mathrm{V}}(TM)$.

Problem 1.24 (Invariant definition of the 1-form θ_L). Show that $\theta_L(v) = dL((i \circ \pi_*)v)$, where $v \in T(TM)$.

Problem 1.25. Give an invariant proof of (1.17).

Problem 1.26. Prove that the path $\gamma(t)$ in M is a trajectory for the Lagrangian system (M, L) if and only if

$$i_{\dot{\gamma}'(t)}(d\theta_L) + dE_L(\gamma'(t)) = 0,$$

where $\dot{\gamma}'(t)$ is the velocity vector of the path $\gamma'(t)$ in TM.

Problem 1.27. Show that for a charged particle in an electromagnetic field, considered in Example 1.5,

$$\boldsymbol{p} = m\dot{\boldsymbol{r}} + \frac{e}{c}\boldsymbol{A} \quad \text{and} \quad H(\boldsymbol{p}, \boldsymbol{r}) = \frac{1}{2m}\left(\boldsymbol{p} - \frac{e}{c}\boldsymbol{A}\right)^2 + e\varphi(\boldsymbol{r}).$$

Problem 1.28. Suppose that for a Lagrangian system $(\mathbb{R}^n, L)$ the Legendre transform τ_L is a diffeomorphism and let H be the corresponding Hamiltonian. Prove that for fixed $\boldsymbol{q}$ and $\dot{\boldsymbol{q}}$ the function $\boldsymbol{p}\dot{\boldsymbol{q}} - H(\boldsymbol{p}, \boldsymbol{q})$ has a single critical point at $\boldsymbol{p} = \dfrac{\partial L}{\partial \dot{\boldsymbol{q}}}$.

Problem 1.29. Give an example of a non-degenerate Lagrangian system (M, L) such that the Legendre transform $\tau_L : TM \to T^*M$ is one-to-one but not onto.

2. Hamiltonian Mechanics

2.1. Hamilton's equations. With every function $H : T^*M \to \mathbb{R}$ on the phase space T^*M there are associated Hamilton's equations — a first-order system of ordinary differential equations, which in the standard coordinates on T^*U has the form

$$(2.1) \qquad \dot{\boldsymbol{p}} = -\frac{\partial H}{\partial \boldsymbol{q}}, \quad \dot{\boldsymbol{q}} = \frac{\partial H}{\partial \boldsymbol{p}}.$$

The corresponding vector field X_H on T^*U,

$$X_H = \sum_{i=1}^{n}\left(\frac{\partial H}{\partial p_i}\frac{\partial}{\partial q^i} - \frac{\partial H}{\partial q^i}\frac{\partial}{\partial p_i}\right) = \frac{\partial H}{\partial \boldsymbol{p}}\frac{\partial}{\partial \boldsymbol{q}} - \frac{\partial H}{\partial \boldsymbol{q}}\frac{\partial}{\partial \boldsymbol{p}},$$

gives rise to a well-defined vector field X_H on T^*M, called the *Hamiltonian vector field*. Suppose now that the vector field X_H on T^*M is complete, i.e., its integral curves exist for all times. The corresponding one-parameter group $\{g_t\}_{t \in \mathbb{R}}$ of diffeomorphisms of T^*M generated by X_H is called the *Hamiltonian phase flow*. It is defined by $g_t(p, q) = (p(t), q(t))$, where $p(t)$, $q(t)$ is a solution of Hamilton's equations satisfying $p(0) = p$, $q(0) = q$.

Liouville's canonical 1-form θ on T^*M defines a 2-form $\omega = d\theta$. In standard coordinates on T^*M it is given by

$$\omega = \sum_{i=1}^{n} dp_i \wedge dq^i = d\boldsymbol{p} \wedge d\boldsymbol{q},$$

and is a non-degenerate 2-form. The form ω is called the *canonical symplectic form* on T^*M. The symplectic form ω defines an isomorphism $J : T^*(T^*M) \to T(T^*M)$ between tangent and cotangent bundles to T^*M. For every $(p, q) \in T^*M$ the linear mapping $J^{-1} : T_{(p,q)}T^*M \to T^*_{(p,q)}T^*M$ is given by

$$\omega(u_1, u_2) = J^{-1}(u_2)(u_1), \quad u_1, u_2 \in T_{(p,q)}T^*M.$$

The mapping J induces the isomorphism between the infinite-dimensional vector spaces $\mathcal{A}^1(T^*M)$ and $\mathrm{Vect}(T^*M)$, which is linear over $C^\infty(T^*M)$. If ϑ is a 1-form on T^*M, then the corresponding vector field $J(\vartheta)$ on T^*M satisfies

$$\omega(X, J(\vartheta)) = \vartheta(X), \quad X \in \mathrm{Vect}(T^*M),$$

and $J^{-1}(X) = -i_X\omega$. In particular, in standard coordinates,

$$J(d\boldsymbol{p}) = \frac{\partial}{\partial \boldsymbol{q}} \quad \text{and} \quad J(d\boldsymbol{q}) = -\frac{\partial}{\partial \boldsymbol{p}},$$

so that $X_H = J(dH)$.

Theorem 2.1. *The Hamiltonian phase flow on T^*M preserves the canonical symplectic form.*

Proof. We need to prove that $(g_t)^*\omega = \omega$. Since g_t is a one-parameter group of diffeomorphisms, it is sufficient to show that

$$\frac{d}{dt}(g_t)^*\omega \bigg|_{t=0} = \mathcal{L}_{X_H}\omega = 0,$$

where $\mathcal{L}_{X_H}$ is the Lie derivative along the vector field X_H. Since for every vector field X,

$$\mathcal{L}_X(df) = d(X(f)),$$

we compute

$$\mathcal{L}_{X_H}(dp_i) = -d\left(\frac{\partial H}{\partial q^i}\right) \quad \text{and} \quad \mathcal{L}_{X_H}(dq^i) = d\left(\frac{\partial H}{\partial p_i}\right),$$

so that

$$\mathcal{L}_{X_H}\omega = \sum_{i=1}^{n} \left(\mathcal{L}_{X_H}(dp_i) \wedge dq^i + dp_i \wedge \mathcal{L}_{X_H}(dq^i)\right)$$

$$= \sum_{i=1}^{n} \left(-d\left(\frac{\partial H}{\partial q^i}\right) \wedge dq^i + dp_i \wedge d\left(\frac{\partial H}{\partial p_i}\right)\right) = -d(dH) = 0. \quad \square$$

Corollary 2.2. $\mathcal{L}_{X_H}(\theta) = d(-H + \theta(X_H))$, *where θ is Liouville's canonical 1-form.*

The canonical symplectic form ω on T^*M defines the volume form $\frac{\omega^n}{n!} = \frac{1}{n!} \underbrace{\omega \wedge \cdots \wedge \omega}_{n}$ on T^*M, called *Liouville's volume form*.

Corollary 2.3 (Liouville's theorem). *The Hamiltonian phase flow on T^*M preserves Liouville's volume form.*

The restriction of the symplectic form ω on T^*M to the configuration space M is 0. Generalizing this property, we get the following notion.

Definition. A submanifold $\mathscr{L}$ of the phase space T^*M is called a *Lagrangian submanifold* if $\dim \mathscr{L} = \dim M$ and $\omega|_{\mathscr{L}} = 0$.

It follows from Theorem 2.1 that the image of a Lagrangian submanifold under the Hamiltonian phase flow is a Lagrangian submanifold.

Problem 2.1. Verify that X_H is a well-defined vector field on T^*M.

Problem 2.2. Show that if all level sets of the Hamiltonian H are compact submanifolds of T^*M, then the Hamiltonian vector field X_H is complete.

Problem 2.3. Let $\pi : T^*M \to M$ be the canonical projection, and let $\mathscr{L}$ be a Lagrangian submanifold. Show that if the mapping $\pi|_{\mathscr{L}} : \mathscr{L} \to M$ is a diffeomorphism, then $\mathscr{L}$ is a graph of a smooth function on M. Give examples when for some $t > 0$ the corresponding projection of $g_t(\mathscr{L})$ onto M is no longer a diffeomorphism.

2.2. The action functional in the phase space. With every function H on the phase space T^*M there is an associated 1-form

$$\theta - H\,dt = \boldsymbol{p}\,d\boldsymbol{q} - H\,dt$$

on the *extended phase space* $T^*M \times \mathbb{R}$, called the *Poincaré-Cartan form*. Let $\gamma : [t_0, t_1] \to T^*M$ be a smooth parametrized path in T^*M such that $\pi(\gamma(t_0)) = q_0$ and $\pi(\gamma(t_1)) = q_1$, where $\pi : T^*M \to M$ is the canonical projection. By definition, the lift of a path γ to the extended phase space $T^*M \times \mathbb{R}$ is a path $\sigma : [t_0, t_1] \to T^*M \times \mathbb{R}$ given by $\sigma(t) = (\gamma(t), t)$, and a path σ in $T^*M \times \mathbb{R}$ is called an *admissible* path if it is a lift of a path γ in T^*M. The space of admissible paths in $T^*M \times \mathbb{R}$ is denoted by $\tilde{P}(T^*M)_{q_0,t_0}^{q_1,t_1}$. A variation of an admissible path σ is a smooth family of admissible paths σ_ε, where $\varepsilon \in [-\varepsilon_0, \varepsilon_0]$ and $\sigma_0 = \sigma$, and the corresponding infinitesimal variation is

$$\delta\sigma = \left.\frac{\partial \sigma_\varepsilon}{\partial \varepsilon}\right|_{\varepsilon=0} \in T_\sigma \tilde{P}(T^*M)_{q_0,t_0}^{q_1,t_1}$$

(cf. Section 1.2). The principle of the least action in the phase space is the following statement.

Theorem 2.4 (Poincaré). *The admissible path σ in $T^*M \times \mathbb{R}$ is an extremal for the action functional*

$$S(\sigma) = \int_\sigma (\boldsymbol{p}d\boldsymbol{q} - Hdt) = \int_{t_0}^{t_1} (\boldsymbol{p}\dot{\boldsymbol{q}} - H)dt$$

if and only if it is a lift of a path $\gamma(t) = (\boldsymbol{p}(t), \boldsymbol{q}(t))$ in T^*M, where $\boldsymbol{p}(t)$ and $\boldsymbol{q}(t)$ satisfy canonical Hamilton's equations

$$\dot{\boldsymbol{p}} = -\frac{\partial H}{\partial \boldsymbol{q}}, \quad \dot{\boldsymbol{q}} = \frac{\partial H}{\partial \boldsymbol{p}}.$$

Proof. As in the proof of Theorem 1.1, for an admissible family $\sigma_\varepsilon(t) = (\boldsymbol{p}(t,\varepsilon), \boldsymbol{q}(t,\varepsilon), t)$ we compute using integration by parts,

$$\left.\frac{d}{d\varepsilon}\right|_{\varepsilon=0} S(\sigma_\varepsilon) = \sum_{i=1}^{n} \int_{t_0}^{t_1} \left(\dot{q}^i \delta p_i - \dot{p}_i \delta q^i - \frac{\partial H}{\partial q^i}\delta q^i - \frac{\partial H}{\partial p_i}\delta p_i \right) dt$$

$$+ \sum_{i=1}^{n} p_i\, \delta q^i \Big|_{t_0}^{t_1}.$$

Since $\delta\boldsymbol{q}(t_0) = \delta\boldsymbol{q}(t_1) = 0$, the path σ is critical if and only if $\boldsymbol{p}(t)$ and $\boldsymbol{q}(t)$ satisfy canonical Hamilton's equations (2.1). $\qquad\square$

Remark. For a Lagrangian system (M, L), every path $\gamma(t) = (\boldsymbol{q}(t))$ in the configuration space M connecting points q_0 and q_1 defines an admissible path $\hat{\gamma}(t) = (\boldsymbol{p}(t), \boldsymbol{q}(t), t)$ in the phase space T^*M by setting $\boldsymbol{p} = \dfrac{\partial L}{\partial \dot{\boldsymbol{q}}}$. If the Legendre transform $\tau_L : TM \to T^*M$ is a diffeomorphism, then

$$S(\hat{\gamma}) = \int_{t_0}^{t_1} (\boldsymbol{p}\dot{\boldsymbol{q}} - H)dt = \int_{t_0}^{t_1} L(\gamma'(t), t)dt.$$

Thus the principle of the least action in a configuration space — Hamilton's principle — follows from the principle of the least action in a phase space. In fact, in this case the two principles are equivalent (see Problem 1.28).

From Corollary 1.5 we immediately get the following result.

Corollary 2.5. *Solutions of canonical Hamilton's equations lying on the hypersurface $H(\boldsymbol{p}, \boldsymbol{q}) = E$ are extremals of the functional $\int_\sigma \boldsymbol{p}d\boldsymbol{q}$ in the class of admissible paths σ lying on this hypersurface.*

Corollary 2.6 (Maupertuis' principle). *The trajectory $\gamma = (\boldsymbol{q}(\tau))$ of a closed Lagrangian system (M, L) connecting points q_0 and q_1 and having energy E is the extremal of the functional*

$$\int_\gamma \boldsymbol{p}d\boldsymbol{q} = \int_\gamma \frac{\partial L}{\partial \dot{\boldsymbol{q}}}(\boldsymbol{q}(\tau), \dot{\boldsymbol{q}}(\tau))\dot{\boldsymbol{q}}(\tau)d\tau$$

on the space of all paths in the configuration space M connecting points q_0 and q_1 and parametrized such that $H(\frac{\partial L}{\partial \dot{q}}(\tau), q(\tau)) = E$.

The functional

$$S_0(\gamma) = \int_\gamma pdq$$

is called the *abbreviated action*[15].

Proof. Every path $\gamma = q(\tau)$, parametrized such that $H(\frac{\partial L}{\partial \dot{q}}, q) = E$, lifts to an admissible path $\sigma = (\frac{\partial L}{\partial \dot{q}}(\tau), q(\tau), \tau)$, $a \leq \tau \leq b$, lying on the hypersurface $H(p, q) = E$. $\qquad\square$

Problem 2.4 (Jacobi). On a Riemannian manifold (M, ds^2) consider a Lagrangian system with $L(q, v) = \frac{1}{2}\|v\|^2 - V(q)$. Let $E > V(q)$ for all $q \in M$. Show that the trajectories of a closed Lagrangian system (M, L) with total energy E are geodesics for the Riemannian metric $d\hat{s}^2 = (E - V(q))ds^2$ on M.

2.3. The action as a function of coordinates. Consider a non-degenerate Lagrangian system (M, L) and denote by $\gamma(t; q_0, v_0)$ the solution of Euler-Lagrange equations

$$\frac{d}{dt}\frac{\partial L}{\partial \dot{q}} - \frac{\partial L}{\partial q} = 0$$

with the initial conditions $\gamma(t_0) = q_0 \in M$ and $\dot{\gamma}(t_0) = v_0 \in T_{q_0}M$. Suppose that there exist a neighborhood $V_0 \subset T_{v_0}M$ of v_0 and $t_1 > t_0$ such that for all $v \in V_0$ the extremals $\gamma(t; q_0, v)$, which start at time t_0 at q_0, do not intersect in the extended configuration space $M \times \mathbb{R}$ for times $t_0 < t < t_1$. Such extremals are said to form a *central field* which includes the extremal $\gamma_0(t) = \gamma(t; q_0, v_0)$. The existence of the central field of extremals is equivalent to the condition that for every $t_0 < t < t_1$ there is a neighborhood $U_t \subset M$ of $\gamma_0(t) \in M$ such that the mapping

$$(2.2) \qquad V_0 \ni v \mapsto q(t) = \gamma(t; q_0, v) \in U_t$$

is a diffeomorphism. Basic theorems in the theory of ordinary differential equations guarantee that for t_1 sufficiently close to t_0 every extremal $\gamma(t)$ for $t_0 < t < t_1$ can be included into the central field. In standard coordinates the mapping (2.2) is given by $\dot{q} \mapsto q(t) = \gamma(t; q_0, \dot{q})$.

For the central field of extremals $\gamma(t; q_0, \dot{q})$, $t_0 < t < t_1$, we define the *action as a function of coordinates and time* (or, *classical action*) by

$$S(q, t; q_0, t_0) = \int_{t_0}^{t} L(\gamma'(\tau))d\tau,$$

---

[15]The accurate formulation of Maupertuis' principle is due to Euler and Lagrange.

where $\gamma(\tau)$ is the extremal from the central field that connects $\boldsymbol{q}_0$ and $\boldsymbol{q}$. For given $\boldsymbol{q}_0$ and t_0, the classical action is defined for $t \in (t_0, t_1)$ and $\boldsymbol{q} \in \bigcup_{t_0 < t < t_1} U_t$. For a fixed energy E,

$$(2.3) \qquad S(\boldsymbol{q}, t; \boldsymbol{q}_0, t_0) = S_0(\boldsymbol{q}, t; \boldsymbol{q}_0, t_0) - E(t - t_0),$$

where S_0 is the abbreviated action from the previous section.

Theorem 2.7. *The differential of the classical action $S(\boldsymbol{q}, t)$ with fixed initial point is given by*

$$dS = \boldsymbol{p}\,d\boldsymbol{q} - H\,dt,$$

where $\boldsymbol{p} = \dfrac{\partial L}{\partial \dot{\boldsymbol{q}}}(\boldsymbol{q}, \dot{\boldsymbol{q}})$ and $H = \boldsymbol{p}\dot{\boldsymbol{q}} - L(\boldsymbol{q}, \dot{\boldsymbol{q}})$ are determined by the velocity $\dot{\boldsymbol{q}}$ of the extremal $\gamma(\tau)$ at time t.

Proof. Let $\boldsymbol{q}_\varepsilon$ be a path in M passing through $\boldsymbol{q}$ at $\varepsilon = 0$ with the tangent vector $\boldsymbol{v} \in T_{\boldsymbol{q}}M \simeq \mathbb{R}^n$, and for ε small enough let $\gamma_\varepsilon(\tau)$ be the family of extremals from the central field satisfying $\gamma_\varepsilon(t_0) = \boldsymbol{q}_0$ and $\gamma_\varepsilon(t) = \boldsymbol{q}_\varepsilon$. For the infinitesimal variation $\delta\gamma$ we have $\delta\gamma(t_0) = 0$ and $\delta\gamma(t) = \boldsymbol{v}$, and for fixed t we get from the formula for variation with the free ends (1.2) that

$$dS(\boldsymbol{v}) = \frac{\partial L}{\partial \dot{\boldsymbol{q}}}\boldsymbol{v}.$$

This shows that $\dfrac{\partial S}{\partial \boldsymbol{q}} = \boldsymbol{p}$. Setting $\boldsymbol{q}(t) = \gamma(t)$, we obtain

$$\frac{d}{dt}S(\boldsymbol{q}(t), t) = \frac{\partial S}{\partial \boldsymbol{q}}\dot{\boldsymbol{q}} + \frac{\partial S}{\partial t} = L,$$

so that $\dfrac{\partial S}{\partial t} = L - \boldsymbol{p}\dot{\boldsymbol{q}} = -H$. $\qquad\qquad\qquad\qquad\qquad\qquad\square$

Corollary 2.8. *The classical action satisfies the following nonlinear partial differential equation*

$$(2.4) \qquad \frac{\partial S}{\partial t} + H\left(\frac{\partial S}{\partial \boldsymbol{q}}, \boldsymbol{q}\right) = 0.$$

This equation is called the *Hamilton-Jacobi equation*. Hamilton's equations (2.1) can be used for solving the Cauchy problem

$$(2.5) \qquad S(\boldsymbol{q}, t)|_{t=0} = s(\boldsymbol{q}), \quad s \in C^\infty(M),$$

for Hamilton-Jacobi equation (2.4) by the method of characteristics. Namely, assume the existence of the Hamiltonian phase flow g_t on T^*M and consider the Lagrangian submanifold

$$\mathscr{L} = \left\{(\boldsymbol{p}, \boldsymbol{q}) \in T^*M : \boldsymbol{p} = \frac{\partial s(\boldsymbol{q})}{\partial \boldsymbol{q}}\right\},$$

a graph of the 1-form ds on M — a section of the cotangent bundle $\pi : T^*M \to M$. The mapping $\pi|_{\mathscr{L}}$ is one-to-one and for sufficiently small t the restriction of the projection π to the Lagrangian submanifold $\mathscr{L}_t = g_t(\mathscr{L})$ remains to be one-to-one. In other words, there is $t_1 > 0$ such that for all $0 \leq t < t_1$ the mapping $\pi_t = \pi \circ g_t \circ (\pi|_{\mathscr{L}})^{-1} : M \to M$ is a diffeomorphism, and the extremals $\gamma(\tau, \boldsymbol{q}_0, \dot{\boldsymbol{q}}_0)$ in the extended configuration space $M \times \mathbb{R}$, where $\dot{\boldsymbol{q}}_0 = \dfrac{\partial H}{\partial \boldsymbol{p}}(\boldsymbol{p}_0, \boldsymbol{q}_0)$ and $(\boldsymbol{p}_0, \boldsymbol{q}_0) \in \mathscr{L}$, do not intersect. Such extremals are called the *characteristics* of the Hamilton-Jacobi equation.

Proposition 2.1. *For $0 \leq t < t_1$ the solution $S(\boldsymbol{q}, t)$ to the Cauchy problem* (2.4)–(2.5) *is given by*

$$S(\boldsymbol{q}, t) = s(\boldsymbol{q}_0) + \int_0^t L(\gamma'(\tau))d\tau.$$

Here $\gamma(\tau)$ is the characteristic with $\gamma(t) = \boldsymbol{q}$ and with the starting point $\boldsymbol{q}_0 = \gamma(0)$ which is uniquely determined by $\boldsymbol{q}$.

Proof. As in the proof of Theorem 2.7 we use formula (1.2), where now $\boldsymbol{q}_0$ depends on $\boldsymbol{q}$, and obtain

$$\frac{\partial S}{\partial \boldsymbol{q}}(\boldsymbol{q}) = \frac{\partial s}{\partial \boldsymbol{q}_0}(\boldsymbol{q}_0)\frac{\partial \boldsymbol{q}_0}{\partial \boldsymbol{q}} + \frac{\partial L}{\partial \dot{\boldsymbol{q}}}(\boldsymbol{q}, \dot{\boldsymbol{q}}) - \frac{\partial L}{\partial \dot{\boldsymbol{q}}_0}(\boldsymbol{q}_0, \dot{\boldsymbol{q}}_0)\frac{\partial \boldsymbol{q}_0}{\partial \boldsymbol{q}} = \boldsymbol{p},$$

since $\dfrac{\partial s}{\partial \boldsymbol{q}_0}(\boldsymbol{q}_0) = \boldsymbol{p}_0 = \dfrac{\partial L}{\partial \dot{\boldsymbol{q}}_0}(\boldsymbol{q}_0, \dot{\boldsymbol{q}}_0)$. Setting $\boldsymbol{q}(t) = \gamma(t)$ we get

$$\frac{d}{dt}S(\boldsymbol{q}(t), t) = \frac{\partial S}{\partial \boldsymbol{q}}\dot{\boldsymbol{q}} + \frac{\partial S}{\partial t} = L(\boldsymbol{q}, \dot{\boldsymbol{q}}),$$

so that

$$\frac{\partial S}{\partial t} = -H(\boldsymbol{p}, \boldsymbol{q}),$$

and S satisfies the Hamilton-Jacobi equation. $\qquad\square$

We can also consider the action $S(\boldsymbol{q}, t; \boldsymbol{q}_0, t_0)$ as a function of both variables $\boldsymbol{q}$ and $\boldsymbol{q}_0$. The analog of Theorem 2.7 is the following statement.

Proposition 2.2. *The differential of the classical action as a function of initial and final points is given by*

$$dS = \boldsymbol{p}d\boldsymbol{q} - \boldsymbol{p}_0 d\boldsymbol{q}_0 - H(\boldsymbol{p}, \boldsymbol{q})dt + H(\boldsymbol{p}_0, \boldsymbol{q}_0)dt_0.$$

Problem 2.5. Prove that the solution to the Cauchy problem for the Hamilton-Jacobi equation is unique.

2.4. Classical observables and Poisson bracket. Smooth real-valued functions on the phase space T^*M are called *classical observables*. The vector space $C^\infty(T^*M)$ is an $\mathbb{R}$-algebra — an associative algebra over $\mathbb{R}$ with a unit given by the constant function 1, and with a multiplication given by the point-wise product of functions. The commutative algebra $C^\infty(T^*M)$ is called the *algebra of classical observables*. Assuming that the Hamiltonian phase flow g_t exists for all times, the time evolution of every observable $f \in C^\infty(T^*M)$ is given by

$$f_t(p,q) = f(g_t(p,q)) = f(p(t),q(t)), \quad (p,q) \in TM.$$

Equivalently, the time evolution is described by the differential equation

$$\frac{df_t}{dt} = \left.\frac{df_{s+t}}{ds}\right|_{s=0} = \left.\frac{d(f_t \circ g_s)}{ds}\right|_{s=0} = X_H(f_t)$$

$$= \sum_{i=1}^n \left(\frac{\partial H}{\partial p_i}\frac{\partial f_t}{\partial q^i} - \frac{\partial H}{\partial q^i}\frac{\partial f_t}{\partial p_i} \right) = \frac{\partial H}{\partial \boldsymbol{p}}\frac{\partial f_t}{\partial \boldsymbol{q}} - \frac{\partial H}{\partial \boldsymbol{q}}\frac{\partial f_t}{\partial \boldsymbol{p}},$$

called Hamilton's equation for classical observables. Setting

$$(2.6) \qquad \{f,g\} = X_f(g) = \frac{\partial f}{\partial \boldsymbol{p}}\frac{\partial g}{\partial \boldsymbol{q}} - \frac{\partial f}{\partial \boldsymbol{q}}\frac{\partial g}{\partial \boldsymbol{p}}, \quad f,g \in C^\infty(T^*M),$$

we can rewrite Hamilton's equation in the concise form

$$(2.7) \qquad \frac{df}{dt} = \{H,f\},$$

where it is understood that (2.7) is a differential equation for a family of functions f_t on T^*M with the initial condition $f_t(p,q)|_{t=0} = f(p,q)$. The properties of the bilinear mapping

$$\{\,,\,\} : C^\infty(T^*M) \times C^\infty(T^*M) \to C^\infty(T^*M)$$

are summarized below.

Theorem 2.9. *The mapping $\{\,,\,\}$ satisfies the following properties.*

(i) *(Relation with the symplectic form)*

$$\{f,g\} = \omega(J(df), J(dg)) = \omega(X_f, X_g).$$

(ii) *(Skew-symmetry)*

$$\{f,g\} = -\{g,f\}.$$

(iii) *(Leibniz rule)*

$$\{fg,h\} = f\{g,h\} + g\{f,h\}.$$

(iv) *(Jacobi identity)*

$$\{f,\{g,h\}\} + \{g,\{h,f\}\} + \{h,\{f,g\}\} = 0$$

for all $f,g,h \in C^\infty(T^*M)$.

Proof. Property (i) immediately follows from the definitions of ω and J in Section 2.1. Properties (ii)-(iii) are obvious. The Jacobi identity could be verified by a direct computation using (2.6), or by the following elegant argument. Observe that $\{f, g\}$ is a bilinear form in the first partial derivatives of f and g, and every term in the left-hand side of the Jacobi identity is a linear homogenous function of second partial derivatives of f, g, and h. Now the only terms in the Jacobi identity which could actually contain second partial derivatives of a function h are the following:

$$\{f, \{g, h\}\} + \{g, \{h, f\}\} = (X_f X_g - X_g X_f)(h).$$

However, this expression does not contain second partial derivatives of h since it is a commutator of two differential operators of the first order which is again a differential operator of the first order! $\qquad\square$

The observable $\{f, g\}$ is called the *canonical Poisson bracket* of the observables f and g. The Poisson bracket map $\{\ ,\ \} : C^\infty(T^*M) \times C^\infty(T^*M) \to C^\infty(T^*M)$ turns the algebra of classical observables $C^\infty(T^*M)$ into a Lie algebra with a Lie bracket given by the Poisson bracket. It has an important property that the Lie bracket is a bi-derivation with respect to the multiplication in $C^\infty(T^*M)$. The algebra of classical observables $C^\infty(T^*M)$ is an example of the *Poisson algebra* — a commutative algebra over $\mathbb{R}$ carrying a structure of a Lie algebra with the property that the Lie bracket is a derivation with respect to the algebra product.

In Lagrangian mechanics, a function I on TM is an integral of motion for the Lagrangian system (M, L) if it is constant along the trajectories. In Hamiltonian mechanics, an observable I — a function on the phase space T^*M — is called an integral of motion (first integral) for Hamilton's equations (2.1) if it is constant along the Hamiltonian phase flow. According to (2.7), this is equivalent to the condition

$$\{H, I\} = 0.$$

It is said that the observables H and I are *in involution* (*Poisson commute*).

2.5. Canonical transformations and generating functions.

Definition. A diffeomorphism g of the phase space T^*M is called a *canonical transformation*, if it preserves the canonical symplectic form ω on T^*M, i.e., $g^*(\omega) = \omega$. By Theorem 2.1, the Hamiltonian phase flow g_t is a one-parameter group of canonical transformations.

Proposition 2.3. *Canonical transformations preserve Hamilton's equations.*

Proof. From $g^*(\omega) = \omega$ it follows that the mapping $J : T^*(T^*M) \to T(T^*M)$ satisfies

$$(2.8) \qquad\qquad g_* \circ J \circ g^* = J.$$

Indeed, for all $X, Y \in \mathrm{Vect}(M)$ we have[16]

$$\omega(X, Y) = g^*(\omega)(X, Y) = \omega(g_*(X), g_*(Y)) \circ g,$$

so that for every 1-form ϑ on M,

$$\omega(X, J(g^*(\vartheta))) = g^*(\vartheta)(X) = \vartheta(g_*(X)) \circ g = \omega(g_*(X), J(\vartheta)) \circ g,$$

which gives $g_*(J(g^*(\vartheta))) = J(\vartheta)$. Using (2.8), we get

$$g_*(X_H) = g_*(J(dH)) = J((g^*)^{-1}(dH)) = X_K,$$

where $K = H \circ g^{-1}$. Thus the canonical transformation g maps trajectories of the Hamiltonian vector field X_H into the trajectories of the Hamiltonian vector field X_K. $\qquad\qquad\square$

Remark. In classical terms, Proposition 2.3 means that canonical Hamilton's equations

$$\dot{p} = -\frac{\partial H}{\partial q}(p, q), \quad \dot{q} = \frac{\partial H}{\partial p}(p, q)$$

in new coordinates $(P, Q) = g(p, q)$ continue to have the canonical form

$$\dot{P} = -\frac{\partial K}{\partial Q}(P, Q), \quad \dot{Q} = \frac{\partial K}{\partial P}(P, Q)$$

with the old Hamiltonian function $K(P, Q) = H(p, q)$.

Consider now the classical case $M = \mathbb{R}^n$. For a canonical transformation $(P, Q) = g(p, q)$ set $P = P(p, q)$ and $Q = Q(p, q)$. Since $dP \wedge dQ = dp \wedge dq$ on $T^*M \simeq \mathbb{R}^{2n}$, the 1-form $pdq - PdQ$ — the difference between the canonical Liouville 1-form and its pullback by the mapping g — is closed. From the Poincaré lemma it follows that there exists a function $F(p, q)$ on $\mathbb{R}^{2n}$ such that

$$(2.9) \qquad\qquad pdq - PdQ = dF(p, q).$$

Now assume that at some point (p_0, q_0) the $n \times n$ matrix $\dfrac{\partial P}{\partial p} = \left\{\dfrac{\partial P_i}{\partial p_j}\right\}_{i,j=1}^{n}$ is non-degenerate. By the inverse function theorem, there exists a neighborhood U of (p_0, q_0) in $\mathbb{R}^{2n}$ for which the functions P, q are coordinate functions. The function

$$S(P, q) = F(p, q) + PQ$$

---

[16]Since g is a diffeomorphism, $g_* X$ is a well-defined vector field on M.

is called a *generating function* of the canonical transformation g in U. It follows from (2.9) that in new coordinates $\boldsymbol{P}, \boldsymbol{q}$ on U,

$$\boldsymbol{p} = \frac{\partial S}{\partial \boldsymbol{q}}(\boldsymbol{P}, \boldsymbol{q}) \quad \text{and} \quad \boldsymbol{Q} = \frac{\partial S}{\partial \boldsymbol{P}}(\boldsymbol{P}, \boldsymbol{q}).$$

The converse statement below easily follows from the implicit function theorem.

Proposition 2.4. *Let $S(\boldsymbol{P}, \boldsymbol{q})$ be a function in some neighborhood U of a point $(\boldsymbol{P}_0, \boldsymbol{q}_0) \in \mathbb{R}^{2n}$ such that the $n \times n$ matrix*

$$\frac{\partial^2 S}{\partial \boldsymbol{P} \partial \boldsymbol{q}}(\boldsymbol{P}_0, \boldsymbol{q}_0) = \left\{ \frac{\partial^2 S}{\partial P_i \partial q^j}(\boldsymbol{P}_0, \boldsymbol{q}_0) \right\}_{i,j=1}^{n}$$

is non-degenerate. Then S is a generating function of a local (i.e., defined in some neighborhood of $(\boldsymbol{P}_0, \boldsymbol{q}_0)$ in $\mathbb{R}^{2n}$) canonical transformation.

Suppose there is a canonical transformation $(\boldsymbol{P}, \boldsymbol{Q}) = g(\boldsymbol{p}, \boldsymbol{q})$ such that $H(\boldsymbol{p}, \boldsymbol{q}) = K(\boldsymbol{P})$ for some function K. Then in the new coordinates Hamilton's equations take the form

$$(2.10) \qquad \dot{\boldsymbol{P}} = 0, \quad \dot{\boldsymbol{Q}} = \frac{\partial K}{\partial \boldsymbol{P}},$$

and are trivially integrated:

$$\boldsymbol{P}(t) = \boldsymbol{P}(0), \quad \boldsymbol{Q}(t) = \boldsymbol{Q}(0) + t\frac{\partial K}{\partial \boldsymbol{P}}(\boldsymbol{P}(0)).$$

Assuming that the matrix $\dfrac{\partial \boldsymbol{P}}{\partial \boldsymbol{p}}$ is non-degenerate, the generating function $S(\boldsymbol{P}, \boldsymbol{q})$ satisfies the differential equation

$$(2.11) \qquad H\left(\frac{\partial S}{\partial \boldsymbol{q}}(\boldsymbol{P}, \boldsymbol{q}), \boldsymbol{q} \right) = K(\boldsymbol{P}),$$

where after the differentiation one should substitute $\boldsymbol{q} = \boldsymbol{q}(\boldsymbol{P}, \boldsymbol{Q})$, defined by the canonical transformation g^{-1}. The differential equation (2.11) for fixed $\boldsymbol{P}$, as it follows from (2.3), coincides with the Hamilton-Jacobi equation for the abbreviated action $S_0 = S - Et$ where $E = K(\boldsymbol{P})$,

$$H\left(\frac{\partial S_0}{\partial \boldsymbol{q}}(\boldsymbol{P}, \boldsymbol{q}), \boldsymbol{q} \right) = E.$$

Theorem 2.10 (Jacobi). *Suppose that there is a function $S(\boldsymbol{P}, \boldsymbol{q})$ which depends on n parameters $\boldsymbol{P} = (P_1, \ldots, P_n)$, satisfies the Hamilton-Jacobi equation (2.11) for some function $K(\boldsymbol{P})$, and has the property that the $n \times n$ matrix $\dfrac{\partial^2 S}{\partial \boldsymbol{P} \partial \boldsymbol{q}}$ is non-degenerate. Then Hamilton's equations*

$$\dot{\boldsymbol{p}} = -\frac{\partial H}{\partial \boldsymbol{q}}, \quad \dot{\boldsymbol{q}} = \frac{\partial H}{\partial \boldsymbol{p}}$$

can be solved explicitly, and the functions $\boldsymbol{P}(\boldsymbol{p},\boldsymbol{q}) = (P_1(\boldsymbol{p},\boldsymbol{q}),\ldots,P_n(\boldsymbol{p},\boldsymbol{q}))$, defined by the equations $\boldsymbol{p} = \dfrac{\partial S}{\partial \boldsymbol{q}}(\boldsymbol{P},\boldsymbol{q})$, are integrals of motion in involution.

Proof. Set $\boldsymbol{p} = \dfrac{\partial S}{\partial \boldsymbol{q}}(\boldsymbol{P},\boldsymbol{q})$ and $\boldsymbol{Q} = \dfrac{\partial S}{\partial \boldsymbol{P}}(\boldsymbol{P},\boldsymbol{q})$. By the inverse function theorem, $g(\boldsymbol{p},\boldsymbol{q}) = (\boldsymbol{P},\boldsymbol{Q})$ is a local canonical transformation with the generating function S. It follows from (2.11) that $H(\boldsymbol{p}(\boldsymbol{P},\boldsymbol{Q}),\boldsymbol{q}(\boldsymbol{P},\boldsymbol{Q})) = K(\boldsymbol{P})$, so that Hamilton's equations take the form (2.10). Since $\omega = d\boldsymbol{P}\wedge d\boldsymbol{Q}$, integrals of motion $P_1(\boldsymbol{p},\boldsymbol{q}),\ldots,P_n(\boldsymbol{p},\boldsymbol{q})$ are in involution. $\qquad\square$

The solution of the Hamilton-Jacobi equation satisfying conditions in Theorem 2.10 is called the *complete integral*. At first glance it seems that solving the Hamilton-Jacobi equation, which is a nonlinear partial differential equation, is a more difficult problem then solving Hamilton's equations, which is a system of ordinary differential equations. It is quite remarkable that for many problems of classical mechanics one can find the complete integral of the Hamilton-Jacobi equation by the method of separation of variables. By Theorem 2.10, this solves the corresponding Hamilton's equations.

Problem 2.6. Find the generating function for the identity transformation $\boldsymbol{P} = \boldsymbol{p}, \boldsymbol{Q} = \boldsymbol{q}$.

Problem 2.7. Prove Proposition 2.4.

Problem 2.8. Suppose that the canonical transformation $g(\boldsymbol{p},\boldsymbol{q}) = (\boldsymbol{P},\boldsymbol{Q})$ is such that locally $(\boldsymbol{Q},\boldsymbol{q})$ can be considered as new coordinates (canonical transformations with this property are called *free*). Prove that $S_1(\boldsymbol{Q},\boldsymbol{q}) = F(\boldsymbol{p},\boldsymbol{q})$, also called a generating function, satisfies

$$\boldsymbol{p} = \frac{\partial S_1}{\partial \boldsymbol{q}} \quad \text{and} \quad \boldsymbol{P} = -\frac{\partial S_1}{\partial \boldsymbol{Q}}.$$

Problem 2.9. Find the complete integral for the case of a particle in $\mathbb{R}^3$ moving in a central field.

2.6. Symplectic manifolds. The notion of a symplectic manifold is a generalization of the example of a cotangent bundle T^*M.

Definition. A non-degenerate, closed 2-form ω on a manifold $\mathscr{M}$ is called a *symplectic form*, and the pair $(\mathscr{M},\omega)$ is called a *symplectic manifold*.

Since a symplectic form ω is non-degenerate, a symplectic manifold $\mathscr{M}$ is necessarily even-dimensional, $\dim\mathscr{M} = 2n$. The nowhere vanishing $2n$-form ω^n defines a canonical orientation on $\mathscr{M}$, and as in the case $\mathscr{M} = T^*M$, $\dfrac{\omega^n}{n!}$ is called Liouville's volume form. We also have the general notion of a Lagrangian submanifold.

Definition. A submanifold $\mathscr{L}$ of a symplectic manifold $(\mathscr{M}, \omega)$ is called a *Lagrangian submanifold*, if $\dim \mathscr{L} = \frac{1}{2} \dim \mathscr{M}$ and the restriction of the symplectic form ω to $\mathscr{L}$ is 0.

Symplectic manifolds form a category. A morphism between $(\mathscr{M}_1, \omega_1)$ and $(\mathscr{M}_2, \omega_2)$, also called a *symplectomorphism*, is a mapping $f : \mathscr{M}_1 \to \mathscr{M}_2$ such that $\omega_1 = f^*(\omega_2)$. When $\mathscr{M}_1 = \mathscr{M}_2$ and $\omega_1 = \omega_2$, the notion of a symplectomorphism generalizes the notion of a canonical transformation. The direct product of symplectic manifolds $(\mathscr{M}_1, \omega_1)$ and $(\mathscr{M}_2, \omega_2)$ is a symplectic manifold

$$(\mathscr{M}_1 \times \mathscr{M}_2, \pi_1^*(\omega_1) + \pi_2^*(\omega_2)),$$

where π_1 and π_2 are, respectively, projections of $\mathscr{M}_1 \times \mathscr{M}_2$ onto the first and second factors in the Cartesian product.

Besides cotangent bundles, another important class of symplectic manifolds is given by Kähler manifolds[17]. Recall that a complex manifold $\mathscr{M}$ is a Kähler manifold if it carries the Hermitian metric whose imaginary part is a closed $(1,1)$-form. In local complex coordinates $z = (z^1, \dots, z^n)$ on $\mathscr{M}$ the Hermitian metric is written as

$$h = \sum_{\alpha,\beta=1}^{n} h_{\alpha\bar{\beta}}(z, \bar{z}) dz^\alpha \otimes d\bar{z}^\beta.$$

Correspondingly,

$$g = \operatorname{Re} h = \frac{1}{2} \sum_{\alpha,\beta=1}^{n} h_{\alpha\bar{\beta}}(z, \bar{z})(dz^\alpha \otimes d\bar{z}^\beta + d\bar{z}^\beta \otimes dz^\alpha)$$

is the Riemannian metric on $\mathscr{M}$ and

$$\omega = -\operatorname{Im} h = \frac{i}{2} \sum_{\alpha,\beta=1}^{n} h_{\alpha\bar{\beta}}(z, \bar{z}) dz^\alpha \wedge d\bar{z}^\beta$$

is the symplectic form on $\mathscr{M}$ (considered as a $2n$-dimensional real manifold).

The simplest compact Kähler manifold is $\mathbb{C}P^1 \simeq S^2$ with the symplectic form given by the area 2-form of the Hermitian metric of Gaussian curvature 1 — the round metric on the 2-sphere. In terms of the local coordinate z associated with the stereographic projection $\mathbb{C}P^1 \simeq \mathbb{C} \cup \{\infty\}$,

$$\omega = 2i \, \frac{dz \wedge d\bar{z}}{(1 + |z|^2)^2}.$$

Similarly, the natural symplectic form on the complex projective space $\mathbb{C}P^n$ is the symplectic form of the Fubini-Study metric. By pull-back, it defines symplectic forms on complex projective varieties.

---

[17]Needless to say, not every symplectic manifold admits a complex structure, not to mention a Kähler structure.

The simplest non-compact Kähler manifold is the n-dimensional complex vector space $\mathbb{C}^n$ with the standard Hermitian metric. In complex coordinates $\boldsymbol{z} = (z^1, \ldots, z^n)$ on $\mathbb{C}^n$ it is given by

$$h = d\boldsymbol{z} \otimes d\bar{\boldsymbol{z}} = \sum_{\alpha=1}^{n} dz^\alpha \otimes d\bar{z}^\alpha.$$

In terms of real coordinates $(\boldsymbol{x}, \boldsymbol{y}) = (x^1, \ldots, x^n, y^1, \ldots, y^n)$ on $\mathbb{R}^{2n} \simeq \mathbb{C}^n$, where $\boldsymbol{z} = \boldsymbol{x} + i\boldsymbol{y}$, the corresponding symplectic form $\omega = -\operatorname{Im} h$ has the canonical form

$$\omega = \frac{i}{2} d\boldsymbol{z} \wedge d\bar{\boldsymbol{z}} = \sum_{\alpha=1}^{n} dx^\alpha \wedge dy^\alpha = d\boldsymbol{x} \wedge d\boldsymbol{y}.$$

This example naturally leads to the following definition.

Definition. A *symplectic vector space* is a pair (V, ω), where V is a vector space over $\mathbb{R}$ and ω is a non-degenerate, skew-symmetric bilinear form on V.

It follows from basic linear algebra that every symplectic vector space V has a *symplectic basis* — a basis $e^1, \ldots, e^n, f_1, \ldots, f_n$ of V, where $2n = \dim V$, such that

$$\omega(e^i, e^j) = \omega(f_i, f_j) = 0 \quad \text{and} \quad \omega(e^i, f_j) = \delta^i_j, \quad i, j = 1, \ldots, n.$$

In coordinates $(\boldsymbol{p}, \boldsymbol{q}) = (p_1, \ldots, p_n, q^1, \ldots, q^n)$ corresponding to this basis, $V \simeq \mathbb{R}^{2n}$ and

$$\omega = d\boldsymbol{p} \wedge d\boldsymbol{q} = \sum_{i=1}^{n} dp_i \wedge dq^i.$$

Thus every symplectic vector space is isomorphic to a direct product of the phase planes $\mathbb{R}^2$ with the canonical symplectic form $dp \wedge dq$. Introducing complex coordinates $\boldsymbol{z} = \boldsymbol{p} + i\boldsymbol{q}$, we get the isomorphism $V \simeq \mathbb{C}^n$, so that every symplectic vector space admits a Kähler structure.

It is a basic fact of symplectic geometry that every symplectic manifold is locally isomorphic to a symplectic vector space.

Theorem 2.11 (Darboux' theorem). *Let $(\mathcal{M}, \omega)$ be a $2n$-dimensional symplectic manifold. For every point $x \in \mathcal{M}$ there is a neighborhood U of x with local coordinates $(\boldsymbol{p}, \boldsymbol{q}) = (p_1, \ldots, p_n, q^1, \ldots, q^n)$ such that on U*

$$\omega = d\boldsymbol{p} \wedge d\boldsymbol{q} = \sum_{i=1}^{n} dp_i \wedge dq^i.$$

Coordinates $\boldsymbol{p}, \boldsymbol{q}$ are called *canonical coordinates* (*Darboux coordinates*). The proof proceeds by induction on n with the two main steps stated as Problems 2.13 and 2.14.

A non-degenerate 2-form ω for every $x \in \mathcal{M}$ defines an isomorphism $J : T_x^*\mathcal{M} \to T_x\mathcal{M}$ by

$$\omega(u_1, u_2) = J^{-1}(u_2)(u_1), \quad u_1, u_2 \in T_x\mathcal{M}.$$

Explicitly, for every $X \in \mathrm{Vect}(\mathcal{M})$ and $\vartheta \in \mathcal{A}^1(\mathcal{M})$ we have

$$\omega(X, J(\vartheta)) = \vartheta(X) \quad \text{and} \quad J^{-1}(X) = -i_X(\omega)$$

(cf. Section 2.1). In local coordinates $\boldsymbol{x} = (x^1, \dots, x^{2n})$ for the coordinate chart (U, φ) on $\mathcal{M}$, the 2-form ω is given by

$$\omega = \tfrac{1}{2} \sum_{i,j=1}^{2n} \omega_{ij}(\boldsymbol{x}) \, dx^i \wedge dx^j,$$

where $\{\omega_{ij}(\boldsymbol{x})\}_{i,j=1}^{2n}$ is a non-degenerate, skew-symmetric matrix-valued function on $\varphi(U)$. Denoting the inverse matrix by $\{\omega^{ij}(\boldsymbol{x})\}_{i,j=1}^{2n}$, we have

$$J(dx^i) = -\sum_{j=1}^{2n} \omega^{ij}(\boldsymbol{x}) \frac{\partial}{\partial x^j}, \quad i = 1, \dots, 2n.$$

Definition. A *Hamiltonian system* is a pair consisting of a symplectic manifold $(\mathcal{M}, \omega)$, called a *phase space*, and a smooth real-valued function H on $\mathcal{M}$, called a *Hamiltonian*. The motion of points on the phase space is described by the vector field

$$X_H = J(dH),$$

called a *Hamiltonian vector field*.

The trajectories of a Hamiltonian system $((\mathcal{M}, \omega), H)$ are the integral curves of a Hamiltonian vector field X_H on $\mathcal{M}$. In canonical coordinates $(\boldsymbol{p}, \boldsymbol{q})$ they are described by the canonical Hamilton's equations (2.1),

$$\dot{\boldsymbol{p}} = -\frac{\partial H}{\partial \boldsymbol{q}}, \quad \dot{\boldsymbol{q}} = \frac{\partial H}{\partial \boldsymbol{p}}.$$

Suppose now that the Hamiltonian vector field X_H on $\mathcal{M}$ is complete. The *Hamiltonian phase flow* on $\mathcal{M}$ associated with a Hamiltonian H is a one-parameter group $\{g_t\}_{t \in \mathbb{R}}$ of diffeomorphisms of $\mathcal{M}$ generated by X_H. The following statement generalizes Theorem 2.1.

Theorem 2.12. *The Hamiltonian phase flow preserves the symplectic form.*

Proof. It is sufficient to show that $\mathcal{L}_{X_H}\omega = 0$. Using Cartan's formula

$$\mathcal{L}_X = i_X \circ d + d \circ i_X$$

and $d\omega = 0$, we get for every $X \in \mathrm{Vect}(\mathcal{M})$,

$$\mathcal{L}_X \omega = (d \circ i_X)(\omega).$$

Since $i_X(\omega)(Y) = \omega(X, Y)$, we have for $X = X_H$ and every $Y \in \mathrm{Vect}(\mathcal{M})$ that

$$i_{X_H}(\omega)(Y) = \omega(J(dH), Y) = -dH(Y).$$

Thus $i_{X_H}(\omega) = -dH$, and the statement follows from $d^2 = 0$. $\square$

Corollary 2.13. *A vector field X on $\mathcal{M}$ is a Hamiltonian vector field if and only if the 1-form $i_X(\omega)$ is exact.*

Definition. A vector field X on a symplectic manifold $(\mathcal{M}, \omega)$ is called a *symplectic vector field* if the 1-form $i_X(\omega)$ is closed, which is equivalent to $\mathscr{L}_X \omega = 0$.

The commutative algebra $C^\infty(\mathcal{M})$, with a multiplication given by the point-wise product of functions, is called the *algebra of classical observables*. Assuming that the Hamiltonian phase flow g_t exists for all times, the time evolution of every observable $f \in C^\infty(\mathcal{M})$ is given by

$$f_t(x) = f(g_t(x)), \quad x \in \mathcal{M},$$

and is described by the differential equation

$$\frac{df_t}{dt} = X_H(f_t)$$

— Hamilton's equation for classical observables. Hamilton's equations for observables on $\mathcal{M}$ have the same form as Hamilton's equations on $\mathcal{M} = T^*M$, considered in Section 2.3. Since

$$X_H(f) = df(X_H) = \omega(X_H, J(df)) = \omega(X_H, X_f),$$

we have the following.

Definition. A Poisson bracket on the algebra $C^\infty(\mathcal{M})$ of classical observables on a symplectic manifold $(\mathcal{M}, \omega)$ is a bilinear mapping $\{\,,\,\} : C^\infty(\mathcal{M}) \times C^\infty(\mathcal{M}) \to C^\infty(\mathcal{M})$, defined by

$$\{f, g\} = \omega(X_f, X_g), \quad f, g \in C^\infty(\mathcal{M}).$$

Now Hamilton's equation takes the concise form

$$(2.12) \qquad \frac{df}{dt} = \{H, f\},$$

understood as a differential equation for a family of functions f_t on $\mathcal{M}$ with the initial condition $f_t|_{t=0} = f$. In local coordinates $\boldsymbol{x} = (x^1, \ldots, x^{2n})$ on $\mathcal{M}$,

$$\{f, g\}(\boldsymbol{x}) = -\sum_{i,j=1}^{2n} \omega^{ij}(\boldsymbol{x}) \frac{\partial f(\boldsymbol{x})}{\partial x^i} \frac{\partial g(\boldsymbol{x})}{\partial x^j}.$$

Theorem 2.14. *The Poisson bracket $\{\,,\,\}$ on a symplectic manifold $(\mathcal{M}, \omega)$ is skew-symmetric and satisfies Leibniz rule and the Jacobi identity.*

Proof. The first two properties are obvious. It follows from the definition of a Poisson bracket and the formula

$$[X_f, X_g](h) = (X_g X_f - X_f X_g)(h) = \{g, \{f, h\}\} - \{f, \{g, h\}\}$$

that the Jacobi identity is equivalent to the property

$$(2.13) \qquad [X_f, X_g] = X_{\{f,g\}}.$$

Let X and Y be symplectic vector fields. Using Cartan's formulas we get

$$\begin{aligned}
i_{[X,Y]}(\omega) &= \mathcal{L}_X(i_Y(\omega)) - i_Y(\mathcal{L}_X(\omega)) \\
&= d(i_X \circ i_Y(\omega)) + i_X d(i_Y(\omega)) \\
&= d(\omega(Y, X)) = i_Z(\omega),
\end{aligned}$$

where Z is a Hamiltonian vector field corresponding to $\omega(X, Y) \in C^\infty(\mathcal{M})$. Since the 2-form ω is non-degenerate, this implies $[X, Y] = Z$, so that setting $X = X_f, Y = X_g$ and using $\{f, g\} = \omega(X_f, X_g)$, we get (2.13). $\qquad\square$

From (2.13) we immediately get the following result.

Corollary 2.15. *The subspace* $\mathrm{Ham}(\mathcal{M})$ *of Hamiltonian vector fields on* $\mathcal{M}$ *is a Lie subalgebra of* $\mathrm{Vect}(\mathcal{M})$. *The mapping* $C^\infty(\mathcal{M}) \to \mathrm{Ham}(\mathcal{M})$, *given by* $f \mapsto X_f$, *is a Lie algebra homomorphism with the kernel consisting of locally constant functions on* $\mathcal{M}$.

As in the case $\mathcal{M} = T^*M$ (see Section 2.4), an observable I — a function on the phase space $\mathcal{M}$ — is called an integral of motion (first integral) for the Hamiltonian system $((\mathcal{M}, \omega), H)$ if it is constant along the Hamiltonian phase flow. According to (2.12), this is equivalent to the condition

$$(2.14) \qquad \{H, I\} = 0.$$

It is said that the observables H and I are *in involution* (*Poisson commute*). From the Jacobi identity for the Poisson bracket we get the following result.

Corollary 2.16 (Poisson's theorem). *The Poisson bracket of two integrals of motion is an integral of motion.*

Proof. If $\{H, I_1\} = \{H, I_2\} = 0$, then

$$\{H, \{I_1, I_2\}\} = \{\{H, I_1\}, I_2\} - \{\{H, I_2\}, I_1\} = 0. \qquad\square$$

It follows from Poisson's theorem that integrals of motion form a Lie algebra and, by (2.13), corresponding Hamiltonian vector fields form a Lie subalgebra in $\mathrm{Vect}(\mathcal{M})$. Since $\{I, H\} = dH(X_I) = 0$, the vector fields X_I are tangent to submanifolds $\mathcal{M}_E = \{x \in \mathcal{M} : H(x) - E\}$ — the level sets of the Hamiltonian H. This defines a Lie algebra of integrals of motion for the Hamiltonian system $((\mathcal{M}, \omega), H)$ at the level set $\mathcal{M}_E$.

Let G be a finite-dimensional Lie group that acts on a connected symplectic manifold $(\mathcal{M}, \omega)$ by symplectomorphisms. The Lie algebra $\mathfrak{g}$ of G acts on $\mathcal{M}$ by vector fields

$$X_\xi(f)(x) = \left.\frac{d}{ds}\right|_{s=0} f(e^{-s\xi} \cdot x),$$

and the linear mapping $\mathfrak{g} \ni \xi \mapsto X_\xi \in \mathrm{Vect}(\mathcal{M})$ is a homomorphism of Lie algebras,

$$[X_\xi, X_\eta] = X_{[\xi,\eta]}, \quad \xi, \eta \in \mathfrak{g}.$$

The G-action is called a *Hamiltonian action* if X_ξ are Hamiltonian vector fields, i.e., for every $\xi \in \mathfrak{g}$ there is $\Phi_\xi \in C^\infty(\mathcal{M})$, defined up to an additive constant, such that $X_\xi = X_{\Phi_\xi} = J(d\Phi_\xi)$. It is called a *Poisson action* if there is a choice of functions Φ_ξ such that the linear mapping $\Phi : \mathfrak{g} \to C^\infty(\mathcal{M})$ is a homomorphism of Lie algebras,

$$(2.15) \qquad \{\Phi_\xi, \Phi_\eta\} = \Phi_{[\xi,\eta]}, \quad \xi, \eta \in \mathfrak{g}.$$

Definition. A Lie group G is a *symmetry group* of the Hamiltonian system $((\mathcal{M}, \omega), H)$ if there is a Hamiltonian action of G on $\mathcal{M}$ such that

$$H(g \cdot x) = H(x), \quad g \in G, \ x \in \mathcal{M}.$$

Theorem 2.17 (Noether theorem with symmetries). *If G is a symmetry group of the Hamiltonian system $((\mathcal{M}, \omega), H)$, then the functions Φ_ξ, $\xi \in \mathfrak{g}$, are the integrals of motion. If the action of G is Poisson, the integrals of motion satisfy* (2.15).

Proof. By definition of the Hamiltonian action, for every $\xi \in \mathfrak{g}$,

$$0 = X_\xi(H) = X_{\Phi_\xi}(H) = \{\Phi_\xi, H\}. \qquad \square$$

Corollary 2.18. *Let (M, L) be a Lagrangian system such that the Legendre transform $\tau_L : TM \to T^*M$ is a diffeomorphism. Then if a Lie group G is a symmetry of (M, L), then G is a symmetry group of the corresponding Hamiltonian system $((T^*M, \omega), H = E_L \circ \tau_L^{-1})$, and the corresponding G-action on T^*M is Poisson. In particular, $\Phi_\xi = -I_\xi \circ \tau_L^{-1}$, where I_ξ are Noether integrals of motion for the one-parameter subgroups of G generated by $\xi \in \mathfrak{g}$.*

Proof. Let X be the vector field associated with the one-parameter subgroup $\{e^{s\xi}\}_{s\in\mathbb{R}}$ of diffeomorphisms of M, used in Theorem 1.3, and let X' be its lift to TM. We have[18]

$$(2.16) \qquad\qquad X_\xi = -(\tau_L)_*(X'),$$

---

[18]The negative sign reflects the difference in definitions of X and X_ξ.

and it follows from (1.16) that $\Phi_\xi = i_{X_\xi}(\theta) = \theta(X_\xi)$, where θ is the canonical Liouville 1-form on T^*M. From Cartan's formula and (1.17) we get

$$d\Phi_\xi = d(i_{X_\xi}(\theta)) = -i_{X_\xi}(d\theta) + \mathcal{L}_{X_\xi}(\theta) = -i_{X_\xi}(\omega),$$

so that

$$J(d\Phi_\xi) = -J(i_{X_\xi}(\omega)) = X_\xi,$$

and the G-action is Hamiltonian. Using (1.17) and another Cartan's formula, we obtain

$$\Phi_{[\xi,\eta]} = i_{[X_\xi, X_\eta]}(\theta) = \mathcal{L}_{X_\xi}(i_{X_\eta}(\theta)) + i_{X_\eta}(\mathcal{L}_{X_\xi}(\theta))$$
$$= X_\xi(\Phi_\eta) = \{\Phi_\xi, \Phi_\eta\}. \qquad \square$$

Example 2.1. The Lagrangian

$$L = \tfrac{1}{2}m\dot{\boldsymbol{r}}^2 - V(r)$$

for a particle in $\mathbb{R}^3$ moving in a central field (see Section 1.6) is invariant with respect to the action of the group $SO(3)$ of orthogonal transformations of the Euclidean space $\mathbb{R}^3$. Let u_1, u_2, u_3 be a basis for the Lie algebra $\mathfrak{so}(3)$ corresponding to the rotations with the axes given by the vectors of the standard basis e_1, e_2, e_3 for $\mathbb{R}^3$ (see Example 1.10 in Section 1.4). These generators satisfy the commutation relations

$$[u_i, u_j] = \varepsilon_{ijk} u_k,$$

where $i, j, k = 1, 2, 3$, and ε_{ijk} is a totally anti-symmetric tensor, $\varepsilon_{123} = 1$. Corresponding Noether integrals of motion are given by $\Phi_{u_i} = -M_i$, where

$$M_1 = (\boldsymbol{r} \times \boldsymbol{p})_1 = r_2 p_3 - r_3 p_2,$$
$$M_2 = (\boldsymbol{r} \times \boldsymbol{p})_2 = r_3 p_1 - r_1 p_3,$$
$$M_3 = (\boldsymbol{r} \times \boldsymbol{p})_3 = r_1 p_2 - r_2 p_1$$

are components of the angular momentum vector $\boldsymbol{M} = \boldsymbol{r} \times \boldsymbol{p}$. (Here it is convenient to lower the indices of the coordinates r_i by the Euclidean metric on $\mathbb{R}^3$.) For the Hamiltonian

$$H = \frac{\boldsymbol{p}^2}{2m} + V(r)$$

we have

$$\{H, M_i\} = 0.$$

According to Theorem 2.17 and Corollary 2.18, Poisson brackets of the components of the angular momentum satisfy

$$\{M_i, M_j\} = -\varepsilon_{ijk} M_k,$$

which is also easy to verify directly using (2.6),

$$\{f, g\}(\boldsymbol{p}, \boldsymbol{r}) = \frac{\partial f}{\partial \boldsymbol{p}} \frac{\partial g}{\partial \boldsymbol{r}} - \frac{\partial f}{\partial \boldsymbol{r}} \frac{\partial g}{\partial \boldsymbol{p}}.$$

Example 2.2 (Kepler's problem). For every $\alpha \in \mathbb{R}$ the Lagrangian system on $\mathbb{R}^3$ with

$$L = \tfrac{1}{2}m\dot{\boldsymbol{r}}^2 + \frac{\alpha}{r}$$

has three extra integrals of motion — the components W_1, W_2, W_3 of the Laplace-Runge-Lenz vector, given by

$$\boldsymbol{W} = \frac{\boldsymbol{p}}{m} \times \boldsymbol{M} - \frac{\alpha \boldsymbol{r}}{r}$$

(see Section 1.6). Using Poisson brackets from the previous example, together with $\{r_i, M_j\} = -\varepsilon_{ijk}r_k$ and $\{p_i, M_j\} = -\varepsilon_{ijk}p_k$, we get by a straightforward computation,

$$\{W_i, M_j\} = -\varepsilon_{ijk}W_k \quad \text{and} \quad \{W_i, W_j\} = \frac{2H}{m}\varepsilon_{ijk}M_k,$$

where $H = \dfrac{\boldsymbol{p}^2}{2m} - \dfrac{\alpha}{r}$ is the Hamiltonian of Kepler's problem.

The Hamiltonian system $((\mathscr{M}, \omega), H)$, $\dim \mathscr{M} = 2n$, is called *completely integrable* if it has n independent integrals of motion $F_1 = H, \ldots, F_n$ in involution. The former condition means that $dF_1(x), \ldots, dF_n(x) \in T_x^*\mathscr{M}$ are linearly independent for almost all $x \in \mathscr{M}$. Hamiltonian systems with one degree of freedom such that dH has only finitely many zeros are completely integrable. Complete separation of variables in the Hamilton-Jacobi equation (see Section 2.5) provides other examples of completely integrable Hamiltonian systems.

Let $((\mathscr{M}, \omega), H)$ be a completely integrable Hamiltonian system. Suppose that the level set $\mathscr{M}_f = \{x \in \mathscr{M} : F_1(x) = f_1, \ldots, F_n(x) = f_n\}$ is compact and tangent vectors $JdF_1, \ldots, JdF_n$ are linearly independent for all $x \in \mathscr{M}_f$. Then by the Liouville-Arnold theorem, in a neighborhood of $\mathscr{M}_f$ there exist so-called *action-angle variables*: coordinates $\boldsymbol{I} = (I_1, \ldots, I_n) \in \mathbb{R}^n_+ = (\mathbb{R}_{>0})^n$ and $\boldsymbol{\varphi} = (\varphi_1, \ldots, \varphi_n) \in T^n = (\mathbb{R}/2\pi\mathbb{Z})^n$ such that $\omega = d\boldsymbol{I} \wedge d\boldsymbol{\varphi}$ and $H = H(I_1, \ldots, I_n)$. According to Hamilton's equations,

$$\dot{I}_i = 0 \quad \text{and} \quad \dot{\varphi}_i = \omega_i = \frac{\partial H}{\partial I_i}, \quad i = 1, \ldots, n,$$

so that action variables are constants, and angle variables change uniformly, $\varphi_i(t) = \varphi_i(0) + \omega_i t$, $i = 1, \ldots, n$. The classical motion is almost-periodic with the frequencies $\omega_1, \ldots, \omega_n$.

Problem 2.10. Show that a symplectic manifold $(\mathscr{M}, \omega)$ admits an *almost complex structure*: a bundle map $\mathscr{J} : T\mathscr{M} \to T\mathscr{M}$ such that $\mathscr{J}^2 = -\mathrm{id}$.

Problem 2.11. Give an example of a symplectic manifold which admits a complex structure but not a Kähler structure.

Problem 2.12 (Coadjoint orbits). Let G be a finite-dimensional Lie group, let $\mathfrak{g}$ be its Lie algebra, and let $\mathfrak{g}^*$ be the dual vector space to $\mathfrak{g}$. For $u \in \mathfrak{g}^*$ let $\mathcal{M} = \mathcal{O}_u$ be the orbit of u under the coadjoint action of G on $\mathfrak{g}^*$. Show that the formula

$$\omega(u_1, u_2) = u([x_1, x_2]),$$

where $u_1 = \mathrm{ad}^* x_1(u), u_2 = \mathrm{ad}^* x_2(u) \in \mathcal{O}_u$, and ad^* stands for the coadjoint action of a Lie algebra $\mathfrak{g}$ on $\mathfrak{g}^*$, gives rise to a well-defined 2-form on $\mathcal{M}$, which is closed and non-degenerate. (The 2-form ω is called the *Kirillov-Kostant* symplectic form.)

Problem 2.13. Let $(\mathcal{M}, \omega)$ be a symplectic manifold. For $x \in \mathcal{M}$ choose a function q^1 on $\mathcal{M}$ such that $q^1(x) = 0$ and dq^1 does not vanish at x, and set $X = -X_{q^1}$. Show that there is a neighborhood U of $x \in \mathcal{M}$ and a function p_1 on U such that $X(q^1) = 1$ on U, and there exist coordinates $p_1, q^1, z^1, \ldots, z^{2n-2}$ on U such that

$$X = \frac{\partial}{\partial p_1} \quad \text{and} \quad Y = X_{p_1} = \frac{\partial}{\partial q^1}.$$

Problem 2.14. Continuing Problem 2.13, show that the 2-form $\omega - dp_1 \wedge dq^1$ on U depends only on coordinates $z^1, \ldots, z^{2n-2}$ and is non-degenerate.

Problem 2.15. Do the computation in Example 2.2 and show that the Lie algebra of the integrals $M_1, M_2, M_3, W_1, W_2, W_3$ in Kepler's problem at $H(\boldsymbol{p}, \boldsymbol{r}) = E$ is isomorphic to the Lie algebra $\mathfrak{so}(4)$, if $E < 0$, to the Euclidean Lie algebra $\mathfrak{e}(3)$, if $E = 0$, and to the Lie algebra $\mathfrak{so}(1, 3)$, if $E > 0$.

Problem 2.16. Find the action-angle variables for a particle with one degree of freedom, when the potential $V(x)$ is a convex function on $\mathbb{R}$ satisfying $\lim_{|x| \to \infty} V(x) = \infty$. (*Hint:* Define $I = \oint p\,dx$, where integration goes over the closed orbit with $H(p, x) = E$.)

Problem 2.17. Show that a Hamiltonian system describing a particle in $\mathbb{R}^3$ moving in a central field is completely integrable, and find the action-angle variables.

Problem 2.18 (Symplectic quotients). For a Poisson action of a Lie group G on a symplectic manifold $(\mathcal{M}, \omega)$, define the *moment map* $P : \mathcal{M} \to \mathfrak{g}^*$ by

$$P(x)(\xi) = \Phi_\xi(x), \ \xi \in \mathfrak{g}, \ x \in \mathcal{M},$$

where $\mathfrak{g}$ is the Lie algebra of G. For every $p \in \mathfrak{g}^*$ such that a stabilizer G_p of p acts freely and properly on $\mathcal{M}_p = P^{-1}(p)$ (such p is called *the regular value* of the moment map), the quotient $M_p = G_p \backslash \mathcal{M}_p$ is called a *reduced phase space*. Show that M_p is a symplectic manifold with the symplectic form uniquely characterized by the condition that its pull-back to $\mathcal{M}_p$ coincides with the restriction to $\mathcal{M}_p$ of the symplectic form ω.

2.7. Poisson manifolds.

The notion of a Poisson manifold generalizes the notion of a symplectic manifold.

Definition. A *Poisson manifold* is a manifold $\mathcal{M}$ equipped with a *Poisson structure* — a skew-symmetric bilinear mapping

$$\{\,,\,\} : C^\infty(\mathcal{M}) \times C^\infty(\mathcal{M}) \to C^\infty(\mathcal{M})$$

which satisfies the Leibniz rule and Jacobi identity.

Equivalently, $\mathcal{M}$ is a Poisson manifold if the algebra $\mathcal{A} = C^\infty(\mathcal{M})$ of classical observables is a Poisson algebra — a Lie algebra such that the Lie bracket is a bi-derivation with respect to the multiplication in $\mathcal{A}$ (a pointwise product of functions). It follows from the derivation property that in local coordinates $\boldsymbol{x} = (x^1, \ldots, x^N)$ on $\mathcal{M}$, the Poisson bracket has the form

$$\{f, g\}(\boldsymbol{x}) = \sum_{i,j=1}^{N} \eta^{ij}(\boldsymbol{x}) \frac{\partial f(\boldsymbol{x})}{\partial x^i} \frac{\partial g(\boldsymbol{x})}{\partial x^j}.$$

The 2-tensor $\eta^{ij}(\boldsymbol{x})$, called a *Poisson tensor*, defines a global section η of the vector bundle $T\mathcal{M} \wedge T\mathcal{M}$ over $\mathcal{M}$.

The evolution of classical observables on a Poisson manifold is given by Hamilton's equations, which have the same form as (2.12),

$$\frac{df}{dt} = X_H(f) = \{H, f\}.$$

The phase flow g_t for a complete Hamiltonian vector field $X_H = \{H, \cdot\}$ defines the *evolution operator* $U_t : \mathcal{A} \to \mathcal{A}$ by

$$U_t(f)(x) = f(g_t(x)), \quad f \in \mathcal{A}.$$

Theorem 2.19. *Suppose that every Hamiltonian vector field on a Poisson manifold $(\mathcal{M}, \{\ ,\ \})$ is complete. Then for every $H \in \mathcal{A}$, the corresponding evolution operator U_t is an automorphism of the Poisson algebra $\mathcal{A}$, i.e.,*

$$(2.17) \qquad U_t(\{f, g\}) = \{U_t(f), U_t(g)\} \quad \text{for all} \quad f, g \in \mathcal{A}.$$

Conversely, if a skew-symmetric bilinear mapping $\{\ ,\ \} : C^\infty(\mathcal{M}) \times C^\infty(\mathcal{M}) \to C^\infty(\mathcal{M})$ is such that $X_H = \{H, \cdot\}$ are complete vector fields for all $H \in \mathcal{A}$, and corresponding evolution operators U_t satisfy (2.17), then $(\mathcal{M}, \{\ ,\ \})$ is a Poisson manifold.

Proof. Let $f_t = U_t(f)$, $g_t = U_t(g)$, and[19] $h_t = U_t(\{f, g\})$. By definition,

$$\frac{d}{dt}\{f_t, g_t\} = \{\{H, f_t\}, g_t\} + \{f_t, \{H, g_t\}\} \quad \text{and} \quad \frac{dh_t}{dt} = \{H, h_t\}.$$

If $(\mathcal{M}, \{\ ,\ \})$ is a Poisson manifold, then it follows from the Jacobi identity that

$$\{\{H, f_t\}, g_t\} + \{f_t, \{H, g_t\}\} = \{H, \{f_t, g_t\}\},$$

so that h_t and $\{f_t, g_t\}$ satisfy the same differential equation (2.12). Since these functions coincide at $t = 0$, (2.17) follows from the uniqueness theorem for the ordinary differential equations.

Conversely, we get the Jacobi identity for the functions f, g, and H by differentiating (2.17) with respect to t at $t = 0$. $\square$

---

[19] Here g_t is not the phase flow!

Corollary 2.20. *A global section η of $T\mathcal{M} \wedge T\mathcal{M}$ is a Poisson tensor if and only if*

$$\mathcal{L}_{X_f}\eta = 0 \quad \text{for all} \quad f \in \mathcal{A}.$$

Definition. The *center* of a Poisson algebra $\mathcal{A}$ is

$$\mathcal{Z}(\mathcal{A}) = \{f \in \mathcal{A} : \{f, g\} = 0 \quad \text{for all} \quad g \in \mathcal{A}\}.$$

A Poisson manifold $(\mathcal{M}, \{\ ,\ \})$ is called *non-degenerate* if the center of a Poisson algebra of classical observables $\mathcal{A} = C^\infty(\mathcal{M})$ consists only of locally constant functions ($\mathcal{Z}(\mathcal{A}) = \mathbb{R}$ for connected $\mathcal{M}$).

Equivalently, a Poisson manifold $(\mathcal{M}, \{\ ,\ \})$ is non-degenerate if the Poisson tensor η is non-degenerate everywhere on $\mathcal{M}$, so that $\mathcal{M}$ is necessarily an even-dimensional manifold. A non-degenerate Poisson tensor for every $x \in \mathcal{M}$ defines an isomorphism $J : T_x^*\mathcal{M} \to T_x\mathcal{M}$ by

$$\eta(u_1, u_2) = u_2(J(u_1)), \ u_1, u_2 \in T_x^*\mathcal{M}.$$

In local coordinates $\boldsymbol{x} = (x^1, \ldots, x^N)$ for the coordinate chart (U, φ) on $\mathcal{M}$, we have

$$J(dx^i) = \sum_{j=1}^N \eta^{ij}(\boldsymbol{x})\frac{\partial}{\partial x^j}, \quad i = 1, \ldots, N.$$

Poisson manifolds form a category. A morphism between $(\mathcal{M}_1, \{\ ,\ \}_1)$ and $(\mathcal{M}_2, \{\ ,\ \}_2)$ is a mapping $\varphi : \mathcal{M}_1 \to \mathcal{M}_2$ of smooth manifolds such that

$$\{f \circ \varphi, g \circ \varphi\}_1 = \{f, g\}_2 \circ \varphi \quad \text{for all} \quad f, g \in C^\infty(\mathcal{M}_2).$$

A direct product of Poisson manifolds $(\mathcal{M}_1, \{\ ,\ \}_1)$ and $(\mathcal{M}_2, \{\ ,\ \}_2)$ is a Poisson manifold $(\mathcal{M}_1 \times \mathcal{M}_2, \{\ ,\ \})$ defined by the property that natural projection maps $\pi_1 : \mathcal{M}_1 \times \mathcal{M}_2 \to \mathcal{M}_1$ and $\pi_2 : \mathcal{M}_1 \times \mathcal{M}_2 \to \mathcal{M}_2$ are Poisson mappings. For $f \in C^\infty(\mathcal{M}_1 \times \mathcal{M}_2)$ and $(x_1, x_2) \in \mathcal{M}_1 \times \mathcal{M}_2$ denote, respectively, by $f_{x_2}^{(1)}$ and $f_{x_1}^{(2)}$ restrictions of f to $\mathcal{M} \times \{x_2\}$ and $\{x_1\} \times \mathcal{M}_2$. Then for $f, g \in C^\infty(\mathcal{M}_1 \times \mathcal{M}_2)$,

$$\{f, g\}(x_1, x_2) = \{f_{x_2}^{(1)}, g_{x_2}^{(1)}\}_1(x_1) + \{f_{x_1}^{(2)}, g_{x_1}^{(2)}\}_2(x_2).$$

Non-degenerate Poisson manifolds form a subcategory of the category of Poisson manifolds.

Theorem 2.21. *The category of symplectic manifolds is (anti-) isomorphic to the category of non-degenerate Poisson manifolds.*

Proof. According to Theorem 2.14, every symplectic manifold carries a Poisson structure. Its non-degeneracy follows from the non-degeneracy of a symplectic form. Conversely, let $(\mathcal{M}, \{\ ,\ \})$ be a non-degenerate Poisson manifold. Define the 2-form ω on $\mathcal{M}$ by

$$\omega(X, Y) = J^{-1}(Y)(X), \quad X, Y \in \text{Vect}(\mathcal{M}),$$

where the isomorphism $J : T^*\mathscr{M} \to T\mathscr{M}$ is defined by the Poisson tensor η. In local coordinates $\boldsymbol{x} = (x^1, \ldots, x^N)$ on $\mathscr{M}$,

$$\omega = - \sum_{1 \leq i < j \leq N} \eta_{ij}(\boldsymbol{x})\, dx^i \wedge dx^j,$$

where $\{\eta_{ij}(\boldsymbol{x})\}_{i,j=1}^N$ is the inverse matrix to $\{\eta^{ij}(\boldsymbol{x})\}_{i,j=1}^N$. The 2-form ω is skew-symmetric and non-degenerate. For every $f \in \mathcal{A}$ let $X_f = \{f, \cdot\}$ be the corresponding vector field on $\mathscr{M}$. The Jacobi identity for the Poisson bracket $\{\, ,\, \}$ is equivalent to $\mathcal{L}_{X_f}\eta = 0$ for every $f \in \mathcal{A}$, so that

$$\mathcal{L}_{X_f}\omega = 0.$$

Since $X_f = Jdf$, we have $\omega(X, Jdf) = df(X)$ for every $X \in \mathrm{Vect}(\mathscr{M})$, so that

$$\omega(X_f, X_g) = \{f, g\}.$$

By Cartan's formula,

$$d\omega(X, Y, Z) = \tfrac{1}{3}\left(\mathcal{L}_X\omega(Y, Z) - \mathcal{L}_Y\omega(X, Z) + \mathcal{L}_Z\omega(X, Y)\right.$$
$$\left. -\omega([X, Y], Z) + \omega([X, Z], Y) - \omega([Y, Z], X)\right),$$

where $X, Y, Z \in \mathrm{Vect}(\mathscr{M})$. Now setting $X = X_f, Y = X_g, Z = X_h$, we get

$$d\omega(X_f, X_g, X_h) = \tfrac{1}{3}\left(\omega(X_h, [X_f, X_g]) + \omega(X_f, [X_g, X_h]) + \omega(X_g, [X_h, X_f])\right)$$
$$= \tfrac{1}{3}\left(\omega(X_h, X_{\{f,g\}}) + \omega(X_f, X_{\{g,h\}}) + \omega(X_g, X_{\{h,f\}})\right)$$
$$= \tfrac{1}{3}\left(\{h, \{f, g\}\} + \{f, \{g, h\}\} + \{g, \{h, f\}\}\right)$$
$$= 0.$$

The exact 1-forms df, $f \in \mathcal{A}$, generate the vector space of 1-forms $\mathcal{A}^1(\mathscr{M})$ as a module over $\mathcal{A}$, so that Hamiltonian vector fields $X_f = Jdf$ generate the vector space $\mathrm{Vect}(\mathscr{M})$ as a module over $\mathcal{A}$. Thus $d\omega = 0$ and $(\mathscr{M}, \omega)$ is a symplectic manifold associated with the Poisson manifold $(\mathscr{M}, \{\, ,\, \})$. It follows from the definitions that Poisson mappings of non-degenerate Poisson manifolds correspond to symplectomorphisms of associated symplectic manifolds. $\qquad\square$

Remark. One can also prove this theorem by a straightforward computation in local coordinates $\boldsymbol{x} = (x^1, \ldots, x^N)$ on $\mathscr{M}$. Just observe that the condition

$$\frac{\partial \eta_{ij}(\boldsymbol{x})}{\partial x^l} + \frac{\partial \eta_{jl}(\boldsymbol{x})}{\partial x^i} + \frac{\partial \eta_{li}(\boldsymbol{x})}{\partial x^j} = 0, \quad i, j, l = 1, \ldots, N,$$

which is a coordinate form of $d\omega = 0$, follows from the condition

$$\sum_{j=1}^N \left(\eta^{ij}(\boldsymbol{x})\frac{\partial \eta^{kl}(\boldsymbol{x})}{\partial x^j} + \eta^{lj}(\boldsymbol{x})\frac{\partial \eta^{ik}(\boldsymbol{x})}{\partial x^j} + \eta^{kj}(\boldsymbol{x})\frac{\partial \eta^{li}(\boldsymbol{x})}{\partial x^j}\right) = 0,$$

which is a coordinate form of the Jacobi identity, by multiplying it three times by the inverse matrix $\eta_{ij}(\boldsymbol{x})$ using

$$\sum_{p=1}^{N}\left(\eta^{ip}(\boldsymbol{x})\frac{\partial\eta_{pk}(\boldsymbol{x})}{\partial x^j}+\frac{\partial\eta^{ip}(\boldsymbol{x})}{\partial x^j}\eta_{pk}(\boldsymbol{x})\right)=0.$$

Remark. Let $\mathscr{M}=T^*\mathbb{R}^n$ with the Poisson bracket $\{\,,\,\}$ given by the canonical symplectic form $\omega=d\boldsymbol{p}\wedge d\boldsymbol{q}$, where $(\boldsymbol{p},\boldsymbol{q})=(p_1,\dots,p_n,q^1,\dots,q^n)$ are coordinate functions on $T^*\mathbb{R}^n$. The non-degeneracy of the Poisson manifold $(T^*\mathbb{R}^n,\{\,,\,\})$ can be formulated as the property that the only observable $f\in C^\infty(T^*\mathbb{R}^n)$ satisfying

$$\{f,p_1\}=\cdots=\{f,p_n\}=0,\quad\{f,q^1\}=\cdots=\{f,q^n\}=0$$

is $f(\boldsymbol{p},\boldsymbol{q})=\text{const.}$

Problem 2.19 (Dual space to a Lie algebra). Let $\mathfrak{g}$ be a finite-dimensional Lie algebra with a Lie bracket $[\,,\,]$, and let $\mathfrak{g}^*$ be its dual space. For $f,g\in C^\infty(\mathfrak{g}^*)$ define

$$\{f,g\}(u)=u\left([df,dg]\right),$$

where $u\in\mathfrak{g}^*$ and $T_u^*\mathfrak{g}^*\simeq\mathfrak{g}$. Prove that $\{\,,\,\}$ is a Poisson bracket. (It was introduced by Sophus Lie and is called a *linear*, or *Lie-Poisson* bracket.) Show that this bracket is degenerate and determine the center of $\mathcal{A}=C^\infty(\mathfrak{g}^*)$.

Problem 2.20. A Poisson bracket $\{\,,\,\}$ on $\mathscr{M}$ restricts to a Poisson bracket $\{\,,\,\}_0$ on a submanifold $\mathscr{N}$ if the inclusion $\imath:\mathscr{N}\to\mathscr{M}$ is a Poisson mapping. Show that the Lie-Poisson bracket on $\mathfrak{g}^*$ restricts to a non-degenerate Poisson bracket on a coadjoint orbit, associated with the Kirillov-Kostant symplectic form.

Problem 2.21 (Lie-Poisson groups). A finite-dimensional Lie group is called a *Lie-Poisson group* if it has a structure of a Poisson manifold $(G,\{\,,\,\})$ such that the group multiplication $G\times G\to G$ is a Poisson mapping, where $G\times G$ is a direct product of Poisson manifolds. Using a basis $\partial_1,\dots,\partial_n$ of left-invariant vector fields on G corresponding to a basis $x_1,\dots,x_n$ of the Lie algebra $\mathfrak{g}$, the Poisson bracket $\{\,,\,\}$ can be written as

$$\{f_1,f_2\}(g)=\sum_{i,j=1}^{n}\eta^{ij}(g)\partial_i f_1\partial_j f_2,$$

where the 2-tensor $\eta^{ij}(g)$ defines a mapping $\eta:G\to\Lambda^2\mathfrak{g}$ by $\eta(g)=\sum_{i,j=1}^{n}\eta^{ij}(g)x_i\otimes x_j$. Show that the bracket $\{\,,\,\}$ equips G with a Lie-Poisson structure if and only if the following conditions are satisfied: (i) for all $g\in G$,

$$\xi^{ijk}(g)=\sum_{l=1}^{n}\left(\eta^{il}(g)\partial_l\eta^{jk}(g)+\eta^{jl}(g)\partial_l\eta^{ki}(g)+\eta^{kl}(g)\partial_l\eta^{ij}(g)\right)$$

$$+\sum_{l,p=1}^{n}\left(c_{lp}^i\eta^{pj}(g)\eta^{kl}(g)+c_{lp}^j\eta^{pk}(g)\eta^{il}(g)+c_{lp}^k\eta^{pi}(g)\eta^{jl}(g)\right)=0,$$

where $[x_i,x_j]=\sum_{k=1}^{n}c_{ij}^k x_k$; (ii) the mapping η is a group 1-cocycle with the adjoint action on $\Lambda^2\mathfrak{g}$, i.e., $\eta(g_1g_2)=\text{Ad}^{-1}g_2\cdot\eta(g_1)+\eta(g_2)$, $g_1,g_2\in G$.

Problem 2.22. Show that the second condition in the previous problem trivially holds when η is a coboundary, $\eta(g) = -r + \mathrm{Ad}^{-1}g \cdot r$ for some $r = \sum_{i,j=1}^{n} r^{ij} x_i \otimes x_j \in \Lambda^2 \mathfrak{g}$, and then the first condition is satisfied if and only if the element

$$\xi(r) = [r_{12}, r_{13} + r_{23}] + [r_{13}, r_{23}] \in \Lambda^3 \mathfrak{g}$$

is invariant under the adjoint action of $\mathfrak{g}$ on $\Lambda^3 \mathfrak{g}$. Here $r_{12} = \sum_{i,j=1}^{n} r^{ij} x_i \otimes x_j \otimes 1$, $r_{13} = \sum_{i,j=1}^{n} r^{ij} x_i \otimes 1 \otimes x_j$, and $r_{23} = \sum_{i,j=1}^{n} r^{ij} 1 \otimes x_i \otimes x_j$ are corresponding elements in the universal enveloping algebra $U\mathfrak{g}$ of a Lie algebra $\mathfrak{g}$. In particular, G is a Lie-Poisson group if $\xi(r) = 0$, which is called the *classical Yang-Baxter equation*.

Problem 2.23. Suppose that $r = \sum_{i,j=1}^{n} r^{ij} x_i \otimes x_j \in \Lambda^2 \mathfrak{g}$ is such that the matrix $\{r^{ij}\}$ is non-degenerate, and let $\{r_{ij}\}$ be the inverse matrix. Show that r satisfies the classical Yang-Baxter equation if and only if the map $c : \Lambda^2 \mathfrak{g} \to \mathbb{C}$, defined by $c(x, y) = \sum_{i,j=1}^{n} r_{ij} u^i v^j$, where $x = \sum_{i=1}^{n} u^i x_i$, $y = \sum_{i=1}^{n} v^i x_i$, is a non-degenerate Lie algebra 2-cocycle, i.e., it satisfies

$$c(x, [y, z]) + c(z, [x, y]) + c(y, [z, x]) = 0, \quad x, y, z \in \mathfrak{g}.$$

2.8. Hamilton's and Liouville's representations. To complete the formulation of classical mechanics, we need to describe the process of *measurement*. In physics, by a measurement of a classical system one understands the result of a physical experiment which gives numerical values for classical observables. The experiment consists of creating certain conditions for the system, and it is always assumed that these conditions can be repeated over and over. The conditions of the experiment define a *state* of the system if repeating these conditions results in probability distributions for the values of all observables of the system.

Mathematically, a state μ on the algebra $\mathcal{A} = C^\infty(\mathscr{M})$ of classical observables on the phase space $\mathscr{M}$ is the assignment

$$\mathcal{A} \ni f \mapsto \mu_f \in \mathscr{P}(\mathbb{R}),$$

where $\mathscr{P}(\mathbb{R})$ is a set of probability measures on $\mathbb{R}$ — Borel measures on $\mathbb{R}$ such that the total measure of $\mathbb{R}$ is 1. For every Borel subset $E \subseteq \mathbb{R}$ the quantity $0 \leq \mu_f(E) \leq 1$ is a probability that in the state μ the value of the observable f belongs to E. By definition, the *expectation value* of an observable f in the state μ is given by the Lebesgue-Stieltjes integral

$$\mathsf{E}_\mu(f) = \int_{-\infty}^{\infty} \lambda d\mu_f(\lambda),$$

where $\mu_f(\lambda) = \mu_f((-\infty, \lambda))$ is a distribution function of the measure $d\mu_f$. The correspondence $f \mapsto \mu_f$ should satisfy the following natural properties.

S1. $|\mathsf{E}_\mu(f)| < \infty$ for $f \in \mathcal{A}_0$ — the subalgebra of bounded observables.

S2. $\mathsf{E}_\mu(1) = 1$, where 1 is the unit in $\mathcal{A}$.

S3. For all $a, b \in \mathbb{R}$ and $f, g \in \mathcal{A}$,

$$\mathsf{E}_\mu(af + bg) = a\mathsf{E}_\mu(f) + b\mathsf{E}_\mu(g),$$

if both $\mathsf{E}_\mu(f)$ and $\mathsf{E}_\mu(g)$ exist.

S4. If $f_1 = \varphi \circ f_2$ with smooth $\varphi : \mathbb{R} \to \mathbb{R}$, then for every Borel subset $E \subseteq \mathbb{R}$,

$$\mu_{f_1}(E) = \mu_{f_2}(\varphi^{-1}(E)).$$

It follows from property **S4** and the definition of the Lebesgue-Stieltjes integral that

$$(2.18) \qquad \mathsf{E}_\mu(\varphi(f)) = \int_{-\infty}^{\infty} \varphi(\lambda) d\mu_f(\lambda).$$

In particular, $\mathsf{E}_\mu(f^2) \geq 0$ for all $f \in \mathcal{A}$, so that the states define normalized, positive, linear functionals on the subalgebra $\mathcal{A}_0$.

Assuming that the functional E_μ extends to the space of bounded, piecewise continuous functions on $\mathcal{M}$, and satisfies (2.18) for measurable functions φ, one can recover the distribution function from the expectation values by the formula

$$\mu_f(\lambda) = \mathsf{E}_\mu\left(\theta(\lambda - f)\right),$$

where $\theta(x)$ is the Heavyside step function,

$$\theta(x) = \begin{cases} 1, & x > 0, \\ 0, & x \leq 0. \end{cases}$$

Indeed, let χ be the characteristic function of the interval $(-\infty, \lambda) \subset \mathbb{R}$. Using (2.18) and the definition of the Lebesgue-Stieltjes integral we get

$$\mathsf{E}_\mu(\theta(\lambda - f)) = \int_{-\infty}^{\infty} \chi(s) d\mu_f(s) = \mu_f((-\infty, \lambda)) = \mu_f(\lambda).$$

Every probability measure $d\mu$ on $\mathcal{M}$ defines the state on $\mathcal{A}$ by assigning[20] to every observable f a probability measure $\mu_f = f_*(\mu)$ on $\mathbb{R}$ — a pushforward of the measure $d\mu$ on $\mathcal{M}$ by the mapping $f : \mathcal{M} \to \mathbb{R}$. It is defined by $\mu_f(E) = \mu(f^{-1}(E))$ for every Borel subset $E \subseteq \mathbb{R}$, and has the distribution function

$$\mu_f(\lambda) = \mu(f^{-1}(-\infty, \lambda)) = \int_{\mathcal{M}_\lambda(f)} d\mu,$$

where $\mathcal{M}_\lambda(f) = \{x \in \mathcal{M} : f(x) < \lambda\}$. It follows from the Fubini theorem that

$$(2.19) \qquad E_\mu(f) = \int_{-\infty}^{\infty} \lambda d\mu_f(\lambda) = \int_{\mathcal{M}} f d\mu.$$

---

[20]There should be no confusion in denoting the state and the measure by μ.

It turns out that probability measures on $\mathcal{M}$ are essentially the only examples of states. Namely, for a locally compact topological space $\mathcal{M}$ the Riesz-Markov theorem asserts that for every positive, linear functional l on the space $C_{\mathrm{c}}(\mathcal{M})$ of continuous functions on $\mathcal{M}$ with compact support, there is a unique regular Borel measure $d\mu$ on $\mathcal{M}$ such that

$$l(f) = \int_{\mathcal{M}} f\, d\mu \quad \text{for all} \quad f \in C_{\mathrm{c}}(\mathcal{M}).$$

This leads to the following definition of states in classical mechanics.

Definition. The set of states $\mathcal{S}$ for a Hamiltonian system with the phase space $\mathcal{M}$ is the convex set $\mathscr{P}(\mathcal{M})$ of all probability measures on $\mathcal{M}$. The states corresponding to Dirac measures $d\mu_x$ supported at points $x \in \mathcal{M}$ are called *pure states*, and the phase space $\mathcal{M}$ is also called the *space of states*[21]. All other states are called *mixed states*. A process of measurement in classical mechanics is the correspondence

$$\mathcal{A} \times \mathcal{S} \ni (f, \mu) \mapsto \mu_f = f_*(\mu) \in \mathscr{P}(\mathbb{R}),$$

which to every observable $f \in \mathcal{A}$ and state $\mu \in \mathcal{S}$ assigns a probability measure μ_f on $\mathbb{R}$ — a push-forward of the measure $d\mu$ on $\mathcal{M}$ by f. For every Borel subset $E \subseteq \mathbb{R}$ the quantity $0 \leq \mu_f(E) \leq 1$ is the probability that for a system in the state μ the result of a measurement of the observable f is in the set E. The expectation value of an observable f in a state μ is given by (2.19).

In physics, pure states are characterized by having the property that a measurement of every observable always gives a well-defined result. Mathematically this can be expressed as follows. Let

$$\sigma_\mu^2(f) = \mathsf{E}_\mu\left((f - \mathsf{E}_\mu(f))^2\right) = \mathsf{E}_\mu(f^2) - \mathsf{E}_\mu(f)^2 \geq 0$$

be the *variance* of the observable f in the state μ.

Lemma 2.1. *Pure states are the only states in which every observable has zero variance.*

Proof. It follows from the Cauchy-Bunyakovski-Schwarz inequality that $\sigma_\mu^2(f) = 0$ if only if f is constant on the support of a probability measure $d\mu$. $\qquad\square$

In particular, a *mixture* of pure states $d\mu_x$ and $d\mu_y$, $x, y \in \mathcal{M}$, is a mixed state with

$$d\mu = \alpha\, d\mu_x + (1 - \alpha)d\mu_y, \quad 0 < \alpha < 1,$$

so that $\sigma_\mu^2(f) > 0$ for every observable f such that $f(x) \neq f(y)$.

---

[21]The space of pure states, to be precise.

For a system consisting of few interacting particles (say, a motion of planets in celestial mechanics) it is possible to measure all coordinates and momenta, so one considers only pure states. Mixed states necessarily appear for *macroscopic* systems, when it is impossible to measure all coordinates and momenta[22].

Remark. As a topological space, the space of states $\mathscr{M}$ can be reconstructed from the algebra $\mathcal{A}$ of classical observables. Namely, suppose for simplicity that $\mathscr{M}$ is compact. Then the $\mathbb{C}$-algebra $\mathcal{C} = C(\mathscr{M})$ of complex-valued continuous functions on $\mathscr{M}$ — the completion of the complexification of the $\mathbb{R}$-algebra $\mathcal{A}$ of classical observables with respect to the sup-norm — is a commutative $\mathbb{C}^*$-algebra. This means that $C(\mathscr{M})$ is a Banach space with respect to the norm $\|f\| = \sup_{x \in \mathscr{M}} |f(x)|$, has a $\mathbb{C}$-algebra structure (associative algebra over $\mathbb{C}$ with a unit) given by the point-wise product of functions such that $\|f \cdot g\| \leq \|f\|\|g\|$, and has a complex anti-linear anti-involution: a map ${}^* : \mathcal{C} \to \mathcal{C}$ given by the complex conjugation $f^*(x) = \overline{f(x)}$ and satisfying $\|f \cdot f^*\| = \|f\|^2$. Then the Gelfand-Naimark theorem asserts that every commutative $\mathbb{C}^*$-algebra $\mathcal{C}$ is isomorphic to the algebra $C(\mathscr{M})$ of continuous functions on its spectrum — the set of maximal ideals of $\mathcal{C}$ — a compact topological space with the topology induced by the weak topology on $\mathcal{C}^*$, the dual space of $\mathcal{C}$.

We conclude our exposition of classical mechanics by presenting two equivalent ways of describing the dynamics — the time evolution of a Hamiltonian system $((\mathscr{M}, \{\,,\,\}), H)$ with the algebra of observables $\mathcal{A} = C^\infty(\mathscr{M})$ and the set of states $\mathcal{S} = \mathscr{P}(\mathscr{M})$. In addition, we assume that the Hamiltonian phase flow g_t exists for all times, and that the phase space $\mathscr{M}$ carries a volume form dx invariant under the phase flow[23].

Hamilton's Description of Dynamics. States do not depend on time, and time evolution of observables is given by Hamilton's equations of motion,

$$\frac{d\mu}{dt} = 0, \quad \mu \in \mathcal{S}, \quad \text{and} \quad \frac{df}{dt} = \{H, f\}, \quad f \in \mathcal{A}.$$

The expectation value of an observable f in the state μ at time t is given by

$$\mathsf{E}_\mu(f_t) = \int_\mathscr{M} f \circ g_t \, d\mu = \int_\mathscr{M} f(g_t(x)) \rho(x) dx,$$

where $\rho(x) = \dfrac{d\mu}{dx}$ is the Radon-Nikodim derivative. In particular, the expectation value of f in the pure state $d\mu_x$ corresponding to the point $x \in \mathscr{M}$ is $f(g_t(x))$. Hamilton's picture is commonly used for mechanical systems consisting of few interacting particles.

---

[22]Typically, a macroscopic system consists of $N \sim 10^{23}$ molecules. Macroscopic systems are studied in classical statistical mechanics.

[23]It is Liouville's volume form when the Poisson structure on $\mathscr{M}$ is non-degenerate.

Liouville's Description of Dynamics. The observables do not depend on time

$$\frac{df}{dt} = 0, \quad f \in \mathcal{A},$$

and states $d\mu(x) = \rho(x)dx$ satisfy Liouville's equation

$$\frac{d\rho}{dt} = -\{H, \rho\}, \quad \rho(x)dx \in \mathcal{S}.$$

Here the Radon-Nikodim derivative $\rho(x) = \dfrac{d\mu}{dx}$ and Liouville's equation are understood in the distributional sense. The expectation value of an observable f in the state μ at time t is given by

$$\mathsf{E}_{\mu_t}(f) = \int_{\mathcal{M}} f(x)\rho(g_{-t}(x))dx.$$

Liouville's picture, where states are described by the distribution functions $\rho(x)$ — positive distributions on $\mathcal{M}$ corresponding to probability measures $\rho(x)dx$ — is commonly used in statistical mechanics. The equality

$$\mathsf{E}_{\mu}(f_t) = \mathsf{E}_{\mu_t}(f) \quad \text{for all} \quad f \in \mathcal{A}, \quad \mu \in \mathcal{S},$$

which follows from the invariance of the volume form dx and the change of variables, expresses the equivalence between Liouville's and Hamilton's descriptions of the dynamics.

3. Notes and references

Classical references are the textbooks [**Arn89**] and [**LL76**], which are written, respectively, from mathematics and physics perspectives. The elegance of [**LL76**] is supplemented by the attention to detail in [**Gol80**], another physics classic. A brief overview of Hamiltonian formalism necessary for quantum mechanics can be found in [**FY80**]. The treatise [**AM78**] and the encyclopaedia surveys [**AG90**], [**AKN97**] provide a comprehensive exposition of classical mechanics, including the history of the subject, and the references to classical works and recent contributions. The textbook [**Ste83**], monographs [**DFN84**], [**DFN85**], and lecture notes [**Bry95**] contain all the necessary material from differential geometry and the theory of Lie groups, as well as references to other sources. In addition, the lectures [**God69**] also provide an introduction to differential geometry and classical mechanics. In particular, [**God69**] and [**Bry95**] discuss the role the second tangent bundle plays in Lagrangian mechanics (see also the monograph [**YI73**] and [**Cra83**]). For a brief review of integration theory, including the Riesz-Markov theorem, see [**RS80**]; for the proof of the Gelfand-Naimark theorem and more details on $\mathbb{C}^*$-algebras, see [**Str05**] and references therein.

Our exposition follows the traditional outline in [**LL76**] and [**Arn89**], which starts with the Lagrangian formalism and introduces Hamiltonian formalism through the Legendre transform. As in [**Arn89**], we made special emphasis on precise mathematical formulations. Having graduate students and research mathematicians as a main audience, we have the advantage to use freely the calculus of differential forms and vector fields on smooth manifolds. This differs from the presentation in [**Arn89**], which is oriented at undergraduate students and needs to introduce this material in the main text. Since the goal of this chapter was to present only those basics of classical mechanics which are fundamental for the formulation of quantum mechanics, we have omitted many important topics, including mechanical–geometrical optics analogy, theory of oscillations, rotation of a rigid body, perturbation theory, etc. The interested reader can find this material in [**LL76**] and [**Arn89**] and in the above-mentioned monographs. Completely integrable Hamiltonian systems were also only briefly mentioned at the end of Section 2.6. We refer the reader to [**AKN97**] and references therein for a comprehensive exposition, and to the monograph [**FT07**] for the so-called Lax pair method in the theory of integrable systems, especially for the case of infinitely many degrees of freedom.

In Section 2.7, following [**FY80**], [**DFN84**], [**DFN85**], we discussed Poisson manifolds and Poisson algebras. These notions, usually not emphasized in standard exposition of classical mechanics, are fundamental for understanding the meaning of quantization — a passage from classical mechanics to quantum mechanics. We also have included in Sections 1.6 and 2.7 the treatment of the Laplace-Runge-Lenz vector, whose components are extra integrals of motions for the Kepler problem[24]. Though briefly mentioned in [**LL76**], the Laplace-Runge-Lenz vector does not actually appear in many textbooks, with the exception of [**Gol80**] and [**DFN84**]. In Section 2.7, following [**FY80**], we also included Theorem 2.19, which clarifies the meaning of the Jacobi identity, and presented Hamilton's and Liouville's descriptions of the dynamics.

Most of the problems in this chapter are fairly standard and are taken from various sources, mainly from [**Arn89**], [**LL76**], [**Bry95**], [**DFN84**], and [**DFN85**]. Other problems indicate interesting relations with representation theory and symplectic geometry. Thus Problems 2.12 and 2.20 introduce the reader to the orbit method [**Kir04**], and Problem 2.18 — to the method of symplectic reduction (see [**Arn89**], [**Bry95**] and references therein). Problems 2.21 – 2.23 introduce the reader to the theory of Lie-Poisson groups (see [**Dri86**], [**Dri87**], [**STS85**], and [**Tak90**] for an elementary exposition).

---

[24]We will see in Chapter 3 that these extra integrals are responsible for the hidden SO(4) symmetry of the hydrogen atom.

Chapter 2

Basic Principles of Quantum Mechanics

We recall the standard notation and basic facts from the theory of self-adjoint operators on Hilbert spaces. Let $\mathcal{H}$ be a separable Hilbert space with an inner product $(\,,\,)$, which is complex-linear with respect to the first argument, and let A be a linear operator in $\mathcal{H}$ with the domain $D(A) \subseteq \mathcal{H}$ — a linear subset of $\mathcal{H}$. An operator A is called *closed* if its graph $\Gamma(A) = \{(\varphi, A\varphi) \in \mathcal{H} \times \mathcal{H} : \varphi \in D(A)\}$ is a closed subspace in $\mathcal{H} \times \mathcal{H}$. If the domain of A is dense[1] in $\mathcal{H}$, i.e., $\overline{D(A)} = \mathcal{H}$, the domain $D(A^*)$ of the *adjoint operator* A^* consists of $\varphi \in \mathcal{H}$ such that there is $\eta \in \mathcal{H}$ with the property that

$$(A\psi, \varphi) = (\psi, \eta) \quad \text{for all} \quad \psi \in D(A),$$

and the operator A^* is defined by $A^*\varphi = \eta$. An operator A is called *symmetric* if

$$(A\psi, \varphi) = (\psi, A\varphi) \quad \text{for all} \quad \varphi, \psi \in D(A).$$

By definition, the *regular set* of a closed operator A with a dense domain $D(A)$ is the set

$$\rho(A) = \{\lambda \in \mathbb{C} \mid A - \lambda I : D(A) \to \mathcal{H} \text{ is a bijection with a bounded inverse}[2]\},$$

and for $\lambda \in \rho(A)$, the bounded operator $R_\lambda(A) = (A - \lambda I)^{-1}$ is called the *resolvent* of A at λ. The regular set $\rho(A) \subset \mathbb{C}$ is open and its complement $\sigma(A) = \mathbb{C} \setminus \rho(A)$ is the *spectrum* of A. The subset $\sigma_p(A)$ of $\sigma(A)$ consisting of eigenvalues of A of finite multiplicity is called the *point spectrum*.

---

[1] We consider only linear operators with dense domains.
[2] By the closed graph theorem, the last condition is redundant.

An operator A is *self-adjoint* (or *Hermitian*) if $A = A^*$. Equivalently, A is symmetric and $D(A) = D(A^*)$, and for such operators $\sigma(A) \subset \mathbb{R}$. A symmetric operator A is called *essentially self-adjoint* if its closure $\bar{A} = A^{**}$ is self-adjoint. For a symmetric operator A the following conditions are equivalent:

(i) A is essentially self-adjoint.

(ii) $\mathrm{Ker}(A^* + iI) = \mathrm{Ker}(A^* - iI) = \{0\}$.

(iii) $\overline{\mathrm{Im}(A + iI)} = \overline{\mathrm{Im}(A - iI)} = \mathscr{H}$.

A symmetric operator A with $D(A) = \mathscr{H}$ is bounded and self-adjoint. An operator A is positive[3] if $(A\varphi, \varphi) \geq 0$ for all $\phi \in D(A)$, which we denote by $A \geq 0$. Positive operators satisfy the Cauchy-Bunyakovski-Schwarz inequality

$$(0.1) \qquad |(A\varphi, \psi)|^2 \leq (A\varphi, \varphi)(A\psi, \psi) \quad \text{for all} \quad \varphi, \psi \in D(A).$$

In particular, $(A\varphi, \varphi) = 0$ implies that $A\varphi = 0$. Every bounded positive operator is self-adjoint[4]. We denote by $\mathscr{L}(\mathscr{H})$ the $\mathbb{C}^*$-algebra of bounded linear operators on $\mathscr{H}$ with the operator norm $\| \cdot \|$ and with the anti-involution $*$ given by the operator adjoint. Operator $A \in \mathscr{L}(\mathscr{H})$ is called *compact* if it maps bounded sets in $\mathscr{H}$ into pre-compact sets[5]. The vector space $\mathscr{C}(\mathscr{H})$ of compact operators on $\mathscr{H}$ is a closed two-sided ideal[6] of the $\mathbb{C}^*$-algebra $\mathscr{L}(\mathscr{H})$.

An operator $A \in \mathscr{C}(\mathscr{H})$ is of *trace class* if

$$\|A\|_1 = \sum_{n=1}^{\infty} \mu_n(A) < \infty,$$

where $\mu_n(A)$ are singular values of A: $\mu_n(A) = \sqrt{\lambda_n(A)} \geq 0$, where $\lambda_n(A)$ are eigenvalues of A^*A. Equivalently, an operator $A \in \mathscr{L}(\mathscr{H})$ is of trace class if and only if for every orthonormal basis $\{e_n\}_{n=1}^{\infty}$ for $\mathscr{H}$

$$\sum_{n=1}^{\infty} |(Ae_n, e_n)| < \infty.$$

Since a permutation of an orthonormal basis is again an orthonormal basis, this condition can be replaced by

$$\sum_{n=1}^{\infty} (Ae_n, e_n) < \infty$$

---

[3]Non-negative, to be precise.

[4]This is true only for complex Hilbert spaces.

[5]The sets with compact closure.

[6]Ideal $\mathscr{C}(\mathscr{H})$ is the only closed two-sided ideal of $\mathscr{L}(\mathscr{H})$.

for every orthonormal basis $\{e_n\}_{n=1}^{\infty}$ for $\mathscr{H}$. The *trace* of a trace class operator A is defined by

$$\mathrm{Tr}\, A = \sum_{n=1}^{\infty} (Ae_n, e_n),$$

and does not depend on the choice of an orthonormal basis $\{e_n\}_{n=1}^{\infty}$ for $\mathscr{H}$. A positive operator $A \in \mathscr{L}(\mathscr{H})$ is of trace class if

$$\sum_{n=1}^{\infty} (Ae_n, e_n) < \infty$$

for some orthonormal basis $\{e_n\}_{n=1}^{\infty}$ for $\mathscr{H}$. The space $\mathscr{S}_1$ of trace class operators in $\mathscr{H}$ is a Banach algebra with the norm $\|A\|_1 = \mathrm{Tr}\, \sqrt{A^*A}$, and is a two-sided ideal (von Neumann-Schatten ideal) in the $\mathbb{C}^*$-algebra $\mathscr{L}(\mathscr{H})$.

The property

$$\mathrm{Tr}\, AB = \mathrm{Tr}\, BA \quad \text{for all} \quad A \in \mathscr{S}_1, \ B \in \mathscr{L}(\mathscr{H})$$

is called the cyclic property of the trace. If $A \in \mathscr{S}_1$, the mapping $l_A(B) = \mathrm{Tr}\, AB$, $B \in \mathscr{L}(\mathscr{H})$, is a continuous linear functional on $\mathscr{L}(\mathscr{H})$, so that $\mathscr{S}_1 \subset \mathscr{L}(\mathscr{H})^*$ — the dual Banach space to $\mathscr{L}(\mathscr{H})$. However, $\mathscr{S}_1 \neq \mathscr{L}(\mathscr{H})^*$, but rather $\mathscr{S}_1 = \mathscr{C}(\mathscr{H})^*$, and the mapping $A \mapsto l_A$ is an isomorphism of Banach spaces. Similarly, the mapping $B \mapsto l_B$ gives an isomorphism $\mathscr{L}(\mathscr{H}) = \mathscr{S}_1^*$.

An operator $A \in \mathscr{L}(\mathscr{H})$ is called a *Hilbert-Schmidt operator* if $AA^* \in \mathscr{S}_1$. Equivalently, an operator $A \in \mathscr{L}(\mathscr{H})$ is a Hilbert-Schmidt operator if and only if for some orthonormal basis $\{e_n\}_{n=1}^{\infty}$ for $\mathscr{H}$

$$\sum_{n=1}^{\infty} \|Ae_n\|^2 < \infty.$$

The vector space $\mathscr{S}_2$ of Hilbert-Schmidt operators in $\mathscr{H}$ is a Hilbert space with the inner product $(A, B)_2 = \mathrm{Tr}\, AB^*$. The Hilbert-Schmidt space $\mathscr{S}_2$ is also a two-sided ideal in the $\mathbb{C}^*$-algebra $\mathscr{L}(\mathscr{H})$.

1. Observables, states, and dynamics

Quantum mechanics studies the microworld — the physical laws at an atomic scale — that cannot be adequately described by classical mechanics. The properties of the microworld are so different from our everyday experiences that it is no surprise that its laws seem to contradict the common sense. Thus classical mechanics and classical electrodynamics cannot explain the stability of atoms and molecules. Neither can these theories reconcile different properties of light, its wave-like behavior in interference and diffraction phenomena, and its particle-like behavior in photo-electric emission and scattering by free photons. The fundamental difference between the microworld and the perceived world around us is that in the microworld every experiment results in interaction with the system and thus disturbs its properties, whereas in classical physics it is always assumed that one can neglect the disturbances the measurement brings upon a system. This imposes a limitation on our powers of observation and leads to a conclusion that there exist observables which cannot be measured simultaneously.

We will not discuss here these and other basic experimental facts, referring the interested reader to physics textbooks. Nor will we follow the historic path of the theory. Instead, we show how to formulate quantum mechanics using the general notions of states, observables, and time evolution, described in Section 2.8 in Chapter 1. There we have seen that commutativity of the algebra of observables $\mathcal{A}$ results in its realization as an algebra of functions on the topological space — the space of states — and thus brings us to the realm of classical mechanics. Therefore in order to get a realization of observables and states which is different from classical mechanics, we must assume that the $\mathbb{C}^*$-algebra associated with observables is no longer commutative. A fundamental example of a non-commutative $\mathbb{C}^*$-algebra is given by the algebra of all bounded operators on a complex Hilbert space, and it turns out that it is this algebra which plays a fundamental role in quantum mechanics!

Here we formulate the basic principles of quantum mechanics using precise mathematical language. At this point it should be noted that one can not verify directly the principles lying at the foundation of quantum mechanics. Nevertheless, the validity of quantum mechanics, whenever it is applicable, is continuously being confirmed by numerous experimental facts which perfectly agree with the predictions of the theory[7].

1.1. Mathematical formulation. The following axioms constitute the basis of quantum mechanics.

A1. With every *quantum system* there is associated an infinite-dimensional separable complex Hilbert space $\mathscr{H}$, in physics terminology called the *space of states*[8]. The Hilbert space of a *composite quantum system* is a tensor product of Hilbert spaces of component systems.

A2. The set of *observables* $\mathscr{A}$ of a quantum system with the Hilbert space $\mathscr{H}$ consists of all self-adjoint operators on $\mathscr{H}$. The subset $\mathscr{A}_0 = \mathscr{A} \cap \mathscr{L}(\mathscr{H})$ of bounded observables is a vector space over $\mathbb{R}$.

A3. The set of *states* $\mathscr{S}$ of a quantum system with a Hilbert space $\mathscr{H}$ consists of all positive (and hence self-adjoint) trace class operators M with

---

[7]This refers to non-relativistic phenomena at the atomic scale.

[8]The space of pure states, to be precise.

$\operatorname{Tr} M = 1$. *Pure states* are projection operators onto one-dimensional subspaces of $\mathscr{H}$. For $\psi \in \mathscr{H}$, $\|\psi\| = 1$, the corresponding projection onto $\mathbb{C}\psi$ is denoted by P_ψ. All other states are called *mixed states*[9].

A4. A process of measurement is the correspondence

$$\mathscr{A} \times \mathscr{S} \ni (A, M) \mapsto \mu_A \in \mathscr{P}(\mathbb{R}),$$

which to every observable $A \in \mathscr{A}$ and state $M \in \mathscr{S}$ assigns a probability measure μ_A on $\mathbb{R}$. For every Borel subset $E \subseteq \mathbb{R}$, the quantity $0 \leq \mu_A(E) \leq 1$ is the probability that for a quantum system in the state M the result of a measurement of the observable A belongs to E. The expectation value (the mean-value) of the observable $A \in \mathscr{A}$ in the state $M \in \mathscr{S}$ is

$$\langle A | M \rangle = \int_{-\infty}^{\infty} \lambda d\mu_A(\lambda),$$

where $\mu_A(\lambda) = \mu_A((-\infty, \lambda))$ is a distribution function for the probability measure μ_A.

The set of states $\mathscr{S}$ is a convex set. According to the Hilbert-Schmidt theorem on the canonical decomposition for compact self-adjoint operators, for every $M \in \mathscr{S}$ there exists an orthonormal set $\{\psi_n\}_{n=1}^N$ in $\mathscr{H}$ (finite or infinite, in the latter case $N = \infty$) such that

$$(1.1) \qquad M = \sum_{n=1}^N \alpha_n P_{\psi_n} \quad \text{and} \quad \operatorname{Tr} M = \sum_{n=1}^N \alpha_n = 1,$$

where $\alpha_n > 0$ are non-zero eigenvalues of M. Thus every mixed state is a convex linear combination of pure states. The following result characterizes the pure states.

Lemma 1.1. *A state $M \in \mathscr{S}$ is a pure state if and only if it cannot be represented as a non-trivial convex linear combination in $\mathscr{S}$.*

Proof. Suppose that

$$P_\psi = aM_1 + (1 - a)M_2, \quad 0 < a < 1,$$

and let $\mathscr{H} = \mathbb{C}\psi \oplus \mathscr{H}_1$ be the orthogonal sum decomposition. Since M_1 and M_2 are positive operators, for $\varphi \in \mathscr{H}_1$ we have

$$a(M_1 \varphi, \varphi) \leq (P_\psi \varphi, \varphi) = 0,$$

so that $(M_1 \varphi, \varphi) = 0$ for all $\varphi \in \mathscr{H}_1$ and by (0.1) we get $M_1|_{\mathscr{H}_1} = 0$. Since M_1 is self-adjoint, it leaves the complementary subspace $\mathbb{C}\psi$ invariant, and from $\operatorname{Tr} M_1 = 1$ it follows that $M_1 = P_\psi$. Therefore, $M_1 = M_2 = P_\psi$. $\qquad\square$

---

[9]In physics terminology, the operator M is called the *density operator*.

Explicit construction of the correspondence $\mathscr{A} \times \mathscr{S} \to \mathscr{P}(\mathbb{R})$ is based on the general spectral theorem of von Neumann, which emphasizes the fundamental role the self-adjoint operators play in quantum mechanics.

Definition. A *projection-valued measure* on $\mathbb{R}$ is a mapping $\mathsf{P} : \mathscr{B}(\mathbb{R}) \to \mathscr{L}(\mathscr{H})$ of the σ-algebra $\mathscr{B}(\mathbb{R})$ of Borel subsets of $\mathbb{R}$ into the algebra of bounded operators on $\mathscr{H}$, satisfying the following properties.

PM1. For every Borel subset $E \subseteq \mathbb{R}$, $\mathsf{P}(E)$ is an orthogonal projection, i.e., $\mathsf{P}(E) = \mathsf{P}(E)^2$ and $\mathsf{P}(E) = \mathsf{P}(E)^*$.

PM2. $\mathsf{P}(\emptyset) = 0$, $\mathsf{P}(\mathbb{R}) = I$, the identity operator on $\mathscr{H}$.

PM3. For every disjoint union of Borel subsets,

$$E = \coprod_{n=1}^{\infty} E_n, \quad \mathsf{P}(E) = \lim_{n \to \infty} \sum_{i=1}^{n} \mathsf{P}(E_i)$$

in the strong topology on $\mathscr{L}(\mathscr{H})$.

Remark. Similarly, a projection-valued measure on $\mathbb{R}^n$ is a mapping $\mathsf{P} : \mathscr{B}(\mathbb{R}^n) \to \mathscr{L}(\mathscr{H})$, satisfying the same properties **PM1-PM3**.

It follows from **PM1-PM3** that

$$(1.2) \qquad \mathsf{P}(E_1)\mathsf{P}(E_2) = \mathsf{P}(E_1 \cap E_2) \quad \text{for all } E_1, E_2 \in \mathscr{B}(\mathbb{R}).$$

With every projection-valued measure P on $\mathbb{R}$ we associate a projection-valued function

$$\mathsf{P}(\lambda) = \mathsf{P}((-\infty, \lambda)),$$

called the *projection-valued resolution of the identity*. It is characterized by the following properties.

PD1.
$$\mathsf{P}(\lambda)\mathsf{P}(\mu) = \mathsf{P}(\min\{\lambda, \mu\}).$$

PD2.
$$\lim_{\lambda \to -\infty} \mathsf{P}(\lambda) = 0, \quad \lim_{\lambda \to \infty} \mathsf{P}(\lambda) = I.$$

PD3.
$$\lim_{\substack{\mu \to \lambda \\ \mu < \lambda}} \mathsf{P}(\mu) = \mathsf{P}(\lambda).$$

For every $\varphi \in \mathscr{H}$ the resolution of the identity $\mathsf{P}(\lambda)$ defines a distribution function $(\mathsf{P}(\lambda)\varphi, \varphi)$ of the bounded measure on $\mathbb{R}$ (probability measure when $\|\varphi\| = 1$). By the polarization identity

$$(\mathsf{P}(\lambda)\varphi, \psi) = \tfrac{1}{4}\left\{ (\mathsf{P}(\lambda)(\varphi + \psi), \varphi + \psi) - (\mathsf{P}(\lambda)(\varphi - \psi), \varphi - \psi) \right.$$
$$\left. + i(\mathsf{P}(\lambda)(\varphi + i\psi), \varphi + i\psi) - i(\mathsf{P}(\lambda)(\varphi - i\psi), \varphi - i\psi) \right\},$$

so that $(\mathsf{P}(\lambda)\varphi, \psi)$ corresponds to a complex measure on $\mathbb{R}$ — a complex linear combination of measures.

A measurable function f on $\mathbb{R}$ is said to be finite almost everywhere (a.e.) with respect to the projection-valued measure P if it is finite a.e. with respect to all measures $(\mathsf{P}\psi, \psi)$, $\psi \in \mathscr{H}$. For separable $\mathscr{H}$ a theorem of von Neumann states that for every projection-valued measure P there exists $\varphi \in \mathscr{H}$ such that a function f is finite a.e. with respect to P if and only if it is finite a.e. with respect to the measure $(\mathsf{P}\varphi, \varphi)$.

The next statement is the celebrated general spectral theorem of von Neumann.

Theorem 1.1 (J. von Neumann). *For every self-adjoint operator A on the Hilbert space $\mathscr{H}$ there exists a unique projection-valued resolution of the identity $\mathsf{P}(\lambda)$, satisfying the following properties.*

(i)
$$D(A) = \left\{ \varphi \in \mathscr{H} : \int_{-\infty}^{\infty} \lambda^2 d(\mathsf{P}(\lambda)\varphi, \varphi) < \infty \right\},$$

and for every $\varphi \in D(A)$

$$A\varphi = \int_{-\infty}^{\infty} \lambda \, d\mathsf{P}(\lambda)\varphi,$$

defined as a limit of Riemann-Stieltjes sums in the strong topology on $\mathscr{H}$. The support of the corresponding projection-valued measure P coincides with the spectrum of the operator A: $\lambda \in \sigma(A)$ if and only if $\mathsf{P}_A((\lambda - \varepsilon, \lambda + \varepsilon)) \neq 0$ for all $\varepsilon > 0$.

(ii) *For every continuous function f on $\mathbb{R}$, $f(A)$ is a linear operator on $\mathscr{H}$ with a dense domain*

$$D(f(A)) = \left\{ \varphi \in \mathscr{H} : \int_{-\infty}^{\infty} |f(\lambda)|^2 d(\mathsf{P}(\lambda)\varphi, \varphi) < \infty \right\},$$

defined for $\varphi \in D(f(A))$ as

$$f(A)\varphi = \int_{-\infty}^{\infty} f(\lambda)d\mathsf{P}(\lambda)\varphi,$$

and understood as in part (i). *The operator $f(A)$ satisfies*

$$f(A)^* = \bar{f}(A),$$

where $\bar{f}$ is the complex conjugate function to f, and the operator $f(A)$ is bounded if and only if the function f is bounded on $\sigma(A)$. For bounded on $\sigma(A)$ continuous functions f and g,

$$f(A)g(A)\varphi = \int_{-\infty}^{\infty} f(\lambda)g(\lambda)d\mathsf{P}(\lambda)\varphi, \quad \varphi \in \mathscr{H}.$$

(iii) *For every measurable function f on $\mathbb{R}$, finite a.e. with respect to the projection-valued measure P, $f(A)$ is a linear operator on $\mathscr{H}$ defined as in (ii), where the integral for $f(A)\varphi$ is now understood in the weak sense: for $\varphi \in D(f(A))$ and every $\psi \in \mathscr{H}$,*

$$(f(A)\varphi, \psi) = \int_{-\infty}^{\infty} f(\lambda) d(\mathsf{P}(\lambda)\varphi, \psi)$$

— a Lebesgue-Stieltjes integral with respect to a complex measure. The correspondence $f \mapsto f(A)$ satisfies the same properties as in (ii), where the integrals are understood in the weak sense.

(iv) *A bounded operator B commutes with A, that is, $B(D(A)) \subset D(A)$ and $AB = BA$ on $D(A)$, if and only if it commutes with $\mathsf{P}(\lambda)$ for all λ and, therefore, B commutes with every operator $f(A)$.*

(v) *For every projection-valued resolution of identity $\mathsf{P}(\lambda)$ the operator A on $\mathscr{H}$, defined as in part (i), is self-adjoint.*

Using the spectral theorem, the correspondence $(A, M) \mapsto \mu_A$, postulated in **A4**, can be explicitly described as follows.

A5. The probability measure μ_A on $\mathbb{R}$, which defines the correspondence $\mathscr{A} \times \mathscr{S} \to \mathscr{P}(\mathbb{R})$, is given by the *Born-von Neumann formula*

$$(1.3) \qquad \mu_A(E) = \operatorname{Tr} \mathsf{P}_A(E) M, \quad E \in \mathscr{B}(\mathbb{R}),$$

where P_A is a projection-valued measure on $\mathbb{R}$ associated with the self-adjoint operator A.

Remark. The probability measure μ_A on $\mathbb{R}$ can be considered as a "quantum push-forward" of the state M by the observable A (cf. the discussion in Section 2.8 in Chapter 1).

From the Hilbert-Schmidt decomposition (1.1) we get

$$\mu_A(E) = \sum_{n=1}^{N} \alpha_n (\mathsf{P}_A(E)\psi_n, \psi_n) = \sum_{n=1}^{N} \alpha_n \|\mathsf{P}_A(E)\psi_n\|^2 \leq \sum_{n=1}^{N} \alpha_n = 1,$$

so that indeed $0 \leq \mu_A(E) \leq 1$. Denote by $\mu_A(\lambda)$ the distribution function of the probability measure μ_A, $\mu_A(\lambda) = (\mathsf{P}_A(\lambda)\psi, \psi)$ for $M = P_\psi$.

Proposition 1.1. *Suppose that an observable $A \in \mathscr{A}$ and a state $M \in \mathscr{S}$ are such that $\langle A|M \rangle < \infty$ and $\operatorname{Im} M \subseteq D(A)$. Then $AM \in \mathscr{S}_1$ and*

$$\langle A|M \rangle = \operatorname{Tr} AM.$$

In particular, if $M = P_\psi$ and $\psi \in D(A)$, then

$$\langle A|M \rangle = (A\psi, \psi) \quad and \quad \langle A^2|M \rangle = \|A\psi\|^2.$$

Proof. Let $\{e_n\}_{n=1}^{\infty}$ be an orthonormal basis for $\mathscr{H}$. Since

$$\mu_A(E) = \operatorname{Tr} \mathsf{P}_A(E)M = \sum_{n=1}^{\infty} (\mathsf{P}_A(E)Me_n, e_n),$$

we get for every $E \in \mathscr{B}(\mathbb{R})$,

$$\mu_A(E) = \sum_{n=1}^{\infty} \mu_n(E),$$

where μ_n are finite measures on $\mathbb{R}$ defined by $\mu_n(E) = (\mathsf{P}_A(E)Me_n, e_n)$. Since

$$\int_{\mathbb{R}} f d\mu_A = \sum_{n=1}^{\infty} \int_{\mathbb{R}} f d\mu_n$$

for every function f integrable with respect to the measure μ_A, it follows from the spectral theorem that

$$\sum_{n=1}^{\infty} (AMe_n, e_n) = \sum_{n=1}^{\infty} \int_{-\infty}^{\infty} \lambda d\mu_n(\lambda) = \int_{-\infty}^{\infty} \lambda d\mu_A(\lambda) < \infty.$$

Thus $AM \in \mathscr{S}_1$ and $\langle A|M\rangle = \operatorname{Tr} AM$. In particular, when $M = P_\psi$ and $\psi \in D(A)$,

$$\langle A|M\rangle = \int_{-\infty}^{\infty} \lambda d(\mathsf{P}_A(\lambda)\psi, \psi) = (A\psi, \psi).$$

Finally, from the spectral theorem and the change of variables formula we get

$$\|A\psi\|^2 = \int_{-\infty}^{\infty} \lambda^2 d(\mathsf{P}_A(\lambda)\psi, \psi) = \int_{0}^{\infty} \lambda d(\mathsf{P}_{A^2}(\lambda)\psi, \psi) = \langle A^2|M\rangle. \qquad \square$$

Corollary 1.2. *If* $\langle A|M\rangle, \langle A^2|M\rangle < \infty$, *then* $AM \in \mathscr{S}_1$ *and* $\langle A|M\rangle = \operatorname{Tr} AM$.

Proof. Since

$$\int_{-\infty}^{\infty} \lambda^2 d\mu_n(\lambda) \leq \int_{-\infty}^{\infty} \lambda^2 d\mu_A(\lambda) = \langle A^2|M\rangle < \infty,$$

we get that $e_n \in D(AM)$, and the result follows from the proof of Proposition 1.1. $\qquad \square$

Remark. It is convenient to approximate an unbounded self-adjoint operator A by bounded operators $A_n = f_n(A)$, where $f_n = \chi_{[-n,n]}$ — a characteristic function of the interval $[-n, n]$. Assuming that $\langle A|M\rangle$ exists, we have

$$\langle A|M\rangle = \int_{-\infty}^{\infty} \lambda d\mu_A(\lambda) = \lim_{n \to \infty} \int_{-n}^{n} \lambda d\mu_A(\lambda) = \lim_{n \to \infty} \langle A_n|M\rangle.$$

Definition. Self-adjoint operators A and B commute if the corresponding projection-valued measures P_A and P_B commute,

$$\mathsf{P}_A(E_1)\mathsf{P}_B(E_2) = \mathsf{P}_B(E_2)\mathsf{P}_A(E_1) \quad \text{for all} \quad E_1, E_2 \in \mathscr{B}(\mathbb{R}).$$

The following results, which follow from the spectral theorem, are very useful in applications.

Proposition 1.2. *The following statements are equivalent.*

(i) *Self-adjoint operators A and B commute.*

(ii) *For all $\lambda, \mu \in \mathbb{C}$, $\operatorname{Im}\lambda, \operatorname{Im}\mu \neq 0$,*

$$R_\lambda(A)R_\mu(B) = R_\mu(B)R_\lambda(A).$$

(iii) *For all $u, v \in \mathbb{R}$,*

$$e^{iuA}e^{ivB} = e^{ivB}e^{iuA}.$$

(iv) *For all $u \in \mathbb{R}$, the operators e^{iuA} and B commute.*

Slightly abusing notation[10], we will often write $[A, B] = AB - BA = 0$ for commuting self-adjoint operators A and B.

Proposition 1.3. *Let $\boldsymbol{A} = \{A_1, \ldots, A_n\}$ be a finite set of self-adjoint, pairwise commuting operators on $\mathscr{H}$. There exists a unique projection-valued measure $\mathsf{P}_{\boldsymbol{A}}$ on the Borel subsets of $\mathbb{R}^n$ having the following properties.*

(i) *For every $\boldsymbol{E} = E_1 \times \cdots \times E_n \in \mathscr{B}(\mathbb{R}^n)$,*

$$\mathsf{P}_{\boldsymbol{A}}(\boldsymbol{E}) = \mathsf{P}_{A_1}(E_1)\ldots\mathsf{P}_{A_n}(E_n).$$

(ii) *In the strong operator topology,*

$$A_k = \int_{\mathbb{R}^n} \lambda_k d\mathsf{P}_{\boldsymbol{A}}, \quad k = 1, \ldots, n,$$

where λ_k is the k-th coordinate function on $\mathbb{R}^n$, $\lambda_k(x_1, \ldots, x_n) = x_k$.

(iii) *For every measurable function f on $\mathbb{R}^n$, finite a.e. with respect to the projection-valued measure $\mathsf{P}_{\boldsymbol{A}}$, $f(A_1, \ldots, A_n)$ is a linear operator on $\mathscr{H}$ defined by*

$$f(A_1, \ldots, A_n) = \int_{\mathbb{R}^n} f d\mathsf{P}_{\boldsymbol{A}},$$

where the integral is understood in the weak operator topology. The correspondence $f \mapsto f(A_1, \ldots, A_n)$ satisfies the same properties as in part (ii) of the spectral theorem.

---

[10] In general, for unbounded self-adjoint operators A and B the commutator $[A, B] = AB - BA$ is not necessarily closed, i.e., it could be defined only for $\varphi = 0$.

The support of the projection-valued measure $\mathsf{P_A}$ on $\mathbb{R}^n$ is called the joint spectrum of the commutative family $\boldsymbol{A} = \{A_1, \ldots, A_n\}$.

Remark. According to von Neumann's theorem on a generating operator, for every commutative family $\boldsymbol{A}$ of self-adjoint operators (not necessarily finite) on a separable Hilbert space $\mathscr{H}$ there is a generating operator — a self-adjoint operator R on $\mathscr{H}$ such that all operators in $\boldsymbol{A}$ are functions of R.

It seems natural that simultaneous measurement of a finite set of observables $\boldsymbol{A} = \{A_1, \ldots, A_n\}$ in the state $M \in \mathscr{S}$ should be described by the probability measure $\mu_{\boldsymbol{A}}$ on $\mathbb{R}^n$ given by the following generalization of the Born-von Neumann formula:

$$(1.4) \quad \mu_{\boldsymbol{A}}(\boldsymbol{E}) = \mathrm{Tr}(\mathsf{P}_{A_1}(E_1) \ldots \mathsf{P}_{A_n}(E_n)M), \quad \boldsymbol{E} = E_1 \times \cdots \times E_n \in \mathscr{B}(\mathbb{R}^n).$$

However, formula (1.4) defines a probability measure on $\mathbb{R}^n$ if and only if $\mathsf{P}_{A_1}(E_1) \ldots \mathsf{P}_{A_n}(E_n)$ defines a projection-valued measure on $\mathbb{R}^n$. Since a product of orthogonal projections is an orthogonal projection only when the projection operators commute, we conclude that the operators $A_1, \ldots, A_n$ should form a commutative family. This agrees with the requirement that simultaneous measurement of several observables should be independent of the order of the measurements of individual observables. We summarize these arguments as the following axiom.

A6. A finite set of observables $\boldsymbol{A} = \{A_1, \ldots, A_n\}$ can be measured simultaneously (*simultaneously measured observables*) if and only if they form a commutative family. Simultaneous measurement of the commutative family $\boldsymbol{A} \subset \mathscr{A}$ in the state $M \in \mathscr{S}$ is described by the probability measure $\mu_{\boldsymbol{A}}$ on $\mathbb{R}^n$ given by

$$\mu_{\boldsymbol{A}}(\boldsymbol{E}) = \mathrm{Tr}\, \mathsf{P}_{\boldsymbol{A}}(\boldsymbol{E})M, \quad \boldsymbol{E} \in \mathscr{B}(\mathbb{R}^n),$$

where $\mathsf{P}_{\boldsymbol{A}}$ is the projection-valued measure from Proposition 1.3. Explicitly, $\mathsf{P}_{\boldsymbol{A}}(\boldsymbol{E}) = \mathsf{P}_{A_1}(E_1) \ldots \mathsf{P}_{A_n}(E_n)$ for $\boldsymbol{E} = E_1 \times \cdots \times E_n \in \mathscr{B}(\mathbb{R}^n)$. For every Borel subset $\boldsymbol{E} \subseteq \mathbb{R}^n$ the quantity $0 \leq \mu_{\boldsymbol{A}}(\boldsymbol{E}) \leq 1$ is the probability that for a quantum system in the state M the result of the simultaneous measurement of observables $A_1, \ldots, A_n$ belongs to $\boldsymbol{E}$.

The axioms **A1-A6** are known as Dirac-von Neumann axioms.

Problem 1.1. Prove property (1.2).

Problem 1.2. Prove that the state M is a pure state if and only if $\mathrm{Tr}\, M^2 = 1$.

Problem 1.3. Prove that the Born-von Neumann formula (1.3) defines a probability measure on $\mathbb{R}$, i.e., μ_A is a σ-additive function on $\mathscr{B}(\mathbb{R})$.

Problem 1.4. Prove all the remaining statements in this section.

1.2. Heisenberg's uncertainty relations. The variance of the observable A in the state M, which measures the mean deviation of A from its expectation value, is defined by

$$\sigma_M^2(A) = \langle (A - \langle A|M\rangle I)^2|M\rangle = \langle A^2|M\rangle - \langle A|M\rangle^2 \geq 0,$$

provided the expectation values $\langle A^2|M\rangle$ and $\langle A|M\rangle$ exist. It follows from Proposition 1.1 that for $M = P_\psi$, where $\psi \in D(A)$,

$$\sigma_M^2(A) = \|(A - \langle A|M\rangle I)\psi\|^2 = \|A\psi\|^2 - (A\psi, \psi)^2.$$

Lemma 1.2. *For $A \in \mathscr{A}$ and $M \in \mathscr{S}$ the variance $\sigma_M(A) = 0$ if and only if $\operatorname{Im} M$ is an eigenspace for the operator A with the eigenvalue $a = \langle A|M\rangle$. In particular, if $M = P_\psi$, then ψ is an eigenvector of A, $A\psi = a\psi$.*

Proof. It follows from the spectral theorem that

$$\sigma_M^2(A) = \int_{-\infty}^{\infty} (\lambda - a)^2 d\mu_A(\lambda),$$

so that $\sigma_M(A) = 0$ if and only if the probability measure μ_A is supported at the point $a \in \mathbb{R}$, i.e., $\mu_A(\{a\}) = 1$. Since $\mu_A(\{a\}) = \operatorname{Tr} P_A(\{a\})M$ and $\operatorname{Tr} M = 1$, we conclude that this is equivalent to $\operatorname{Im} M$ being an invariant subspace for $P_A(\{a\})$, and it follows from the spectral theorem that $\operatorname{Im} M$ is an eigenspace for A with the eigenvalue a. $\square$

Now we formulate generalized *Heisenberg's uncertainty relations*.

Proposition 1.4 (H. Weyl). *Let $A, B \in \mathscr{A}$ and let $M = P_\psi$ be the pure state such that $\psi \in D(A) \cap D(B)$ and $A\psi, B\psi \in D(A) \cap D(B)$. Then*

$$\sigma_M^2(A)\sigma_M^2(B) \geq \tfrac{1}{4}\langle i[A, B]|M\rangle^2.$$

The same inequality holds for all $M \in \mathscr{S}$, where by definition $\langle i[A, B]|M\rangle = \lim_{n\to\infty}\langle i[A_n, B_n]|M\rangle$.

Proof. Let $M = P_\psi$. Since

$$[A - \langle A|M\rangle I, B - \langle B|M\rangle I] = [A, B],$$

it is sufficient to prove the inequality

$$\langle A^2|M\rangle\langle B^2|M\rangle \geq \tfrac{1}{4}\langle i[A, B]|M\rangle^2.$$

We have for all $\alpha \in \mathbb{R}$,

$$\|(A + i\alpha B)\psi\|^2 = \alpha^2(B\psi, B\psi) - i\alpha(A\psi, B\psi) + i\alpha(B\psi, A\psi) + (A\psi, A\psi)$$

$$= \alpha^2(B^2\psi, \psi) + \alpha(i[A, B]\psi, \psi) + (A^2\psi, \psi) \geq 0,$$

so that necessarily $4(A^2\psi, \psi)(B^2\psi, \psi) \geq (i[A, B]\psi, \psi)^2$.

The same argument works for the mixed states. Since

$$\sigma_M^2(A)\sigma_M^2(B) = \lim_{n\to\infty} \sigma_M^2(A_n)\sigma_M^2(B_n)$$

(see the remark in the previous section), it is sufficient to prove the inequality
for bounded A and B. Then using the cyclic property of the trace we have
for all $\alpha \in \mathbb{R}$,

$$0 \le \operatorname{Tr}((A + i\alpha B)M(A + i\alpha B)^*) = \operatorname{Tr}((A + i\alpha B)M(A - i\alpha B))$$

$$= \alpha^2 \operatorname{Tr} BMB + i\alpha \operatorname{Tr} BMA - i\alpha \operatorname{Tr} AMB + \operatorname{Tr} AMA$$

$$= \alpha^2 \operatorname{Tr} B^2 M + \alpha \operatorname{Tr}(i[A, B]M) + \operatorname{Tr} A^2 M,$$

so that $4 \operatorname{Tr} A^2 M \operatorname{Tr} B^2 M \ge \operatorname{Tr}(i[A, B]M)^2$. $\square$

Heisenberg's uncertainty relations provide a quantitative expression of
the fact that even in a pure state non-commuting observables cannot be
measured simultaneously. This shows a fundamental difference between the
process of measurement in classical mechanics and in quantum mechanics.

1.3. Dynamics. The set $\mathscr{A}$ of quantum observables does not form an algebra with respect to an operator product[11]. Nevertheless, a real vector space
$\mathscr{A}_0$ of bounded observables has a Lie algebra structure with the Lie bracket

$$i[A, B] = i(AB - BA), \quad A, B \in \mathscr{A}_0.$$

Remark. In fact, the $\mathbb{C}^*$-algebra $\mathscr{L}(\mathscr{H})$ of bounded operators on $\mathscr{H}$ has
a structure of a complex Lie algebra with the Lie bracket given by a com-
mutator $[A, B] = AB - BA$. It satisfies the Leibniz rule

$$[AB, C] = A[B, C] + [A, C]B,$$

so that the Lie bracket is a derivation of the $\mathbb{C}^*$-algebra $\mathscr{L}(\mathscr{H})$.

In analogy with classical mechanics, we postulate that the time evolution
of a quantum system with the space of states $\mathscr{H}$ is completely determined
by a special observable $H \in \mathscr{A}$, called a *Hamiltonian operator* (Hamiltonian
for brevity). As in classical mechanics, the Lie algebra structure on $\mathscr{A}_0$ leads
to corresponding *quantum equations of motion*.

Specifically, the analog of Hamilton's picture in classical mechanics (see
Section 2.8 in Chapter 1) is the *Heisenberg picture* in quantum mechanics,
where the states do not depend on time

$$\frac{dM}{dt} = 0, \quad M \in \mathscr{S},$$

---

[11]The product of two non-commuting self-adjoint operators is not self-adjoint.

and bounded observables satisfy the *Heisenberg equation of motion*

$$(1.5) \qquad \frac{dA}{dt} = \{H, A\}_\hbar, \quad A \in \mathscr{A}_0,$$

where

$$(1.6) \qquad \{\ ,\ \}_\hbar = \frac{i}{\hbar}[\ ,\ ]$$

is the *quantum bracket* — the $\hbar$-dependent Lie bracket on $\mathscr{A}_0$. The positive number $\hbar$, called the *Planck constant*, is one of the fundamental constants in physics[12].

The Heisenberg equation (1.5) is well defined when $H \in \mathscr{A}_0$. Indeed, let $U(t)$ be a strongly continuous one-parameter group of unitary operators associated with a bounded self-adjoint operator H,

$$(1.7) \qquad U(t) = e^{-\frac{i}{\hbar}tH}, \quad t \in \mathbb{R}.$$

It satisfies the differential equation

$$(1.8) \qquad i\hbar\frac{dU(t)}{dt} = HU(t) = U(t)H,$$

so that the solution $A(t)$ of the Heisenberg equation of motion with the initial condition $A(0) = A \in \mathscr{A}_0$ is given by

$$(1.9) \qquad A(t) = U(t)^{-1}AU(t).$$

In general, a strongly one-parameter group of unitary operators (1.7), associated with a self-adjoint operator H, satisfies differential equation (1.8) only on $D(H)$ in a strong sense, that is, applied to $\varphi \in D(H)$. The quantum dynamics is defined by the same formula (1.9), and in this sense all quantum observables satisfy the Heisenberg equation of motion (1.5). The *evolution operator* $U_t : \mathscr{A} \to \mathscr{A}$ is defined by $U_t(A) = A(t) = U(t)^{-1}AU(t)$, and is an automorphism of the Lie algebra $\mathscr{A}_0$ of bounded observables. This is a quantum analog of the statement that the evolution operator in classical mechanics is an automorphism of the Poisson algebra of classical observables (see Theorem 2.19 in Section 2.7 of Chapter 1).

By Stone's theorem, every strongly-continuous one-parameter group of unitary operators[13] $U(t)$ is of the form (1.7), where

$$D(H) = \left\{\varphi \in \mathscr{H} : \lim_{t \to 0}\frac{U(t) - I}{t}\varphi \text{ exists}\right\} \quad \text{and} \quad H\varphi = i\hbar\lim_{t \to 0}\frac{U(t) - I}{t}\varphi.$$

The domain $D(H)$ of the self-adjoint operator H, called the *infinitesimal generator* of $U(t)$, is an invariant linear subspace for all operators $U(t)$.

---

[12]The Planck constant has a physical dimension of the action (energy $\times$ time). Its value $\hbar = 1.054 \times 10^{-27}$ erg $\times$ sec, which is determined from the experiment, manifests that quantum mechanics is a microscopic theory.

[13]According to a theorem of von Neumann, on a separable Hilbert space every weakly measurable one-parameter group of unitary operators is strongly continuous.

We summarize the presented arguments as the following axiom.

A7 (Heisenberg's Picture). The dynamics of a quantum system is described by the strongly continuous one-parameter group $U(t)$ of unitary operators. Quantum states do not depend on time, $\mathscr{S} \ni M \mapsto M(t) = M \in \mathscr{S}$, and time dependence of quantum observables is given by the evolution operator U_t,

$$\mathscr{A} \ni A \mapsto A(t) = U_t(A) = U(t)^{-1}AU(t) \in \mathscr{A}.$$

Infinitesimally, the evolution of quantum observables is described by the Heisenberg equation of motion (1.5), where the Hamiltonian operator H is the infinitesimal generator of $U(t)$.

The analog of Liouville's picture in classical mechanics (see Section 2.8 in Chapter 1) is *Schrödinger's picture* in quantum mechanics, defined as follows.

A8 (Schrödinger's Picture). The dynamics of a quantum system is described by the strongly continuous one-parameter group $U(t)$ of unitary operators. Quantum observables do not depend on time, $\mathscr{A} \ni A \mapsto A(t) = A \in \mathscr{A}$, and time dependence of states is given by the inverse of the evolution operator $U_t^{-1} = U_{-t}$,

$$(1.10) \qquad \mathscr{S} \ni M \mapsto M(t) = U_{-t}(M) = U(t)MU(t)^{-1} \in \mathscr{S}.$$

Infinitesimally, the evolution of quantum states is described by the Schrödinger equation of motion

$$(1.11) \qquad \frac{dM}{dt} = -\{H, M\}_h, \quad M \in \mathscr{S},$$

where the Hamiltonian operator H is the infinitesimal generator of $U(t)$.

Proposition 1.5. *Heisenberg and Schrödinger descriptions of dynamics are equivalent.*

Proof. Let $\mu_{A(t)}$ and $(\mu_t)_A$ be, respectively, probability measures on $\mathbb{R}$ associated with $(A(t), M) \in \mathscr{A} \times \mathscr{S}$ and $(A, M(t)) \in \mathscr{A} \times \mathscr{S}$ according to **A3**-**A4**, where $A(t) = U_t(A)$ and $M(t) = U_{-t}(M)$. We need to show that $\mu_{A(t)} = (\mu_t)_A$. It follows from the spectral theorem that $\mathsf{P}_{A(t)} = U(t)^{-1}\mathsf{P}_A U(t)$, so that using the Born-von Neumann formula (1.3) and the cyclic property of the trace, we get for $E \in \mathscr{B}(\mathbb{R})$,

$$\mu_{A(t)}(E) = \mathrm{Tr}\,\mathsf{P}_{A(t)}(E)M = \mathrm{Tr}(U(t)^{-1}\mathsf{P}_A(E)U(t)M)$$

$$= \mathrm{Tr}(\mathsf{P}_A(E)U(t)MU(t)^{-1}) = \mathrm{Tr}\,\mathsf{P}_A(E)M(t) = (\mu_t)_A(E). \qquad \square$$

Corollary 1.3. $\langle A(t)|M \rangle = \langle A|M(t) \rangle.$

In analogy with classical mechanics (see Section 1.4 of Chapter 1), we have the following definition.

Definition. An observable $A \in \mathscr{A}$ is a *quantum integral of motion* (or a *constant of motion*) for a quantum system with the Hamiltonian H if in Heisenberg's picture

$$\frac{dA(t)}{dt} = 0.$$

It follows from Proposition 1.2 that $A \in \mathscr{A}$ is an integral of motion if and only if it commutes with the Hamiltonian H, so that, in agreement with (1.5),

$$\{H, A\}_\hbar = 0.$$

This is a quantum analog of the Poisson commutativity property, given by formula (2.14) in Section 2.6 of Chapter 1.

It follows from (1.11) that the time evolution of a pure state $M = P_\psi$ is given by $M(t) = P_{\psi(t)}$, where $\psi(t) = U(t)\psi$. Since $D(H)$ is invariant under $U(t)$, the vector $\psi(t) = U(t)\psi$ satisfies the *time-dependent Schrödinger equation*

$$(1.12) \qquad\qquad i\hbar\frac{d\psi}{dt} = H\psi$$

with the initial condition $\psi(0) = \psi$.

Definition. A state $M \in \mathscr{S}$ is called *stationary* for a quantum system with Hamiltonian H if in Schrödinger's picture

$$\frac{dM(t)}{dt} = 0.$$

The state M is stationary if and only if $[M, U(t)] = 0$ for all t, and by Proposition 1.2 this is equivalent to

$$\{H, M\}_\hbar = 0,$$

in agreement with (1.11). The following simple result is fundamental.

Lemma 1.3. *The pure state $M = P_\psi$ is stationary if and only if ψ is an eigenvector for H,*

$$H\psi = \lambda\psi,$$

and in this case

$$\psi(t) = e^{-\frac{i}{\hbar}\lambda t}\psi.$$

Proof. It follows from $U(t)P_\psi = P_\psi U(t)$ that ψ is a common eigenvector for unitary operators $U(t)$ for all t, $U(t)\psi = c(t)\psi$, $|c(t)| = 1$. Since $U(t)$ is a strongly continuous one-parameter group of unitary operators, the continuous function $c(t) = (U(t)\psi, \psi)$ satisfies the equation $c(t_1 + t_2) = c(t_1)c(t_2)$ for all $t_1, t_2 \in \mathbb{R}$, so that $c(t) = e^{-\frac{i}{\hbar}\lambda t}$ for some $\lambda \in \mathbb{R}$. Thus by Stone's theorem $\psi \in D(H)$ and $H\psi = \lambda\psi$. $\qquad\square$

In physics terminology, the eigenvectors of H are called *bound states*. The corresponding eigenvalues are called *energy levels* and are usually denoted by E. The eigenvalue equation $H\psi = E\psi$ is called the *stationary Schrödinger equation*.

Problem 1.5. Show that if an observable A is such that for every state M the expectation value $\langle A|M(t)\rangle$ does not depend on t, then A is a quantum integral of motion. (This is the definition of integrals of motion in the Schrödinger picture.)

Problem 1.6. Show that a solution of the initial value problem for the time-dependent Schrödinger equation (1.12) is given by

$$\psi(t) = \int_{-\infty}^{\infty} e^{-\frac{i}{\hbar}t\lambda}\,d\mathsf{P}(\lambda)\psi,$$

where $\mathsf{P}(\lambda)$ is the resolution of identity for the Hamiltonian H.

Problem 1.7. Let D be a linear subspace of $\mathscr{H}$, consisting of *Gårding vectors*

$$\psi_f = \int_{-\infty}^{\infty} f(s)U(s)\psi\,ds, \quad f \in \mathscr{S}(\mathbb{R}), \ \ \psi \in \mathscr{H},$$

where $\mathscr{S}(\mathbb{R})$ is the Schwartz space of rapidly decreasing functions on $\mathbb{R}$. Prove that D is dense in $\mathscr{H}$ and is invariant for $U(t)$ and for the Hamiltonian H. (*Hint:* Show that $U(t)\psi_f = \psi_{f_t} \in D$, where $f_t(s) = f(s-t)$, and deduce $H\psi_f = \frac{\hbar}{i}\psi_{f'}$.)

2. Quantization

To study the quantum system one needs to describe its Hilbert space of states $\mathscr{H}$ and the Hamiltonian H — a self-adjoint operator in $\mathscr{H}$ which defines the evolution of a system. When the quantum system has a classical analog, the procedure of constructing the corresponding Hilbert space $\mathscr{H}$ and the Hamiltonian H is called *quantization*.

Definition. Quantization of a classical system $((\mathscr{M},\{\ ,\ \}),H_\mathrm{c})$ with the Hamiltonian function[14] H_c is a one-to-one mapping $\mathsf{Q}_\hbar : \mathcal{A} \to \mathscr{A}$ from the set of classical observables $\mathcal{A} = C^\infty(\mathscr{M})$ to the set $\mathscr{A}$ of quantum observables — the set of self-adjoint operators on a Hilbert space $\mathscr{H}$. The map $\mathsf{Q}_\hbar$ depends on the parameter $\hbar > 0$, and its restriction to the subspace of bounded classical observables $\mathcal{A}_0$ is a linear mapping to the subspace $\mathscr{A}_0$ of bounded quantum observables, which satisfies the properties

$$\lim_{\hbar \to 0} \tfrac{1}{2}\mathsf{Q}_\hbar^{-1}\big(\mathsf{Q}_\hbar(f_1)\mathsf{Q}_\hbar(f_2) + \mathsf{Q}_\hbar(f_2)\mathsf{Q}_\hbar(f_1)\big) = f_1 f_2$$

and

$$\lim_{\hbar \to 0} \mathsf{Q}_\hbar^{-1}\big(\{\mathsf{Q}_\hbar(f_1),\mathsf{Q}_\hbar(f_2)\}_\hbar\big) = \{f_1, f_2\} \quad \text{for all} \quad f_1, f_2 \in \mathcal{A}_0.$$

---

[14]Notation H_c is used to distinguish the Hamiltonian function in classical mechanics from the Hamiltonian operator H in quantum mechanics.

The latter property is the celebrated *correspondence principle* of Niels Bohr. In particular, $H_c \mapsto Q_\hbar(H_c) = H$ — the Hamiltonian operator for a quantum system.

Remark. In physics literature the correspondence principle is often stated in the form

$$[\,,\,] \simeq \frac{\hbar}{i}\{\,,\,\} \quad \text{as} \quad \hbar \to 0.$$

Quantum mechanics is different from classical mechanics, so that the correspondence $f \mapsto Q_\hbar(f)$ cannot be an isomorphism between the Lie algebras of bounded classical and quantum observables with respect to classical and quantum brackets. It becomes an isomorphism only in the limit $\hbar \to 0$ when, according to the correspondence principle, quantum mechanics turns into classical mechanics. Since quantum mechanics provides a more accurate and refined description than classical mechanics, quantization of a classical system may not be unique.

Definition. Two quantizations $Q_\hbar^{(1)}$ and $Q_\hbar^{(2)}$ of a given classical system $((\mathscr{M}, \{\,,\,\}), H_c)$ are said to be equivalent if there exists a linear mapping $\mathscr{U}_\hbar : \mathcal{A} \to \mathcal{A}$ such that $Q_\hbar^{(2)} = Q_\hbar^{(1)} \circ \mathscr{U}_\hbar$ and $\lim_{\hbar \to 0} \mathscr{U}_\hbar = \mathrm{id}$.

For many "real world" quantum systems — the systems describing actual physical phenomena — the corresponding Hamiltonian H does not depend on a choice of equivalent quantization, and is uniquely determined by the classical Hamiltonian function H_c.

2.1. Heisenberg commutation relations. The simplest classical system with one degree of freedom is described by the phase space $\mathbb{R}^2$ with coordinates p, q and the Poisson bracket $\{\,,\,\}$, associated with the canonical symplectic form $\omega = dp \wedge dq$. In particular, the Poisson bracket between classical observables p and q — momentum and coordinate of a particle — has the following simple form:

$$(2.1) \qquad\qquad\qquad \{p, q\} = 1.$$

It is another postulate of quantum mechanics that under quantization classical observables p and q correspond to quantum observables P and Q — self-adjoint operators on a Hilbert space $\mathscr{H}$, satisfying the following properties.

CR1. There is a dense linear subset $D \subset \mathscr{H}$ such that $P : D \to D$ and $Q : D \to D$.

CR2. For all $\psi \in D$,

$$(PQ - QP)\psi = -i\hbar\psi.$$

CR3. Every bounded operator on $\mathscr{H}$ which commutes with P and Q is a multiple of the identity operator I.

Property **CR2** is called the *Heisenberg commutation relation* for one degree of freedom. In terms of the quantum bracket (1.6) it takes the form

$$(2.2) \qquad \{P, Q\}_\hbar = I,$$

which is exactly the same as the Poisson bracket (2.1). Property **CR3** is a quantum analog of the classical property that the Poisson manifold $(\mathbb{R}^2, \{\ ,\ \})$ is non-degenerate: every function which Poisson commutes with p and q is a constant (see the last remark in Section 2.7 of Chapter 1).

The operators P and Q are called, respectively, the *momentum operator* and the *coordinate operator*. The correspondence $p \mapsto P$, $q \mapsto Q$ with P and Q satisfying **CR1-CR3** is the cornerstone for the quantization of classical systems. The validity of (2.2), as well as of quantum mechanics as a whole, is confirmed by the agreement of the theory with numerous experiments.

Remark. It is tempting to extend the correspondence $p \mapsto P$, $q \mapsto Q$ to all observables by defining the mapping $f(p, q) \mapsto f(P, Q)$. However, this approach to quantization is rather naive: operators P and Q satisfy (2.2) and do not commute, so that one needs to understand how $f(P, Q)$ — a "function of non-commuting variables" — is actually defined. We will address this problem of the ordering of non-commuting operators P and Q in Section 3.3.

It follows from Heisenberg's uncertainty relations (see Proposition 1.4), that for any pure state $M = P_\psi$ with $\psi \in D$,

$$\sigma_M(P)\sigma_M(Q) \geq \frac{\hbar}{2}.$$

This is a fundamental result saying that it is impossible to measure the coordinate and the momentum of a quantum particle simultaneously: the more accurate the measurement of one quantity is, the less accurate the value of the other is. It is often said that a quantum particle has no observed path, so that "quantum motion" differs dramatically from the motion in classical mechanics.

It is now straightforward to consider a classical system with n degrees of freedom, described by the phase space $\mathbb{R}^{2n}$ with coordinates $\boldsymbol{p} = (p_1, \ldots, p_n)$ and $\boldsymbol{q} = (q^1, \ldots, q^n)$, and the Poisson bracket $\{\ ,\ \}$, associated with the canonical symplectic form $\omega = d\boldsymbol{p} \wedge d\boldsymbol{q}$. The Poisson brackets between classical observables $\boldsymbol{p}$ and $\boldsymbol{q}$ — momenta and coordinates of a particle — have the form

$$(2.3) \qquad \{p_k, p_l\} = 0, \quad \{q^k, q^l\} = 0, \quad \{p_k, q^l\} = \delta_k^l, \quad k, l = 1, \ldots, n.$$

Corresponding momenta and coordinate operators $\boldsymbol{P} = (P_1, \ldots, P_n)$ and $\boldsymbol{Q} = (Q^1, \ldots, Q^n)$ are self-adjoint operators that have a common invariant dense linear subset $D \subset \mathscr{H}$, and on D satisfy the following commutation relations:

$$(2.4) \quad \{P_k, P_l\}_\hbar = 0, \quad \{Q^k, Q^l\}_\hbar = 0, \quad \{P_k, Q^l\}_\hbar = \delta_k^l I, \quad k, l = 1, \ldots, n.$$

These relations are called *Heisenberg commutation relations* for n degrees of freedom. The analog of **CR3** is the property that every bounded operator on $\mathscr{H}$ which commutes with all operators $\boldsymbol{P}$ and $\boldsymbol{Q}$ is a multiple of the identity operator I.

The fundamental algebraic structure associated with Heisenberg commutation relations is the so-called *Heisenberg algebra*.

Definition. The Heisenberg algebra $\mathfrak{h}_n$ with n degrees of freedom is a Lie algebra with the generators $e^1, \ldots, e^n, f_1, \ldots, f_n, c$ and the relations

$$(2.5) \quad [e^k, c] = 0, \quad [f_k, c] = 0, \quad [e^k, f_l] = \delta_l^k c, \quad k, l = 1, \ldots, n.$$

The invariant definition is the following. Let (V, ω) be a $2n$-dimensional symplectic vector space considered as an abelian Lie algebra, and let $\mathfrak{g}$ be a one-dimensional central extension of V by a Lie algebra 2-cocycle given by the bilinear form ω. This means that there is an exact sequence of vector spaces

$$(2.6) \qquad\qquad 0 \to \mathbb{R} \to \mathfrak{g} \to V \to 0,$$

and the Lie bracket in $\mathfrak{g}$ is defined by

$$(2.7) \qquad\qquad [x, y] = \omega(\bar{x}, \bar{y})c,$$

where $\bar{x}, \bar{y}$ are the images in V of elements $x, y \in \mathfrak{g}$, and c is the image of 1 under the embedding $\mathbb{R} \hookrightarrow \mathfrak{g}$, called the central element of $\mathfrak{g}$. A choice of a symplectic basis $e^1, \ldots, e^n, f_1, \ldots, f_n$ for V (see Section 2.6 of Chapter 1) establishes the isomorphism $\mathfrak{g} \simeq \mathfrak{h}_n$, and relations (2.5) are obtained from the Lie bracket (2.7).

By Ado's theorem, the Heisenberg algebra $\mathfrak{h}_n$ is isomorphic to a Lie subalgebra of a matrix algebra over $\mathbb{R}$. Explicitly, it is realized as a nilpotent subalgebra of the Lie algebra $\mathfrak{gl}_{n+2}$ of $(n+2) \times (n+2)$ matrices with the elements

$$(2.8) \qquad \sum_{k=1}^{n} (u^k f_k + v_k e^k) + \alpha c = \begin{pmatrix} 0 & u^1 & u^2 & \cdots & u^n & \alpha \\ 0 & 0 & 0 & \cdots & 0 & v_1 \\ 0 & 0 & 0 & \cdots & 0 & v_2 \\ \vdots & \vdots & \vdots & \ddots & \vdots & \vdots \\ 0 & 0 & 0 & \cdots & 0 & v_n \\ 0 & 0 & 0 & \cdots & 0 & 0 \end{pmatrix}.$$

Remark. The faithful representation $\mathfrak{h}_n \to \mathfrak{gl}_{n+2}$, given by (2.8), is clearly reducible: the subspace $V = \{\boldsymbol{x} = (x_1, \dots, x_{n+2}) \in \mathbb{R}^{n+2} : x_{n+2} = 0\}$ is an invariant subspace for $\mathfrak{h}_n$ with the central element c acting by zero. However, this representation is not decomposable: the vector space $\mathbb{R}^{n+2}$ cannot be written as a direct sum of V and a one-dimensional invariant subspace for $\mathfrak{h}_n$. This explains why the central element c is not represented by a diagonal matrix with the first $n+1$ zeros, but rather has a special form given by (2.8).

Analytically, Heisenberg commutation relations (2.5) correspond to an irreducible unitary representation of the Heisenberg Lie algebra $\mathfrak{h}_n$. Recall that a unitary representation ρ of $\mathfrak{h}_n$ in the Hilbert space $\mathscr{H}$ is the linear mapping $\rho : \mathfrak{h}_n \to i\mathscr{A}$ — the space of skew-Hermitian operators in $\mathscr{H}$ — such that all self-adjoint operators $i\rho(x)$, $x \in \mathfrak{h}_n$, have a common invariant dense linear subset $D \subset \mathscr{H}$ and satisfy

$$\rho([x, y])\varphi = (\rho(x)\rho(y) - \rho(y)\rho(x))\varphi, \quad x, y \in \mathfrak{h}_n, \ \varphi \in D.$$

Formally applying Schur's lemma we say that the representation ρ is irreducible if every bounded operator which commutes with all operators $i\rho(x)$ is a multiple of the identity operator I. Then Heisenberg commutation relations (2.5) define an irreducible unitary representation ρ of the Heisenberg Lie algebra $\mathfrak{h}_n$ in the Hilbert space $\mathscr{H}$ by setting

$$(2.9) \qquad \rho(f_k) = -iP_k, \quad \rho(e^k) = -iQ^k, \quad k = 1, \dots, n, \quad \rho(c) = -i\hbar I.$$

Since the operators P^k and Q_k are necessarily unbounded (see Problem 2.1), the condition

$$P_k P_l \varphi = P_l P_k \varphi \quad \text{for all} \quad \varphi \in D$$

does not necessarily imply (see Problem 2.2) that self-adjoint operators P_k and P_l commute in the sense of the definition in Section 1.1. To avoid such "pathological" representations, we will assume that ρ is an *integrable* representation, i.e., it can be integrated (in a precise sense specified below) to an irreducible unitary representation of the *Heisenberg group* $\mathbf{H}_n$ — a connected, simply-connected Lie group with the Lie algebra $\mathfrak{h}_n$.

Explicitly, the Heisenberg group is a unipotent subgroup of the Lie algebra $\mathrm{SL}(n+2, \mathbb{R})$ with the elements

$$g = \begin{pmatrix} 1 & u^1 & u^2 & \cdots & u^n & \alpha \\ 0 & 1 & 0 & \cdots & 0 & v_1 \\ 0 & 0 & 1 & \cdots & 0 & v_2 \\ \vdots & \vdots & \vdots & \ddots & \vdots & \vdots \\ 0 & 0 & 0 & \cdots & 1 & v_n \\ 0 & 0 & 0 & \cdots & 0 & 1 \end{pmatrix}.$$

The exponential map $\exp : \mathfrak{h}_n \to \mathbf{H}_n$ is onto, and the Heisenberg group $\mathbf{H}_n$ is generated by two n-parameter abelian subgroups

$$\exp \boldsymbol{u} X = \exp\left(\sum_{k=1}^{n} u^k f_k\right), \quad \exp \boldsymbol{v} Y = \exp\left(\sum_{k=1}^{n} v_k e^k\right), \quad \boldsymbol{u}, \boldsymbol{v} \in \mathbb{R}^n,$$

and a one-parameter center $\exp \alpha c$, which satisfy the relations

$$(2.10) \quad \exp \boldsymbol{u} X \exp \boldsymbol{v} Y = \exp(-\boldsymbol{u}\boldsymbol{v}c) \exp \boldsymbol{v} Y \exp \boldsymbol{u} X, \quad \boldsymbol{u}\boldsymbol{v} = \sum_{k=0}^{n} u^k v_k.$$

Indeed, it follows from (2.5) that $[\boldsymbol{u} X, \boldsymbol{v} Y] = -\boldsymbol{u}\boldsymbol{v}c$ is a central element, so that using the Baker-Campbell-Hausdorff formula we obtain

$$\exp \boldsymbol{u} X \exp \boldsymbol{v} Y = \exp(-\tfrac{1}{2}\boldsymbol{u}\boldsymbol{v}c) \exp(\boldsymbol{u} X + \boldsymbol{v} Y),$$

$$\exp \boldsymbol{v} Y \exp \boldsymbol{u} X = \exp(\tfrac{1}{2}\boldsymbol{u}\boldsymbol{v}c) \exp(\boldsymbol{u} X + \boldsymbol{v} Y).$$

In the matrix realization, the exponential map is given by the matrix exponential and we get $e^{\boldsymbol{u} X} = I + \boldsymbol{u} X$, $e^{\boldsymbol{v} Y} = I + \boldsymbol{v} Y$, and $e^{\alpha c} = I + \alpha c$, where I is the $(n+2) \times (n+2)$ identity matrix.

Let R be an irreducible unitary representation of the Heisenberg group $\mathbf{H}_n$ in the Hilbert space $\mathscr{H}$ — a strongly continuous group homomorphism $R : \mathbf{H}_n \to \mathscr{U}(\mathscr{H})$, where $\mathscr{U}(\mathscr{H})$ is the group of unitary operators in $\mathscr{H}$. By Schur's lemma, $R(e^{\alpha c}) = e^{-i\lambda\alpha} I$, $\lambda \in \mathbb{R}$. Suppose now that $\lambda = \hbar$, and define two strongly continuous n-parameter abelian groups of unitary operators

$$U(\boldsymbol{u}) = R(\exp \boldsymbol{u} X), \quad V(\boldsymbol{v}) = R(\exp \boldsymbol{v} Y), \quad \boldsymbol{u}, \boldsymbol{v} \in \mathbb{R}^n.$$

Then it follows from (2.10) that unitary operators $U(\boldsymbol{u})$ and $V(\boldsymbol{v})$ satisfy commutation relations

$$(2.11) \qquad\qquad U(\boldsymbol{u})V(\boldsymbol{v}) = e^{i\hbar\boldsymbol{u}\boldsymbol{v}} V(\boldsymbol{v})U(\boldsymbol{u}),$$

called *Weyl relations*. Let $\boldsymbol{P} = (P_1, \ldots, P_n)$ and $\boldsymbol{Q} = (Q^1, \ldots, Q^n)$ be, respectively, infinitesimal generators of the subgroups $U(\boldsymbol{u})$ and $V(\boldsymbol{v})$, given by the Stone theorem,

$$P_k = i \left.\frac{\partial U(\boldsymbol{u})}{\partial u^k}\right|_{\boldsymbol{u}=0} \quad \text{and} \quad Q^k = i \left.\frac{\partial V(\boldsymbol{v})}{\partial v_k}\right|_{\boldsymbol{v}=0}, \quad k = 1, \ldots, n.$$

Taking the second partial derivatives of Weyl relations (2.11) at the origin $\boldsymbol{u} = \boldsymbol{v} = 0$ and using the solution of Problem 1.7 in the previous section, we easily obtain the following result.

Lemma 2.1. *Let $R : \mathbf{H}_n \to \mathscr{U}(\mathscr{H})$ be an irreducible unitary representation of the Heisenberg group $\mathbf{H}_n$ in $\mathscr{H}$ such that $R(e^{\alpha c}) = e^{-i\hbar\alpha} I$, and let $\boldsymbol{P} = (P_1, \ldots, P_n)$ and $\boldsymbol{Q} = (Q^1, \ldots, Q^n)$ be, respectively, infinitesimal generators of the strongly continuous n-parameter abelian subgroups $U(\boldsymbol{u})$ and $V(\boldsymbol{v})$.*

Then formulas (2.9) define an irreducible unitary representation ρ of the Heisenberg algebra $\mathfrak{h}_n$ in $\mathscr{H}$.

The representation ρ in Lemma 2.1 is called the differential of a representation R, and is denoted by dR. The irreducible unitary representation ρ of $\mathfrak{h}_n$ is called *integrable* if $\rho = dR$ for some irreducible unitary representation R of $\mathbf{H}_n$.

Remark. Not every irreducible unitary representation of the Heisenberg algebra is integrable, so that Weyl relations cannot be obtained from the Heisenberg commutation relations. However, the following heuristic argument (which ignores the subtleties of dealing with unbounded operators) is commonly used in physics textbooks. Consider the case of one degree of freedom and start with

$$\{P, Q\}_\hbar = I.$$

Since the quantum bracket satisfies the Leibniz rule we have, for a "suitable" function f,

$$\{f(P), Q\}_\hbar = f'(P).$$

In particular, choosing $f(P) = e^{-iuP} = U(u)$, we obtain

$$U(u)Q - QU(u) = \hbar u U(u) \quad \text{or} \quad U(u)QU(u)^{-1} = Q + \hbar u I.$$

This implies, for a "suitable" function g,

$$U(u)g(Q) = g(Q + \hbar u I)U(u),$$

and setting $g(Q) = e^{-ivQ} = V(v)$, we get the Weyl relation.

We will prove in Section 3.1 that all integrable irreducible unitary representations of the Heisenberg algebra $\mathfrak{h}_n$ with the same action of the central element c are unitarily equivalent. This justifies the following mathematical formulation of the Heisenberg commutation relations for n degrees of freedom.

A9 (Heisenberg's Commutation Relations). Momenta and coordinate operators $\boldsymbol{P} = (P_1, \ldots, P_n)$ and $\boldsymbol{Q} = (Q^1, \ldots, Q^n)$ for a quantum particle with n degrees of freedom are defined by formulas (2.9), where ρ is an integrable irreducible unitary representation of the Heisenberg algebra $\mathfrak{h}_n$ with the property $\rho(c) = -i\hbar I$.

Problem 2.1. Prove that there are no bounded operators on the Hilbert space $\mathscr{H}$ satisfying $[A, B] = I$.

Problem 2.2. Give an example of self adjoint operators A and B which have a common invariant dense linear subset $D \subset \mathscr{H}$ such that $AB\varphi = BA\varphi$ for all $\varphi \in D$, but e^{iA} and e^{iB} do not commute.

Problem 2.3. Prove Lemma 2.1. (*Hint:* As in Problem 1.7, let D be the linear set of Gårding vectors

$$\psi_f = \int_{\mathbb{R}^{2n}} f(\boldsymbol{u}, \boldsymbol{v}) U(\boldsymbol{u}) V(\boldsymbol{v}) \psi \, d^n \boldsymbol{u} d^n \boldsymbol{v}, \quad f \in \mathscr{S}(\mathbb{R}^{2n}), \ \psi \in \mathscr{H},$$

where $\mathscr{S}(\mathbb{R}^{2n})$ is the Schwartz space of rapidly decreasing functions on $\mathbb{R}^{2n}$.)

2.2. Coordinate and momentum representations. We start with the case of one degree of freedom and consider two natural realizations of the Heisenberg commutation relation. They are defined by the property that one of the self-adjoint operators P and Q is "diagonal" (i.e., is a multiplication by a function operator in the corresponding Hilbert space).

In the *coordinate representation*, $\mathscr{H} = L^2(\mathbb{R}, dq)$ is the L^2-space on the configuration space $\mathbb{R}$ with the coordinate q, which is a Lagrangian subspace of $\mathbb{R}^2$ defined by the equation $p = 0$. Set

$$D(Q) = \left\{ \varphi \in \mathscr{H} : \int_{-\infty}^{\infty} q^2 |\varphi(q)|^2 dq < \infty \right\}$$

and for $\varphi \in D(Q)$ define the operator Q as a "multiplication by q operator",

$$(Q\varphi)(q) = q\varphi(q), \ q \in \mathbb{R},$$

justifying the name coordinate representation. The coordinate operator Q is obviously self-adjoint and its projection-valued measure is given by

(2.12) $$(\mathsf{P}(E)\varphi)(q) = \chi_E(q)\varphi(q),$$

where χ_E is the characteristic function of a Borel subset $E \subseteq \mathbb{R}$. Therefore $\operatorname{supp} \mathsf{P} = \mathbb{R}$ and $\sigma(Q) = \mathbb{R}$.

Recall that a self-adjoint operator A has an absolutely continuous spectrum if for every $\psi \in \mathscr{H}$, $\|\psi\| = 1$, the probability measure ν_ψ,

$$\nu_\psi(E) = (\mathsf{P}_A(E)\psi, \psi), \quad E \in \mathscr{B}(\mathbb{R}),$$

is absolutely continuous with respect to the Lebesgue measure on $\mathbb{R}$.

Lemma 2.2. *The coordinate operator Q has an absolutely continuous spectrum $\mathbb{R}$, and every bounded operator B which commutes with Q is a function of Q, $B = f(Q)$ with $f \in L^\infty(\mathbb{R})$.*

Proof. It follows from (2.12) that $\nu_\psi(E) = \int_E |\psi(q)|^2 dq$, which proves the first statement. Now a bounded operator B on $\mathscr{H}$ commutes with Q if and only if $B\mathsf{P}(E) = \mathsf{P}(E)B$ for all $E \in \mathscr{B}(\mathbb{R})$, and using (2.12) we get

(2.13) $$B(\chi_E \varphi) = \chi_E B(\varphi).$$

Choosing in (2.13) $E = E_1$ and $\varphi = \chi_{E_2}$, where E_1 and E_2 have finite Lebesgue measure, we obtain

$$B(\chi_{E_1} \cdot \chi_{E_2}) = B(\chi_{E_1 \cap E_2}) = \chi_{E_1} B(\chi_{E_2}) = \chi_{E_2} B(\chi_{E_1}),$$

so that denoting $f_E = B(\chi_E)$ we get $\operatorname{supp} f_E \subseteq E$, and

$$f_{E_1}\big|_{E_1 \cap E_2} = f_{E_2}\big|_{E_1 \cap E_2}$$

for all $E_1, E_2 \in \mathscr{B}(\mathbb{R})$ with finite Lebesgue measure. Thus there exists a measurable function f on $\mathbb{R}$ such that $f|_E = f_E|_E$ for every $E \in \mathscr{B}(\mathbb{R})$ with finite Lebesgue measure. The linear subspace spanned by all $\chi_E \in L^2(\mathbb{R})$ is dense in $L^2(\mathbb{R})$ and the operator B is continuous, so that we get

$$(B\varphi)(q) = f(q)\varphi(q) \quad \text{for all} \quad \varphi \in L^2(\mathbb{R}).$$

Since B is a bounded operator, $f \in L^\infty(\mathbb{R})$ and $\|B\| = \|f\|_\infty$. $\qquad\square$

Remark. By the Schwartz kernel theorem, the operator B can be represented by an integral operator with a distributional kernel $K(q, q')$. Then the commutativity $BQ = QB$ implies that, in the distributional sense,

$$(q - q')K(q, q') = 0,$$

so that K is "proportional" to the Dirac delta-function, i.e.,

$$K(q, q') = f(q)\delta(q - q'),$$

with some $f \in L^\infty(\mathbb{R})$. This argument is usually given in the physics textbooks.

Remark. The operator Q has no eigenvectors — the eigenvalue equation

$$Q\varphi = \lambda\varphi$$

has no solutions in $L^2(\mathbb{R})$. However, in the distributional sense, this equation for every $\lambda \in \mathbb{R}$ has a unique (up to a constant factor) solution $\varphi_\lambda(q) = \delta(q - \lambda)$, and these "generalized eigenfunctions", also called eigenfunctions of the continuous spectrum, combine to a Schwarz kernel of the identity operator I on $L^2(\mathbb{R})$. This reflects the fact that operator Q is diagonal in the coordinate representation.

Remark. Normalization of the eigenfunctions of the continuous spectrum $\varphi_\lambda(q)$ can be also determined by the condition that for every $\lambda \in \mathbb{R}$ the function

$$\Phi_\lambda(q) = \int_{\lambda_0}^{\lambda} \varphi_\mu(q)\,d\mu, \quad \Phi_\lambda \in L^2(\mathbb{R}),$$

satisfies

$$(2.14) \qquad\qquad \lim_{\Delta \to 0} \frac{1}{\Delta}\|\Phi_{\lambda+\Delta} - \Phi_\lambda\|^2 = 1.$$

Here $\lambda_0 \in \mathbb{R}$ is fixed and does not enter (2.14). Indeed, in our case $\Phi_\lambda = \chi_{(\lambda_0, \lambda)}$ — the characteristic function of the interval (λ_0, λ) — so that $\|\Phi_{\lambda+\Delta} - \Phi_\lambda\|^2 = \Delta$.

For a pure state $M = P_\psi$, $\|\psi\| = 1$, the corresponding probability measure μ_Q on $\mathbb{R}$ is given by

$$\mu_Q(E) = \nu_\psi(E) = \int_E |\psi(q)|^2 dq, \quad E \in \mathscr{B}(\mathbb{R}).$$

Physically, this is interpreted that in the state P_ψ with the "wave function" $\psi(q)$, the probability of finding a quantum particle between q and $q + dq$ is $|\psi(q)|^2 dq$. In other words, the modulus square of a wave function is the probability distribution for the coordinate of a quantum particle.

The corresponding momentum operator P is given by a differential operator

$$P = \frac{\hbar}{i}\frac{d}{dq}$$

with $D(P) = W^{1,2}(\mathbb{R})$ — the Sobolev space of absolutely continuous functions f on $\mathbb{R}$ such that f and its derivative f' (defined a.e.) are in $L^2(\mathbb{R})$. The operator P is self-adjoint and it is straightforward to verify that on $D = C_c^\infty(\mathbb{R})$, the space of smooth functions on $\mathbb{R}$ with compact support,

$$QP - PQ = i\hbar I.$$

Remark. The operator P on $\mathscr{H}$ has no eigenvectors — the eigenvalue equation

$$P\varphi = p\,\varphi, \; p \in \mathbb{R},$$

has a solution

$$\varphi(q) = \mathrm{const} \times e^{\frac{i}{\hbar}pq}$$

which does not belong to $L^2(\mathbb{R})$. We will see later that the family of normalized eigenfunctions of the continuous spectrum

$$\varphi_p(q) = \frac{1}{\sqrt{2\pi\hbar}}e^{\frac{i}{\hbar}pq}$$

combines to a Schwartz kernel of the inverse $\hbar$-dependent Fourier transform operator, which diagonalizes the momentum operator P. In the distributional sense,

$$\int_{-\infty}^{\infty} \varphi_p(q)\overline{\varphi_{p'}(q)}dq = \delta(p - p').$$

Remark. As for the case of the coordinate operator, the normalization of the eigenfunctions of the continuous spectrum $\varphi_p(q)$ of the momentum operator can be determined from the condition (2.14). Indeed,

$$\Phi_p(q) = \int_{p_0}^{p} ce^{\frac{i}{\hbar}kq}dk = \frac{c\hbar}{iq}\left(e^{\frac{i}{\hbar}pq} - e^{\frac{i}{\hbar}p_0q}\right),$$

so that

$$\Phi_{p+\Delta}(q) - \Phi_p(q) = \frac{2c\hbar}{q}e^{\frac{i}{\hbar}(p+\frac{1}{2}\Delta)q}\sin\frac{\Delta q}{2\hbar}.$$

Using the elementary integral, we obtain

$$\frac{1}{\Delta}\|\Phi_{p+\Delta}(q) - \Phi_p(q)\|^2 = 2c^2\hbar \int_{-\infty}^{\infty} \frac{\sin^2 q}{q^2}\,dq = 2\pi c^2\hbar,$$

so that $c = \frac{1}{\sqrt{2\pi\hbar}}$.

Proposition 2.1. *The coordinate representation defines an irreducible, unitary, integrable representation of the Heisenberg algebra.*

Proof. To show that the coordinate representation is integrable, let $U(u) = e^{-iuP}$ and $V(v) = e^{-ivQ}$ be the corresponding one-parameter groups of unitary operators. Clearly, $(V(v)\varphi)\psi(q) = e^{-ivq}\varphi(q)$ and it easily follows from the Stone theorem (or by the definition of a derivative) that $(U(u)\varphi)(q) = \varphi(q - \hbar u)$, so that unitary operators $U(u)$ and $V(v)$ satisfy the Weyl relation (2.11). Such a realization of the Weyl relation is called the *Schrödinger representation*.

To prove that the coordinate representation is irreducible, let B be a bounded operator commuting with P and Q. By Lemma 2.2, $T = f(Q)$ for some $f \in L^\infty(\mathbb{R})$. Now commutativity between T and P implies that

$$TU(u) = U(u)T \quad \text{for all} \quad u \in \mathbb{R},$$

which is equivalent to $f(q - \hbar u) = f(q)$ for all $q, u \in \mathbb{R}$, so that $f = \mathrm{const}$ a.e. on $\mathbb{R}$. $\qquad\square$

To summarize, the coordinate representation is characterized by the property that the coordinate operator Q is a multiplication by q operator and the momentum operator P is a differentiation operator,

$$Q = q \quad \text{and} \quad P = \frac{\hbar}{i}\frac{d}{dq}.$$

Similarly, *momentum representation* is defined by the property that the momentum operator P is a multiplication by p operator. Namely let $\mathscr{H} = L^2(\mathbb{R}, dp)$ be the Hilbert L^2-space on the "momentum space" $\mathbb{R}$ with the coordinate p, which is a Lagrangian subspace of $\mathbb{R}^2$ defined by the equation $q = 0$. The coordinate and momentum operators are given by

$$\hat{Q} = i\hbar\frac{d}{dp} \quad \text{and} \quad \hat{P} = p,$$

and satisfy the Heisenberg commutation relation. As the coordinate representation, the momentum representation is an irreducible, unitary, integrable representation of the Heisenberg algebra. In the momentum representation, the modulus square of the wave function $\psi(p)$ of a pure state $M = P_\psi$,

$\|\psi\| = 1$, is the probability distribution for the momentum of the quantum particle, i.e., the probability that a quantum particle has momentum between p and $p + dp$ is $|\psi(p)|^2 dp$.

Let $\mathscr{F}_\hbar : L^2(\mathbb{R}) \to L^2(\mathbb{R})$ be the $\hbar$-dependent Fourier transform operator, defined by

$$\hat{\varphi}(p) = \mathscr{F}_\hbar(\varphi)(p) = \frac{1}{\sqrt{2\pi\hbar}} \int_{-\infty}^{\infty} e^{-\frac{i}{\hbar}pq} \varphi(q) dq.$$

Here the integral is understood as the limit $\hat{\varphi} = \lim_{n\to\infty} \hat{\varphi}_n$ in the strong topology on $L^2(\mathbb{R})$, where

$$\hat{\varphi}_n(p) = \frac{1}{\sqrt{2\pi\hbar}} \int_{-n}^{n} e^{-\frac{i}{\hbar}pq} \varphi(q) dq.$$

By Plancherel's theorem, $\mathscr{F}_\hbar$ is a unitary operator on $L^2(\mathbb{R})$,

$$\mathscr{F}_\hbar \mathscr{F}_\hbar^* = \mathscr{F}_\hbar^* \mathscr{F}_\hbar = I,$$

and

$$\hat{Q} = \mathscr{F}_\hbar Q \mathscr{F}_\hbar^{-1}, \quad \hat{P} = \mathscr{F}_\hbar P \mathscr{F}_\hbar^{-1},$$

so that coordinate and momentum representations are unitarily equivalent. In particular, since the operator $\hat{P}$ is obviously self-adjoint, this immediately shows that the operator P is self-adjoint.

For n degrees of freedom, the coordinate representation is defined by setting $\mathscr{H} = L^2(\mathbb{R}^n, d^n\boldsymbol{q})$, where $d^n\boldsymbol{q} = dq^1 \cdots dq^n$ is the Lebesgue measure on $\mathbb{R}^n$, and

$$\boldsymbol{Q} = \boldsymbol{q} = (q^1, \ldots, q^n), \quad \boldsymbol{P} = \frac{\hbar}{i} \frac{\partial}{\partial \boldsymbol{q}} = \left(\frac{\hbar}{i} \frac{\partial}{\partial q^1}, \ldots, \frac{\hbar}{i} \frac{\partial}{\partial q^n} \right).$$

Here $\mathbb{R}^n$ is the configuration space with coordinates $\boldsymbol{q}$ — a Lagrangian subspace of $\mathbb{R}^{2n}$ defined by the equations $\boldsymbol{p} = 0$. The coordinate and momenta operators are self-adjoint and satisfy Heisenberg commutation relations. Projection-valued measures for the operators Q^k are given by

$$(\mathsf{P}_k(E)\varphi)(\boldsymbol{q}) = \chi_{\lambda_k^{-1}(E)}(\boldsymbol{q})\varphi(\boldsymbol{q}),$$

where $E \in \mathscr{B}(\mathbb{R})$ and $\lambda_k : \mathbb{R}^n \to \mathbb{R}$ is a canonical projection onto the k-th component, $k = 1, \ldots, n$. Correspondingly, the projection-valued measure P for the commutative family $\boldsymbol{Q} = (Q^1, \ldots, Q^n)$ (see Proposition 1.3) is defined on the Borel subsets $\boldsymbol{E} \subseteq \mathbb{R}^n$ by

$$(\mathsf{P}(\boldsymbol{E})\varphi)(\boldsymbol{q}) = \chi_{\boldsymbol{E}}(\boldsymbol{q})\varphi(\boldsymbol{q}).$$

The family $\boldsymbol{Q}$ has absolutely continuous joint spectrum $\mathbb{R}^n$.

Coordinate operators $Q^1, \ldots, Q^n$ form a *complete system of commuting observables*. This means, by definition, that none of these operators is a function of the other operators, and that every bounded operator commuting with $Q^1, \ldots, Q^n$ is a function of $Q^1, \ldots, Q^n$, i.e., is a multiplication by $f(\boldsymbol{q})$

operator for some $f \in L^\infty(\mathbb{R}^n)$. The proof repeats verbatim the proof of Lemma 2.2. For a pure state $M = P_\psi$, $\|\psi\| = 1$, the modulus square $|\psi(\boldsymbol{q})|^2$ of the wave function is the density of a joint distribution function $\mu_{\boldsymbol{Q}}$ for the commutative family $\boldsymbol{Q}$, i.e., the probability of finding a quantum particle in a Borel subset $\boldsymbol{E} \subseteq \mathbb{R}^n$ is given by

$$\mu_{\boldsymbol{Q}}(\boldsymbol{E}) = \int_{\boldsymbol{E}} |\psi(\boldsymbol{q})|^2 d^n \boldsymbol{q}.$$

The coordinate representation defines an irreducible, unitary, integrable representation of the Heisenberg algebra $\mathfrak{h}_n$. Indeed, n-parameter groups of unitary operators $U(\boldsymbol{u}) = e^{-i\boldsymbol{u}\boldsymbol{P}}$ and $V(\boldsymbol{v}) = e^{-i\boldsymbol{v}\boldsymbol{Q}}$ are given by

$$(U(\boldsymbol{u})\varphi)(\boldsymbol{q}) = \varphi(\boldsymbol{q} - \hbar\boldsymbol{u}), \quad (V(\boldsymbol{v})\varphi)(\boldsymbol{q}) = e^{-i\boldsymbol{v}\boldsymbol{q}}\varphi(\boldsymbol{q}),$$

and satisfy Weyl relations (2.11). The same argument as in the proof of Proposition 2.1 shows that this representation of the Heisenberg group $\mathbf{H}_n$, called the *Schrödinger representation for n degrees of freedom*, is irreducible.

In the momentum representation, $\mathscr{H} = L^2(\mathbb{R}^n, d^n\boldsymbol{p})$, where $d^n\boldsymbol{p} = dp_1 \cdots dp_n$ is the Lebesgue measure on $\mathbb{R}^n$, and

$$\hat{\boldsymbol{Q}} = i\hbar\frac{\partial}{\partial\boldsymbol{p}} = \left(i\hbar\frac{\partial}{\partial p_1}, \dots, i\hbar\frac{\partial}{\partial p_n}\right), \quad \hat{\boldsymbol{P}} = \boldsymbol{p} = (p_1, \dots, p_n).$$

Here $\mathbb{R}^n$ is the momentum space with coordinates $\boldsymbol{p}$ — a Lagrangian subspace of $\mathbb{R}^{2n}$ defined by the equations $\boldsymbol{q} = 0$.

The coordinate and momentum representations are unitarily equivalent by the Fourier transform. As in the case $n = 1$, the Fourier transform $\mathscr{F}_\hbar : L^2(\mathbb{R}^n) \to L^2(\mathbb{R}^n)$ is a unitary operator defined by

$$\hat{\varphi}(\boldsymbol{p}) = \mathscr{F}_\hbar(\varphi)(\boldsymbol{p}) = (2\pi\hbar)^{-n/2} \int_{\mathbb{R}^n} e^{-\frac{i}{\hbar}\boldsymbol{p}\boldsymbol{q}} \varphi(\boldsymbol{q}) d^n\boldsymbol{q}$$

$$= \lim_{N\to\infty} (2\pi\hbar)^{-n/2} \int_{|\boldsymbol{q}|\leq N} e^{-\frac{i}{\hbar}\boldsymbol{p}\boldsymbol{q}} \varphi(\boldsymbol{q}) d^n\boldsymbol{q},$$

where the limit is understood in the strong topology on $L^2(\mathbb{R}^n)$. As in the case $n = 1$, we have

$$\hat{Q}_k = \mathscr{F}_\hbar Q_k \mathscr{F}_\hbar^{-1}, \quad \hat{P}_k = \mathscr{F}_\hbar P_k \mathscr{F}_\hbar^{-1}, \quad k = 1, \dots, n.$$

In particular, since operators $\hat{P}_1, \dots, \hat{P}_n$ are obviously self-adjoint, this immediately shows that $P_1, \dots, P_n$ are also self-adjoint.

Remark. Following Dirac, physicists denote a vector $\psi \in \mathscr{H}$ by a *ket vector* $|\psi\rangle$, a vector $\varphi \in \mathscr{H}^*$ in the dual space to $\mathscr{H}$ ($\mathscr{H}^* \simeq \mathscr{H}$ is a complex antilinear isomorphism) by a *bra vector* $\langle\varphi|$, and their inner product by $\langle\varphi|\psi\rangle$. In standard mathematics notation,

$$(\psi, \varphi) = \langle\varphi|\psi\rangle \quad \text{and} \quad (A\psi, \varphi) = \langle\varphi|A|\psi\rangle,$$

where A is a linear operator. From a physics point of view, Dirac's notation is intuitive and convenient for working with coordinate and momentum representations. Denoting by $|\boldsymbol{q}\rangle = \delta(\boldsymbol{q}-\boldsymbol{q}')$ and $|\boldsymbol{p}\rangle = (2\pi\hbar)^{-n/2}e^{\frac{i}{\hbar}\boldsymbol{pq}}$ the set of generalized common eigenfunctions for the operators $\boldsymbol{Q}$ and $\boldsymbol{P}$, respectively, we formally get

$$\boldsymbol{Q}|\boldsymbol{q}\rangle = \boldsymbol{q}|\boldsymbol{q}\rangle, \quad \boldsymbol{P}|\boldsymbol{p}\rangle = \boldsymbol{p}|\boldsymbol{p}\rangle,$$

where operators $\boldsymbol{Q}$ act on $\boldsymbol{q}'$, and

$$\langle\boldsymbol{q}|\psi\rangle = \int_{\mathbb{R}^n} \delta(\boldsymbol{q} - \boldsymbol{q}')\psi(\boldsymbol{q}')d^n\boldsymbol{q}' = \psi(\boldsymbol{q}),$$

$$\langle\boldsymbol{p}|\psi\rangle = (2\pi\hbar)^{-n/2}\int_{\mathbb{R}^n} e^{-\frac{i}{\hbar}\boldsymbol{pq}}\psi(\boldsymbol{q})d^n\boldsymbol{q} = \hat{\psi}(\boldsymbol{p}),$$

as well as $\langle\boldsymbol{q}|\boldsymbol{q}'\rangle = \delta(\boldsymbol{q} - \boldsymbol{q}')$, $\langle\boldsymbol{p}|\boldsymbol{p}'\rangle = \delta(\boldsymbol{p} - \boldsymbol{p}')$. Though in our exposition we are not using Dirac's notation, these formulas would help the interested reader "translate" the notation used in physics textbooks to standard mathematics notation.

Remark. We will show in Section 3.2 that every Lagrangian subspace of the symplectic vector space $\mathbb{R}^{2n}$ with the canonical symplectic form $\omega = d\boldsymbol{p}\wedge d\boldsymbol{q}$ gives rise to an integrable, unitary, irreducible representation of the Heisenberg algebra $\mathfrak{h}_n$. This is the simplest example of the *real polarization*, which for a given symplectic manifold $(\mathcal{M},\omega)$ is defined as an integrable distribution $\{\mathscr{L}_x\}_{x\in\mathcal{M}}$ of Lagrangian subspaces $\mathscr{L}_x$ of tangent spaces $T_x\mathcal{M}$. The notion of a polarization plays a fundamental role in *geometric quantization*: it allows us to construct (under certain conditions) the Hilbert space of states $\mathscr{H}$ associated with the classical phase space $(\mathcal{M},\omega)$. In the linear case $\mathcal{M} = \mathbb{R}^{2n}$ every Lagrangian subspace $\mathscr{L}$ in $\mathbb{R}^{2n}$ gives rise to a real polarization by using the identification $T_x\mathbb{R}^{2n} \simeq \mathbb{R}^{2n}$. In particular, for the coordinate representation, $\mathscr{L}$ is given by the equation $\boldsymbol{q} = 0$, and for the momentum representation — by the equation $\boldsymbol{p} = 0$. The corresponding Hilbert $\mathscr{H}$ space consists of functions on $\mathbb{R}^{2n}$ which are constant along the fibers of the polarization.

Problem 2.4. Give an example of a non-integrable representation of the Heisenberg algebra.

Problem 2.5. Prove that there exists $\varphi \in \mathscr{H} = L^2(\mathbb{R}, dq)$ such that the vectors $\mathsf{P}(E)\varphi$, $E \in \mathscr{B}(\mathbb{R})$, where P is a projection-valued measure for the coordinate operator Q, are dense in $\mathscr{H}$.

Problem 2.6. Find the generating operator for the commutative family $\boldsymbol{Q} = (Q^1,\ldots,Q^n)$. Does it have a physical interpretation?

Problem 2.7. Find the projection-valued measure for the commutative family $\boldsymbol{P} = (P_1,\ldots,P_n)$ in the coordinate representation.

2.3. Free quantum particle. A free classical particle with one degree of freedom is described by the phase space $\mathbb{R}^2$ with coordinates p, q and the Poisson bracket (2.1), and by the Hamiltonian function

$$(2.15) \qquad H_{\mathrm{c}}(p, q) = \frac{p^2}{2m}.$$

The Hamiltonian operator of a free quantum particle with one degree of freedom is

$$H_0 = \frac{P^2}{2m},$$

and in coordinate representation is given by

$$H_0 = -\frac{\hbar^2}{2m}\frac{d^2}{dq^2}.$$

It is a self-adjoint operator on $\mathscr{H} = L^2(\mathbb{R}, dq)$ with $D(H_0) = W^{2,2}(\mathbb{R})$ — the Sobolev space of functions in $L^2(\mathbb{R})$, whose generalized first and second derivatives are in $L^2(\mathbb{R})$.

The operator H_0 is positive with absolutely continuous spectrum $[0, \infty)$ of multiplicity two. Indeed, let $\mathfrak{H}_0 = L^2(\mathbb{R}_{>0}, \mathbb{C}^2; d\sigma)$ be the Hilbert space of $\mathbb{C}^2$-valued measurable functions Ψ on the semi-line $\mathbb{R}_{>0} = (0, \infty)$, which are square-integrable with respect to the measure $d\sigma(\lambda) = \sqrt{\frac{m}{2\lambda}}\, d\lambda$,

$$\mathfrak{H}_0 = \left\{ \Psi(\lambda) = \begin{pmatrix} \psi_1(\lambda) \\ \psi_2(\lambda) \end{pmatrix} : \|\Psi\|^2 = \int_0^\infty (|\psi_1(\lambda)|^2 + |\psi_2(\lambda)|^2)d\sigma(\lambda) < \infty \right\}.$$

It follows from the unitarity of the Fourier transform that the operator $\mathscr{U}_0 : L^2(\mathbb{R}, dq) \to \mathfrak{H}_0$,

$$\mathscr{U}_0(\psi)(\lambda) = \Psi(\lambda) = \begin{pmatrix} \hat{\psi}(\sqrt{2m\lambda}) \\ \hat{\psi}(-\sqrt{2m\lambda}) \end{pmatrix},$$

is unitary, $\mathscr{U}_0^* \mathscr{U}_0 = I$ and $\mathscr{U}_0 \mathscr{U}_0^* = I_0$, where I and I_0 are, respectively, identity operators in $\mathscr{H}$ and $\mathfrak{H}_0$. The operator $\mathscr{U}_0$ establishes the isomorphism $L^2(\mathbb{R}, dq) \simeq \mathfrak{H}_0$, and since in the momentum representation H_0 is a multiplication by $\frac{1}{2m}p^2$ operator, the operator $\mathscr{U}_0 H_0 \mathscr{U}_0^{-1}$ is a multiplication by λ operator in $\mathfrak{H}_0$.

Remark. The Hamiltonian operator H_0 has no eigenvectors — the eigenvalue equation

$$H_0 \psi = \lambda \psi$$

has no solutions in $L^2(\mathbb{R})$. However, for every $\lambda = \frac{1}{2m}k^2 > 0$ this differential equation has two linear independent bounded solutions

$$\psi_k^{(\pm)}(q) = \frac{1}{\sqrt{2\pi\hbar}}\, e^{\pm\frac{i}{\hbar}kq}, \quad k > 0.$$

In the distributional sense, these eigenfunctions of the continuous spectrum combine to a Schwartz kernel of the unitary operator $\mathscr{U}_0$, which establishes the isomorphism between $\mathscr{H} = L^2(\mathbb{R}, dq)$ and the Hilbert space $\mathfrak{H}_0$, where H_0 acts as a multiplication by λ operator. The normalization of the eigenfunctions of the continuous spectrum is also determined by the condition (2.14):

$$\lim_{\Delta \to 0} \frac{1}{\Delta} \left\| \Psi_{k+\Delta}^{(\pm)} - \Psi_k^{(\pm)} \right\|^2 = 1, \quad \lim_{\Delta \to 0} \frac{1}{\Delta} \left(\Psi_{k+\Delta}^{(+)} - \Psi_k^{(+)}, \Psi_{k+\Delta}^{(-)} - \Psi_k^{(-)} \right) = 0,$$

where $\Psi_k^{(\pm)}(q) = \int_{k_0}^k \psi_p^{(\pm)}(q)dp$.

The Cauchy problem for the Schrödinger equation for a free particle,

$$(2.16) \qquad\qquad i\hbar \frac{d\psi(t)}{dt} = H_0 \psi(t), \quad \psi(0) = \psi,$$

is easily solved by the Fourier transform. Indeed, in the momentum representation it takes the form

$$i\hbar \frac{\partial \hat{\psi}(p,t)}{\partial t} = \frac{p^2}{2m} \hat{\psi}(p,t), \quad \hat{\psi}(p,0) = \hat{\psi}(p),$$

so that

$$\hat{\psi}(p,t) = e^{-\frac{ip^2}{2m\hbar}t} \hat{\psi}(p).$$

In the coordinate representation, the solution of (2.16) is given by

$$(2.17) \quad \psi(q,t) = \frac{1}{\sqrt{2\pi\hbar}} \int_{-\infty}^{\infty} e^{\frac{i}{\hbar}pq} \hat{\psi}(p,t)dp = \frac{1}{\sqrt{2\pi\hbar}} \int_{-\infty}^{\infty} e^{\frac{i}{\hbar}\chi(p,q,t)t} \hat{\psi}(p)dp,$$

where

$$\chi(p,q,t) = -\frac{p^2}{2m} + \frac{pq}{t}.$$

Formula (2.17) describes the motion of a quantum particle, and admits the following physical interpretation. Let initial condition ψ in (2.16) be such that its Fourier transform $\hat{\psi} = \mathcal{F}_\hbar(\psi)$ is a smooth function supported in a neighborhood U_0 of $p_0 \in \mathbb{R} \setminus \{0\}$, $0 \notin U_0$, and

$$\int_{-\infty}^{\infty} |\hat{\psi}(p)|^2 dp = 1.$$

Such states are called "wave packets". Then for every compact subset $E \subset \mathbb{R}$ we have

$$(2.18) \qquad\qquad \lim_{|t| \to \infty} \int_E |\psi(q,t)|^2 dq = 0.$$

Since

$$\int_{-\infty}^{\infty} |\psi(q,t)|^2 dq = 1$$

for all t, it follows from (2.18) that the particle leaves every compact subset of $\mathbb{R}$ as $|t| \to \infty$ and the quantum motion is infinite. To prove (2.18), observe that the function $\chi(p,q,t)$ — the "phase" in integral representation (2.17)

— has the property that $|\frac{\partial \chi}{\partial p}| > C > 0$ for all $p \in U_0$, $q \in E$ and large enough $|t|$. Integrating by parts we get

$$\psi(q,t) = \frac{1}{\sqrt{2\pi\hbar}} \int_{U_0} e^{\frac{i}{\hbar}\chi(p,q,t)t} \hat{\psi}(p)\,dp$$

$$= -\frac{1}{it}\sqrt{\frac{\hbar}{2\pi}} \int_{U_0} \frac{\partial}{\partial p} \left(\frac{\hat{\psi}(p)}{\frac{\partial \chi(p,q,t)}{\partial p}} \right) e^{\frac{i}{\hbar}\chi(p,q,t)t}\,dp,$$

so that uniformly on E,

$$\psi(q,t) = O(|t|^{-1}) \quad \text{as} \quad |t| \to \infty.$$

By repeated integration by parts, we obtain that for every $n \in \mathbb{N}$, uniformly on E,

$$\psi(q,t) = O(|t|^{-n}),$$

so that $\psi(q,t) = O(|t|^{-\infty})$.

To describe the motion of a free quantum particle in unbounded regions, we use the stationary phase method. In its simplest form it is stated as follows.

The Method of Stationary Phase. Let $f, g \in C^\infty(\mathbb{R})$, where f is real-valued and g has compact support, and suppose that f has a single non-degenerate critical point x_0, i.e., $f'(x_0) = 0$ and $f''(x_0) \neq 0$. Then

$$\int_{-\infty}^{\infty} e^{iNf(x)} g(x)\,dx = \left(\frac{2\pi}{N|f''(x_0)|} \right)^{\frac{1}{2}} e^{iNf(x_0) + \frac{i\pi}{4}\operatorname{sgn} f''(x_0)} g(x_0) + O\left(\frac{1}{N}\right)$$

as $N \to \infty$.

Applying the stationary phase method to the integral representation (2.17) (and setting $N = t$), we find that the critical point of $\chi(p,q,t)$ is $p_0 = \frac{mq}{t}$ with $\chi''(p_0) = -\frac{1}{m} \neq 0$, and

$$\psi(q,t) = \sqrt{\frac{m}{t}} \hat{\psi}\left(\frac{mq}{t}\right) e^{\frac{imq^2}{2\hbar t} - \frac{\pi i}{4}} + O(t^{-1})$$

$$= \psi_0(q,t) + O(t^{-1}) \quad \text{as} \quad t \to \infty.$$

Thus as $t \to \infty$, the wave function $\psi(q,t)$ is supported on $\frac{t}{m}U_0$ — a domain where the probability of finding a particle is asymptotically different from zero. At large t the points in this domain move with constant velocities $v = \frac{p}{m}$, $p \in U_0$. In this sense, the classical relation $p = mv$ remains valid in the quantum picture. Moreover, the asymptotic wave function ψ_0 satisfies

$$\int_{-\infty}^{\infty} |\psi_0(q,t)|^2\,dq = \sqrt{\frac{m}{t}} \int_{-\infty}^{\infty} \left| \hat{\psi}\left(\frac{mq}{t}\right) \right|^2\,dq = 1,$$

and, therefore, describes the asymptotic probability distribution. Similarly, setting $N = -|t|$, we can describe the behavior of the wave function $\psi(q, t)$ as $t \to -\infty$.

Remark. We have $\lim_{|t| \to \infty} \psi(t) = 0$ in the weak topology on $\mathscr{H}$. Indeed, for every $\varphi \in \mathscr{H}$ we get by Parseval's identity for the Fourier integrals,

$$(\psi(t), \varphi) = \int_{-\infty}^{\infty} \hat{\psi}(p)\overline{\hat{\varphi}(p)}e^{-\frac{ip^2 t}{2m\hbar}}\, dp,$$

and the integral goes to zero as $|t| \to \infty$ by the Riemann-Lebesgue lemma.

A free classical particle with n degrees of freedom is described by the phase space $\mathbb{R}^{2n}$ with coordinates $\boldsymbol{p} = (p_1, \ldots, p_n)$ and $\boldsymbol{q} = (q^1, \ldots, q^n)$, the Poisson bracket (2.3), and the Hamiltonian function

$$H_{\mathrm{c}}(\boldsymbol{p}, \boldsymbol{q}) = \frac{\boldsymbol{p}^2}{2m} = \frac{1}{2m}(p_1^2 + \cdots + p_n^2).$$

The Hamiltonian operator of a free quantum particle with n degrees of freedom is

$$H_0 = \frac{\boldsymbol{P}^2}{2m} = \frac{1}{2m}(P_1^2 + \cdots + P_n^2),$$

and in the coordinate representation is

$$H_0 = -\frac{\hbar^2}{2m}\Delta,$$

where

$$\Delta = \left(\frac{\partial}{\partial \boldsymbol{q}}\right)^2 = \left(\frac{\partial}{\partial q^1}\right)^2 + \cdots + \left(\frac{\partial}{\partial q^n}\right)^2$$

is the Laplace operator[15] in the Cartesian coordinates on $\mathbb{R}^n$. The Hamiltonian H_0 is a self-adjoint operator on $\mathscr{H} = L^2(\mathbb{R}^n, d^n\boldsymbol{q})$ with $D(H_0) = W^{2,2}(\mathbb{R}^n)$ — the Sobolev space on $\mathbb{R}^n$. In the momentum representation,

$$H_0 = \frac{\boldsymbol{p}^2}{2m}$$

— a multiplication by a function operator on $\mathscr{H} = L^2(\mathbb{R}^n, d^n\boldsymbol{p})$.

The operator H_0 is positive with absolutely continuous spectrum $[0, \infty)$ of infinite multiplicity. Namely, let $S^{n-1} = \{\boldsymbol{n} \in \mathbb{R}^n : \boldsymbol{n}^2 = 1\}$ be the $(n-1)$-dimensional unit sphere in $\mathbb{R}^n$, let $d\boldsymbol{n}$ be the measure on S^{n-1} induced by the Lebesgue measure on $\mathbb{R}^n$, and let

$$\mathfrak{h} = \{f : S^{n-1} \to \mathbb{C} : \|f\|_{\mathfrak{h}}^2 = \int_{S^{n-1}} |f(\boldsymbol{n})|^2 d\boldsymbol{n} < \infty\}.$$

---

[15] It is the negative of the Laplace-Beltrami operator of the standard Euclidean metric on $\mathbb{R}^n$.

Let $\mathfrak{H}_0^{(n)} = L^2(\mathbb{R}_{>0}, \mathfrak{h}; d\sigma_n)$ be the Hilbert space of $\mathfrak{h}$-valued measurable functions[16] Ψ on $\mathbb{R}_{>0} = (0, \infty)$, square-integrable on $\mathbb{R}_{>0}$ with respect to the measure $d\sigma_n(\lambda) = (2m\lambda)^{\frac{n}{2}}\frac{d\lambda}{2\lambda}$,

$$\mathfrak{H}_0^{(n)} = \left\{ \Psi : \mathbb{R}_{>0} \to \mathfrak{h}, \ \|\Psi\|^2 = \int_0^\infty \|\Psi(\lambda)\|_{\mathfrak{h}}^2 \, d\sigma_n(\lambda) < \infty \right\}.$$

When $n = 1$, $\mathfrak{H}_0^{(1)} = \mathfrak{H}_0$ — the corresponding Hilbert space for one degree of freedom. The operator $\mathscr{U}_0 : L^2(\mathbb{R}^n, d^n\boldsymbol{q}) \to \mathfrak{H}_0^{(n)}$,

$$\mathscr{U}_0(\psi)(\lambda) = \Psi(\lambda), \quad \Psi(\lambda)(\boldsymbol{n}) = \hat{\psi}(\sqrt{2m\lambda}\,\boldsymbol{n}),$$

is unitary and establishes the isomorphism $L^2(\mathbb{R}^n, d^n\boldsymbol{q}) \simeq \mathfrak{H}_0^{(n)}$. In the momentum representation H_0 is a multiplication by $\frac{1}{2m}\boldsymbol{p}^2$ operator, so that the operator $\mathscr{U}_0 H_0 \mathscr{U}_0^{-1}$ is a multiplication by λ operator in $\mathfrak{H}_0^{(n)}$.

Remark. As in the case $n = 1$, the Hamiltonian operator H_0 has no eigenvectors — the eigenvalue equation

$$H_0\psi = \lambda\psi$$

has no solutions in $L^2(\mathbb{R}^n)$. However, for every $\lambda > 0$ this differential equation has infinitely many linearly independent bounded solutions

$$\psi_{\boldsymbol{n}}(\boldsymbol{q}) = (2\pi\hbar)^{-\frac{n}{2}} e^{\frac{i}{\hbar}\sqrt{2m\lambda}\,\boldsymbol{n}\boldsymbol{q}},$$

parametrized by the unit sphere S^{n-1}. These solutions do not belong to $L^2(\mathbb{R}^n)$, but in the distributional sense they combine to a Schwartz kernel of the unitary operator $\mathscr{U}_0$, which establishes the isomorphism between $\mathscr{H} = L^2(\mathbb{R}^n, d^n\boldsymbol{q})$ and the Hilbert space $\mathfrak{H}_0^{(n)}$, where H_0 acts as a multiplication by λ operator.

As in the case $n = 1$, the Schrödinger equation for free particle,

$$i\hbar\frac{d\psi(t)}{dt} = H_0\psi(t), \quad \psi(0) = \psi,$$

is solved by the Fourier transform

$$\psi(\boldsymbol{q}, t) = (2\pi\hbar)^{-n/2} \int_{\mathbb{R}^n} e^{\frac{i}{\hbar}(\boldsymbol{p}\boldsymbol{q} - \frac{\boldsymbol{p}^2}{2m}t)} \hat{\psi}(\boldsymbol{p}) d^n\boldsymbol{p}.$$

For a wave packet, an initial condition ψ such that its Fourier transform $\hat{\psi} = \mathscr{F}_\hbar(\psi)$ is a smooth function supported on a neighborhood U_0 of $\boldsymbol{p}_0 \in \mathbb{R}^n \setminus \{0\}$ such that $0 \notin U_0$ and

$$\int_{\mathbb{R}^n} |\hat{\psi}(\boldsymbol{p})|^2 d^n\boldsymbol{p} = 1,$$

---

[16] That is, for every $f \in \mathfrak{h}$ the function (f, Ψ) is measurable on $\mathbb{R}_{>0}$.

the quantum particle leaves every compact subset of $\mathbb{R}^n$ and the motion is infinite. Asymptotically as $|t| \to \infty$, the wave function $\psi(\boldsymbol{q}, t)$ is different from 0 only when $\boldsymbol{q} = \frac{\boldsymbol{p}}{m}t$, $\boldsymbol{p} \in U_0$.

Problem 2.8. Find the asymptotic wave function for a free quantum particle with n degrees of freedom.

2.4. Examples of quantum systems. Here we describe quantum systems that correspond to the classical Lagrangian systems introduced in Section 1.3 of Chapter 1. In Hamiltonian formulation, the phase space of these systems, except for the last example, is a symplectic vector space $\mathbb{R}^{2n}$ with canonical coordinates $\boldsymbol{p}, \boldsymbol{q}$ and symplectic form $\omega = d\boldsymbol{p} \wedge d\boldsymbol{q}$.

Example 2.1 (Newtonian particle). According to Section 1.7 in Chapter 1, a classical particle in $\mathbb{R}^n$ moving in a potential field $V(\boldsymbol{q})$ is described by the Hamiltonian function

$$H_{\mathrm{c}}(\boldsymbol{p}, \boldsymbol{q}) = \frac{\boldsymbol{p}^2}{2m} + V(\boldsymbol{q}).$$

Assume that the Hamiltonian operator for the quantum system is given by

$$H = \frac{\boldsymbol{P}^2}{2m} + V,$$

with some operator V, so that coordinate and momenta operators satisfy Heisenberg equations of motion

(2.19) $$\dot{\boldsymbol{P}} = \{H, \boldsymbol{P}\}_\hbar, \ \dot{\boldsymbol{Q}} = \{H, \boldsymbol{Q}\}_\hbar.$$

To determine V, we require that the classical relation $\dot{\boldsymbol{q}} = \dfrac{\boldsymbol{p}}{m}$ between the velocity and the momentum of a particle is preserved under the quantization, i.e.,

$$\dot{\boldsymbol{Q}} = \frac{\boldsymbol{P}}{m}.$$

Since $\{\boldsymbol{P}^2, \boldsymbol{Q}\}_\hbar = 2\boldsymbol{P}$, it follows from (2.19) that this condition is equivalent to

$$[V, Q_k] = 0, \quad k = 1, \dots, n.$$

It follows from Section 2.2 that V is a function of commuting operators $Q_1, \dots, Q_n$, and the natural choice[17] is $V = V(\boldsymbol{Q})$. Thus the Hamiltonian operator of a Newtonian particle is

$$H = \frac{\boldsymbol{P}^2}{2m} + V(\boldsymbol{Q}),$$

---

[17]Confirmed by the agreement of the theory with the experiments.

in agreement with $H = H_c(\boldsymbol{P}, \boldsymbol{Q})$[18]. In coordinate representation the Hamiltonian is the *Schrödinger operator*

$$(2.20) \qquad H = -\frac{\hbar^2}{2m}\Delta + V(\boldsymbol{q})$$

with the real-valued potential $V(\boldsymbol{q})$.

Remark. The sum of two unbounded, self-adjoint operators is not necessarily self-adjoint, and one needs to describe admissible potentials $V(\boldsymbol{q})$ for which H is a self-adjoint operator on $L^2(\mathbb{R}^n, d^n\boldsymbol{q})$. If the potential $V(\boldsymbol{q})$ is a real-valued, locally integrable function on $\mathbb{R}^n$, then the differential operator (2.20) defines a symmetric operator H with the domain $C_0^2(\mathbb{R}^n)$ — twice continuously differentiable functions on $\mathbb{R}^n$ with compact support. Potentials for which the symmetric operator H has no self-adjoint extensions are obviously non-physical. It may also happen that H has several self-adjoint extensions[19]. These extensions are specified by some boundary conditions at infinity and there are no physical principles distinguishing between them. The only physical case is when the symmetric operator H admits a unique self-adjoint extension, that is, when H is essentially self-adjoint. In Chapter 3 we present necessary conditions for the essential self-adjointness. Here we only mention the von Neumann criterion that if A is a closed operator and $\overline{D(A)} = \mathscr{H}$, then $H = A^*A$ is a positive self-adjoint operator.

Example 2.2 (Interacting quantum particles). In Lagrangian formalism, a closed classical system of N interacting particles on $\mathbb{R}^3$ was described in Example 1.2 in Section 1.3 of Chapter 1. In Hamiltonian formalism, it is described by the canonical coordinates $\boldsymbol{r} = (\boldsymbol{r}_1, \ldots, \boldsymbol{r}_N)$, the canonical momenta $\boldsymbol{p} = (\boldsymbol{p}_1, \ldots, \boldsymbol{p}_N)$, $\boldsymbol{r}_a, \boldsymbol{p}_a \in \mathbb{R}^3$, and by the Hamiltonian function

$$(2.21) \qquad H_c(\boldsymbol{p}, \boldsymbol{r}) = \sum_{a=1}^{N} \frac{\boldsymbol{p}_a^2}{2m_a} + V(\boldsymbol{r}),$$

where m_a is the mass of the a-th particle, $a = 1, \ldots, N$ (see Section 1.7 in Chapter 1). The corresponding Hamiltonian operator H in the coordinate representation has the form

$$(2.22) \qquad H = -\sum_{a=1}^{N} \frac{\hbar^2}{2m_a}\Delta_a + V(\boldsymbol{r}).$$

In particular, when

$$V(\boldsymbol{r}) = \sum_{1 \le a < b \le N} V(\boldsymbol{r}_a - \boldsymbol{r}_b),$$

---

[18]In the special case $H_c(\boldsymbol{p}, \boldsymbol{q}) = f(\boldsymbol{p}) + g(\boldsymbol{q})$ the problem of the ordering of non-commuting operators $\boldsymbol{P}$ and $\boldsymbol{Q}$ does not arise.

[19]This is the case when the defect indices of H are equal and are non-zero.

the Schrödinger operator (2.22) describes the N-body problem in quantum mechanics. The fundamental quantum system is the complex atom, formed by a nucleus of charge Ne and mass M, and by N electrons of charge $-e$ and mass m. Denoting by $\boldsymbol{R} \in \mathbb{R}^3$ the position of the nucleus, and by $\boldsymbol{r}_1, \ldots, \boldsymbol{r}_N$ the positions of the electrons and assuming that the interaction is given by the Coulomb attraction, we get for the Hamiltonian function (2.21)

$$H_{\mathrm{c}}(\boldsymbol{P}, \boldsymbol{p}, \boldsymbol{R}, \boldsymbol{r}) = \frac{\boldsymbol{P}^2}{2M} + \sum_{a=1}^{N} \frac{\boldsymbol{p}_a^2}{2m} - \sum_{a=1}^{N} \frac{Ne^2}{|\boldsymbol{R} - \boldsymbol{r}_a|} + \sum_{1 \leq a < b \leq N} \frac{e^2}{|\boldsymbol{r}_a - \boldsymbol{r}_b|},$$

where $\boldsymbol{P}$ is the canonical momentum of the nucleus. The corresponding Schrödinger operator H in the coordinate representation has the form[20]

$$H = -\frac{\hbar^2}{2M}\Delta - \sum_{a=1}^{N} \frac{\hbar^2}{2m}\Delta_a - \sum_{a=1}^{N} \frac{Ne^2}{|\boldsymbol{R} - \boldsymbol{r}_a|} + \sum_{1 \leq a < b \leq N} \frac{e^2}{|\boldsymbol{r}_a - \boldsymbol{r}_b|}.$$

In the simplest case of the hydrogen atom, when $N = 1$ and the nucleus consists of a single proton[21], the Hamiltonian is

$$H = -\frac{\hbar^2}{2M}\Delta_p - \frac{\hbar^2}{2m}\Delta_e - \frac{e^2}{|\boldsymbol{r}_p - \boldsymbol{r}_e|},$$

where $\boldsymbol{r}_p$ is the position of the proton and $\boldsymbol{r}_e$ is the position of the electron. As the first approximation, the proton can be considered as infinitely heavy, so that the hydrogen atom is described by an electron in an attractive Coulomb field $-e^2/|\boldsymbol{r}|$, where now $\boldsymbol{r} = \boldsymbol{r}_e - \boldsymbol{r}_p$. The corresponding Hamiltonian operator takes the form

$$(2.23) \qquad H = -\frac{\hbar^2}{2m}\Delta - \frac{e^2}{|\boldsymbol{r}|}.$$

We will solve the Schrödinger equation with this Hamiltonian H and determine its energy levels in Section 5.1 of Chapter 3.

Example 2.3 (Charged particle in an electromagnetic field). A classical particle of charge e and mass m moving in the time-independent electromagnetic field with scalar and vector potentials $\varphi(\boldsymbol{r})$ and $\boldsymbol{A}(\boldsymbol{r})$, $\boldsymbol{r} \in \mathbb{R}^3$, is described by the Hamiltonian function

$$H_{\mathrm{c}}(\boldsymbol{p}, \boldsymbol{r}) = \frac{1}{2m}\left(\boldsymbol{p} - \frac{e}{c}\boldsymbol{A}\right)^2 + e\varphi(\boldsymbol{r})$$

(see Problem 1.27 in Section 1.7 of Chapter 1). The corresponding classical velocity vector $\boldsymbol{v} = \{H_{\mathrm{c}}, \boldsymbol{r}\}$ is given by

$$\boldsymbol{v} = \boldsymbol{p} - \frac{e}{c}\boldsymbol{A},$$

---

[20] Ignoring the fact that electron has spin, see Chapter 4.

[21] In the case of hydrogen-1 or protium; it includes one or more neutrons for deuterium, tritium, and other isotopes.

and its components $\boldsymbol{v} = (v_1, v_2, v_3)$ have non-vanishing Poisson brackets:

$$\{v_1, v_2\} = -\frac{e}{m^2 c} B_3, \quad \{v_2, v_3\} = -\frac{e}{m^2 c} B_1, \quad \{v_3, v_1\} = -\frac{e}{m^2 c} B_2,$$

where $\boldsymbol{B} = (B_1, B_2, B_3)$ are components of the magnetic field $\boldsymbol{B} = \operatorname{curl} \boldsymbol{A}$.

The Hamiltonian operator of a quantum particle is

$$(2.24) \qquad H = \frac{1}{2m} \left(\boldsymbol{P} - \frac{e}{c} \boldsymbol{A} \right)^2 + e\varphi(\boldsymbol{r})$$

— the Schrödinger operator of a charged particle in an electromagnetic field. The corresponding quantum velocity vector $\boldsymbol{V} = \{H, \boldsymbol{Q}\}_\hbar$ is given by the same formula as in the classical case,

$$\boldsymbol{V} = \boldsymbol{P} - \frac{e}{c} \boldsymbol{A},$$

and its components $\boldsymbol{V} = (V_1, V_2, V_3)$ have non-vanishing quantum brackets:

$$\{V_1, V_2\}_\hbar = -\frac{e}{m^2 c} B_3, \quad \{V_2, V_3\}_\hbar = -\frac{e}{m^2 c} B_1, \quad \{V_3, V_1\}_\hbar = -\frac{e}{m^2 c} B_2.$$

Thus in the presence of a magnetic field the three components of a quantum velocity operator no longer commute and cannot be measured simultaneously.

Example 2.4 (Free quantum particle on a Riemannian manifold). The phase space of a classical particle of mass $m = 1$ moving on a Riemannian manifold (M, g) is the cotangent bundle T^*M, and the corresponding Hamiltonian function is given by

$$H_{\mathrm{c}}(\boldsymbol{p}, \boldsymbol{x}) = \tfrac{1}{2} g^{\mu\nu}(\boldsymbol{x}) p_\mu p_\nu,$$

where $g^{\mu\nu}(\boldsymbol{x})$ is the inverse of the metric tensor $g_{\mu\nu}(\boldsymbol{x})$, and

$$(\boldsymbol{p}, \boldsymbol{x}) = (p_1, \dots, p_n, x^1, \dots, x^n)$$

are standard coordinates on T^*M (see Section 1.7 of Chapter 1). The Hilbert space of the quantum system is $\mathscr{H} = L^2(M, d\mu)$, where $d\mu = \sqrt{g(\boldsymbol{x})} d^n \boldsymbol{x}$, $g(\boldsymbol{x}) = \det(g_{\mu\nu}(\boldsymbol{x}))$, is the measure associated with the density of the Riemannian metric (Riemannian volume form if M is oriented). When canonical coordinates $(\boldsymbol{p}, \boldsymbol{x})$ are only locally defined on T^*M, it is not possible to construct corresponding operators $\boldsymbol{P}$ and $\boldsymbol{Q}$. Still, one can always define the Hamiltonian operator by

$$(2.25) \qquad H = \frac{\hbar^2}{2} \Delta_g, \quad \text{where} \quad \Delta_g = -\frac{1}{\sqrt{g(\boldsymbol{x})}} \frac{\partial}{\partial x^\mu} \left(\sqrt{g(\boldsymbol{x})} \, g^{\mu\nu} \frac{\partial}{\partial x^\nu} \right)$$

is the Laplace-Beltrami operator of the Riemannian metric g on M. Note that in this case there is a non-trivial problem of the ordering of non-commuting operators in the quantization of $H_{\mathrm{c}}(\boldsymbol{p}, \boldsymbol{x})$, which arises if in a coordinate chart on M one replaces canonical coordinates p and x by P and Q. The formula

$$\Delta_g = -g^{\mu\nu}(x)\frac{\partial^2}{\partial x^\mu \partial x^\nu} + g^{\mu\sigma}(x)\Gamma^\nu_{\mu\sigma}(x)\frac{\partial}{\partial x^\nu}$$

shows that local expressions defined by

$$(2.26) \qquad H_c(P, Q) = \tfrac{1}{2}\big(g^{\mu\nu}(Q)P_\mu P_\nu + i\hbar g^{\mu\sigma}(Q)\Gamma^\nu_{\mu\sigma}(Q)P_\nu\big)$$

combine to a well-defined self-adjoint operator H on $\mathscr{H}$ given by (2.25). Thus even when $M = \mathbb{R}^n$ and canonical coordinates (p, x) are globally defined on $T^*\mathbb{R}^n$, the correct formula for $H_c(P, Q)$ — the one which extends to general Riemannian manifolds — is given by (2.26), where the second term represents the "quantum correction" to the naive expression $g^{\mu\nu}(Q)P_\mu P_\nu$.

2.5. Old quantum mechanics. The formulation of quantum mechanics presented here goes back to 1925-1927, and is due to Heisenberg, Schrödinger, Born, Jordan, and Dirac. It replaced the old quantum theory, proposed in 1913 by Bohr, which was based on Rutherford's planetary model of the atom. In the old theory, the energy levels of a one-dimensional quantum system correspond to the closed orbits of the associated classical Hamiltonian system which satisfy the *Bohr-Wilson-Sommerfeld quantization rule* (BWS rule)

$$\oint pdq = 2\pi\hbar(n + \tfrac{1}{2}),$$

where n is a non-negative integer, and integration goes over the closed orbit in the phase space $\mathbb{R}^2$. Bohr-Wilson-Sommerfeld quantization rules also apply to completely integrable Hamiltonian systems with several degrees of freedom (see Section 2.6 of Chapter 1). Namely, let $F_1 = H_c, \ldots, F_N$ be N independent integrals of motion in involution. The BWS quantization rules are

$$\oint_\gamma pdq = 2\pi\hbar(n_\gamma + \tfrac{1}{4}\operatorname{ind}\gamma),$$

where integration goes over all 1-cycles γ in the Lagrangian submanifold $\Lambda = \{(p, q) \in \mathbb{R}^{2N} : H_c(p, q) = E, F_2(p, q) = E_2, \ldots, F_N(p, q) = E_N\}$, and $\operatorname{ind}\gamma \in \mathbb{Z}$ is the so-called *Maslov index* of a cycle γ in Λ. In the one-dimensional case the closed orbit is topologically a circle and its Maslov index is 2. It will be shown in Section 6.3 of Chapter 3 that, in general, Bohr-Wilson-Sommerfeld quantization rules only give asymptotics of the energy levels as $\hbar \to 0$. However, for integrable systems with extra symmetry, such as the harmonic oscillator and the Kepler problem, the BWS quantization rules determine the energy levels exactly. We will show this in the next section for the harmonic oscillator, and in Section 5.1 of Chapter 3 — for the Kepler problem.

2.6. Harmonic oscillator. The simplest classical system with one degree of freedom, besides the free particle, is the harmonic oscillator. It is described by the phase space $\mathbb{R}^2$ with the canonical coordinates p, q, and the Hamiltonian function

$$(2.27) \qquad H_{\mathrm{c}}(p, q) = \frac{p^2}{2m} + \frac{m\omega^2 q^2}{2}$$

(see Sections 1.5 and 1.7 in Chapter 1). Hamilton's equations

$$\dot{p} = \{H_{\mathrm{c}}, p\} = -m\omega^2 q, \quad \dot{q} = \{H_{\mathrm{c}}, q\} = \frac{p}{m}$$

with the initial conditions p_0, q_0 are readily solved,

$$(2.28) \qquad p(t) = p_0 \cos \omega t - m\omega q_0 \sin \omega t,$$

$$(2.29) \qquad q(t) = q_0 \cos \omega t + \frac{1}{m\omega} p_0 \sin \omega t,$$

and describe the harmonic motion. As in Section 2.6 of Chapter 1, it is convenient to introduce complex coordinates on the phase space on $\mathbb{R}^2 \simeq \mathbb{C}$,

$$(2.30) \qquad z = \frac{1}{\sqrt{2\omega}} \left(\omega q + ip \right), \quad \bar{z} = \frac{1}{\sqrt{2\omega}} \left(\omega q - ip \right).$$

We have

$$(2.31) \qquad \{z, \bar{z}\} = \frac{i}{m}, \quad H_{\mathrm{c}}(z, \bar{z}) = m\omega |z|^2,$$

so that Hamilton's equations decouple,

$$\dot{z} = \{H_{\mathrm{c}}, z\} = -i\omega z, \quad \dot{\bar{z}} = \{H_{\mathrm{c}}, \bar{z}\} = i\omega \bar{z},$$

and are trivially solved,

$$(2.32) \qquad z(t) = e^{-i\omega t} z_0, \quad \bar{z} = e^{i\omega t} \bar{z}_0.$$

Here

$$z_0 = \frac{1}{\sqrt{2\omega}} \left(\omega q_0 + ip_0 \right), \quad \bar{z}_0 = \frac{1}{\sqrt{2\omega}} \left(\omega q_0 - ip_0 \right).$$

For the quantum system, the corresponding Hamiltonian operator is

$$H = \frac{P^2}{2m} + \frac{m\omega^2 Q^2}{2},$$

and in the coordinate representation $\mathscr{H} = L^2(\mathbb{R}, dq)$ it is a Schrödinger operator with a quadratic potential,

$$H = -\frac{\hbar^2}{2m} \frac{d^2}{dq^2} + \frac{m\omega^2 q^2}{2}.$$

The quantum harmonic oscillator is the simplest non-trivial quantum system, besides the free particle, whose Schrödinger equation can be solved explicitly. It appears in all problems involving quantized oscillations, namely in molecular and crystalline vibrations. The exact solution of the harmonic oscillator, described below, has remarkable[22] algebraic and analytic properties.

Temporarily set $m = 1$ and consider the operators

$$(2.33) \qquad a = \frac{1}{\sqrt{2\omega\hbar}}\left(\omega Q + iP\right), \quad a^* = \frac{1}{\sqrt{2\omega\hbar}}\left(\omega Q - iP\right),$$

which are quantum analogs of complex coordinates (2.30). The operators a and a^* are defined on $W^{1,2}(\mathbb{R}) \cap \widehat{W}^{1,2}(\mathbb{R})$, where $\widehat{W}^{1,2}(\mathbb{R}) = \mathscr{F}(W^{1,2}(\mathbb{R}))$, and it is easy to show that a^* is the adjoint operator to a and $a^{**} = a$, so that a is a closed operator. From the Heisenberg commutation relation (2.2) we get the *canonical commutation relation*

$$(2.34) \qquad\qquad\qquad [a, a^*] = I$$

on $W^{2,2}(\mathbb{R}) \cap \widehat{W}^{2,2}(\mathbb{R})$. Indeed,

$$aa^* = \frac{P^2 + \omega^2 Q^2}{2\omega\hbar} + \frac{i\omega}{2\omega\hbar}[P, Q] = \frac{P^2 + \omega^2 Q^2}{2\omega\hbar} + \frac{1}{2}I,$$

and

$$a^*a = \frac{P^2 + \omega^2 Q^2}{2\omega\hbar} - \frac{i\omega}{2\omega\hbar}[P, Q] = \frac{P^2 + \omega^2 Q^2}{2\omega\hbar} - \frac{1}{2}I,$$

so that (2.34) holds on $W^{2,2}(\mathbb{R}) \cap \widehat{W}^{2,2}(\mathbb{R})$, where $\widehat{W}^{2,2}(\mathbb{R}) = \mathscr{F}(W^{2,2}(\mathbb{R}))$, and

$$H = \omega\hbar\left(a^*a + \tfrac{1}{2}I\right) = \omega\hbar\left(aa^* - \tfrac{1}{2}I\right).$$

In particular, it follows from the von Neumann criterion that the Hamiltonian operator H is self-adjoint.

The operators a, a^* and $N = a^*a$ satisfy the commutation relations

$$(2.35) \qquad\qquad [N, a] = -a, \quad [N, a^*] = a^*, \quad [a, a^*] = I.$$

These commutation relations correspond to the irreducible unitary representation of a four-dimensional solvable Lie algebra $\tilde{\mathfrak{h}}$ associated with the Heisenberg algebra $\mathfrak{h} = \mathfrak{h}_1$, introduced in Section 2.1. Namely, $\tilde{\mathfrak{h}}$ is a Lie algebra with the generators e, f, h, and c, where e, f, c satisfy the relations of the Heisenberg algebra $\mathfrak{h}$, and

$$[h, e] = -f, \quad [h, f] = \omega^2 e, \quad [h, c] = 0.$$

The irreducible integrable representation ρ of the Heisenberg algebra $\mathfrak{h}$ (see (2.9) in Section 2.1) extends to a unitary representation of the Lie algebra

---

[22]The algebraic structure of the exact solution of the harmonic oscillator plays a fundamental role in quantum electrodynamics and in quantum field theory in general.

$\tilde{\mathfrak{h}}$ by setting

$$\rho(h) = -i\omega N = \frac{P^2 + \omega^2 Q^2}{2i\hbar} + \frac{i\omega}{2} I.$$

Remark. In invariant terms, the Lie algebra $\tilde{\mathfrak{h}}$ is a one-dimensional right extension of the Heisenberg algebra $\mathfrak{h}$,

$$0 \to \mathfrak{h} \to \tilde{\mathfrak{h}} \to \mathbb{R} \to 0.$$

It is defined by the $\mathfrak{h}$-valued Lie algebra 1-cocycle $r \in Z^1(\mathfrak{h}, \mathfrak{h})$ — a derivation of $\mathfrak{h}$, given on generators by

$$r(e) = f, \quad r(f) = -\omega^2 e, \quad r(c) = 0.$$

Explicitly, if $h \in \tilde{\mathfrak{h}}$ is such that $\bar{h} = 1 \in \mathbb{R}$ under the projection $\tilde{\mathfrak{h}} \to \mathbb{R}$, then identifying elements in $\mathfrak{h}$ with their images under the embedding $\mathfrak{h} \hookrightarrow \tilde{\mathfrak{h}}$, we have

$$[x + \alpha h, y + \beta h] = [x, y] - \alpha r(y) + \beta r(y), \quad x, y \in \mathfrak{h}.$$

It is due to this Lie-algebraic structure of commutation relations (2.35) that Heisenberg equations of motion for the harmonic oscillator can be solved exactly. Namely, we have

$$\dot{a} = \{H, a\}_\hbar = -i\omega a, \quad \dot{a}^* = \{H, a^*\}_\hbar = i\omega a^*,$$

so that

$$a(t) = e^{-i\omega t} a_0, \quad a^*(t) = e^{i\omega t} a_0^*.$$

Comparing with (2.32) we see that solutions of classical and quantum equations of motion for the harmonic oscillator have the same form!

Next, using commutation relations (2.35) and positivity of the operator N, we will solve the eigenvalue problem for the Hamiltonian H of the harmonic oscillator explicitly by finding its energy levels and corresponding eigenvectors. We will prove that the eigenvectors form a complete system of vectors in $\mathscr{H}$, so that the spectrum of the Hamiltonian H is the point spectrum. This is a quantum mechanical analog of the fact that classical motion of the harmonic oscillator is always finite.

The algebraic part of the exact solution is the following fundamental result.

Proposition 2.2. *Suppose that there exists a non-zero $\psi \in D(a^n) \cap D((a^*)^n)$, $n = 1, 2, \dots$, such that*

$$H\psi = \lambda\psi.$$

Then the following statements hold.

(i) *There exists $\psi_0 \in \mathscr{H}$, $\|\psi_0\| = 1$, such that*

$$H\psi_0 = \tfrac{1}{2}\hbar\omega\psi_0.$$

(ii) *The vectors*

$$\psi_n = \frac{(a^*)^n}{\sqrt{n!}}\psi_0 \in \mathcal{H}, \quad n = 0, 1, 2, \ldots,$$

are orthonormal eigenvectors for H with the eigenvalues $\hbar\omega(n+\frac{1}{2})$,

$$H\psi_n = \hbar\omega(n + \tfrac{1}{2})\psi_n.$$

(iii) *Restriction of the operator H to the Hilbert space $\mathcal{H}_0$ — a closed subspace of $\mathcal{H}$, spanned by the orthonormal set $\{\psi_n\}_{n=0}^{\infty}$ — is essentially self-adjoint.*

Proof. Rewriting commutation relations (2.35) as

$$Na = a(N - I) \quad \text{and} \quad Na^* = a^*(N + I),$$

and putting $\lambda = \hbar\omega(\mu + \frac{1}{2})$, we get for all $n \geq 0$,

$$(2.36) \qquad Na^n\psi = (\mu - n)a^n\psi \quad \text{and} \quad N(a^*)^n\psi = (\mu + n)(a^*)^n\psi.$$

Since $N \geq 0$ on $D(N)$, it follows from the first equation in (2.36) that there exists $n_0 \geq 0$ such that $a^{n_0}\psi \neq 0$ but $a^{n_0+1}\psi = 0$. Setting $\psi_0 = \dfrac{a^{n_0}\psi}{\|a^{n_0}\psi\|} \in \mathcal{H}$ we get

$$(2.37) \qquad a\psi_0 = 0 \quad \text{and} \quad N\psi_0 = 0.$$

Since $H = \hbar\omega(N + \frac{1}{2}I)$, this proves part (i). To prove part (ii), we use commutation relations

$$(2.38) \qquad [a, (a^*)^n] = n(a^*)^{n-1},$$

which follow from (2.34) and the Leibniz rule. Using (2.37)-(2.38), we get

$$(2.39) \qquad a^*\psi_n = \sqrt{n+1}\,\psi_{n+1}, \quad a\psi_n = \sqrt{n}\,\psi_{n-1},$$

so that

$$\|\psi_n\|^2 = \frac{1}{\sqrt{n}}(a^*\psi_{n-1}, \psi_n) = \frac{1}{\sqrt{n}}(\psi_{n-1}, a\psi_n) = \|\psi_{n-1}\|^2 = \cdots = \|\psi_0\|^2 = 1.$$

From the second equation in (2.36) it follows that $N\psi_n = n\psi_n$, so ψ_n are normalized eigenvectors of H with the eigenvalues $\hbar\omega(n + \frac{1}{2})$. The eigenvectors ψ_n are orthogonal since the corresponding eigenvalues are distinct and the operator H is symmetric. Finally, part (iii) immediately follows from the fact that, according to part (ii), the subspaces $\mathrm{Im}\,(H \pm iI)|_{\mathcal{H}_0}$ are dense in $\mathcal{H}_0$, which is the criterion of essential self-adjointness. $\qquad\square$

Remark. Since the coordinate representation of the Heisenberg commutation relations is irreducible, it is tempting to conclude, using Proposition 2.2, that $\mathcal{H}_0 = \mathcal{H}$. Namely, it follows from the construction that the linear span of vectors ψ_n — a dense subspace of $\mathcal{H}_0$ — is invariant for the operators P and Q. However, this does not immediately imply that the projection operator Π_0 onto the subspace $\mathcal{H}_0$ commutes with self-adjoint operators P and Q in the sense of the definition in Section 1.1.

Using the coordinate representation, we can immediately show the existence of the vector ψ_0 in Proposition 2.2, and prove that $\mathcal{H}_0 = \mathcal{H}$. Indeed, equation $a\psi_0 = 0$ becomes a first order linear differential equation

$$\left(\hbar \frac{d}{dq} + \omega q \right) \psi_0 = 0,$$

so that

$$\psi_0(q) = \sqrt[4]{\frac{\omega}{\pi\hbar}}\, e^{-\frac{\omega}{2\hbar} q^2},$$

and

$$\|\psi_0\|^2 = \sqrt{\frac{\omega}{\pi\hbar}} \int_{-\infty}^{\infty} e^{-\frac{\omega}{\hbar} q^2}\, dq = 1.$$

The vector ψ_0 is called the *ground state* for the harmonic oscillator. Correspondingly, the eigenfunctions

$$\psi_n(q) = \frac{1}{\sqrt{n!}} \left(\frac{1}{\sqrt{2\omega\hbar}} \left(\omega q - \hbar \frac{d}{dq} \right) \right)^n \psi_0$$

are of the form $P_n(q) e^{-\frac{\omega}{2\hbar} q^2}$, where $P_n(q)$ are polynomials of degree n. The following result guarantees that the functions $\{\psi_n\}_{n=0}^{\infty}$ form an orthonormal basis in $L^2(\mathbb{R}, dq)$.

Lemma 2.3. *The functions $q^n e^{-q^2}$, $n = 0, 1, 2, \ldots$, are complete in $L^2(\mathbb{R}, dq)$.*

Proof. Let $f \in L^2(\mathbb{R}, dq)$ is such that

$$\int_{-\infty}^{\infty} f(q) q^n e^{-q^2}\, dq = 0, \quad n = 0, 1, 2, \ldots.$$

The integral

$$F(z) = \int_{-\infty}^{\infty} f(q) e^{iqz - q^2}\, dq$$

is absolutely convergent for all $z \in \mathbb{C}$ and, therefore, defines an entire function. We have

$$F^{(n)}(0) = i^n \int_{-\infty}^{\infty} f(q) q^n e^{-q^2}\, dq = 0, \quad n = 0, 1, 2, \ldots,$$

so that $F(z) = 0$ for all $z \subset \mathbb{C}$. This implies the function $g(q) = f(q) e^{-q^2} \in L^1(\mathbb{R}) \cap L^2(\mathbb{R})$ satisfies $\mathscr{F}(g) = 0$, where $\mathscr{F}$ is the "ordinary" ($\hbar = 1$) Fourier transform. Thus we conclude that $g = 0$. $\qquad\square$

The polynomials P_n are expressed through classical Hermite-Tchebyscheff polynomials H_n, defined by

$$H_n(q) = (-1)^n e^{q^2} \frac{d^n}{dq^n} e^{-q^2}, \quad n = 0, 1, 2, \ldots.$$

Namely, using the identity

$$e^{\frac{q^2}{2}} \frac{d^n}{dq^n} e^{-q^2} = -\left(q - \frac{d}{dq}\right) \left[e^{\frac{q^2}{2}} \frac{d^{n-1}}{dq^{n-1}} e^{-q^2}\right]$$

$$= \cdots = (-1)^n \left(q - \frac{d}{dq}\right)^n e^{-\frac{q^2}{2}}$$

we obtain

$$\psi_n(q) = \sqrt[4]{\frac{\omega}{\pi\hbar}} \frac{1}{\sqrt{2^n n!}} e^{-\frac{\omega}{2\hbar}q^2} H_n\left(\sqrt{\frac{\omega}{\hbar}} q\right).$$

We summarize the obtained results as follows.

Theorem 2.1. *The Hamiltonian*

$$H = -\frac{\hbar^2}{2m} \frac{d^2}{dq^2} + \frac{m\omega^2 q^2}{2}$$

of the quantum harmonic oscillator with one degree of freedom is a self-adjoint operator on $\mathscr{H} = L^2(\mathbb{R}, dq)$ with the domain $D(H) = W^{2,2}(\mathbb{R}) \cap \widehat{W}^{2,2}(\mathbb{R})$. The operator H has pure point spectrum

$$H\psi_n = \lambda_n \psi_n, \quad n = 0, 1, 2, \ldots,$$

with the eigenvalues $\lambda_n = \hbar\omega(n + \frac{1}{2})$. Corresponding eigenfunctions ψ_n form an orthonormal basis for $\mathscr{H}$ and are given by

$$(2.40) \qquad \psi_n(q) = \sqrt[4]{\frac{m\omega}{\pi\hbar}} \frac{1}{\sqrt{2^n n!}} e^{-\frac{m\omega}{2\hbar}q^2} H_n\left(\sqrt{\frac{m\omega}{\hbar}} q\right),$$

where $H_n(q)$ are classical Hermite-Tchebyscheff polynomials.

Proof. Consider the operator H defined on the Schwartz space $\mathscr{S}(\mathbb{R})$ of rapidly decreasing functions. Since the operator H is symmetric and has a complete system of eigenvectors in $\mathscr{S}(\mathbb{R})$, the subspaces $\mathrm{Im}(H \pm iI)$ are dense in $\mathscr{H}$, so that H is essentially self-adjoint. The proof that its self-adjoint closure (which we continue to denote by H) has the domain $W^{2,2}(\mathbb{R}) \cap \widehat{W}^{2,2}(\mathbb{R})$ is left to the reader. $\qquad \square$

Remark. The energy levels of the harmonic oscillator can be also obtained from the Bohr-Wilson-Sommerfeld quantization rule. Indeed, the corresponding classical orbit with the energy E is an ellipse in the phase plane given by the equations $q(t) = A\cos(\omega t + \alpha)$, $p(t) = -m\omega A\sin(\omega t + \alpha)$, where $E = \frac{1}{2}m\omega^2 A^2$ (see Section 1.5 in Chapter 1). Thus

$$\oint p\,dq = \iint dp \wedge dq = \pi m\omega A^2 = \frac{2\pi}{\omega}E,$$

and the BWS quantization rule gives the energy levels $E_n = \hbar\omega(n + \frac{1}{2})$.

Remark. Since the energy levels of the Hamiltonian H are equidistant by $\hbar\omega$, the quantum harmonic oscillator describes the system of identical "quanta" with the energy $\hbar\omega$. The ground state $|0\rangle = \psi_0$, in Dirac's notation, is the *vacuum state* with no quanta present and with the vacuum energy $\frac{1}{2}\hbar\omega$, and the states $|n\rangle = \psi_n$ consist of n quanta with the energy $\hbar\omega(n+\frac{1}{2})$. According to (2.39), the operator a^* adds one quantum to the state $|n\rangle$ and is called a *creation operator*, and the operator a destroys one quantum in the state $|n\rangle$ and is called an *annihilation operator*.

An example of the harmonic oscillator illustrates the dramatic difference between the motion in quantum mechanics and in classical mechanics. The classical motion in the potential field $V(q) = \frac{1}{2}m\omega^2 q^2$ is finite: a particle with energy E moves in the region $|\omega q| \leq \sqrt{\frac{2E}{m}}$, whereas there is always a non-zero probability of finding a quantum particle outside the classical region. Thus for the ground state energy $E = \frac{1}{2}\hbar\omega$ this probability is

$$\int_{|q| \geq \sqrt{\frac{\hbar}{m\omega}}} |\psi_0(q)|^2 dq = \frac{2}{\sqrt{\pi}} \int_1^\infty e^{-x^2} dx \simeq 0.1572992070.$$

The classical harmonic oscillator with n degrees of freedom is described by the phase space $\mathbb{R}^{2n}$ with the canonical coordinates $\boldsymbol{p}, \boldsymbol{q}$, and the Hamiltonian function

$$H_c(\boldsymbol{p}, \boldsymbol{q}) = \frac{\boldsymbol{p}^2}{2m} + \sum_{j=1}^n \frac{m\omega_j^2 q_j^2}{2},$$

where $\omega_1, \ldots, \omega_n > 0$ (see Sections 1.3 and 1.7 in Chapter 1).

The corresponding Hamiltonian operator is

$$H = \frac{\boldsymbol{P}^2}{2m} + \sum_{j=1}^n \frac{m\omega_j^2 Q_j^2}{2}$$

and in the coordinate representation $\mathscr{H} = L^2(\mathbb{R}^n, d^n\boldsymbol{q})$ is a Schrödinger operator with quadratic potential,

$$H = -\frac{\hbar^2}{2m}\Delta + \sum_{j=1}^n \frac{m\omega_j^2 q_j^2}{2}.$$

The Hamiltonian H is a self-adjoint operator with $D(H) = W^{2,2}(\mathbb{R}^n) \cap \widehat{W}^{2,2}(\mathbb{R}^n)$ and a pure point spectrum. The corresponding eigenfunctions

$$\psi_{\boldsymbol{k}}(\boldsymbol{q}) = \psi_{k_1}(q_1) \ldots \psi_{k_n}(q_n),$$

where $\boldsymbol{k} = (k_1, \ldots, k_n)$ and $\psi_{k_j}(q_j)$ are eigenfunctions (2.40) with $\omega = \omega_j$, form an orthonormal basis for $L^2(\mathbb{R}^n, d^n\boldsymbol{q})$. The corresponding energy levels are given by

$$\lambda_{\boldsymbol{k}} = \hbar\omega_1(k_1 + \tfrac{1}{2}) + \cdots + \hbar\omega_n(k_n + \tfrac{1}{2}).$$

The spectrum of H is simple if and only if $\hbar\omega_1, \ldots, \hbar\omega_n$ are linearly independent over $\mathbb{Z}$. The highest degeneracy case is $\omega_1 = \cdots = \omega_n = \omega$, when the multiplicity of the eigenvalue

$$\lambda_{\boldsymbol{k}} = \hbar\omega \sum_{j=1}^{n}(k_j + \tfrac{1}{2})$$

is the partition function $p_n(|\boldsymbol{k}|)$ — the number of representations of the integer $|\boldsymbol{k}| = k_1 + \cdots + k_n$ as a sum of n non-negative integers. Setting $m = 1$ and introducing the operators[23]

$$(2.41) \quad a_j = \frac{1}{\sqrt{2\omega\hbar}}\left(\omega Q_j + iP_j\right), \quad a_j^* = \frac{1}{\sqrt{2\omega\hbar}}\left(\omega Q_j - iP_j\right), \quad j = 1, \ldots, n,$$

we get canonical commutation relations for creation and annihilation operators for n degrees of freedom,

$$(2.42) \qquad [a_j, a_l] = 0, \quad [a_j^*, a_l^*] = 0, \quad [a_j, a_l^*] = \delta_{jl}I, \quad j, l = 1, \ldots, n,$$

which generalize relation (2.34) for the one degree of freedom. The operators a_j, a_j^* and $N_j = a_j^* a_j$, $j = 1, \ldots, n$, satisfy commutation relations

$$(2.43) \qquad [N_j, a_l] = -\delta_{jl}a_l, \quad [N_j, a_l^*] = \delta_{jl}a_l^*, \quad j, l = 1, \ldots, n.$$

In particular, the operator

$$N = \sum_{j=1}^{n} N_j = \sum_{j=1}^{n} a_j^* a_j$$

satisfies

$$[N, a_j] = -a_j, \quad [N, a_j^*] = a_j^*, \quad j = 1, \ldots, n,$$

and $H = \hbar\omega(N + \tfrac{n}{2}I)$.

Commutation relations (2.42) and (2.43) correspond to the irreducible unitary representation of the solvable Lie algebra $\tilde{\mathfrak{h}}_n$ with $3n + 1$ generators e_j, f_j, h_j, and c, where e_j, f_j, c satisfy the relations in the Heisenberg algebra $\mathfrak{h}_n$, and

$$[h_j, e_l] = -\delta_{jl}f_l, \quad [h_j, f_l] = \delta_{jl}\omega_l^2 e_l, \quad [h_j, c] = 0, \quad j, l = 1, \ldots, n.$$

---

[23]Here, using the standard Euclidean metric on $\mathbb{R}^n$, we lowered the indices for Q^j.

Problem 2.9. Show that $\langle H|M \rangle \geq \frac{1}{2}\hbar\omega$ for every $M \in \mathscr{S}$, where H is the Hamiltonian of the harmonic oscillator with one degree of freedom.

Problem 2.10. Let $q(t) = A\cos(\omega t + \alpha)$ be the classical trajectory of the harmonic oscillator with $m = 1$ and the energy $E = \frac{1}{2}\omega^2 A^2$, and let μ_α be the probability measure on $\mathbb{R}$ supported at the point $q(t)$. Show that the convex linear combination of the measures μ_α, $0 \leq \alpha \leq 2\pi$, is the probability measure on $\mathbb{R}$ with the distribution function $\mu(q) = \frac{\theta(A^2 - q^2)}{\pi\sqrt{A^2 - q^2}}$, where $\theta(q)$ is the Heaviside step function.

Problem 2.11. Show that when $n \to \infty$ and $\hbar \to 0$ such that $\hbar\omega(n + \frac{1}{2}) = \frac{1}{2}\omega^2 A^2$ remains fixed, the envelope of the distribution function $|\psi_n(q)|^2$ on the interval $|q| \leq A$ coincides with the classical distribution function $\mu(q)$ from the previous problem. (*Hint:* Prove the integral representation

$$e^{-q^2} H_n(q) = \frac{2^{n+1}}{\sqrt{\pi}} \int_0^\infty e^{-y^2} y^n \cos(2qy - \tfrac{1}{2}n\pi)dy,$$

and derive the asymptotic formula

$$\psi_n(q) = \sqrt{\frac{2}{\pi}}\,\frac{1}{\sqrt[4]{A^2 - q^2}} \cos\left\{\frac{\omega}{2\hbar}\left(A^2 \sin^{-1}\frac{q}{A} + q\sqrt{A^2 - q^2} - \frac{1}{2}A^2\pi\right) + O(1)\right\}$$

when $\hbar \to 0$ and $\hbar(n + \frac{1}{2}) = \frac{1}{2}\omega A^2$, $|q| < A$.)

Problem 2.12. Complete the proof of Theorem 2.1.

Problem 2.13 (The N-representation theorem). Let $\psi \in \mathscr{S}(\mathbb{R})$. Show that the L^2-convergent expansion $\psi = \sum_{n=0}^\infty c_n\psi_n$, where $c_n = (\psi, \psi_n)$, converges in $\mathscr{S}(\mathbb{R})$. (*Hint*: Use $N\psi_n = n\psi_n$.)

Problem 2.14. Show that the operators $E_{ij} = a_i^* a_j$, $i, j = 1, \ldots, n$, satisfy the commutation relations of the Lie algebra $\mathrm{sl}(n, \mathbb{C})$.

2.7. Holomorphic representation and Wick symbols. Let

$$\ell^2 = \left\{ c = \{c_n\}_{n=0}^\infty \;:\; \|c\|^2 = \sum_{n=0}^\infty |c_n|^2 < \infty \right\}$$

be the Hilbert ℓ^2-space. The choice of an orthonormal basis $\{\psi_n\}_{n=0}^\infty$ for $L^2(\mathbb{R}, dq)$, given by the eigenfunctions (2.40) of the Schrödinger operator for the harmonic oscillator, establishes the Hilbert space isomorphism $L^2(\mathbb{R}, dq) \simeq \ell^2$,

$$L^2(\mathbb{R}, dq) \ni \psi = \sum_{n=0}^\infty c_n\psi_n \mapsto c = \{c_n\}_{n=0}^\infty \in \ell^2,$$

where

$$c_n = (\psi, \psi_n) = \int_{-\infty}^\infty \psi(q)\psi_n(q)dq,$$

since the functions ψ_n are real-valued. Using (2.39) we get

$$a^*\psi = \sum_{n=0}^{\infty} c_n a^* \psi_n = \sum_{n=0}^{\infty} \sqrt{n+1}\, c_n \psi_{n+1} = \sum_{n=1}^{\infty} \sqrt{n}\, c_{n-1}\psi_n, \quad \psi \in D(a^*),$$

and

$$a\psi = \sum_{n=0}^{\infty} c_n a \psi_n = \sum_{n=1}^{\infty} \sqrt{n}\, c_n \psi_{n-1} = \sum_{n=0}^{\infty} \sqrt{n+1}\, c_{n+1}\psi_n, \quad \psi \in D(a),$$

so that in ℓ^2 creation and annihilation operators a^* and a are represented by the following semi-infinite matrices:

$$a = \begin{pmatrix} 0 & \sqrt{1} & 0 & 0 & \cdots \\ 0 & 0 & \sqrt{2} & 0 & \cdots \\ 0 & 0 & 0 & \sqrt{3} & \cdots \\ 0 & 0 & 0 & 0 & \cdots \\ \vdots & \vdots & \vdots & \vdots & \ddots \end{pmatrix}, \quad a^* = \begin{pmatrix} 0 & 0 & 0 & 0 & \cdots \\ \sqrt{1} & 0 & 0 & 0 & \cdots \\ 0 & \sqrt{2} & 0 & 0 & \cdots \\ 0 & 0 & \sqrt{3} & 0 & \cdots \\ \vdots & \vdots & \vdots & \vdots & \ddots \end{pmatrix}.$$

As a result,

$$N = a^* a = \begin{pmatrix} 0 & 0 & 0 & 0 & \cdots \\ 0 & 1 & 0 & 0 & \cdots \\ 0 & 0 & 2 & 0 & \cdots \\ 0 & 0 & 0 & 3 & \cdots \\ \vdots & \vdots & \vdots & \vdots & \ddots \end{pmatrix},$$

so that the Hamiltonian of the harmonic oscillator is represented by a diagonal matrix,

$$H = \hbar\omega(N + \tfrac{1}{2}) = \mathrm{diag}\{\tfrac{1}{2}\hbar\omega, \tfrac{3}{2}\hbar\omega, \tfrac{5}{2}\hbar\omega, \dots\}.$$

This representation of the Heisenberg commutation relations is called the *representation by occupation numbers*, and has the property that in this representation the Hamiltonian H of the harmonic operator is diagonal.

Another representation where H is diagonal is constructed as follows. Let $\mathscr{D}$ be the space of entire functions $f(z)$ with the inner product

$$(2.44) \qquad (f,g) = \frac{1}{\pi} \int_{\mathbb{C}} f(z)\overline{g(z)} e^{-|z|^2} d^2 z,$$

where $d^2 z = \tfrac{i}{2} dz \wedge d\bar{z}$ is the Lebesgue measure on $\mathbb{C} \simeq \mathbb{R}^2$. It is easy to check that $\mathscr{D}$ is a Hilbert space with the orthonormal basis

$$f_n(z) = \frac{z^n}{\sqrt{n!}}, \quad n = 0, 1, 2, \dots.$$

The correspondence

$$\ell^2 \ni c = \{c_n\}_{n=0}^{\infty} \mapsto f(z) = \sum_{n=0}^{\infty} c_n f_n(z) \in \mathscr{D}$$

establishes the Hilbert space isomorphism $\ell^2 \simeq \mathscr{D}$. The realization of a Hilbert space $\mathscr{H}$ as the Hilbert space $\mathscr{D}$ of entire functions is called a *holomorphic representation*. In the holomorphic representation,

$$a^* = z, \quad a = \frac{d}{dz}, \quad \text{and} \quad H = \hbar\omega\left(z\frac{d}{dz} + \frac{1}{2}\right),$$

and it is very easy to show that a^* is the adjoint operator to a. The mapping

$$\mathscr{H} \ni \psi = \sum_{n=0}^{\infty} c_n\psi_n \mapsto f(z) = \sum_{n=0}^{\infty} c_n f_n(z) \in \mathscr{D}$$

establishes the isomorphism between the coordinate and holomorphic representations. It follows from the formula for the generating function for Hermite-Tchebyscheff polynomials,

$$\sum_{n=0}^{\infty} H_n(q)\frac{z^n}{n!} = e^{2qz - z^2},$$

that the corresponding unitary operator $U : \mathscr{H} \to \mathscr{D}$ is an integral operator

$$U\psi(z) = \int_{-\infty}^{\infty} U(z,q)\psi(q)dq$$

with the kernel

$$(2.45) \qquad U(z,q) = \sum_{n=0}^{\infty} \psi_n(q)f_n(z) = \sqrt[4]{\frac{m\omega}{\pi\hbar}}\, e^{\frac{m\omega}{2\hbar}q^2 - \left(\sqrt{\frac{m\omega}{\hbar}}q - \frac{1}{\sqrt{2}}z\right)^2}.$$

Another useful realization is a representation in the Hilbert space $\bar{\mathscr{D}}$ of anti-holomorphic functions $f(\bar{z})$ on $\mathbb{C}$ with the inner product

$$(f,g) = \frac{1}{\pi}\int_{\mathbb{C}} f(\bar{z})\overline{g(\bar{z})}e^{-|z|^2}d^2z,$$

given by

$$a^* = \bar{z}, \quad a = \frac{d}{d\bar{z}}.$$

Remark. Holomorphic and anti-holomorphic representations correspond to different choices of a *complex polarization* of the symplectic manifold $\mathbb{R}^2$. By definition, a complex polarization of a symplectic manifold $(\mathscr{M}, \omega)$ is an integrable distribution on $\mathscr{M}$ of complex Lagrangian subspaces of the complexified vector spaces $T_x\mathscr{M} \otimes_{\mathbb{R}} \mathbb{C}$. As the notion of real polarization, the notion of complex polarization plays a fundamental role in geometric quantization. In particular, holomorphic representation corresponds to the complex Lagrangian subspace in $T_x\mathbb{R}^2 \otimes_{\mathbb{R}} \mathbb{C} \simeq \mathbb{C}^2$ given by the equation $z = 0$, where z and $\bar{z}$ are complex coordinates on $\mathbb{C}^2$. (Note that here $\bar{z}$ is not a complex conjugate to z!) The equation $\bar{z} = 0$ defines a complex Lagrangian subspace which corresponds to the anti-holomorphic representation.

The anti-holomorphic representation is used to introduce the so-called *Wick symbols* of the operators. Namely, let A be an operator in $\bar{\mathscr{D}}$ which is a polynomial with constant coefficients in creation and annihilation operators a^* and a. Using the commutation relation (2.34), we can move all operators a^* to the left of the operators a, and represent A in the *Wick normal form* as follows:

$$(2.46) \qquad A = \sum_{l,m} A_{lm}(a^*)^l a^m.$$

By definition, the Wick symbol $A(\bar{z}, z)$ of the operator A is

$$(2.47) \qquad A(\bar{z}, z) = \sum_{l,m} A_{lm}\bar{z}^l z^m.$$

It is a restriction of a polynomial $A(v, z)$ in variables v and z to $v = \bar{z}$.

In order to define Wick symbols of bounded operators in $\bar{\mathscr{D}}$, we consider the family of *coherent states* (or *Poisson vectors*) $\Phi_v \in \bar{\mathscr{D}}$, $v \in \mathbb{C}$, defined by

$$\Phi_v(\bar{z}) = e^{v\bar{z}}, \quad z \in \mathbb{C}.$$

They satisfy the properties

$$(2.48) \qquad a\Phi_v = v\Phi_v \quad \text{and} \quad f(\bar{v}) = (f, \Phi_v), \quad f \in \bar{\mathscr{D}}, \ v \in \mathbb{C}.$$

Indeed, the first property is trivial, whereas the "reproducing property" immediately follows from the formula

$$(2.49) \qquad \Phi_v(\bar{z}) = \sum_{n=0}^{\infty} f_n(v)\bar{f}_n(\bar{z}),$$

where $\bar{f}_n(\bar{z}) = \overline{f_n(z)}$, $n = 0, 1, 2, \ldots$, is the orthonormal basis for $\bar{\mathscr{D}}$.

We also have

$$(2.50) \qquad (f, g) = \frac{1}{\pi} \int_{\mathbb{C}} (f, \Phi_v)\overline{(g, \Phi_v)}e^{-|v|^2} d^2v.$$

Now for the operator A in the Wick normal form (2.46) we get, using the first property in (2.48),

$$(A\Phi_z, \Phi_{\bar{v}}) = \sum_{l,m} A_{lm}((a^*)^l a^m \Phi_z, \Phi_{\bar{v}}) = \sum_{l,m} A_{lm}(a^m \Phi_z, a^l \Phi_{\bar{v}})$$
$$= A(v, z)(\Phi_z, \Phi_{\bar{v}}).$$

Therefore,

$$A(v, z) = \frac{(A\Phi_z, \Phi_{\bar{v}})}{(\Phi_z, \Phi_{\bar{v}})} = e^{-vz}(A\Phi_z, \Phi_{\bar{v}}),$$

since by the reproducing property, $(\Phi_z, \Phi_{\bar{v}}) = \Phi_z(v) = e^{vz}$.

Definition. The Wick symbol $A(\bar{z}, z)$ of a bounded operator A in the Hilbert space $\bar{\mathscr{D}}$ is a restriction to $v = \bar{z}$ of an entire function $A(v, z)$ in variables v and z, defined by

$$A(v, z) = e^{-vz}(A\Phi_z, \Phi_{\bar{v}}).$$

In the next theorem, we summarize the properties of the Wick symbols.

Theorem 2.2. *Wick symbols of the bounded operators on $\bar{\mathscr{D}}$ have the following properties.*

    (i) *If $A(\bar{z}, z)$ is the Wick symbol of an operator A, then for the Wick symbol of the operator A^* we have $A^*(\bar{z}, z) = \overline{A(\bar{z}, z)}$ and*

$$A^*(v, z) = \overline{A(\bar{z}, \bar{v})}.$$

    (ii) *For $f \in \bar{\mathscr{D}}$,*

$$(Af)(\bar{z}) = \frac{1}{\pi} \int_{\mathbb{C}} A(\bar{z}, v) f(\bar{v}) e^{-v(\bar{v} - \bar{z})} d^2 v.$$

    (iii) *A real-analytic function $A(\bar{z}, z)$ is a Wick symbol of a bounded operator A in $\bar{\mathscr{D}}$ if and only if it is a restriction to $v = \bar{z}$ of an entire function $A(v, z)$ in variables v and z with the property that for every $f \in \bar{\mathscr{D}}$ the integral in part (ii) is absolutely convergent and defines a function in $\bar{\mathscr{D}}$.*

    (iv) *If $A_1(\bar{z}, z)$ and $A_2(\bar{z}, z)$ are the Wick symbols of operators A_1 and A_2, then the Wick symbol of the operator $A = A_1 A_2$ is given by*

$$A(\bar{z}, z) = \frac{1}{\pi} \int_{\mathbb{C}} A_1(\bar{z}, v) A_2(\bar{v}, z) e^{-(v - z)(\bar{v} - \bar{z})} d^2 v.$$

Proof. We have

$$A^*(v, z) = e^{-vz}(A^*\Phi_z, \Phi_{\bar{v}}) = e^{-vz}(\Phi_z, A\Phi_{\bar{v}}) = e^{-vz}\overline{(A\Phi_{\bar{v}}, \Phi_z)} = \overline{A(\bar{z}, \bar{v})},$$

which proves (i). To prove (ii), we use the reproducing property to get

$$(Af)(\bar{z}) = (Af, \Phi_z) = (f, A^*\Phi_z) = \frac{1}{\pi} \int_{\mathbb{C}} f(\bar{v})\overline{(A^*\Phi_z)(\bar{v})} e^{-|v|^2} d^2 v.$$

Using the reproducing property once again we have

$$(A^*\Phi_z)(\bar{v}) = (A^*\Phi_z, \Phi_v) = A^*(\bar{v}, z)(\Phi_z, \Phi_v) = \overline{e^{v\bar{z}} A(\bar{z}, v)},$$

which proves (ii). Property (iii) follows from the definition and the uniform boundness principle, which is needed to show that the operator A on $\bar{\mathscr{D}}$, defined by the integral in (ii), is bounded. We leave the standard details to the reader. To prove (iv), by using (2.50) and (i) we get

$$A(\bar{z}, z) = e^{-|z|^2}(A_1 A_2 \Phi_z, \Phi_z) = e^{-|z|^2}(A_2 \Phi_z, A_1^* \Phi_z)$$

$$= \frac{1}{\pi} \int_{\mathbb{C}} (A_2 \Phi_z, \Phi_v)\overline{(A_1^* \Phi_z, \Phi_v)} e^{-(|v|^2+|z|^2)} d^2 v$$

$$= \frac{1}{\pi} \int_{\mathbb{C}} A_1(\bar{z}, v) A_2(\bar{v}, z) e^{-(v-z)(\bar{v}-\bar{z})} d^2 v. \qquad \square$$

Remark. Properties (i) and (iv) remain valid for the polynomials in a^* and a — operators of the form (2.46).

A *matrix symbol* $\tilde{A}(\bar{z}, z)$ of a bounded operator A in the Hilbert space $\bar{\mathscr{D}}$ is a restriction to $v = \bar{z}$ of an entire function $\tilde{A}(v, z)$ in variables v and z, defined by the following absolutely convergent series:

$$(2.51) \qquad \tilde{A}(v, z) = \sum_{m,n=0}^{\infty} (A\bar{f}_m, \bar{f}_n) f_n(v) f_m(z).$$

The matrix and Wick symbols are related as follows.

Lemma 2.4. *For a bounded operator A in the Hilbert space $\bar{\mathscr{D}}$,*

$$\tilde{A}(v, z) = e^{vz} A(v, z).$$

Proof. Using (2.49) we get

$$\tilde{A}(v, z) = \frac{1}{\pi} \sum_{m,n=0}^{\infty} f_n(v) f_m(z) \int_{\mathbb{C}} (A\bar{f}_m)(\bar{u}) f_n(u) e^{-|u|^2} d^2 u$$

$$= \frac{1}{\pi} \int_{\mathbb{C}} (A\Phi_z)(\bar{u})\overline{\Phi_{\bar{v}}(\bar{u})} e^{-|u|^2} d^2 u = (A\Phi_z, \Phi_{\bar{v}}) = e^{vz} A(v, z).$$

Changing the order of summation and integration is justified by the absolute convergence. $\qquad \square$

Corollary 2.3. *If $\tilde{A}_1(\bar{z}, z)$ and $\tilde{A}_2(\bar{z}, z)$ are matrix symbols of operators A_1 and A_2, then the matrix symbol of the operator $A = A_1 A_2$ is given by*

$$\tilde{A}(\bar{z}, z) = \frac{1}{\pi} \int_{\mathbb{C}} \tilde{A}_1(\bar{z}, v) \tilde{A}_2(\bar{v}, z) e^{-|v|^2} d^2 v.$$

Proof. The proof immediately follows from part (iv) of Theorem 2.2 and Lemma 2.4. $\qquad \square$

It is straightforward to generalize these constructions to n degrees of freedom. The Hilbert space $\mathscr{D}_n$ defining the holomorphic representation is the space of entire functions $f(\boldsymbol{z})$ of n complex variables $\boldsymbol{z} = (z_1, \ldots, z_n)$ with the inner product

$$(f, g) = \frac{1}{\pi^n} \int_{\mathbb{C}^n} f(\boldsymbol{z}) \overline{g(\boldsymbol{z})} e^{-|\boldsymbol{z}|^2} d^{2n}\boldsymbol{z} < \infty,$$

where $|\boldsymbol{z}|^2 = z_1^2 + \cdots + z_n^2$ and $d^{2n}\boldsymbol{z} = d^2 z_1 \cdots d^2 z_n$ is the Lebesgue measure on $\mathbb{C}^n \simeq \mathbb{R}^{2n}$. The functions

$$f_{\boldsymbol{m}}(\boldsymbol{z}) = \frac{z_1^{m_1} \cdots z_n^{m_n}}{\sqrt{m_1! \ldots m_n!}}, \qquad m_1, \ldots, m_n = 0, 1, 2, \ldots,$$

where $\boldsymbol{m} = (m_1, \ldots, m_n)$ is a multi-index, form an orthonormal basis for $\mathscr{D}_n$. Corresponding creation and annihilation operators are given by

$$a_j^* = z_j, \quad a_j = \frac{\partial}{\partial z_j}, \quad j = 1, \ldots, n.$$

The Hilbert space $\bar{\mathscr{D}}_n$ of anti-holomorphic functions $f(\bar{\boldsymbol{z}})$ on $\mathbb{C}^n$ is defined by the inner product

$$(2.52) \qquad (f, g) = \frac{1}{\pi^n} \int_{\mathbb{C}^n} f(\bar{\boldsymbol{z}}) \overline{g(\bar{\boldsymbol{z}})} e^{-|\boldsymbol{z}|^2} d^{2n}\boldsymbol{z} < \infty,$$

and the creation and annihilation operators are given by

$$a_j^* = \bar{z}_j, \quad a_j = \frac{\partial}{\partial \bar{z}_j}, \quad j = 1, \ldots, n.$$

The coherent states are $\Phi_{\boldsymbol{v}}(\bar{\boldsymbol{z}}) = e^{\boldsymbol{v}\bar{\boldsymbol{z}}}$, where $\boldsymbol{v}\bar{\boldsymbol{z}} = v_1 \bar{z}_1 + \cdots + v_n \bar{z}_n$, and satisfy the reproducing property

$$f(\bar{\boldsymbol{v}}) = (f, \Phi_{\boldsymbol{v}}), \quad f \in \bar{\mathscr{D}}_n, \ \boldsymbol{v} \in \mathbb{C}^n.$$

The Wick symbol $A(\bar{\boldsymbol{z}}, \boldsymbol{z})$ of a bounded operator A on $\bar{\mathscr{D}}_n$ is defined as a restriction to $\boldsymbol{v} = \bar{\boldsymbol{z}}$ of an entire function $A(\boldsymbol{v}, \boldsymbol{z})$ of $2n$ variables $\boldsymbol{v} = (v_1, \ldots, v_n)$ and $\boldsymbol{z} = (z_1, \ldots, z_n)$, given by

$$A(\boldsymbol{v}, \boldsymbol{z}) = e^{-\boldsymbol{v}\boldsymbol{z}} (A\Phi_{\boldsymbol{z}}, \Phi_{\bar{\boldsymbol{v}}}).$$

We have

$$(Af)(\bar{\boldsymbol{z}}) = \frac{1}{\pi^n} \int_{\mathbb{C}^n} A(\bar{\boldsymbol{z}}, \boldsymbol{v}) f(\bar{\boldsymbol{v}}) e^{-\boldsymbol{v}(\bar{\boldsymbol{v}} - \bar{\boldsymbol{z}})} d^{2n}\boldsymbol{v}, \quad f \in \bar{\mathscr{D}}_n,$$

and the Wick symbol $A(\bar{\boldsymbol{z}}, \boldsymbol{z})$ of the operator $A = A_1 A_2$ is given by

$$A(\bar{\boldsymbol{z}}, \boldsymbol{z}) = \frac{1}{\pi^n} \int_{\mathbb{C}^n} A_1(\bar{\boldsymbol{z}}, \boldsymbol{v}) A_2(\bar{\boldsymbol{v}}, \boldsymbol{z}) e^{-(\boldsymbol{v} - \boldsymbol{z})(\bar{\boldsymbol{v}} - \bar{\boldsymbol{z}})} d^{2n}\boldsymbol{v},$$

where $A_1(\bar{\boldsymbol{z}}, \boldsymbol{z})$ and $A_2(\bar{\boldsymbol{z}}, \boldsymbol{z})$ are the Wick symbols of the operators A_1 and A_2.

The matrix symbol $\tilde{A}(\bar{z}, z)$ of a bounded operator A on $\bar{\mathscr{D}}_n$ is defined as a restriction to $v = \bar{z}$ of an entire function $\tilde{A}(v, z)$ of $2n$ variables $v = (v_1, \ldots, v_n)$ and $z = (z_1, \ldots, z_n)$, given by the absolutely convergent series

$$\tilde{A}(v, z) = \sum_{k,m=0}^{\infty} (A\bar{f}_k, \bar{f}_m) f_m(v) f_k(z),$$

where $k = (k_1, \ldots, k_n)$, $m = (m_1, \ldots, m_n)$ are multi-indices, and $\bar{f}_m(\bar{z}) = \overline{f_m(z)}$. The matrix and Wick symbols of a bounded operator A are related by

$$\tilde{A}(v, z) = e^{vz} A(v, z).$$

Remark. The anti-holomorphic representation is very useful in quantum mechanics and especially in quantum field theory, where it is called holomorphic representation (with respect to the variables $\bar{z}$). Slightly abusing terminology, in Part 2 we will also call it holomorphic representation.

Problem 2.15. Find an explicit formula for the unitary operator establishing the Hilbert space isomorphism $\bar{\mathscr{D}}_n \simeq L^2(\mathbb{R}^n, d^n q)$.

Problem 2.16. Prove that for the bounded operator A the functions $A(v, z)$ and $\tilde{A}(v, z)$ are entire functions of $2n$ variables.

Problem 2.17. Let A be a trace class operator on $\bar{\mathscr{D}}_n$ with the Wick symbol $A(\bar{z}, z)$. Prove that

$$\operatorname{Tr} A = \frac{1}{\pi^n} \int_{\mathbb{C}^n} A(\bar{z}, z) e^{-|z|^2} d^{2n} z.$$

Problem 2.18. Show that the Wick symbol $A(\bar{z}, z)$ of the product $A = A_l \ldots A_1$ is given by

$$A(\bar{z}, z) = \frac{1}{\pi^l} \int_{\mathbb{C}^l} A_l(\bar{z}, z_{l-1}) \ldots A_1(\bar{z}_1, z) \exp\left\{ \sum_{k=1}^{l} \bar{z}_k (z_{k-1} - z_k) \right\} d^2 z_1 \ldots d^2 z_{l-1},$$

where $z_0 = z_l = z$, $\bar{z}_l = \bar{z}$, and $A_k(\bar{z}, z)$ are the Wick symbols of the operators A_k.

3. Weyl relations

Let R be an irreducible unitary representation of the Heisenberg group $\mathbf{H}_n$ in Hilbert space $\mathscr{H}$. It follows from Schur's lemma that $R(e^{\alpha c}) = e^{i\hbar\alpha} I$ for some $\hbar \in \mathbb{R}$, where I is the identity operator on $\mathscr{H}$. If $\hbar = 0$, the n-parameter abelian groups of unitary operators $U(u) = R(e^{uX})$ and $V(v) = R(e^{vY})$ commute, so using Schur's lemma once again we conclude that there exist $p, q \in \mathbb{R}^n$ such that $U(u) = e^{-iup} I$ and $V(v) = e^{-ivq} I$. Therefore, in this case the irreducible representation R is one-dimensional and is parametrized by a vector $(p, q) \in \mathbb{R}^{2n}$. When $\hbar \neq 0$, unitary operators $U(u)$ and $V(v)$ satisfy the Weyl relations

$$(3.1) \qquad\qquad U(u)V(v) = e^{i\hbar uv} V(v) U(u),$$

which admit the Schrödinger representation, introduced in Section 2.2. It turns out that every unitary irreducible representation of the Heisenberg group $\mathbf{H}_n$ is unitarily equivalent either to a one-dimensional representation with parameters $(\boldsymbol{p}, \boldsymbol{q}) \in \mathbb{R}^{2n}$, or to the Schrödinger representation with some $\hbar \neq 0$. The physically meaningful case corresponds to $\hbar > 0$, and representation with $-\hbar$ is given by the operators $U^{-1}(\boldsymbol{u}) = U(-\boldsymbol{u})$ and $V(\boldsymbol{v})$.

3.1. Stone-von Neumann theorem. Here we prove the following fundamental result.

Theorem 3.1 (Stone-von Neumann theorem). *Every irreducible unitary representation of the Weyl relations for n degrees of freedom,*

$$U(\boldsymbol{u})V(\boldsymbol{v}) = e^{i\hbar\boldsymbol{uv}}V(\boldsymbol{v})U(\boldsymbol{u}),$$

is unitarily equivalent to the Schrödinger representation.

Proof. It is sufficient to consider $n = 1$, the general case $n > 1$ is treated similarly. Set

$$S(u, v) = e^{-\frac{i\hbar uv}{2}} U(u)V(v).$$

The unitary operator $S(u, v)$ satisfies

$$(3.2) \qquad\qquad S(u, v)^* = S(-u, -v),$$

and it follows from the Weyl relation that

$$(3.3) \qquad S(u_1, v_1)S(u_2, v_2) = e^{\frac{i\hbar}{2}(u_1 v_2 - u_2 v_1)}S(u_1 + u_2, v_1 + v_2).$$

Define a linear map $W : L^1(\mathbb{R}^2) \to \mathscr{L}(\mathscr{H})$, called the *Weyl transform*, by

$$W(f) = \frac{1}{2\pi} \int_{\mathbb{R}^2} f(u, v)S(u, v)\,du\,dv.$$

Here the integral is understood in the weak sense: for every $\psi_1, \psi_2 \in \mathscr{H}$,

$$(W(f)\psi_1, \psi_2) = \frac{1}{2\pi} \int_{\mathbb{R}^2} f(u, v)(S(u, v)\psi_1, \psi_2)\,du\,dv.$$

The integral is absolutely convergent for all $\psi_1, \psi_2 \in \mathscr{H}$, and defines a bounded operator $W(f)$ satisfying

$$\|W(f)\| \leq \frac{1}{2\pi}\|f\|_{L^1}.$$

The Weyl transform has the following properties.

WT1. For $f \in L^1(\mathbb{R}^2)$,

$$W(f)^* = W(f^*),$$

where

$$f^*(u, v) = \overline{f(-u, -v)}.$$

WT2. For $f \in L^1(\mathbb{R}^2)$,

$$S(u_1, v_1) W(f) S(u_2, v_2) = W(\tilde{f}),$$

where

$$\tilde{f}(u, v) = e^{\frac{i\hbar}{2}\{(u_1 - u_2)v - (v_1 - v_2)u + u_1 v_2 - u_2 v_1\}} f(u - u_1 - u_2, v - v_1 - v_2).$$

WT3. $\operatorname{Ker} W = \{0\}$.

WT4. For $f_1, f_2 \in L^1(\mathbb{R}^2)$,

$$W(f_1) W(f_2) = W(f_1 *_\hbar f_2),$$

where

$$(f_1 *_\hbar f_2)(u, v) = \frac{1}{2\pi} \int_{\mathbb{R}^2} e^{\frac{i\hbar}{2}(uv' - u'v)} f_1(u - u', v - v') f_2(u', v') du' dv'.$$

The first two properties follow from the definition and (3.2) and (3.3). To prove **WT3**, suppose that $W(f) = 0$. Then for every $\psi_1, \psi_2 \in \mathscr{H}$ and all $u', v' \in \mathbb{R}$ we have

$$0 = (W(f) S(u', v') \psi_1, S(u', v') \psi_2)$$

$$= \frac{1}{2\pi} \int_{\mathbb{R}^2} e^{i\hbar(uv' - u'v)} f(u, v)(S(u, v)\psi_1, \psi_2) du dv.$$

Therefore $f(u, v)(S(u, v)\psi_1, \psi_2) = 0$ a.e. on $\mathbb{R}^2$, so that $f = 0$.

To prove **WT4**, we compute

$$(W(f_1) W(f_2) \psi_1, \psi_2) = (W(f_2) \psi_1, W(f_1)^* \psi_2)$$

$$= \frac{1}{2\pi} \int_{\mathbb{R}^2} f_2(u_2, v_2)(S(u_2, v_2)\psi_1, W(f_1)^* \psi_2) du_2 dv_2$$

$$= \frac{1}{2\pi} \int_{\mathbb{R}^2} f_2(u_2, v_2)(W(f_1) S(u_2, v_2)\psi_1, \psi_2) du_2 dv_2$$

$$= \frac{1}{(2\pi)^2} \int_{\mathbb{R}^2} \int_{\mathbb{R}^2} f_1(u_1 - u_2, v_1 - v_2) f_2(u_2, v_2) \cdot$$

$$\cdot e^{\frac{i\hbar}{2}(u_1 v_2 - u_2 v_1)}(S(u_1, v_1)\psi_1, \psi_2) du_1 dv_1 du_2 dv_2$$

$$= \frac{1}{2\pi} \int_{\mathbb{R}^2} (f_1 *_\hbar f_2)(u, v)(S(u, v)\psi_1, \psi_2) du dv.$$

We have $f_1 *_\hbar f_2 \in L^1(\mathbb{R}^2)$, and it follows from the associativity of the operator product and property **WT3** that the linear mapping

$$*_\hbar : L^1(\mathbb{R}^2) \times L^1(\mathbb{R}^2) \to L^1(\mathbb{R}^2)$$

defines a new associative product on $L^1(\mathbb{R}^2)$,

$$f_1 *_\hbar (f_2 *_\hbar f_3) = (f_1 *_\hbar f_2) *_\hbar f_3 \quad \text{for all} \quad f_1, f_2, f_3 \in L^1(\mathbb{R}^2).$$

When $\hbar = 0$, the product $*_\hbar$ becomes the usual convolution product on $L^1(\mathbb{R}^2)$.

For every non-zero $\psi \in \mathcal{H}$ denote by $\mathcal{H}_\psi$ the closed subspace of $\mathcal{H}$ spanned by the vectors $S(u,v)\psi$ for all $u,v \in \mathbb{R}$. The subspace $\mathcal{H}_\psi$ is invariant for all operators $U(u)$ and $V(v)$. Since the representation of the Weyl relation is irreducible, $\mathcal{H}_\psi = \mathcal{H}$.

There is a vector $\psi_0 \in \mathcal{H}$ for which all inner products of the generating vectors $S(u,v)\psi_0$ of $\mathcal{H}_{\psi_0}$ can be explicitly computed. Namely, let

$$f_0(u,v) = \hbar\, e^{-\frac{\hbar}{4}(u^2+v^2)}$$

and set

$$W_0 = W(f_0).$$

Then $W_0^* = W_0$ and

$$W_0 S(u,v) W_0 = e^{-\frac{\hbar}{4}(u^2+v^2)} W_0.$$

In particular, $W_0^2 = W_0$, so that W_0 is an orthogonal projection. Indeed, using **WT4**, we have

$$W_0 S(u,v) W_0 = W(f_0 *_\hbar \tilde{f_0}),$$

where

$$\tilde{f_0}(u',v') = e^{\frac{i\hbar}{2}(uv'-u'v)} f_0(u'-u, v'-v).$$

By "completing the square" we obtain

$$(f_0 *_\hbar \tilde{f_0})(u',v') = \hbar e^{-\frac{\hbar}{4}(u^2+v^2+u'^2+v'^2)} I(u',v'),$$

where

$$I(u',v') = \frac{\hbar}{2\pi} \int_{\mathbb{R}^2} e^{-\frac{\hbar}{2}\{(u''-u-u'+iv+iv')^2 + (v''-v-v'-iu-iu')^2\}} du''\,dv''.$$

Shifting contours of integration to $\operatorname{Im} u'' = -v-v'$, $\operatorname{Im} v'' = u+u'$ and substituting $\xi = u''-u-u'+iv+iv'$, $\eta = v''-v-v'-iu-iu'$, we get

$$I(u',v') = \frac{\hbar}{2\pi} \int_{\mathbb{R}^2} e^{-\frac{\hbar}{2}(\xi^2+\eta^2)} d\xi\,d\eta = 1.$$

Now let $\mathcal{H}_0 = \operatorname{Im} W_0$ — a non-zero closed subspace of $\mathcal{H}$ by property **WT3**. For every $\psi_1, \psi_2 \in \mathcal{H}_0$ we have $W_0\psi_1 = \psi_1$, $W_0\psi_2 = \psi_2$, and

$$
\begin{aligned}
(S(u_1,v_1)\psi_1, S(u_2,v_2)\psi_2) &= (S(u_1,v_1)W_0\psi_1, S(u_2,v_2)W_0\psi_2) \\
&= (W_0 S(-u_2,-v_2) S(u_1,v_1) W_0\psi_1, \psi_2) \\
&= e^{\frac{i\hbar}{2}(u_1 v_2 - u_2 v_1)} (W_0 S(u_1-u_2, v_1-v_2) W_0\psi_1, \psi_2) \\
&= e^{\frac{i\hbar}{2}(u_1 v_2 - u_2 v_1) - \frac{\hbar}{4}\{(u_1-u_2)^2 + (v_1-v_2)^2\}} (W_0\psi_1, \psi_2) \\
&= e^{\frac{i\hbar}{2}(u_1 v_2 - u_2 v_1) - \frac{\hbar}{4}\{(u_1-u_2)^2 + (v_1-v_2)^2\}} (\psi_1, \psi_2).
\end{aligned}
$$

This implies that the subspace $\mathscr{H}_0$ is one-dimensional. Indeed, for every $\psi_1, \psi_2 \in \mathscr{H}_0$ such that $(\psi_1, \psi_2) = 0$, we have that the corresponding subspaces $\mathscr{H}_{\psi_1}$ and $\mathscr{H}_{\psi_2}$ are orthogonal. Since $\mathscr{H}_\psi = \mathscr{H}$ for every non-zero ψ, at least one of the vectors ψ_1, ψ_2 is 0. Let $\mathscr{H}_0 = \mathbb{C}\,\psi_0$, $\|\psi_0\| = 1$, and set

$$\psi_{\alpha,\beta} = S(\alpha, \beta)\psi_0, \quad \alpha, \beta \in \mathbb{R}.$$

The closure of the linear span of the vectors $\psi_{\alpha,\beta}$ for all $\alpha, \beta \in \mathbb{R}$ is $\mathscr{H}$. We have

$$(\psi_{\alpha,\beta}, \psi_{\gamma,\delta}) = e^{\frac{i\hbar}{2}(\alpha\delta - \beta\gamma) - \frac{\hbar}{4}\{(\alpha-\gamma)^2 + (\beta-\delta)^2\}}$$

and

$$S(u, v)\psi_{\alpha,\beta} = e^{\frac{i\hbar}{2}(u\beta - v\alpha)}\psi_{\alpha+u,\beta+v}.$$

Next we consider the Schrödinger representation of the Weyl relation in $L^2(\mathbb{R}, dq)$:

$$(\mathsf{U}(u)\varphi)(q) = \varphi(q - \hbar u),$$

$$(\mathsf{V}(v)\varphi)(q) = e^{-ivq}\varphi(q),$$

$$(\mathsf{S}(u, v)\varphi)(q) = e^{\frac{i\hbar}{2}uv - ivq}\varphi(q - \hbar u).$$

For the corresponding operator W_0 we have

$$(\mathsf{W}_0\varphi)(q) = \frac{\hbar}{2\pi} \int_{\mathbb{R}^2} e^{-\frac{\hbar}{4}(u^2+v^2)} e^{\frac{i\hbar}{2}uv - ivq}\varphi(q - \hbar u)\,du\,dv$$

$$= \sqrt{\frac{\hbar}{\pi}} \int_{-\infty}^{\infty} e^{-\frac{\hbar}{4}\{u^2 + (u - \frac{2}{\hbar}q)^2\}}\varphi(q - \hbar u)\,du$$

$$= \frac{1}{\sqrt{\hbar\pi}} \int_{-\infty}^{\infty} e^{-\frac{1}{2\hbar}(q^2 + q'^2)}\varphi(q')\,dq',$$

so that W_0 is a projection onto the one-dimensional subspace of $L^2(\mathbb{R}, dq)$ spanned by the Gaussian exponential $\varphi_0(q) = \frac{1}{\sqrt[4]{\pi\hbar}}e^{-\frac{1}{2\hbar}q^2}$, $\|\varphi_0\| = 1$. Let

$$\varphi_{\alpha,\beta} = S(\alpha, \beta)\varphi_0.$$

Since the Schrödinger representation is irreducible, the closed subspace spanned by the functions $\varphi_{\alpha,\beta}$ for all $\alpha, \beta \in \mathbb{R}$ is the whole Hilbert space $L^2(\mathbb{R}, dq)$. (This also follows from the completeness of Hermite-Tchebyscheff functions, since $\varphi_{\alpha,\beta}(q) = \frac{1}{\sqrt[4]{\pi\hbar}}e^{\frac{i\hbar}{2}\alpha\beta - i\beta q - \frac{1}{2\hbar}(q - \hbar\alpha)^2}$.) We have

$$(\varphi_{\alpha,\beta}, \varphi_{\gamma,\delta}) = e^{\frac{i\hbar}{2}(\alpha\delta - \beta\gamma) - \frac{\hbar}{4}\{(\alpha-\gamma)^2 + (\beta-\delta)^2\}}$$

and

$$S(u, v)\varphi_{\alpha,\beta} = e^{\frac{i\hbar}{2}(u\beta - v\alpha)}\varphi_{\alpha+u,\beta+v}.$$

For $\psi = \sum_{i=1}^{n} c_i \psi_{\alpha_i, \beta_i} \in \mathscr{H}$ define

$$\mathscr{U}(\psi) = \sum_{i=1}^{n} c_i \varphi_{\alpha_i, \beta_i} \in L^2(\mathbb{R}, dq).$$

Since inner products between the vectors $\psi_{\alpha, \beta}$ coincide with the inner products between the vectors $\varphi_{\alpha, \beta}$, we have

$$\|\mathscr{U}(\psi)\|_{L^2}^2 = \|\psi\|_{\mathscr{H}}^2,$$

so that $\mathscr{U}$ is a well-defined unitary operator between the subspaces spanned by the vectors $\{\psi_{\alpha, \beta}\}_{\alpha, \beta \in \mathbb{R}}$ and $\{\varphi_{\alpha, \beta}\}_{\alpha, \beta \in \mathbb{R}}$. Thus $\mathscr{U}$ extends to the unitary operator $\mathscr{U} : \mathscr{H} \to L^2(\mathbb{R}, dq)$ and

$$\mathscr{U} \mathsf{S}(\alpha, \beta) = \mathsf{S}(\alpha, \beta) \mathscr{U} \quad \text{for all} \quad \alpha, \beta \in \mathbb{R}. \qquad \square$$

Lemma 2.1 in Section 2.1 is now an easy corollary of the Stone-von Neumann theorem.

Corollary 3.2. *The self-adjoint generators* $\boldsymbol{P} = (P_1, \ldots, P_n)$ *and* $\boldsymbol{Q} = (Q^1, \ldots, Q^n)$ *of the n-parameter abelian groups of unitary operators* $U(\boldsymbol{u})$ *and* $V(\boldsymbol{v})$ *satisfy Heisenberg commutation relations.*

Proof. It follows from the Stone theorem that in the Schrödinger representation the generators $\boldsymbol{P}$ and $\boldsymbol{Q}$ are momenta and coordinate operators. $\qquad \square$

Remark. For n degrees of freedom

$$S(\boldsymbol{u}, \boldsymbol{v}) = e^{-\frac{i\hbar}{2} \boldsymbol{uv}} U(\boldsymbol{u}) V(\boldsymbol{v}) = e^{-\frac{i\hbar}{2} \boldsymbol{uv}} e^{-i\boldsymbol{uP}} e^{-i\boldsymbol{vQ}},$$

and formally using the Baker-Campbell-Hausdorff formula, we get

$$(3.4) \qquad\qquad S(\boldsymbol{u}, \boldsymbol{v}) = e^{-i(\boldsymbol{uP} + \boldsymbol{vQ})}.$$

Thus the Weyl transform

$$W(f) = \frac{1}{(2\pi)^n} \int_{\mathbb{R}^{2n}} f(\boldsymbol{u}, \boldsymbol{v}) e^{-i(\boldsymbol{uP} + \boldsymbol{vQ})} d^n \boldsymbol{u} \, d^n \boldsymbol{v}$$

can be considered as a *non-commutative Fourier transform* — an operator-valued generalization of the "ordinary" Fourier transform

$$\hat{f}(\boldsymbol{p}, \boldsymbol{q}) = \frac{1}{(2\pi)^n} \int_{\mathbb{R}^{2n}} f(\boldsymbol{u}, \boldsymbol{v}) e^{-i(\boldsymbol{up} + \boldsymbol{vq})} d^n \boldsymbol{u} \, d^n \boldsymbol{v}.$$

Remark. The Stone-von Neumann theorem is a very strong result. In particular, it implies that creation and annihilation operators for n degrees of freedom,

$$\boldsymbol{a} = \frac{1}{\sqrt{2\hbar}} (\boldsymbol{Q} + i\boldsymbol{P}), \quad \boldsymbol{a}^* = \frac{1}{\sqrt{2\hbar}} (\boldsymbol{Q} - i\boldsymbol{P}),$$

where $\boldsymbol{P}$ and $\boldsymbol{Q}$ are the corresponding generators[24] satisfying Heisenberg commutation relations, are unitarily equivalent to the corresponding operators in the Schrödinger representation. Thus there always exists a ground state — a vector $\psi_0 \in \mathscr{H}$, annihilated by the operators $\boldsymbol{a} = (a_1, \ldots, a_n)$. The corresponding statement no longer holds for quantum systems with infinitely many degrees of freedom, described by quantum field theory, where one needs to postulate the existence of the ground state (the "physical vacuum").

The Weyl transform in the Schrödinger representation $\mathscr{H} = L^2(\mathbb{R}^n, d^n\boldsymbol{q})$ can be described explicitly as a bounded operator with an integral kernel. Namely,

$$(3.5) \qquad (S(\boldsymbol{u}, \boldsymbol{v})\psi)(\boldsymbol{q}) = e^{\frac{i\hbar}{2}\boldsymbol{uv} - i\boldsymbol{vq}}\psi(\boldsymbol{q} - \hbar\boldsymbol{u}), \quad \psi \in L^2(\mathbb{R}^n, d^n\boldsymbol{q}),$$

and for $\psi_1, \psi_2 \in L^2(\mathbb{R}^n, d^n\boldsymbol{q})$ and $f \in L^1(\mathbb{R}^{2n}) \cap L^2(\mathbb{R}^{2n})$ we have

$$(W(f)\psi_1, \psi_2) = \frac{1}{(2\pi)^n} \int_{\mathbb{R}^{2n}} f(\boldsymbol{u}, \boldsymbol{v})(S(\boldsymbol{u}, \boldsymbol{v})\psi_1, \psi_2) d^n\boldsymbol{u}\, d^n\boldsymbol{v}$$

$$= \frac{1}{(2\pi)^n} \int_{\mathbb{R}^{2n}} \left(\int_{\mathbb{R}^n} e^{\frac{i\hbar}{2}\boldsymbol{uv} - i\boldsymbol{vq}} f(\boldsymbol{u}, \boldsymbol{v})\psi_1(\boldsymbol{q} - \hbar\boldsymbol{u})\overline{\psi_2(\boldsymbol{q})} d^n\boldsymbol{q} \right) d^n\boldsymbol{u}\, d^n\boldsymbol{v}$$

$$= \int_{\mathbb{R}^n} \left(\int_{\mathbb{R}^n} K(\boldsymbol{q}, \boldsymbol{q}')\psi_1(\boldsymbol{q}') d^n\boldsymbol{q}' \right) \overline{\psi_2(\boldsymbol{q})} d^n\boldsymbol{q},$$

where

$$(3.6) \qquad K(\boldsymbol{q}, \boldsymbol{q}') = \frac{1}{(2\pi\hbar)^n} \int_{\mathbb{R}^n} f\left(\tfrac{\boldsymbol{q} - \boldsymbol{q}'}{\hbar}, \boldsymbol{v}\right) e^{-\frac{i}{2}(\boldsymbol{q} + \boldsymbol{q}')\boldsymbol{v}} d^n\boldsymbol{v}.$$

The interchange of the order of integrations is justified by the Fubini theorem. Thus $W(f)$ is an integral operator with the integral kernel $K(\boldsymbol{q}, \boldsymbol{q}')$,

$$(W(f)\psi)(\boldsymbol{q}) = \int_{\mathbb{R}^n} K(\boldsymbol{q}, \boldsymbol{q}')\psi(\boldsymbol{q}') d^n\boldsymbol{q}', \quad \psi \in L^2(\mathbb{R}^n, d^n\boldsymbol{q}).$$

By the Plancherel theorem,

$$\int_{\mathbb{R}^{2n}} |K(\boldsymbol{q}, \boldsymbol{q}')|^2 d^n\boldsymbol{q}\, d^n\boldsymbol{q}' = \int_{\mathbb{R}^{2n}} |f(\boldsymbol{u}, \boldsymbol{v})|^2 d^n\boldsymbol{u}\, d^n\boldsymbol{v},$$

so that $W(f)$ is a Hilbert-Schmidt operator on $\mathscr{H}$. The linear subspace $L^1(\mathbb{R}^{2n}) \cap L^2(\mathbb{R}^{2n})$ is dense in $L^2(\mathbb{R}^{2n})$ and the Weyl transform extends to the isometry $W : L^2(\mathbb{R}^{2n}) \to \mathscr{S}_2$, where $\mathscr{S}_2$ is the Hilbert space of Hilbert-Schmidt operators on $\mathscr{H} = L^2(\mathbb{R}^n, d^n\boldsymbol{q})$. Since every Hilbert-Schmidt operator on $\mathscr{H}$ is a bounded operator with the integral kernel in $L^2(\mathbb{R}^{2n})$, the mapping W is onto. Thus we have proved the following result.

---

[24]Here it is convenient to lower the indices by the Euclidean metric on $\mathbb{R}^n$ and use $\boldsymbol{Q} = (Q_1, \ldots, Q_n)$.

Lemma 3.1. *The Weyl transform W defines the isomorphism $L^2(\mathbb{R}^{2n}) \simeq \mathscr{S}_2$.*

Problem 3.1. Let $\boldsymbol{P} = (P_1, \ldots, P_n)$ and $\boldsymbol{Q} = (Q_1, \ldots, Q_n)$ be self-adjoint operators[25] on $\mathscr{H}$ satisfying Heisenberg commutation relations (2.4) on some dense domain D in $\mathscr{H}$, and such that the symmetric operator $H = \sum_{k=1}^{n}(P_k^2 + Q_k^2)$, defined on D, is essentially self-adjoint. Also suppose that every bounded operator which commutes with $\boldsymbol{P}$ and $\boldsymbol{Q}$ is a multiple of the identity operator in $\mathscr{H}$. Prove that $\boldsymbol{P}$ and $\boldsymbol{Q}$ are unitarily equivalent to momenta and coordinate operators in the Schrödinger representation. (*Hint:* By the Stone-von Neumann theorem, it is sufficient to show that $\boldsymbol{P}$ and $\boldsymbol{Q}$ define an integrable representation of the Heisenberg algebra. One can also prove this result directly by using commutation relations between the operators H, $\boldsymbol{a}^*$, $\boldsymbol{a}$ and the arguments in Section 2.6. This would give another proof of the Stone-von Neumann theorem.)

Problem 3.2. Prove formula (3.4).

3.2. Invariant formulation. Here we present a coordinate-free description of the Schrödinger representation. Let (V, ω) be a finite-dimensional symplectic vector space, $\dim V = 2n$. Recall that (see Section 2.1) the Heisenberg algebra $\mathfrak{g} = \mathfrak{g}(V)$ is a one-dimensional central extension of the abelian Lie algebra V by the skew-symmetric bilinear form ω. As a vector space,

$$\mathfrak{g} = V \oplus \mathbb{R}c,$$

with the Lie bracket

$$[u + \alpha c, v + \beta c] = \omega(u, v)c, \quad u, v \in V, \ \alpha, \beta \in \mathbb{R}.$$

Every Lagrangian subspace ℓ of V defines an abelian subalgebra $\ell \oplus \mathbb{R}c$ of $\mathfrak{g}$. A Lagrangian subspace ℓ' is complementary to ℓ in V if $\ell \oplus \ell' = V$. A choice of the complementary Lagrangian subspace ℓ' to ℓ establishes the isomorphism

$$\mathfrak{g}/(\ell \oplus \mathbb{R}c) \simeq \ell'.$$

The Heisenberg group $G = G(V)$ is a connected and simply-connected Lie group with the Lie algebra $\mathfrak{g}$. By the exponential map, $G \simeq V \oplus \mathbb{R}c$ as a manifold, with the group law

$$\exp(v_1 + \alpha_1 c)\exp(v_2 + \alpha_2 c) = \exp(v_1 + v_2 + (\alpha_1 + \alpha_2 + \tfrac{1}{2}\omega(v_1, v_2))c),$$

where $v_1, v_2 \in V$, $\alpha_1, \alpha_2 \in \mathbb{R}$. A choice of the symplectic basis for V establishes the isomorphism $G(V) \simeq \mathbf{H}_n$ with the matrix Heisenberg group defined in Section 2.1. The volume form $d^{2n}\boldsymbol{v} \wedge dc$, where $d^{2n}\boldsymbol{v} \in \Lambda^{2n}V^*$ is a volume form on V and V^* is the dual vector space to V, defines a bi-invariant Haar measure on G. Every Lagrangian subspace ℓ defines an

---

[25]See previous footnote.

abelian subgroup $L = \exp(\ell \oplus \mathbb{R}c)$ of G, and a choice of the complementary Lagrangian subspace ℓ' establishes the isomorphism

$$(3.7) \qquad\qquad G/L \simeq \ell'.$$

The isomorphism $\Lambda^{2n}V^* \simeq \Lambda^n\ell^* \wedge \Lambda^n\ell'^*$ gives rise to the volume form $d^n\boldsymbol{v}'$ on ℓ' and defines a measure dg on the homogeneous space G/L. The measure dg is invariant under the left G-action and does not depend on the choice of ℓ'.

For a given $\hbar \in \mathbb{R}$, the function $\chi : L \to \mathbb{C}$,

$$\chi(\exp(v + \alpha c)) = e^{i\hbar\alpha}, \quad v \in V,\, \alpha \in \mathbb{R},$$

defines a one-dimensional unitary character of L,

$$\chi(l_1 l_2) = \chi(l_1)\chi(l_2), \quad l_1, l_2 \in L.$$

Definition. The Schrödinger representation S_ℓ of the Heisenberg group $G(V)$, associated with a Lagrangian subspace ℓ of V, is a representation of G induced by the one-dimensional representation χ of the corresponding abelian Lie group L,

$$S_\ell = \mathrm{Ind}_L^G \chi.$$

By definition of the induced representation, the Hilbert space $\mathscr{H}_\ell$ of the representation S_ℓ consists of measurable functions $f : G \to \mathbb{C}$ satisfying the property

$$f(gl) = \chi(l)^{-1} f(g), \quad g \in G,\, l \in L,$$

and such that

$$\|f\|^2 = \int_{G/L} |f(g)|^2 dg < \infty.$$

Corresponding unitary operators $S_\ell(g)$, $g \in G$, are defined by the left translations,

$$(S_\ell(g)f)(g') = f(g^{-1}g'), \quad f \in \mathscr{H}_\ell,\, g \in G.$$

In particular,

$$(S_\ell(\exp \alpha c)f)(g) = f(\exp(-\alpha c)g) = f(g\exp(-\alpha c)) = e^{i\hbar\alpha} f(g),$$

so that $S_\ell(\exp \alpha c) = e^{i\hbar\alpha} I$, where I is the identity operator on $\mathscr{H}_\ell$.

For every Lagrangian subspace ℓ of V, the representation S_ℓ is unitarily equivalent to the Schrödinger representation. This can be explicitly shown as follows. By (3.7), a choice of a complementary Lagrangian subspace ℓ' gives rise to the unique decomposition $g = \exp v' \exp(v + \alpha c)$, and for $f \in \mathscr{H}_\ell$ we get

$$f(g) = f(\exp v' \exp(v + \alpha c)) = e^{-i\hbar\alpha} f(\exp v'), \quad v \in \ell,\, v' \in \ell',\, \alpha \in \mathbb{R}.$$

Thus every $f \in \mathscr{H}_\ell$ is completely determined by its restriction to $\exp \ell' \simeq \ell'$, and the mapping $\mathscr{U}_\ell : \mathscr{H}_\ell \to L^2(\ell', d^n v')$, defined by

$$\mathscr{U}_\ell(f)(v') = f(\exp \tfrac{1}{\hbar} v'), \quad v' \in \ell',$$

is the isomorphism of Hilbert spaces. For the corresponding representation $\mathsf{S}_\ell = \mathscr{U}_\ell \mathsf{S}_\ell \mathscr{U}_\ell^{-1}$ in $L^2(\ell', d^n v')$ we have

$$(\mathsf{S}_\ell(\exp v)\psi)(v') = e^{i\omega(v,v')}\psi(v'), \quad v \in \ell, \ v' \in \ell',$$

$$(\mathsf{S}_\ell(\exp u')\psi)(v') = \psi(v' - \hbar u'), \quad u', v' \in \ell'.$$

Let $e^1, \ldots, e^n, f_1, \ldots, f_n$ be the symplectic basis for V such that

$$\ell = \mathbb{R}e^1 \oplus \cdots \oplus \mathbb{R}e^n \quad \text{and} \quad \ell' = \mathbb{R}f_1 \oplus \cdots \oplus \mathbb{R}f_n,$$

and let $(\boldsymbol{p}, \boldsymbol{q}) = (p_1, \ldots, p_n, q^1, \ldots, q^n)$ be the corresponding coordinates in V (see Section 2.6 in Chapter 1). Then $L^2(\ell', d^n v') \simeq L^2(\mathbb{R}^n, d^n \boldsymbol{q})$ and

$$\mathsf{S}_\ell(\exp \boldsymbol{u}X) = U(\boldsymbol{u}) \quad \text{and} \quad \mathsf{S}_\ell(\exp \boldsymbol{v}Y) = V(\boldsymbol{v}),$$

where $\boldsymbol{u}X = \sum_{k=1}^n u^k f_k$, $\boldsymbol{v}Y = \sum_{k=1}^n v_k e^k$, and $U(\boldsymbol{u}) = e^{-i\boldsymbol{u}\boldsymbol{P}}$, $V(\boldsymbol{v}) = e^{-i\boldsymbol{v}\boldsymbol{Q}}$ are, respectively, n-parameter groups of unitary operators corresponding to momenta and coordinate operators $\boldsymbol{P}$ and $\boldsymbol{Q}$.

The mapping $*_\hbar$ — the new associative product on $L^1(\mathbb{R}^{2n})$ introduced in the last section — can also be described in invariant terms. Let $B_\hbar = G/\Gamma_\hbar$ be the quotient group, where $\Gamma_\hbar = \{\exp(\frac{2\pi n}{\hbar} c) \in G \,|\, n \in \mathbb{Z}\}$ is a discrete central subgroup in G, and let $db = d^{2n} \boldsymbol{v} \wedge d^* \alpha$ be the left-invariant Haar measure on $B_\hbar$, where $d^* \alpha$ is the normalized Haar measure on the circle $\mathbb{R}/\frac{2\pi}{\hbar}\mathbb{Z}$. The Banach space $L^1(B_\hbar, db)$ has an algebra structure with respect to the convolution:

$$(\varphi_1 *_{B_\hbar} \varphi_2)(b) = \int_{B_\hbar} \varphi_1(b_1)\varphi_2(b_1^{-1}b)db_1, \quad \varphi_1, \varphi_2 \in L^1(B_\hbar, db).$$

The correspondence

$$L^1(V, d^{2n}\boldsymbol{v}) \ni f(v) \mapsto \varphi(\exp(v + \alpha c)) = e^{-i\hbar\alpha} f(v) \in L^1(B_\hbar, db)$$

defines the inclusion map $L^1(V, d^{2n}\boldsymbol{v}) \hookrightarrow L^1(B_\hbar, db)$, and the image of $L^1(V, d^{2n}\boldsymbol{v})$ is a subalgebra of $L^1(B_\hbar, db)$ under the convolution. We have for $f_1, f_2 \in L^1(V, d^{2n}\boldsymbol{v})$,

$$(\varphi_1 *_{B_\hbar} \varphi_2)(\exp v) = \int_{B_\hbar} \varphi_1(\exp(u + \alpha c))\varphi_2(\exp(-u - \alpha c)\exp v)d^{2n}\boldsymbol{v}d^*\alpha$$

$$= \int_V \varphi_1(\exp u)\varphi_2(\exp(v - u)\exp(-\tfrac{1}{2}\omega(u,v)c))d^{2n}\boldsymbol{v}$$

$$= \int_V f_1(u)f_2(v - u)e^{\frac{i\hbar}{2}\omega(u,v)}d^{2n}\boldsymbol{v}$$

$$= (f_1 *_\hbar f_2)(v), \quad v \in V.$$

Problem 3.3. Prove that the Weyl transform is the restriction to $L^1(V, d^{2n}\boldsymbol{v})$ of the mapping

$$L^1(B_\hbar, db) \ni \varphi \mapsto \int_{B_\hbar} \varphi(b) S_\ell(b) db \in \mathscr{L}(\mathscr{H}_\ell).$$

Problem 3.4. Let ℓ_1, ℓ_2, ℓ_3 be Lagrangian subspaces of V. Define the *triple Maslov index* $\tau(\ell_1, \ell_2, \ell_3)$ as a signature of the quadratic form Q on $\ell_1 \oplus \ell_2 \oplus \ell_3$, defined by

$$Q(x_1, x_2, x_3) = \omega(x_1, x_2) + \omega(x_2, x_3) + \omega(x_3, x_1).$$

For Lagrangian subspaces ℓ_1, ℓ_2 of V let $\mathscr{F}_{\ell_2, \ell_1}$ be the unitary operator establishing the Hilbert space isomorphism $\mathscr{H}_{\ell_1} \simeq \mathscr{H}_{\ell_2}$ and intertwining representations S_{ℓ_1} and S_{ℓ_2},

$$\mathscr{F}_{\ell_2, \ell_1} S_{\ell_1}(g) = S_{\ell_2}(g) \mathscr{F}_{\ell_2, \ell_1}, \quad g \in G.$$

Prove that $\mathscr{F}_{\ell_1, \ell_3} \mathscr{F}_{\ell_3, \ell_2} \mathscr{F}_{\ell_2, \ell_1} = e^{-\pi i \tau(\ell_1, \ell_2, \ell_3)} I_1$, where I_1 is the identity operator in $\mathscr{H}_{\ell_1}$.

Problem 3.5. Let $\mathrm{Sp}(V)$ be the symplectic group of (V, ω) — the subgroup of $\mathrm{GL}(V)$, preserving the symplectic form ω. The group $\mathrm{Sp}(V)$ acts as a group of automorphisms of the Heisenberg group G by the formula $h \cdot \exp(v + \alpha c) = \exp(h \cdot v + \alpha c)$, $h \in \mathrm{Sp}(V)$, $v \in V$. Let R be an irreducible unitary representation of G in the Hilbert space $\mathscr{H}$. Show that there exist unitary operators $U(h)$, defined up to multiplication by complex numbers of modulus 1, such that

$$U(h) R(g) U(h)^{-1} = R(h \cdot g), \quad h \in \mathrm{Sp}(V), \ g \in G,$$

and that U defines a projective unitary representation of $\mathrm{Sp}(V)$ in $\mathscr{H}$, called the *Shale-Weil representation*. Find a realization of the Hilbert space $\mathscr{H}$ such that the 2-cocycle of the Shale-Weil representation is explicitly computed in terms of the Maslov index.

3.3. Weyl quantization. The Weyl transform, introduced in Section 3.1, defines a quantization of classical systems associated with the phase space $\mathbb{R}^{2n}$ with coordinates $\boldsymbol{p} = (p_1, \dots, p_n)$, $\boldsymbol{q} = (q^1, \dots, q^n)$, and the Poisson bracket $\{\, , \,\}$ associated with the canonical symplectic form $\omega = d\boldsymbol{p} \wedge d\boldsymbol{q}$. Let

$$\Phi = W \circ \mathscr{F}^{-1} : \mathscr{S}(\mathbb{R}^{2n}) \to \mathscr{L}(\mathscr{H}),$$

where W is the Weyl transform and $\mathscr{F}^{-1}$ is the inverse Fourier transform, be the linear mapping of the Schwartz space $\mathscr{S}(\mathbb{R}^{2n})$ of complex-valued functions of rapid decay on $\mathbb{R}^{2n}$ into the Banach space of bounded operators $\mathscr{L}(\mathscr{H})$ on $\mathscr{H} = L^2(\mathbb{R}^n, d^n\boldsymbol{q})$. It is given explicitly by the integral

$$\Phi(f) = \frac{1}{(2\pi)^n} \int_{\mathbb{R}^{2n}} \check{f}(\boldsymbol{u}, \boldsymbol{v}) S(\boldsymbol{u}, \boldsymbol{v}) d^n\boldsymbol{u}\, d^n\boldsymbol{v},$$

understood as a limit of Riemann sums in the uniform topology on $\mathscr{L}(\mathscr{H})$, where

$$\check{f}(\boldsymbol{u}, \boldsymbol{v}) = \mathscr{F}^{-1}(f)(\boldsymbol{u}, \boldsymbol{v}) = \frac{1}{(2\pi)^n} \int_{\mathbb{R}^{2n}} f(\boldsymbol{p}, \boldsymbol{q}) e^{i(\boldsymbol{u}\boldsymbol{p} + \boldsymbol{v}\boldsymbol{q})} d^n\boldsymbol{p}\, d^n\boldsymbol{q},$$

and $S(\boldsymbol{u},\boldsymbol{v}) = e^{-\frac{i\hbar}{2}\boldsymbol{u}\boldsymbol{v}}U(\boldsymbol{u})V(\boldsymbol{v})$. It follows from (3.6) that $\Phi(f)$ is an integral operator: for every $\psi \in L^2(\mathbb{R}^n, d^n\boldsymbol{q})$,

$$(\Phi(f)\psi)(\boldsymbol{q}) = \int_{\mathbb{R}^n} K(\boldsymbol{q},\boldsymbol{q}')\psi(\boldsymbol{q}')d^n\boldsymbol{q}',$$

where

$$K(\boldsymbol{q},\boldsymbol{q}') = \frac{1}{(2\pi\hbar)^n} \int_{\mathbb{R}^n} \check{f}\left(\frac{\boldsymbol{q}-\boldsymbol{q}'}{\hbar}, \boldsymbol{v}\right)e^{-\frac{i}{2}\boldsymbol{v}(\boldsymbol{q}+\boldsymbol{q}')}d^n\boldsymbol{v}$$

$$= \frac{1}{(2\pi\hbar)^n} \int_{\mathbb{R}^n} f\left(\boldsymbol{p}, \frac{\boldsymbol{q}+\boldsymbol{q}'}{2}\right)e^{\frac{i}{\hbar}\boldsymbol{p}(\boldsymbol{q}-\boldsymbol{q}')}d^n\boldsymbol{p}.$$

Since the Schwartz space is self-dual with respect to the Fourier transform, $K \in \mathscr{S}(\mathbb{R}^n \times \mathbb{R}^n) \subset L^2(\mathbb{R}^n \times \mathbb{R}^n)$, so that $\Phi(f)$ is a Hilbert-Schmidt operator. Using the property $S(\boldsymbol{u},\boldsymbol{v})^* = S(-\boldsymbol{u},-\boldsymbol{v})$ we get

$$\Phi(f)^* = \Phi(\bar{f}),$$

so that classical observables — real-valued functions on $\mathbb{R}^{2n}$, correspond to quantum observables — self-adjoint operators in $\mathscr{H}$. It follows from the property **WT3** in Section 3.1 that the mapping Φ is injective; its image $\operatorname{Im}\Phi$ and the inverse mapping Φ^{-1} are explicitly described as follows.

Proposition 3.1. *The subspace* $\operatorname{Im}\Phi \subset \mathscr{L}(\mathscr{H})$ *consists of operators* $B \in \mathscr{S}_1$ *such that corresponding functions* $g(\boldsymbol{u},\boldsymbol{v}) = \hbar^n \operatorname{Tr}(B\,S(\boldsymbol{u},\boldsymbol{v})^{-1})$ *belong to the Schwartz class* $\mathscr{S}(\mathbb{R}^{2n})$; *in this case* $W(g) = B$. *The inverse mapping* $\Phi^{-1} = \mathscr{F} \circ W^{-1}$ *is given by the Weyl's inversion formula*

$$\check{f}(\boldsymbol{u},\boldsymbol{v}) = \hbar^n \operatorname{Tr}(\Phi(f)S(\boldsymbol{u},\boldsymbol{v})^{-1}), \quad f \in \mathscr{S}(\mathbb{R}^{2n}).$$

Proof. First consider the case $n = 1$. Let

$$H = \frac{P^2 + Q^2}{2}$$

be the Hamiltonian of the harmonic oscillator with $m = 1$ and $\omega = 1$. According to Theorem 2.1, the operator H has a complete orthonormal system of real-valued eigenfunctions $\psi_n(q)$, given by Hermite-Tchebyscheff functions, with the eigenvalues $\hbar(n+\frac{1}{2})$, so that the inverse operator $H^{-1} \in \mathscr{S}_2$. The operator $H\Phi(f)$ is an integral operator with the kernel $\frac{1}{2}(-\hbar^2\frac{\partial^2}{\partial q^2} + q^2)K(q,q')$, which is a Schwartz class function on $\mathbb{R}^2$, so that $H\Phi(f) \in \mathscr{S}_2$. The operator

$$\Phi(f) = H^{-1}H\Phi(f)$$

is a product of Hilbert-Schmidt operators and, therefore, is of trace class. Using the orthonormal basis $\{\psi_n(q)\}_{n=0}^{\infty}$ for $\mathscr{H} = L^2(\mathbb{R})$, we get

$$(3.8) \qquad K(q,q') = \sum_{m,n=0}^{\infty} c_{mn}\psi_n(q)\psi_m(q'),$$

where

$$c_{mn} = \int_{\mathbb{R}^2} K(q, q')\psi_n(q)\psi_m(q')dqdq',$$

and the series (3.8) converges in $L^2(\mathbb{R}^2)$. Since $K \in \mathscr{S}(\mathbb{R}^2)$, it follows from the N-representation theorem (see Problem 2.13) that this series is also convergent in the $\mathscr{S}(\mathbb{R}^2)$ topology. It is this fact which allows us to set $q' = q$ in (3.8) and to get the expansion

$$K(q, q) = \sum_{m,n=0}^{\infty} c_{mn}\psi_n(q)\psi_m(q),$$

which converges in $\mathscr{S}(\mathbb{R})$. Thus we obtain

$$\operatorname{Tr}\Phi(f) = \sum_{n=0}^{\infty}(\Phi(f)\psi_n, \psi_n) = \sum_{n=0}^{\infty} c_{nn} = \int_{-\infty}^{\infty} K(q, q)dq,$$

where the interchange of orders of summation and integration is legitimate.

The general case $n > 1$ is similar. We consider the operator

$$H = \frac{\boldsymbol{P}^2 + \boldsymbol{Q}^2}{2}$$

and use the fact that operators $H^n\Phi(f)$ and H^{-n} are Hilbert-Schmidt. To prove the formula

$$\operatorname{Tr}\Phi(f) = \int_{\mathbb{R}^n} K(\boldsymbol{q}, \boldsymbol{q})d^n\boldsymbol{q},$$

we expand the kernel $K(\boldsymbol{q}, \boldsymbol{q}')$ using the orthonormal basis $\{\psi_{\boldsymbol{k}}(\boldsymbol{q})\}_{\boldsymbol{k},=0}^{\infty}$ for $L^2(\mathbb{R}^n)$, where

$$\psi_{\boldsymbol{k}}(\boldsymbol{q}) = \psi_{k_1}(q_1)\cdots\psi_{k_n}(q_n), \quad \boldsymbol{k} = (k_1, \ldots, k_n).$$

From the explicit form of the kernel K we get

$$(3.9) \qquad \operatorname{Tr}\Phi(f) = \frac{1}{(2\pi\hbar)^n} \int_{\mathbb{R}^n} \int_{\mathbb{R}^n} \check{f}(0, \boldsymbol{v})e^{-i\boldsymbol{v}\boldsymbol{q}}d^n\boldsymbol{v}\, d^n\boldsymbol{q} = \hbar^{-n}\check{f}(0, 0),$$

which gives the inversion formula for $\boldsymbol{u} = \boldsymbol{v} = 0$. To get the inversion formula for all $\boldsymbol{u}, \boldsymbol{v} \in \mathbb{R}^n$, it is sufficient to apply (3.9) to the function $f_{\boldsymbol{uv}} = \mathscr{F}(\check{f}_{\boldsymbol{u},\boldsymbol{v}})$, where

$$\check{f}_{\boldsymbol{u},\boldsymbol{v}}(\boldsymbol{u}', \boldsymbol{v}') = \check{f}(\boldsymbol{u} + \boldsymbol{u}', \boldsymbol{v} + \boldsymbol{v}')e^{\frac{i\hbar}{2}(\boldsymbol{v}'\boldsymbol{u} - \boldsymbol{u}'\boldsymbol{v})},$$

and to use $S(\boldsymbol{u}, \boldsymbol{v})^{-1} = S(\boldsymbol{u}, \boldsymbol{v})^* = S(-\boldsymbol{u}, -\boldsymbol{v})$ and the property **WT2** in Section 3.1,

$$W(\check{f})S(-\boldsymbol{u}, -\boldsymbol{v}) = W(\check{f}_{\boldsymbol{u},\boldsymbol{v}}).$$

To complete the proof, we need to show that $\operatorname{Im}\Phi$ consists of all $B \in \mathscr{S}_1$ with the property that the function $g(\boldsymbol{u}, \boldsymbol{v}) = \hbar^n \operatorname{Tr}(B\,S(\boldsymbol{u}, \boldsymbol{v})^{-1})$ belongs to the Schwartz class. By the inversion formula,

$$g(\boldsymbol{u}, \boldsymbol{v}) = \hbar^n \operatorname{Tr}(\Phi(f)\,S(\boldsymbol{u}, \boldsymbol{v})^{-1}),$$

where $f = \mathscr{F}(g)$, so it is sufficient to prove that if

$$\mathrm{Tr}(B\, S(\boldsymbol{u}, \boldsymbol{v})^{-1}) = 0$$

for all $\boldsymbol{u}, \boldsymbol{v} \in \mathbb{R}^n$, then $B = 0$. Indeed, multiplying by $\overline{f(\boldsymbol{u}, \boldsymbol{v})}$ and integrating, we get

$$\mathrm{Tr}\, BW(f)^* = 0$$

for all $f \in \mathscr{S}(\mathbb{R}^{2n})$. Since Schwartz class functions are dense in $L^2(\mathbb{R}^{2n})$, by Lemma 3.1 we get $\mathrm{Tr}\, BB^* = 0$ and, therefore, $B = 0$. $\qquad\square$

Corollary 3.3.

$$\mathrm{Tr}\, \Phi(f) = \frac{1}{(2\pi\hbar)^n} \int_{\mathbb{R}^{2n}} f(\boldsymbol{p}, \boldsymbol{q})d^n\boldsymbol{p}\, d^n\boldsymbol{q}.$$

Remark. By the Schwartz kernel theorem, the operator $S(\boldsymbol{u}, \boldsymbol{v})$ is an integral operator with the distributional kernel

$$e^{\frac{i\hbar}{2}\boldsymbol{u}\boldsymbol{v} - i\boldsymbol{v}\boldsymbol{q}} \delta(\boldsymbol{q} - \boldsymbol{q}' - \hbar\boldsymbol{u}),$$

so that

$$\mathrm{Tr}\, S(\boldsymbol{u}, \boldsymbol{v}) = \left(\frac{2\pi}{\hbar}\right)^n \delta(\boldsymbol{u})\delta(\boldsymbol{v}).$$

Thus, as is customary in physics texts,

$$\mathrm{Tr}\, \Phi(f) = \hbar^{-n} \int_{\mathbb{R}^n} \check{f}(\boldsymbol{u}, \boldsymbol{v})\delta(\boldsymbol{u})\delta(\boldsymbol{v})d^n\boldsymbol{u}\, d^n\boldsymbol{v} = \hbar^{-n}\check{f}(0, 0).$$

Let $\mathcal{A}_0 = \mathscr{S}(\mathbb{R}^{2n}, \mathbb{R}) \subset \mathcal{A}$ be the subalgebra of classical observables on $\mathbb{R}^{2n}$ of rapid decay, and let $\mathscr{A}_0 = \mathscr{A} \cap \mathscr{L}(\mathscr{H})$ be the space of bounded quantum observables.

Proposition 3.2. *The mapping $\mathcal{A}_0 \ni f \mapsto \Phi(f) \in \mathscr{A}_0$ is a quantization, i.e., it satisfies*

$$\lim_{\hbar\to 0} \tfrac{1}{2}\Phi^{-1}\left(\Phi(f_1)\Phi(f_2) + \Phi(f_2)\Phi(f_1)\right) = f_1 f_2$$

and the correspondence principle

$$\lim_{\hbar\to 0} \Phi^{-1}\left(\{\Phi(f_1), \Phi(f_2)\}_\hbar\right) = \{f_1, f_2\}, \quad f_1, f_2 \in \mathcal{A}_0,$$

where

$$\{\Phi(f_1), \Phi(f_2)\}_\hbar = \frac{i}{\hbar}[\Phi(f_1), \Phi(f_2)] \quad and \quad \{f_1, f_2\} - \frac{\partial f_1}{\partial \boldsymbol{p}}\frac{\partial f_2}{\partial \boldsymbol{q}} - \frac{\partial f_1}{\partial \boldsymbol{q}}\frac{\partial f_2}{\partial \boldsymbol{p}}$$

are, respectively, the quantum bracket and the Poisson bracket.

Proof. In terms of the product $*_\hbar$ introduced in Section 3.1 we have

$$\Phi^{-1}(\Phi(f_1)\Phi(f_2)) = \mathscr{F}(\check{f}_1 *_\hbar \check{f}_2),$$

and it follows from the property **WT4** that

$$(3.10) \quad \Phi^{-1}(\Phi(f_1)\Phi(f_2))(\boldsymbol{p},\boldsymbol{q}) = \frac{1}{(2\pi)^{2n}} \int_{\mathbb{R}^{2n}} \int_{\mathbb{R}^{2n}} \check{f}_1(\boldsymbol{u}_1,\boldsymbol{v}_1)\check{f}_2(\boldsymbol{u}_2,\boldsymbol{v}_2) \cdot$$
$$\cdot\, e^{\frac{i\hbar}{2}(\boldsymbol{u}_1\boldsymbol{v}_2 - \boldsymbol{u}_2\boldsymbol{v}_1) - i(\boldsymbol{u}_1+\boldsymbol{u}_2)\boldsymbol{p} - i(\boldsymbol{v}_1+\boldsymbol{v}_2)\boldsymbol{q}} d^n\boldsymbol{u}_1 d^n\boldsymbol{u}_2 d^n\boldsymbol{v}_1 d^n\boldsymbol{v}_2.$$

Using the expansion

$$e^{\frac{i\hbar}{2}(\boldsymbol{u}_1\boldsymbol{v}_2 - \boldsymbol{u}_2\boldsymbol{v}_1)} = 1 + \frac{i\hbar}{2}(\boldsymbol{u}_1\boldsymbol{v}_2 - \boldsymbol{u}_2\boldsymbol{v}_1) + O(\hbar^2(\boldsymbol{u}_1\boldsymbol{v}_2 - \boldsymbol{u}_2\boldsymbol{v}_1)^2)$$

as $\hbar \to 0$, and the basic properties of the Fourier transform,

$$\mathscr{F}(\boldsymbol{u}\check{f}(\boldsymbol{u},\boldsymbol{v})) = i\frac{\partial f}{\partial \boldsymbol{p}}(\boldsymbol{p},\boldsymbol{q}) \quad \text{and} \quad \mathscr{F}(\boldsymbol{v}\check{f}(\boldsymbol{u},\boldsymbol{v})) = i\frac{\partial f}{\partial \boldsymbol{q}}(\boldsymbol{p},\boldsymbol{q}),$$

we get from (3.10) that as $\hbar \to 0$,

$$\Phi^{-1}(\Phi(f_1)\Phi(f_2))(\boldsymbol{p},\boldsymbol{q}) = (f_1 f_2)(\boldsymbol{p},\boldsymbol{q}) - \frac{i\hbar}{2}\{f_1, f_2\}(\boldsymbol{p},\boldsymbol{q}) + O(\hbar^2).$$

Using skew-symmetry of the Poisson bracket completes the proof. $\square$

The quantization associated with the mapping $\Phi = W \circ \mathscr{F}^{-1}$ is called the *Weyl quantization*. The correspondence $f \mapsto \Phi(f)$ can be easily extended to the vector space $\widehat{L^1(\mathbb{R}^{2n})}$ — the image of $L^1(\mathbb{R}^{2n})$ under the Fourier transform, which is a subspace of $C(\mathbb{R}^{2n})$. More generally, for $f \in \mathscr{S}(\mathbb{R}^{2n})'$ — the space of tempered distributions on $\mathbb{R}^{2n}$ — the corresponding kernel

$$(3.11) \qquad K(\boldsymbol{q},\boldsymbol{q}') = \frac{1}{(2\pi\hbar)^n} \int_{\mathbb{R}^n} f(\boldsymbol{p}, \tfrac{\boldsymbol{q}+\boldsymbol{q}'}{2}) e^{\frac{i}{\hbar}\boldsymbol{p}(\boldsymbol{q}-\boldsymbol{q}')} d^n\boldsymbol{p},$$

considered as a tempered distribution on $\mathbb{R}^n \times \mathbb{R}^n$, is a Schwartz kernel of the linear operator

$$\Phi(f) : \mathscr{S}(\mathbb{R}^n) \to \mathscr{S}(\mathbb{R}^n)'.$$

In particular, the constant function $f = 1$ corresponds to the identity operator I with $K(\boldsymbol{q},\boldsymbol{q}') = \delta(\boldsymbol{q} - \boldsymbol{q}')$. In terms of the kernel $K(\boldsymbol{q},\boldsymbol{q}')$ the Weyl inversion formula takes the form

$$(3.12) \qquad f(\boldsymbol{p},\boldsymbol{q}) = \int_{\mathbb{R}^n} K(\boldsymbol{q} - \tfrac{1}{2}\boldsymbol{v}, \boldsymbol{q} + \tfrac{1}{2}\boldsymbol{v}) e^{\frac{i}{\hbar}\boldsymbol{p}\boldsymbol{v}} d^n\boldsymbol{v}.$$

The distribution $f(\boldsymbol{p},\boldsymbol{q})$ defined by (3.12) is called the *Weyl symbol* of the operator in $L^2(\mathbb{R}^n, d^n\boldsymbol{q})$ with the Schwartz kernel $K(\boldsymbol{q},\boldsymbol{q}')$.

Examples below describe classes of distributions f such that the operators $\Phi(f)$ are essentially self-adjoint unbounded operators on the domain $\mathscr{S}(\mathbb{R}^n) \subset L^2(\mathbb{R}^n, d^n\boldsymbol{q})$.

Example 3.1. Let $f = f(q) \in L^p(\mathbb{R}^n)$ for some $1 \le p \le \infty$, or let f be a polynomially bounded function as $|q| \to \infty$. In the distributional sense,

$$\check{f}(u, v) = (2\pi)^{n/2}\delta(u)\check{f}(v)$$

so that

$$K(q, q') = \frac{1}{(\hbar\sqrt{2\pi})^n} \int_{\mathbb{R}^n} \delta(\tfrac{q-q'}{\hbar})\check{f}(v)e^{-\frac{i}{2}v(q+q')}d^n v$$

$$= \delta(q - q')f(\tfrac{q+q'}{2}) = f(q)\delta(q - q').$$

Thus the operator $\Phi(f)$ is a multiplication by $f(q)$ operator on $L^2(\mathbb{R}^n)$. In particular, coordinates q in classical mechanics correspond to coordinate operators Q in quantum mechanics. Similarly, if $f = f(p)$, then $\Phi(f) = f(P)$. In particular, momenta p in classical mechanics correspond to the momenta operators P in quantum mechanics.

Example 3.2. Let

$$H_{\mathrm{c}} = \frac{p^2}{2m} + V(q)$$

be the Hamiltonian function in classical mechanics. Then $H = \Phi(H_{\mathrm{c}})$ is the corresponding Hamiltonian operator in quantum mechanics,

$$H = \frac{P^2}{2m} + V(Q).$$

In Chapter 3 we give necessary conditions for the operator H to be essentially self-adjoint.

Example 3.3. Here we find $f \in \mathscr{S}(\mathbb{R}^{2n})'$ such that

$$\Phi(f) = P_\psi$$

— a pure state P_ψ, where $\psi \in L^2(\mathbb{R}^n)$, $\|\psi\| = 1$. The projection P_ψ is an integral operator with the kernel $\psi(q)\overline{\psi(q')}$, and we get from (3.11),

$$\frac{1}{(2\pi\hbar)^n} \int_{\mathbb{R}^n} f(p, \tfrac{q+q'}{2})e^{\frac{i}{\hbar}p(q-q')}d^n p = \psi(q)\overline{\psi(q')}.$$

Introducing $q_+ = \frac{1}{2}(q + q'), q_- = \frac{1}{2}(q - q')$, we obtain

$$\check{f}(\tfrac{2q_-}{\hbar}, v) = \hbar^n \int_{\mathbb{R}^n} \psi(q_- + q_+)\overline{\psi(q_+ - q_-)}e^{ivq_+}d^n q_+,$$

or

$$\check{f}(u, v) = \hbar^n \int_{\mathbb{R}^n} \psi(q + \tfrac{1}{2}\hbar u)\overline{\psi(q - \tfrac{1}{2}\hbar u)}e^{ivq}d^n q.$$

Assuming that ψ does not depend on $\hbar$, we get in accordance with Corollary 3.3,

$$\check{\rho}(u, v) = \lim_{\hbar \to 0} \frac{1}{(2\pi\hbar)^n}\check{f}(u, v) = \frac{1}{(2\pi)^n} \int_{\mathbb{R}^n} |\psi(q)|^2 e^{ivq}d^n v.$$

Thus in the classical limit $\hbar \to 0$ the pure state P_ψ in quantum mechanics becomes a mixed state in classical mechanics, given by a probability measure $d\mu = \rho(\boldsymbol{p}, \boldsymbol{q}) d^n\boldsymbol{p}\, d^n\boldsymbol{q}$ on $\mathbb{R}^{2n}$ with the density

$$\rho(\boldsymbol{p}, \boldsymbol{q}) = \delta(\boldsymbol{p})|\psi(\boldsymbol{q})|^2.$$

It describes a classical particle at rest ($\boldsymbol{p} = 0$) with the distribution of coordinates given by the probability measure $|\psi(\boldsymbol{q})|^2 d^n\boldsymbol{q}$ on $\mathbb{R}^n$. When $\psi(\boldsymbol{q}) = e^{\frac{i}{\hbar}\boldsymbol{p}_0 \boldsymbol{q}}\varphi(\boldsymbol{q})$, where $\varphi(\boldsymbol{q})$ does not depend on $\hbar$, the corresponding density is $\rho(\boldsymbol{p}, \boldsymbol{q}) = \delta(\boldsymbol{p} - \boldsymbol{p}_0)|\varphi(\boldsymbol{q})|^2$.

Remark. The Weyl quantization can be considered as a way of defining a function $f(\boldsymbol{P}, \boldsymbol{Q})$ of non-commuting operators $\boldsymbol{P} = (P_1, \ldots, P_n)$ and $\boldsymbol{Q} = (Q^1, \ldots, Q^n)$ by setting

$$f(\boldsymbol{P}, \boldsymbol{Q}) = \Phi(f).$$

In particular, if $f(\boldsymbol{p}, \boldsymbol{q}) = g(\boldsymbol{p}) + h(\boldsymbol{q})$, then

$$f(\boldsymbol{P}, \boldsymbol{Q}) = g(\boldsymbol{P}) + h(\boldsymbol{Q}).$$

For $f(\boldsymbol{p}, \boldsymbol{q}) = \boldsymbol{p}\boldsymbol{q} = p_1 q^1 + \cdots + p_n q^n$ we get, using (3.11),

$$f(\boldsymbol{P}, \boldsymbol{Q}) = \frac{\boldsymbol{P}\boldsymbol{Q} + \boldsymbol{Q}\boldsymbol{P}}{2}.$$

This shows that the Weyl quantization symmetrizes products of the non-commuting factors $\boldsymbol{P}$ and $\boldsymbol{Q}$. In general, let f be a polynomial function,

$$(3.13) \qquad f(\boldsymbol{p}, \boldsymbol{q}) = \sum_{|\alpha|, |\beta| \leq N} c_{\alpha\beta}\, \boldsymbol{p}^\alpha \boldsymbol{q}^\beta,$$

where for the multi-indices $\alpha = (\alpha_1, \ldots, \alpha_n)$ and $\beta = (\beta_1, \ldots, \beta_n)$,

$$\boldsymbol{p}^\alpha = p_1^{\alpha_1} \ldots p_n^{\alpha_n}, \quad \boldsymbol{q}^\beta = (q^1)^{\beta_1} \ldots (q^n)^{\beta_n},$$

and $|\alpha| = \alpha_1 + \cdots + \alpha_n$, $\quad |\beta| = \beta_1 + \cdots + \beta_n$. Using (3.11), we get the following formula

$$(3.14) \qquad \Phi(f) = \sum_{|\alpha|, |\beta| \leq N} c_{\alpha\beta} \operatorname{Sym}(\boldsymbol{P}^\alpha \boldsymbol{Q}^\beta).$$

Here $\operatorname{Sym}(\boldsymbol{P}^\alpha \boldsymbol{Q}^\beta)$ is a symmetric product, defined by

$$(3.15) \qquad (\boldsymbol{u}\boldsymbol{P} + \boldsymbol{v}\boldsymbol{Q})^k = \sum_{|\alpha| + |\beta| = k} \frac{k!}{\alpha!\beta!} \boldsymbol{u}^\alpha \boldsymbol{v}^\beta \operatorname{Sym}(\boldsymbol{P}^\alpha \boldsymbol{Q}^\beta),$$

where $\boldsymbol{u}\boldsymbol{P} + \boldsymbol{v}\boldsymbol{Q} = u^1 P_1 + \cdots + u^n P_n + v_1 Q^1 + \cdots + v_n Q^n$ and

$$\alpha! = \alpha_1! \ldots \alpha_n!, \quad \beta! = \beta_1! \ldots \beta_n!.$$

Remark. In addition to the Weyl quantization Φ, consider also the mappings $\Phi_1 : \mathscr{S}(\mathbb{R}^{2n}) \to \mathscr{L}(\mathscr{H})$ and $\Phi_2 : \mathscr{S}(\mathbb{R}^{2n}) \to \mathscr{L}(\mathscr{H})$, defined by

$$\Phi_1(f) = \frac{1}{(2\pi)^n} \int_{\mathbb{R}^{2n}} \check{f}(\boldsymbol{u}, \boldsymbol{v}) e^{\frac{i\hbar}{2}\boldsymbol{uv}} S(\boldsymbol{u}, \boldsymbol{v}) d^n\boldsymbol{u}\, d^n\boldsymbol{v}$$

and

$$\Phi_2(f) = \frac{1}{(2\pi)^n} \int_{\mathbb{R}^{2n}} \check{f}(\boldsymbol{u}, \boldsymbol{v}) e^{-\frac{i\hbar}{2}\boldsymbol{uv}} S(\boldsymbol{u}, \boldsymbol{v}) d^n\boldsymbol{u}\, d^n\boldsymbol{v},$$

where $\check{f} = \mathscr{F}^{-1}(f)$ is the inverse Fourier transform. Though Φ_1 and Φ_2 no longer map $\mathcal{A}_0$ into the real vector space $\mathscr{A}_0$ of bounded quantum observables, they satisfy all the properties in Proposition 3.2. It follows from (3.5) that for $f \in \mathscr{S}(\mathbb{R}^{2n})$ the operators $\Phi_1(f)$ and $\Phi_2(f)$ are integral operators with the integral kernels

$$K_1(\boldsymbol{q}, \boldsymbol{q}') = \frac{1}{(2\pi\hbar)^n} \int_{\mathbb{R}^n} f(\boldsymbol{p}, \boldsymbol{q}') e^{\frac{i}{\hbar}\boldsymbol{p}(\boldsymbol{q}-\boldsymbol{q}')} d^n\boldsymbol{p}$$

and

$$K_2(\boldsymbol{q}, \boldsymbol{q}') = \frac{1}{(2\pi\hbar)^n} \int_{\mathbb{R}^n} f(\boldsymbol{p}, \boldsymbol{q}) e^{\frac{i}{\hbar}\boldsymbol{p}(\boldsymbol{q}-\boldsymbol{q}')} d^n\boldsymbol{p},$$

respectively. As in the case of the Weyl quantization, these formulas extend the mappings $f \mapsto \Phi_1(f)$ and $f \mapsto \Phi_2(f)$ of the space $\mathscr{S}(\mathbb{R}^{2n})'$ of tempered distributions on $\mathbb{R}^{2n}$. In particular, if $f(\boldsymbol{p}, \boldsymbol{q})$ is a polynomial function (3.13), then

$$(3.16) \qquad \Phi_1(f) = \sum_{|\alpha|,|\beta|\leq N} c_{\alpha\beta}\, \boldsymbol{P}^\alpha \boldsymbol{Q}^\beta$$

and

$$(3.17) \qquad \Phi_2(f) = \sum_{|\alpha|,|\beta|\leq N} c_{\alpha\beta}\, \boldsymbol{Q}^\beta \boldsymbol{P}^\alpha.$$

Therefore the mapping $f \mapsto \Phi_1(f)$ is called the **pq-*quantization*, and the mapping $f \mapsto \Phi_2(f)$ — the **qp**-*quantization*. Corresponding inversion formulas are

$$(3.18) \qquad f(\boldsymbol{p}, \boldsymbol{q}) = \int_{\mathbb{R}^n} K_1(\boldsymbol{q}-\boldsymbol{v}, \boldsymbol{q}) e^{\frac{i}{\hbar}\boldsymbol{pv}} d^n\boldsymbol{v}$$

and

$$(3.19) \qquad f(\boldsymbol{p}, \boldsymbol{q}) = \int_{\mathbb{R}^n} K_2(\boldsymbol{q}, \boldsymbol{q}+\boldsymbol{v}) e^{\frac{i}{\hbar}\boldsymbol{pv}} d^n\boldsymbol{v}.$$

The distribution $f(\boldsymbol{p}, \boldsymbol{q})$ defined by (3.18) is called the **pq**-*symbol* of an operator with the Schwartz kernel $K_1(\boldsymbol{q}, \boldsymbol{q}')$, and the distribution $f(\boldsymbol{p}, \boldsymbol{q})$ defined by (3.19) — the **qp**-*symbol* of an operator with the Schwartz kernel $K_2(\boldsymbol{q}, \boldsymbol{q}')$. The **$qp$**-symbols are commonly used in the theory of pseudodifferential operators. It follows from (3.18) and (3.19) that if $f(\boldsymbol{p}, \boldsymbol{q})$ is a **pq**-symbol of the operator $\Phi_1(f)$, then $\overline{f(\boldsymbol{p}, \boldsymbol{q})}$ is a **qp**-symbol of the adjoint operator $\Phi_1(f)^*$.

Problem 3.6. Prove formula (3.14).

Problem 3.7. Prove formulas (3.16)-(3.17).

Problem 3.8. Prove that Weyl, $\boldsymbol{pq}$, and $\boldsymbol{qp}$-quantizations are equivalent. (*Hint:* Find the relations between Weyl, $\boldsymbol{pq}$, and $\boldsymbol{qp}$-symbols of a given operator.)

3.4. The $\star$-product. The Weyl quantization $\Phi : \mathscr{S}(\mathbb{R}^{2n}) \to \mathscr{L}(\mathscr{H})$, studied in the previous section, defines a new bilinear operation

$$\star_\hbar : \mathscr{S}(\mathbb{R}^{2n}) \times \mathscr{S}(\mathbb{R}^{2n}) \to \mathscr{S}(\mathbb{R}^{2n})$$

on $\mathscr{S}(\mathbb{R}^{2n})$ by the formula

$$f_1 \star_\hbar f_2 = \Phi^{-1}(\Phi(f_1)\Phi(f_2)).$$

This operation is called the $\star$-product[26]. According to (3.10),

$$(3.20) \qquad (f_1 \star_\hbar f_2)(\boldsymbol{p}, \boldsymbol{q}) = \frac{1}{(2\pi)^{2n}} \int_{\mathbb{R}^{2n}} \int_{\mathbb{R}^{2n}} \check{f}_1(\boldsymbol{u}_1, \boldsymbol{v}_1)\check{f}_2(\boldsymbol{u}_2, \boldsymbol{v}_2) \cdot$$
$$\cdot e^{\frac{i\hbar}{2}(\boldsymbol{u}_1\boldsymbol{v}_2 - \boldsymbol{u}_2\boldsymbol{v}_1) - i(\boldsymbol{u}_1 + \boldsymbol{u}_2)\boldsymbol{p} - i(\boldsymbol{v}_1 + \boldsymbol{v}_2)\boldsymbol{q}} d^n\boldsymbol{u}_1 d^n\boldsymbol{u}_2 d^n\boldsymbol{v}_1 d^n\boldsymbol{v}_2.$$

The $\star$-product on $\mathscr{S}(\mathbb{R}^{2n})$ has the following properties.

  1. Associativity:

$$f_1 \star_\hbar (f_2 \star_\hbar f_3) = (f_1 \star_\hbar f_2) \star_\hbar f_3.$$

  2. Semi-classical limit:

$$(f_1 \star_\hbar f_2)(\boldsymbol{p}, \boldsymbol{q}) = (f_1 f_2)(\boldsymbol{p}, \boldsymbol{q}) - \tfrac{i\hbar}{2}\{f_1, f_2\}(\boldsymbol{p}, \boldsymbol{q}) + O(\hbar^2) \ \text{ as } \ \hbar \to 0.$$

  3. Property of the unit:

$$f \star_\hbar \mathbf{1} = \mathbf{1} \star_\hbar f,$$

where $\mathbf{1}$ is a function which identically equals 1 on $\mathbb{R}^{2n}$.

  4. The cyclic trace property:

$$\tau(f_1 \star_\hbar f_2) = \tau(f_2 \star_\hbar f_1),$$

where the $\mathbb{C}$-linear map $\tau : \mathscr{S}(\mathbb{R}^{2n}) \to \mathbb{C}$ is defined by

$$\tau(f) = \frac{1}{(2\pi\hbar)^n} \int_{\mathbb{R}^{2n}} f(\boldsymbol{p}, \boldsymbol{q}) d^n\boldsymbol{p}\, d^n\boldsymbol{q}.$$

Property **1** follows from the corresponding property for the product $*_\hbar$ (see Section 3.1), property **2** follows from Proposition 3.2, and properties **3** and **4** directly follow from the definition (3.20). The complex vector space $\mathscr{S}(\mathbb{R}^{2n}) \oplus \mathbb{C}\,\mathbf{1}$ with the bilinear operation $\star_\hbar$ is an associative algebra over $\mathbb{C}$ with unit $\mathbf{1}$ and the cyclic trace τ, satisfying the correspondence principle,

$$\lim_{\hbar \to 0} \tfrac{i}{\hbar}(f_1 \star_\hbar f_2 - f_2 \star_\hbar f_1) = \{f_1, f_2\}.$$

---

[26]Also called *Moyal product* in physics.

Consider the tensor product of Hilbert spaces

$$L^2(\mathbb{R}^{2n}) \otimes L^2(\mathbb{R}^{2n}) \simeq L^2(\mathbb{R}^{2n} \times \mathbb{R}^{2n}),$$

and define the unitary operator $\boldsymbol{U}_1$ on $L^2(\mathbb{R}^{2n}) \otimes L^2(\mathbb{R}^{2n})$ by

$$\boldsymbol{U}_1 = e^{-\frac{i\hbar}{2}\left(\frac{\partial}{\partial \boldsymbol{p}} \otimes \frac{\partial}{\partial \boldsymbol{q}}\right)},$$

where

$$\frac{\partial}{\partial \boldsymbol{p}} \otimes \frac{\partial}{\partial \boldsymbol{q}} = \sum_{k=1}^{n} \frac{\partial}{\partial p_k} \otimes \frac{\partial}{\partial q^k}.$$

It follows from the theory of Fourier transform that for $f_1, f_2 \in \mathscr{S}(\mathbb{R}^{2n})$,

$$\left(\boldsymbol{U}_1(f_1 \otimes f_2)\right)(\boldsymbol{p}_1, \boldsymbol{q}_1, \boldsymbol{p}_2, \boldsymbol{q}_2) = \frac{1}{(2\pi)^{2n}} \int_{\mathbb{R}^{2n}} \int_{\mathbb{R}^{2n}} \check{f}_1(\boldsymbol{u}_1, \boldsymbol{v}_1) \check{f}_2(\boldsymbol{u}_2, \boldsymbol{v}_2) \cdot$$

$$\cdot e^{\frac{i\hbar}{2}\boldsymbol{u}_1\boldsymbol{v}_2 - i\boldsymbol{u}_1\boldsymbol{p}_1 - i\boldsymbol{u}_2\boldsymbol{p}_2 - i\boldsymbol{v}_1\boldsymbol{q}_1 - i\boldsymbol{v}_2\boldsymbol{q}_2} d^n\boldsymbol{u}_1 d^n\boldsymbol{u}_2 d^n\boldsymbol{v}_1 d^n\boldsymbol{v}_2.$$

Similarly, defining the unitary operator $\boldsymbol{U}_2$ on $L^2(\mathbb{R}^{2n}) \otimes L^2(\mathbb{R}^{2n})$ by

$$\boldsymbol{U}_2 = e^{-\frac{i\hbar}{2}\left(\frac{\partial}{\partial \boldsymbol{q}} \otimes \frac{\partial}{\partial \boldsymbol{p}}\right)},$$

we get

$$\left(\boldsymbol{U}_2(f_1 \otimes f_2)\right)(\boldsymbol{p}_1, \boldsymbol{q}_1, \boldsymbol{p}_2, \boldsymbol{q}_2) = \frac{1}{(2\pi)^{2n}} \int_{\mathbb{R}^{2n}} \int_{\mathbb{R}^{2n}} \check{f}_1(\boldsymbol{u}_1, \boldsymbol{v}_1) \check{f}_2(\boldsymbol{u}_2, \boldsymbol{v}_2) \cdot$$

$$\cdot e^{\frac{i\hbar}{2}\boldsymbol{u}_2\boldsymbol{v}_1 - i\boldsymbol{u}_1\boldsymbol{p}_1 - i\boldsymbol{u}_2\boldsymbol{p}_2 - i\boldsymbol{v}_1\boldsymbol{q}_1 - \boldsymbol{v}_2\boldsymbol{q}_2} d^n\boldsymbol{u}_1 d^n\boldsymbol{u}_2 d^n\boldsymbol{v}_1 d^n\boldsymbol{v}_2.$$

Finally, define the unitary operator $\boldsymbol{U}_\hbar$ on $L^2(\mathbb{R}^{2n}) \otimes L^2(\mathbb{R}^{2n})$ by

$$\boldsymbol{U}_\hbar = \boldsymbol{U}_1 \boldsymbol{U}_2^{-1} = e^{-\frac{i\hbar}{2}\left(\frac{\partial}{\partial \boldsymbol{p}} \otimes \frac{\partial}{\partial \boldsymbol{q}} - \frac{\partial}{\partial \boldsymbol{q}} \otimes \frac{\partial}{\partial \boldsymbol{p}}\right)},$$

and denote by $m : \mathscr{S}(\mathbb{R}^{2n}) \otimes \mathscr{S}(\mathbb{R}^{2n}) \to \mathscr{S}(\mathbb{R}^{2n})$ the point-wise product of functions, $(m(f_1 \otimes f_2))(\boldsymbol{p}, \boldsymbol{q}) = f_1(\boldsymbol{p}, \boldsymbol{q}) f_2(\boldsymbol{p}, \boldsymbol{q})$. Then the $\star$-product can be written in the following concise form:

$$(3.21) \qquad f_1 \star_\hbar f_2 = (m \circ \boldsymbol{U}_\hbar)(f_1 \otimes f_2).$$

In analogy with the Poisson bracket $\{\,,\,\}$ on $\mathbb{R}^{2n}$ associated with the canonical symplectic form $\omega = d\boldsymbol{p} \wedge d\boldsymbol{q}$,

$$\{f_1, f_2\} = \frac{\partial f_1}{\partial \boldsymbol{p}} \frac{\partial f_2}{\partial \boldsymbol{q}} - \frac{\partial f_1}{\partial \boldsymbol{q}} \frac{\partial f_2}{\partial \boldsymbol{p}},$$

we introduce the notation

$$\left\{\overset{\otimes}{,}\right\} = \frac{\partial}{\partial \boldsymbol{p}} \otimes \frac{\partial}{\partial \boldsymbol{q}} - \frac{\partial}{\partial \boldsymbol{q}} \otimes \frac{\partial}{\partial \boldsymbol{p}} = \frac{\partial^2}{\partial \boldsymbol{p}_1 \partial \boldsymbol{q}_2} - \frac{\partial^2}{\partial \boldsymbol{q}_1 \partial \boldsymbol{p}_2}.$$

Then it follows from the theory of Fourier transform that formula (3.21) for the $\star$-product can be rewritten as

$$(3.22) \qquad (f_1 \star_\hbar f_2)(\boldsymbol{p}, \boldsymbol{q}) = \left(e^{-\frac{i\hbar}{2}\left\{\overset{\otimes}{,}\right\}} f_1(\boldsymbol{p}_1, \boldsymbol{q}_1) f_2(\boldsymbol{p}_2, \boldsymbol{q}_2)\right)\Big|_{\substack{\boldsymbol{p}_1 = \boldsymbol{p}_2 = \boldsymbol{p} \\ \boldsymbol{q}_1 = \boldsymbol{q}_2 = \boldsymbol{q}}} .$$

Thus we have shown that on $\mathscr{S}(\mathbb{R}^{2n})$ the $\star$-product can be equivalently defined by (3.20), or by (3.21) and (3.22). The latter representation has the advantage that the formal expansion of the exponential $e^{-\frac{i\hbar}{2}\{\overset{\otimes}{,}\}}$ into the power series gives the asymptotic expansion of the $\star$-product as $\hbar \to 0$. Namely, define the bidifferential operators

$$B_k : \mathscr{S}(\mathbb{R}^{2n}) \otimes \mathscr{S}(\mathbb{R}^{2n}) \to \mathscr{S}(\mathbb{R}^{2n})$$

by $B_k = m \circ \{\overset{\otimes}{,}\}^k$ for $k \geq 1$ and $B_0 = m$.

Lemma 3.2. *For every $f_1, f_2 \in \mathscr{S}(\mathbb{R}^{2n})$ and $l \in \mathbb{N}$ there is $C > 0$ such that for all $p, q \in \mathbb{R}^{2n}$*

$$\left| (f_1 \star_\hbar f_2)(p, q) - \sum_{k=0}^{l} \frac{(-i\hbar)^k}{2^k k!} B_k(f_1, f_2)(p, q) \right| \leq C\hbar^{l+1} \quad as \quad \hbar \to 0.$$

Succinctly,

$$(3.23) \qquad (f_1 \star_\hbar f_2)(p, q) = \sum_{k=0}^{\infty} \frac{(-i\hbar)^k}{2^k k!} B_k(f_1, f_2)(p, q) + O(\hbar^\infty).$$

Proof. Expanding the exponential function $e^{\frac{i\hbar}{2}(u_1 v_2 - u_2 v_1)}$ into the power series and repeating the proof of Proposition 3.2 gives the result. $\qquad \square$

Finally, we get another integral representation for the $\star$-product. Applying the Fourier inversion formula to the integral over $d^n u_1 d^n v_1$ in (3.20), we get

$$(f_1 \star_\hbar f_2)(p, q) = \frac{1}{(2\pi)^n} \int_{\mathbb{R}^{2n}} f_1(p - \tfrac{\hbar}{2}v_2, q + \tfrac{\hbar}{2}u_2) \check{f}_2(u_2, v_2) \cdot$$

$$\cdot e^{-iu_2 p - iv_2 q} d^n u_2 d^n v_2$$

$$= \frac{1}{(2\pi)^{2n}} \int_{\mathbb{R}^{2n}} \int_{\mathbb{R}^{2n}} f_1(p - \tfrac{\hbar}{2}v_2, q + \tfrac{\hbar}{2}u_2) f_2(p_2, q_2) \cdot$$

$$\cdot e^{-iu_2 p - iv_2 q + iu_2 p_2 + iv_2 q_2} d^n p_2 d^n q_2 d^n u_2 d^n v_2,$$

and changing variables $p_1 = p - \tfrac{\hbar}{2}v_2$, $q_1 = q + \tfrac{\hbar}{2}u_2$, we obtain

$$(f_1 \star_\hbar f_2)(p, q) = \frac{1}{(\pi\hbar)^{2n}} \int_{\mathbb{R}^{2n}} \int_{\mathbb{R}^{2n}} f_1(p_1, q_1) f_2(p_2, q_2) \cdot$$

$$\cdot e^{\frac{2i}{\hbar}(p_1 q - pq_1 + q_1 p_2 - q_2 p_1 + pq_2 - p_2 q)} d^n p_1 d^n q_1 d^n p_2 d^n q_2.$$

Let $\triangle$ be a Euclidean triangle (a 2-simplex) in the phase space $\mathbb{R}^{2n}$ with the vertices (p, q), (p_1, q_1), and (p_2, q_2). It is easy to see that

$$p_1 q - pq_1 + q_1 p_2 - q_2 p_1 + pq_2 - p_2 q = 2 \int_{\triangle} \omega,$$

which is twice the symplectic area of $\triangle$ — the sum of oriented areas of the projections of $\triangle$ onto two-dimensional planes $(p_1, q^1), \ldots, (p_n, q^n)$. Thus we have the final formula

$$(3.24) \qquad (f_1 \star_\hbar f_2)(\boldsymbol{p}, \boldsymbol{q}) = \frac{1}{(\pi\hbar)^{2n}} \int_{\mathbb{R}^{2n}} \int_{\mathbb{R}^{2n}} f_1(\boldsymbol{p}_1, \boldsymbol{q}_1) f_2(\boldsymbol{p}_2, \boldsymbol{q}_2) \cdot$$
$$\cdot\, e^{\frac{4i}{\hbar} \int_\triangle \omega}\, d^n \boldsymbol{p}_1 d^n \boldsymbol{q}_1 d^n \boldsymbol{p}_2 d^n \boldsymbol{q}_2,$$

which is a composition formula for the Weyl symbols.

Remark. It is instructive to compare formulas (3.23) and (3.24). The latter formula represents the $\star$-product on $\mathscr{S}(\mathbb{R}^{2n})$ as an absolutely convergent integral, and is equivalent to the Weyl quantization. The former formula is an asymptotic expansion of the $\star$-product as $\hbar \to 0$, and does not capture all properties of the Weyl quantization. In general, the power series in (3.23) diverges; for polynomial functions this series becomes a finite sum and gives a formula for the $\star$-product of polynomials.

Problem 3.9 (Composition formula for $\boldsymbol{pq}$-symbols). Let $f_1(\boldsymbol{p}, \boldsymbol{q})$ and $f_2(\boldsymbol{p}, \boldsymbol{q})$ be, respectively, the $\boldsymbol{pq}$-symbols of the operators $\Phi_1(f_1)$ and $\Phi_1(f_2)$. Show that the $\boldsymbol{pq}$-symbol of the operator $\Phi_1(f_1)\Phi_1(f_2)$ is given by

$$f(\boldsymbol{p}, \boldsymbol{q}) = \frac{1}{(2\pi\hbar)^n} \int_{\mathbb{R}^{2n}} f_1(\boldsymbol{p}, \boldsymbol{q}_1) f_2(\boldsymbol{p}_1, \boldsymbol{q}) e^{\frac{i}{\hbar}(\boldsymbol{p}-\boldsymbol{p}_1)(\boldsymbol{q}-\boldsymbol{q}_1)} d^n \boldsymbol{p}_1 d^n \boldsymbol{q}_1.$$

(*Hint:* Use the formula for $\left(m \circ U_1^2\right)(f_1 \otimes f_2)$.)

Problem 3.10 (Composition formula for $\boldsymbol{qp}$-symbols). Let $f_1(\boldsymbol{p}, \boldsymbol{q})$ and $f_2(\boldsymbol{p}, \boldsymbol{q})$ be, respectively, the $\boldsymbol{qp}$-symbols of the operators $\Phi_2(f_1)$ and $\Phi_2(f_2)$. Show that the $\boldsymbol{qp}$-symbol of the operator $\Phi_2(f_1)\Phi_2(f_2)$ is given by

$$f(\boldsymbol{p}, \boldsymbol{q}) = \frac{1}{(2\pi\hbar)^n} \int_{\mathbb{R}^{2n}} f_1(\boldsymbol{p}_1, \boldsymbol{q}) f_2(\boldsymbol{p}, \boldsymbol{q}_1) e^{-\frac{i}{\hbar}(\boldsymbol{p}-\boldsymbol{p}_1)(\boldsymbol{q}-\boldsymbol{q}_1)} d^n \boldsymbol{p}_1 d^n \boldsymbol{q}_1.$$

(*Hint:* Use the formula for $\left(m \circ U_2^{-2}\right)(f_1 \otimes f_2)$.)

Problem 3.11. Using (3.24) prove that the $\star$-product is associative.

Problem 3.12. For classical observable $f(\boldsymbol{p}, \boldsymbol{q})$ define the $\star$-exponential (the analog of the evolution operator) by

$$\exp_\star f = \sum_{n=0}^{\infty} \frac{\hbar^{-n}}{n!} \underbrace{f \star_\hbar f \star_\hbar \cdots \star_\hbar f}_{n}.$$

Compute $\exp_\star(-itH_c)$, where $H_c(\boldsymbol{p}, \boldsymbol{q})$ is the Hamiltonian function (2.27) of the harmonic oscillator.

3.5. Deformation quantization. Here we consider the quantization procedure from a formal algebraic point of view as the deformation theory of associative algebras. Let $\mathcal{A}$ be a $\mathbb{C}$-algebra (or an associative algebra with unit over a field k of characteristic zero) with a bilinear multiplication map

$m_0 : \mathcal{A} \otimes_{\mathbb{C}} \mathcal{A} \to \mathcal{A}$, which we will abbreviate as $a \cdot b = m_0(a, b)$. Denote by $\mathbb{C}[[t]]$ the ring of formal power series in t with coefficients in $\mathbb{C}$,

$$\mathbb{C}[[t]] = \left\{ \sum_{n=0}^{\infty} a_n t^n : a_n \in \mathbb{C} \right\},$$

and let

$$\mathcal{A}_t = \mathbb{C}[[t]] \otimes_{\mathbb{C}} \mathcal{A}$$

be the $\mathbb{C}[[t]]$-algebra of formal power series in t with coefficients in $\mathcal{A}$. The multiplication in $\mathcal{A}_t$ is a $\mathbb{C}[[t]]$-bilinear extension of the multiplication in $\mathcal{A}$, which we continue to denote by m_0. The algebra $\mathcal{A}_t$ is $\mathbb{Z}$-graded,

$$\mathcal{A}_t = \bigoplus_{n=0}^{\infty} A_n,$$

where $A_n = t^n \mathcal{A}$, so that $A_m \cdot A_m \subset A_{m+n}$.

Definition. A *formal deformation* of a $\mathbb{C}$-algebra $\mathcal{A}$ with a multiplication m_0 is an associative algebra $\mathcal{A}_t$ over the ring $\mathbb{C}[[t]]$ with a $\mathbb{C}[[t]]$-bilinear multiplication map $m_t : \mathcal{A}_t \otimes_{\mathbb{C}[[t]]} \mathcal{A}_t \to \mathcal{A}_t$ such that

$$m_t(a, b) = a \cdot b + \sum_{n=1}^{\infty} t^n m_n(a, b)$$

for all $a, b \in \mathcal{A}$, where $m_n : \mathcal{A} \otimes_{\mathbb{C}} \mathcal{A} \to \mathcal{A}$ are bilinear mappings.

It follows from $\mathbb{C}[[t]]$-bilinearity of m_t that the associativity condition is equivalent to

(3.25) $$m_t(m_t(a, b), c) = m_t(a, m_t(b, c))$$

for all $a, b, c \in \mathcal{A}$.

Definition. Two formal deformations m_t and $\tilde{m}_t$ of a $\mathbb{C}$-algebra $\mathcal{A}$ are equivalent if there is a $\mathbb{C}[[t]]$-linear mapping $F_t : \mathcal{A}_t \to \mathcal{A}_t$ such that

(i) for every $a \in \mathcal{A}$,

$$F_t(a) = a + \sum_{n=1}^{\infty} t^n f_n(a),$$

where $f_n : \mathcal{A} \to \mathcal{A}$ are linear mappings;

(ii) for all $a, b \in \mathcal{A}$,

$$F_t(\tilde{m}_t(a, b)) = m_t(F_t(a), F_t(b)).$$

The bilinear maps m_n satisfy infinitely many relations which are obtained by expanding (3.25) into formal power series in t. The first two of them, arising from comparing the coefficients at t and t^2, are

$$a \cdot m_1(b, c) - m_1(a \cdot b, c) + m_1(a, b \cdot c) - m_1(a, b) \cdot c = 0,$$

and

$$a \cdot m_2(b, c) - m_2(a \cdot b, c) + m_2(a, b \cdot c) - m_2(a, b) \cdot c$$
$$= m_1(m_1(a, b), c) - m_1(a, m_1(b, c)).$$

In general

$$a \cdot m_n(b, c) - m_n(a \cdot b, c) + m_n(a, b \cdot c) - m_n(a, b) \cdot c$$
$$= \sum_{j=1}^{n-1} (m_j(m_{n-j}(a, b), c) - m_j(a, m_{n-j}(b, c))).$$

The main tool for understanding these equations and studying the deformation theory of associative algebras is the *Hochschild cohomology*. Namely, let M be an $\mathcal{A}$-bimodule, i.e., a left and right module for the $\mathbb{C}$-algebra $\mathcal{A}$.

Definition. The Hochschild cochain complex $\mathsf{C}^\bullet(\mathcal{A}, M)$ of a $\mathbb{C}$-algebra $\mathcal{A}$ with coefficients in $\mathcal{A}$-bimodule M is defined by the cochains

$$\mathsf{C}^n(\mathcal{A}, M) = \mathrm{Hom}_{\mathbb{C}}(\mathcal{A}^{\otimes n}, M),$$

i.e., n-linear maps $f(a_1, \ldots, a_n)$ on $\mathcal{A}$ with values in M, and the differential $d_n : \mathsf{C}^n(\mathcal{A}, M) \to \mathsf{C}^{n+1}(\mathcal{A}, M)$,

$$(d_n f)(a_1, a_2, \ldots, a_{n+1}) = a_1 \cdot f(a_2, \ldots, a_{n+1})$$
$$+ \sum_{j=1}^{n} (-1)^j f(a_1, \ldots, a_{j-1}, a_j \cdot a_{j+1}, a_{j+2} \ldots, a_{n+1})$$
$$+ (-1)^{n+1} f(a_1, \ldots, a_n) \cdot a_{n+1}.$$

We have $d^2 = 0$, i.e., $d_{n+1} \circ d_n = 0$, and the cohomology $H^\bullet(\mathcal{A}, M)$ of the complex $(\mathsf{C}^\bullet(\mathcal{A}, M), d)$,

$$H^n(\mathcal{A}, M) = \ker d_n / \operatorname{Im} d_{n-1},$$

is called the Hochschild cohomology of the algebra $\mathcal{A}$ with coefficients in the $\mathcal{A}$-bimodule M.

In the deformation theory of associative algebras we have the simplest non-trivial case $M = \mathcal{A}$ with the left and right $\mathcal{A}$-actions given by the multiplication map. The associativity equation (3.25) can be written as

$$(d_2 m_1)(a, b, c) = 0,$$

$$(d_2 m_n)(a, b, c) = \sum_{j-1}^{n-1} (m_j(m_{n-j}(a, b), c) - m_j(a, m_{n-j}(b, c))), \quad a, b, c \in \mathcal{A}.$$

It is quite remarkable that the Hochschild cochain complex $\mathsf{C}^\bullet(\mathcal{A}, \mathcal{A})$ carries an additional structure of a graded Lie algebra, which plays a fundamental role in studying the associativity equation (3.25). Namely, for $f \in \mathsf{C}^m(\mathcal{A}, \mathcal{A})$ and $g \in \mathsf{C}^n(\mathcal{A}, \mathcal{A})$ let

$$(f \circ g)(a_1, \ldots, a_{m+n-1})$$

$$= \sum_{j=0}^{m-1} (-1)^j f(a_1, \ldots, a_j, g(a_{j+1}, \ldots, a_{j+n}), a_{j+n+1}, \ldots, a_{m+n-1}),$$

and define

$$[f, g]_G = f \circ g - (-1)^{(m-1)(n-1)} g \circ f.$$

The linear mapping $[\;,\;]_G : \mathsf{C}^m(\mathcal{A}, \mathcal{A}) \times \mathsf{C}^n(\mathcal{A}, \mathcal{A}) \to \mathbb{C}^{m+n-1}(\mathcal{A}, \mathcal{A})$ satisfies $[f, g]_G = -(-1)^{(m-1)(n-1)}[g, f]_G$, and is called the *Gerstenhaber bracket*[27]. Considering the restriction of a multiplication m_t to $\mathcal{A} \otimes_{\mathbb{C}} \mathcal{A}$ as formal power series in t with coefficients in $\mathsf{C}^{\bullet}(\mathcal{A}, \mathcal{A})$ and using the Gerstenhaber bracket, we can rewrite (3.25) in the succinct form:

$$(3.26) \qquad\qquad [m_t, m_t]_G = 0.$$

Let $\mathcal{A}$ be a commutative $\mathbb{C}$-algebra, and let $\mathcal{A}_t$ be a formal deformation of $\mathcal{A}$. Define the bilinear map $\{\;,\;\} : \mathcal{A} \otimes_{\mathbb{C}} \mathcal{A} \to \mathcal{A}$ by

$$(3.27) \qquad\qquad \{a, b\} = m_1(a, b) - m_1(b, a), \quad a, b \in \mathcal{A}.$$

Lemma 3.3. *A formal deformation $\mathcal{A}_t$ of a commutative $\mathbb{C}$-algebra $\mathcal{A}$ equips $\mathcal{A}$ with a Poisson algebra structure with the bracket $\{\;,\;\}$.*

Proof. Consider the equation $d\mu_1(a, b, c) = 0$. Subtracting from it the equation with a and c interchanged, and using the commutativity of $\mathcal{A}$, we obtain

$$\{a \cdot b, c\} - \{a, b \cdot c\} = a \cdot \{b, c\} - c \cdot \{a, b\}.$$

Interchanging b and c, we get

$$\{a \cdot c, b\} - \{a, b \cdot c\} = a \cdot \{c, b\} - b \cdot \{a, c\},$$

and interchanging further a and c, we obtain

$$\{a \cdot c, b\} - \{c, a \cdot b\} = c \cdot \{a, b\} - b \cdot \{c, a\}.$$

Adding the first and third equations and subtracting the second equation gives

$$\{a \cdot b, c\} = a \cdot \{a, c\} + b \cdot \{a, c\},$$

so that the skew-symmetric cochain $\{\;,\;\} \in \mathsf{C}^2(\mathcal{A}, \mathcal{A})$ satisfies the Leibniz rule. To prove the Jacobi identity, observe that

$$(3.28) \qquad\qquad \{a, b\} = \frac{m_t(a, b) - m_t(b, a)}{t} \ \mathrm{mod} \ t\mathcal{A}_t.$$

---

[27]Together with the cup product of cochains, the Gerstenhaber bracket equips $\mathsf{C}^{\bullet}(\mathcal{A}, \mathcal{A})$ with the structure of a *Gerstenhaber algebra*.

Using associativity condition (3.25) (associativity modulo $t^2 \mathcal{A}_t$ is sufficient) we get

$$\{\{a,b\},c\} + \{\{c,a\},b\} + \{\{b,c\},a\} = \frac{1}{t^2}((m_t(m_t(a,b) - m_t(b,a)),c)$$
$$-m_t(c, m_t(a,b) - m_t(b,a)) + m_t(m_t(c,a) - m_t(a,c),b)$$
$$-m_t(b, m_t(c,a) - m_t(a,c)) + m_t(m_t(b,c) - m_t(c,b),a)$$
$$-m_t(a, m_t(b,c) - m_t(c,b))) \bmod t\mathcal{A}_t = 0. \qquad \square$$

This result motivates the following definition.

Definition. A *deformation quantization* of a Poisson algebra $(\mathcal{A}, \{\ ,\ \})$ with the commutative product $m_0 : \mathcal{A} \otimes_\mathbb{C} \mathcal{A} \to \mathcal{A}$ is a formal deformation $\mathcal{A}_t$ of an algebra $\mathcal{A}$ such that the multiplication map m_t satisfies (3.28).

By Lemma 3.3, every formal deformation $\mathcal{A}_t$ of a commutative algebra $\mathcal{A}$ is a deformation quantization of the Poisson algebra $(\mathcal{A}, \{\ ,\ \})$ with the Poisson bracket given by (3.27).

According to Lemma 3.2, a deformation quantization of the Poisson algebra $(\mathscr{S}(\mathbb{R}^{2n}), \{\ ,\ \})$, where the Poisson bracket is associated with the canonical symplectic form $\omega = d\boldsymbol{p} \wedge d\boldsymbol{q}$, is the algebra $\mathscr{S}(\mathbb{R}^{2n})[[t]]$, where $t = -i\hbar$ and $\hbar$ is considered as a formal parameter, and the multiplication map is given by the $\star$-product. The following statement is a formal algebraic analog of the representation (3.22) for the $\star$-product.

Theorem 3.4 (Universal deformation). *Let $\mathcal{A}$ be a commutative $\mathbb{C}$-algebra with the multiplication map m_0, and let φ_1 and φ_2 be two commuting derivations of $\mathcal{A}$, i.e., linear maps $\varphi_1, \varphi_2 : \mathcal{A} \to \mathcal{A}$ satisfying Leibniz rule and*

$$\varphi_1 \circ \varphi_2 = \varphi_2 \circ \varphi_1.$$

Then

$$\{a,b\} = \varphi_1(a) \cdot \varphi_2(b) - \varphi_2(a) \cdot \varphi_1(b), \quad a,b \in \mathcal{A},$$

defines a Poisson bracket on $\mathcal{A}$, and the formula

$$m_t = m_0 \circ e^{t\Phi}, \quad \Phi = \varphi_1 \otimes \varphi_2,$$

called the universal deformation formula, gives a deformation quantization of the Poisson algebra $(\mathcal{A}, \{\ ,\ \})$.

Proof. The Jacobi identity for the bracket $\{\ ,\ \}$ follows from the commutativity of the derivations φ_1 and φ_2. To show that $\mathcal{A}_t$ is a deformation quantization of the Poisson algebra $(\mathcal{A}, \{\ ,\ \})$, we need to verify the associativity condition (3.25) for the map m_t. Let $\Delta : \mathcal{A} \to \mathcal{A} \otimes_\mathbb{C} \mathcal{A}$ be the coproduct map,

$$\Delta(a) = a \otimes 1 + 1 \otimes a, \quad a \in \mathcal{A}.$$

It extends to a $\mathbb{C}$-linear mapping from $\mathrm{Hom}_{\mathbb{C}}(\mathcal{A}, \mathcal{A})$ to $\mathrm{Hom}_{\mathbb{C}}(\mathcal{A}^{\otimes 2}, \mathcal{A}^{\otimes 2})$, which we continue to denote by Δ, and

$$\Delta(\varphi) = \varphi \otimes \mathrm{id} + \mathrm{id} \otimes \varphi, \quad \varphi \in \mathrm{Hom}_{\mathbb{C}}(\mathcal{A}, \mathcal{A}).$$

To prove (3.25), we observe that the Leibniz rule and the binomial formula yield the following identity for a derivation φ of $\mathcal{A}$:

$$e^{t\varphi} \circ m_0 = m_0 \circ e^{t\Delta(\varphi)}.$$

Using commutativity of φ_1 and φ_2, we obtain

$$m_t(a, m_t(b, c)) = m_0(e^{t\Phi}(a \otimes m_0(e^{t\Phi}(b \otimes c))))$$
$$= (m_0 \circ (\mathrm{id} \otimes m_0))(e^{t(\mathrm{id} \otimes \Delta)(\Phi)} e^{t \, \mathrm{id} \otimes \Phi}(a \otimes b \otimes c))$$
$$= (m_0 \circ (\mathrm{id} \otimes m_0))(e^{t(\mathrm{id} \otimes \Delta)(\Phi) + t \, \mathrm{id} \otimes \Phi}(a \otimes b \otimes c)),$$

where $(\mathrm{id} \otimes \Delta)(\Phi) = \varphi_1 \otimes \Delta(\varphi_2) \in \mathcal{A}^{\otimes 3}$. Similarly,

$$m_t(m_t(a, b), c) = (m_0 \circ (m_0 \otimes \mathrm{id}))(e^{t(\Delta \otimes \mathrm{id})(\Phi) + t\Phi \otimes \mathrm{id}}(a \otimes b \otimes c)),$$

where $(\Delta \otimes \mathrm{id})(\Phi) = \Delta(\varphi_1) \otimes \varphi_2 \in \mathcal{A}^{\otimes 3}$. Since $m_0 \circ (\mathrm{id} \otimes m_0) = m_0 \circ (m_0 \otimes \mathrm{id})$, the associativity of the multiplication m_t is equivalent to

$$(\Delta \otimes \mathrm{id})(\Phi) + \Phi \otimes \mathrm{id} = (\mathrm{id} \otimes \Delta)(\Phi) + \mathrm{id} \otimes \Phi,$$

which is obviously satisfied since $\Phi = \varphi_1 \otimes \varphi_2$. $\square$

In general, $2n$ pair-wise commuting derivations φ_i on a commutative algebra $\mathcal{A}$ define a Poisson bracket on $\mathcal{A}$,

$$\{a, b\} = \sum_{i=1}^{n} (\varphi_i(a) \cdot \varphi_{i+n}(b) - \varphi_{i+n}(a) \cdot \varphi_i(b)),$$

and deformation quantization of the corresponding Poisson algebra $(\mathcal{A}, \{\,,\,\})$ is given by the universal deformation formula

$$m_t = m_0 \circ e^{t \sum_{i=1}^{n} (\varphi_i \otimes \varphi_{i+n} - \varphi_{i+n} \otimes \varphi_i)}.$$

A deformation quantization of a Poisson manifold $(\mathcal{M}, \{\,,\,\})$ (see Section 2.7 in Chapter 1) is, by definition, a deformation quantization of the corresponding Poisson algebra of classical observables $(C^{\infty}(\mathcal{M}), \{\,,\,\})$ with the property that the linear maps m_n, $n \geq 1$, are bidifferential operators[28]. The formal power series expansion (3.22) of the $\star$-product is a deformation quantization of the Poisson manifold $(\mathbb{R}^{2n}, \{\,,\,\})$, where the Poisson bracket corresponding to the canonical symplectic form $\omega = d\boldsymbol{p} \wedge d\boldsymbol{q}$.

---

[28]This property ensures that the unit in Poisson algebra is preserved under the deformation.

Problem 3.13. Show that $C^\bullet(\mathcal{A},\mathcal{A})$ is a *graded Lie algebra* with respect to the Gerstenhaber bracket, i.e., that $[\ ,\ ]_G$ satisfies the *graded Jacobi identity*

$$(-1)^{(m-1)(p-1)}[f,[g,h]] + (-1)^{(n-1)(p-1)}[h,[f,g]] + (-1)^{(m-1)(n-1)}[g,[h,f]] = 0$$

for all $f \in C^m(\mathcal{A},\mathcal{A}), g \in C^n(\mathcal{A},\mathcal{A}), h \in C^p(\mathcal{A},\mathcal{A})$.

Problem 3.14. Verify that $d_n f = [f, m_0]_G$, where $f \in \mathsf{C}^n(\mathcal{A},\mathcal{A})$ and d_n is the Hochschild differential.

Problem 3.15. Show that deformation quantizations associated with Weyl, $\boldsymbol{pq}$, and $\boldsymbol{qp}$-quantizations are equivalent. (In fact, all deformation quantizations of the canonical Poisson manifold $(\mathbb{R}^{2n}, \{\ ,\ \})$ are equivalent.)

Problem 3.16. Let $\mathfrak{g}$ be a finite-dimensional Lie algebra with a basis $x_1, \ldots, x_n$, and let $\mathfrak{g}^*$ be its dual space equipped with the Lie-Poisson bracket $\{\ ,\ \}$ (see Problem 2.20 in Section 2.7 of Chapter 1). For $u \in \mathfrak{g}^*$ and $x = \sum_{i=1}^n \xi^i x_i, y = \sum_{i=1}^n \eta^i x_i \in \mathfrak{g}$ let

$$e^{u(\frac{1}{t}\log e^{tx}e^{ty} - x - y)} = \sum_{\alpha,\beta} a_{\alpha\beta}(u,t)\xi^\alpha \eta^\beta,$$

where $\xi^\alpha = (\xi^1)^{\alpha_1} \ldots (\xi^n)^{\alpha_n}$, $\eta^\beta = (\eta^1)^{\beta_1} \ldots (\eta^n)^{\beta_n}$, be the formal group law — the "Baker-Campbell-Hausdorff series". Show that the product

$$m_t(f_1,f_2) = \sum_{\alpha,\beta} a_{\alpha\beta}(u,t)D^\alpha f_1(u)D^\beta f_2(u), \quad f_1,f_2 \in C^\infty(\mathfrak{g}^*),$$

where for a multi-index $\alpha = (\alpha_1, \ldots, \alpha_n)$ and $f \in C^\infty(\mathfrak{g}^*)$, $D^\alpha f = \frac{\partial^{\alpha_1 + \cdots + \alpha_n} f}{\partial u_1^{\alpha_1} \ldots \partial u_n^{\alpha_n}}$, gives a deformation quantization of the Poisson manifold $(\mathfrak{g}^*, \{\ ,\ \})$.

Problem 3.17. Let $(G, \{\ ,\ \})$ be a Lie-Poisson group, where $\eta(g) = -r + \mathrm{Ad}^{-1} g \cdot r$ and non-degenerate r satisfies the classical Yang-Baxter equation (see Problems 2.21 – 2.23 in Section 2.7 of Chapter 1). Using Baker-Campbell-Hausdorff series for the Lie algebra $\tilde{\mathfrak{g}}$ — the one-dimensional central extension of $\mathfrak{g}$ by the 2-cocycle c (see Problem 2.23 in Section 2.7 of Chapter 1), show that there exists an element $F \in (U\mathfrak{g} \otimes U\mathfrak{g})[[t]]$ of the form $F = 1 - \frac{1}{2}tr + O(t^2)$, satisfying

$$(\Delta \otimes \mathrm{id})(F)(F \otimes 1) = (\mathrm{id} \otimes \Delta)(F)(1 \otimes F),$$

where Δ is the standard coproduct in $U\mathfrak{g}$.

Problem 3.18. Let F_L and F_R be the images of F and F^{-1} under the identifications of the universal enveloping algebra $U\mathfrak{g}$ with the algebras of left and right-invariant differential operators on G (see Problem 2.21 in Section 2.7 of Chapter 1), and let $\mathscr{F} = F_L \circ F_R$. Show that the product $m_t = m_0 \circ \mathscr{F}$ gives a deformation quantization of the Poisson-Lie group $(G, \{\ ,\ \})$ such that

$$\Delta \circ m_t = (m_t \otimes m_t) \circ \Delta,$$

where Δ is the standard coproduct on functions on G, $\Delta(f)(g_1,g_2) = f(g_1 g_2), f \in C^\infty(G)$. The *Hopf algebra* $(C^\infty(G), m_t, \Delta, S)$, where S is the standard antipode on G, $S(f)(g) = f(g^{-1})$, is called the *quantum group* corresponding to the Poisson-Lie group G, and is denoted by G_q, $q = e^t$.

Problem 3.19. Let $R = \sigma(F^{-1})F \in (U\mathfrak{g} \otimes U\mathfrak{g})[[t]]$, where σ is a permutation — an involution of $U\mathfrak{g} \otimes U\mathfrak{g}$, defined by $\sigma(a \otimes b) = b \otimes a$, $a, b \in U\mathfrak{g}$. Show that $R = 1 - tr + O(t^2)$ and satisfies the *quantum Yang-Baxter equation*

$$R_{12}R_{13}R_{23} = R_{23}R_{13}R_{12}$$

(see Problem 2.22 in Section 2.7 in Chapter 1 for notation).

4. Notes and references

Dirac's classical monograph [**Dir47**] is the fundamental text. Other classical references are physics textbooks [**Foc78**] and [**LL58**]. The monograph [**Sak94**] is a popular text for graduate courses in physics departments. Another useful physics reference is the encyclopedic treatise [**Mes99**], which discusses at length the origin of quantum theory and the development of its mathematical formalism. An elementary textbook [**PW35**] provides an introduction to the old quantum theory, including Bohr-Wilson-Sommerfeld quantization rules, and applications of quantum mechanics to chemistry. We refer the interested reader to these sources for physical formulation and the origin of quantum mechanics, including a discussion of basic experimental facts: the double-slit experiment and particle-wave duality. Quantum mechanics does not account for single measurement outcomes in a deterministic way, and the quantum process of measurement admits different interpretations. The accepted one is the so-called Copenhagen interpretation, in which a measurement causes an instantaneous "collapse" of the wave function describing the quantum system.

Monographs [**Rud87, BS87, AG93**], as well as [**RS80, RS75**], contain all the necessary material from the theory of operators in Hilbert spaces; the latter treatise also covers the theory of distributions, Fourier transform, and self-adjointness. The book [**Kir76**] introduces the reader to various methods in the representation theory, and the monograph [**BR86**], written for theoretical physicists, contains all the necessary information on the representation theory of Lie groups and Lie algebras. A succinct introduction to the method of stationary phase and other asymptotic methods is given in the classical text [**Erd56**] and in [**Olv97**]; see also Appendix B in [**BW97**]. Properties of Hermite-Tchebyscheff polynomials[29] can be found in the classical monograph [**Sze75**], as well as in many other reference books on special functions and orthogonal polynomials.

Our exposition in Section 1.1 follows a traditional approach to the mathematical foundation of quantum mechanics based on Dirac-von Neumann axioms. These axioms go back to von Neumann's classical monograph [**vN96**], with more detailed discussion in [**Mac04**]; see also [**BS91**], notes to Chapter VIII of [**RS80**], and references therein. Another mathematical description of quantum mechanics is based

---

[29]Following V.A. Fock [**Foc78**], in Section 2.6 we used the more appropriate name Hermite-Tchebyscheff polynomials instead of the customary Hermite polynomials (see [**Ger50**] for the relevant historic arguments).

on $\mathbb{C}^*$-algebras and the Gelfand-Naimark-Segal reconstruction theorem, and can be found in [**Str05**] and references therein.

Our exposition in Section 2 follows the outline in [**FY80**]; see [**Ber74**] for the general mathematical formulation of the quantization problem and the recent survey [**AE05**]. Having in mind a more advanced audience than for [**FY80**], in Sections 2.1 and 2.2 we included a complete mathematical treatment of Heisenberg commutation relations and coordinate and momentum representations. In Sections 2.2 and 2.3 we use the same normalization of the eigenfunctions of continuous spectrum as in [**Foc78**]; it corresponds to the orthogonality relation in Section 2.2 of Chapter 3. In our exposition in Section 2.6, which is otherwise standard, we include a proof of the completeness of the eigenfunctions $\psi_n(q)$, which is usually only mentioned in the physics textbooks. Holomorphic representation in quantum mechanics was introduced by V.A. Fock in 1932 [**Foc32**], where the scalar product in $\mathscr{D}_n$ was given in terms of the orthonormal basis of monomials $f_m(z)$. In terms of the integral (2.52) this scalar product was defined by V. Bargmann [**Bar61**]; holomorphic representation is also called *Fock-Bargmann representation*. Discussion of Wick symbols of operators in Section 2.7 essentially follows [**BS91**]; more details can be found in the original paper [**Ber71b**]. The method of geometric quantization is only briefly mentioned in remarks in Sections 2.2 and 2.7; the interested reader is referred to monographs [**GS77, Woo92**], lecture notes [**SW76, BW97**] and the survey [**Kir90**].

Our proof of the celebrated Stone-von Neumann theorem in Section 3.1 essentially follows the original paper of von Neumann [**vN31**]. The main instrument of the proof — the Weyl transform — was introduced by Weyl in the classical monograph [**Wey50**] (see [**Ros04**] for the history and generalizations of the Stone-von Neumann theorem). Invariant formulation of the Stone-von Neumann theorem, discussed in Section 3.2, as well as the relation with the triple Maslov index, methaplectic group, Shale-Weil representation, and applications to number theory, can be found in [**LV80**]. We refer to [**BS91**] for the extended discussion of $\boldsymbol{pq}$ and $\boldsymbol{qp}$-quantizations as their relations with the Weyl quantization, as well as for more details on the $\star$-product, introduced in Section 3.4. The beautiful formula (3.24) for the $\star$-product — composition of the Weyl symbols — belongs to Berezin [**Ber71a**].

The notion of the deformation quantization was introduced in [**BFF$^+$78a, BFF$^+$78b**], where the solution of Problem 3.12 can be found. The fundamental theorem that every Poisson manifold admits a deformation quantization was proved by Kontsevich in [**Kon03**]. More on Hochschild cohomology and the deformation theory of associative algebras can be found in the survey [**Vor05**] and references therein. The deformation theory point of view on the evolution of physical theories, like a passage from classical mechanics to quantum mechanics emphasized in Faddeev's lectures [**FY80**], can be found in [**Fla82**] and [**Fad98**].

Most of the problems in this chapter are rather standard and are mainly taken from the references [**RS80, BS91**]. Other require more sophistication and are aimed at introducing the reader to a new topic. Thus the asymptotic expansion in Problem 2.11, which can be found in monograph [**Sze75**], is an example of semi-classical asymptotics, and Problem 3.1, taken from [**Nel59**], gives a criterion when the irreducible unitary representation of the Heisenberg algebra is integrable. Problems 3.17–3.19 introduce the reader to the theory of quantum groups (see [**Jim85, Dri86, Dri87, RTF89**]) and are taken from [**Dri83**] (see also [**Tak90**]). The fundamental result that every Lie-Poisson group admits a deformation quantization described in Problem 3.18 has been proved in [**EK96**] (see also [**Enr01**]).

Chapter 3

Schrödinger Equation

1. General properties

Here we describe the general properties of the Schrödinger operator for a quantum particle in $\mathbb{R}^n$ moving in a potential field. To simplify the notation, in this chapter we set $\hbar = 1$ and $m = \frac{1}{2}$, and use $\boldsymbol{x} = (x_1, \ldots, x_n)$ for Cartesian coordinates on $\mathbb{R}^n$. Recall (see Section 2.4 in Chapter 2) that the corresponding Schrödinger operator is given by the formal differential expression

$$H = -\Delta + V(\boldsymbol{x}),$$

where $V(\boldsymbol{x})$ is a real-valued, measurable function on $\mathbb{R}^n$. Denote by

$$H_0 = -\Delta = -\left(\frac{\partial^2}{\partial x_1^2} + \cdots + \frac{\partial^2}{\partial x_n^2} \right)$$

the Schrödinger operator of a free quantum particle of mass $m = \frac{1}{2}$ on $\mathbb{R}^n$, also called the *kinetic energy operator*, and by V — the multiplication by $V(\boldsymbol{x})$ operator, also called the *potential energy operator*, so that

$$(1.1) \qquad\qquad H = H_0 + V.$$

According to Section 2.3 of Chapter 2, the operator H_0 is self-adjoint and unbounded on $\mathscr{H} = L^2(\mathbb{R}^n)$, with the domain $D(H_0) = W^{2,2}(\mathbb{R}^n)$ and absolutely continuous spectrum $[0, \infty)$, and the operator V is self-adjoint and is bounded if and only if $V(\boldsymbol{x}) \in L^\infty(\mathbb{R}^n)$, when $\|V\| = \|V\|_\infty$. The sum $H_0 + V$ is not necessarily self-adjoint, and the first major mathematical problem of quantum mechanics is to characterize potentials $V(\boldsymbol{x})$ for which the formal differential expression (1.1) uniquely defines a self-adjoint operator H on $\mathscr{H}$. We start by presenting some useful criteria for the self-adjointness.

1.1. Self-adjointness. The main result here is the Kato-Rellich theorem on perturbation of self-adjoint operators.

Definition. Let A and B be densely defined operators in $\mathscr{H}$. The operator B is smaller than A in the sense of Kato if $D(A) \subseteq D(B)$ and there exist $a, b \in \mathbb{R}$ with $a < 1$ such that for all $\psi \in D(A)$,

$$(1.2) \qquad \|B\psi\| \leq a\|A\psi\| + b\|\psi\|.$$

Equivalently, the operator B is smaller than A in the sense of Kato, if there exist $\alpha, \beta \in \mathbb{R}$ with $\alpha < 1$ such that for all $\psi \in D(A)$,

$$(1.3) \qquad \|B\psi\|^2 \leq \alpha\|A\psi\|^2 + \beta\|\psi\|^2.$$

Theorem 1.1 (Kato-Rellich). *If A is a self-adjoint operator with domain $D(A)$ and B is a symmetric operator smaller than A in the sense of Kato, then $H = A + B$ with $D(H) = D(A)$ is a self-adjoint operator.*

Proof. Recall that the operator H is self-adjoint if and only if $\operatorname{Im}(H + \lambda I) = \operatorname{Im}(H - \lambda I) = \mathscr{H}$ for some $\lambda \in i\mathbb{R}$, and hence for all $\lambda \in \mathbb{C} \setminus \sigma(H)$. Since $A = A^*$, for every $\lambda \in i\mathbb{R}$ we have $R_\lambda = (A - \lambda I)^{-1} \in \mathscr{L}(\mathscr{H})$ and $\operatorname{Im} R_\lambda = D(A)$. Now

$$H - \lambda I = (I + BR_\lambda)(A - \lambda I),$$

and in order to prove that $\operatorname{Im}(H - \lambda I) = \mathscr{H}$ it is sufficient to show that for $|\lambda|$ large enough $\|BR_\lambda\| < 1$, since then by Neumann series $I + BR_\lambda$ is an invertible bounded operator and $\operatorname{Im}(I + BR_\lambda) = \mathscr{H}$. Indeed, for all $\varphi \in \mathscr{H}$ we have the inequalities

$$\|R_\lambda \varphi\| \leq \tfrac{1}{|\lambda|}\|\varphi\| \quad \text{and} \quad \|AR_\lambda \varphi\| \leq \|\varphi\|,$$

which follow from the equation $\|(A - \lambda I)\psi\|^2 = \|A\psi\|^2 + |\lambda|^2\|\psi\|^2$ by setting $\varphi = (A - \lambda I)\psi$. Now using (1.2) with $\psi = R_\lambda \varphi$, we get

$$\|BR_\lambda \varphi\| \leq a\|AR_\lambda \varphi\| + b\|R_\lambda \varphi\| \leq (a + \tfrac{b}{|\lambda|})\|\varphi\|,$$

so that $\|BR_\lambda\| < 1$ for large enough $|\lambda|$. $\qquad\qquad\square$

We use the Kato-Rellich criterion for the physically important case when $A = H_0$ — the Schrödinger operator of a free particle in $\mathbb{R}^3$ — and $B = V$ — the multiplication by $V(\boldsymbol{x})$ operator.

Theorem 1.2. *Let $V = V_1 + V_2$, where $V_1(\boldsymbol{x}) \in L^2(\mathbb{R}^3)$ and $V_2(\boldsymbol{x}) \in L^\infty(\mathbb{R}^3)$. Then $H = H_0 + V$ is a self-adjoint operator and $D(H) = W^{2,2}(\mathbb{R}^3)$.*

Proof. It is sufficient to show that V is smaller than H_0 in the sense of Kato on $C_0^\infty(\mathbb{R}^3) \subset W^{2,2}(\mathbb{R}^3)$. Denoting by $\|\cdot\|_\infty$ the norm in $L^\infty(\mathbb{R}^3)$, for $\varphi \in C_0^\infty(\mathbb{R}^3)$ we have

$$\|V\varphi\| \leq \|V_1\|\,\|\varphi\|_\infty + \|V_2\|_\infty\,\|\varphi\|.$$

Let $h(\boldsymbol{p}) = \boldsymbol{p}^2$. Denoting by $\|\cdot\|_1$ the norm in $L^1(\mathbb{R}^3)$ and using the Fourier transform and the Cauchy-Bunyakovski-Schwarz inequality, we get

$$(2\pi)^{3/2}\|\varphi\|_\infty = \sup_{\boldsymbol{x}\in\mathbb{R}^3}\left|\int_{\mathbb{R}^3} e^{i\boldsymbol{px}}\check{\varphi}(\boldsymbol{p})d^3\boldsymbol{p}\right| \le \|\check{\varphi}\|_1$$

$$\le \|(h+1)^{-1}\|\,\|(h+1)\check{\varphi}\| \le C(\|h\check{\varphi}\| + \|\check{\varphi}\|)$$

$$= C(\|H_0\varphi\| + \|\varphi\|).$$

Now replace $\check{\varphi}(\boldsymbol{p})$ by $\check{\varphi}_r(\boldsymbol{p}) = r^3\check{\varphi}(r\boldsymbol{p})$, $r > 0$. Since $\|\check{\varphi}_r\|_\infty = \|\check{\varphi}\|_\infty$, $\|\check{\varphi}_r\|_1 = \|\check{\varphi}\|_1$, $\|\check{\varphi}_r\| = r^{3/2}\|\check{\varphi}\|$, and $\|h\check{\varphi}_r\| = r^{-1/2}\|h\check{\varphi}\|$, we get

$$(2\pi)^{3/2}\|\varphi\|_\infty \le r^{-1/2}C(\|H_0\varphi\| + r^2\|\varphi\|),$$

where $r > 0$ is arbitrary. Choosing r such that $a = r^{-1/2}(2\pi)^{-3/2}C\|V_1\| < 1$ completes the proof. $\qquad\square$

Corollary 1.3. *The Schrödinger operator of a complex atom, considered in Example 2.2 in Section 2.4 of Chapter 2, is essentially self-adjoint on $C_0^\infty(\mathbb{R}^{3(N+1)})$.*

Proof. We consider only the special case of the Hamiltonian of the hydrogen atom,

$$H = -\Delta - \frac{e^2}{r}, \quad r = |\boldsymbol{x}|.$$

Writing $V = \chi_1 V + (1 - \chi_1)V = V_1 + V_2$, where χ_1 is the characteristic function of the unit ball $B_1 = \{x \in \mathbb{R}^3 : |x| \le 1\}$, we have $V_1(\boldsymbol{x}) \in L^2(\mathbb{R}^3)$ and $V_2(\boldsymbol{x}) \in L^\infty(\mathbb{R}^3)$. $\qquad\square$

Another useful criterion applies to real-valued potentials $V(\boldsymbol{x}) \in L^\infty_{\mathrm{loc}}(\mathbb{R}^n)$, the space of locally bounded a.e. functions on $\mathbb{R}^n$. In this case the operator H, defined by the formal differential expression (1.1), is symmetric on $C_0^\infty(\mathbb{R}^n)$ and we have the following result.

Theorem 1.4. *If $V(\boldsymbol{x}) \in L^\infty_{\mathrm{loc}}(\mathbb{R}^n)$ is bounded from below, $V(\boldsymbol{x}) \ge C$ a.e. on $\mathbb{R}^n$, then the Schrödinger operator $H = H_0 + V$ is essentially self-adjoint on $C_0^\infty(\mathbb{R}^n)$.*

In fact, a much more general statement holds.

Theorem 1.5 (Sears). *Suppose that the potential $V(\boldsymbol{x}) \in L^\infty_{\mathrm{loc}}(\mathbb{R}^n)$ for all $\boldsymbol{x} \in \mathbb{R}^n$ satisfies the condition*

$$V(\boldsymbol{x}) \ge -Q(|\boldsymbol{x}|),$$

where $Q(r)$ is an increasing continuous positive function on $[0, \infty)$ such that

$$\int_0^\infty \frac{dr}{\sqrt{Q(r)}} = \infty.$$

Then the Schrödinger operator $H = H_0 + V$ is essentially self-adjoint on $C_0^\infty(\mathbb{R}^n)$.

Problem 1.1. Prove the general case of Corollary 1.3 (*Hint*: Derive the estimate (1.3) for each term in the corresponding potential energy operator.)

Problem 1.2 (Kato's inequality). Let $\psi \in L^1_{\text{loc}}(\mathbb{R}^n)$ be such that $\Delta\psi$, defined in the distributional sense, is represented by a function in $L^1_{\text{loc}}(\mathbb{R}^n)$. Prove that in the distributional sense, $\Delta|u| \geq \operatorname{Re}\left(\dfrac{\bar{u}}{u}\,\Delta u\right)$, where it is assumed that $\dfrac{\overline{u(\boldsymbol{x})}}{u(\boldsymbol{x})} = 0$ if $u(\boldsymbol{x}) = 0$. (The distribution $T \in \mathscr{S}(\mathbb{R}^n)'$ is non-negative if $T(\varphi) \geq 0$ for all non-negative $\varphi \in \mathscr{S}(\mathbb{R}^n)$.)

Problem 1.3. Prove Theorem 1.4 using Kato's inequality. (*Hint*: Show that if $\psi \in L^2(\mathbb{R}^n)$ in the distributional sense satisfies $(-\Delta + V(\boldsymbol{x}) + C + 1)\psi = 0$, then $\psi = 0$.)

Problem 1.4. Prove that one-dimensional Schrödinger operators with unbounded below potentials $V(x) = x$ and $V(x) = -x^2$ are essentially self-adjoint on $C_0^\infty(\mathbb{R})$.

1.2. Characterization of the spectrum. The second major mathematical problem in quantum mechanics is to describe the spectral properties of the Schrödinger operator H. Here we present some general results characterizing the spectrum of H. The first basic result is the following.

Theorem 1.6. *Suppose that $V(\boldsymbol{x}) \in L^\infty_{\text{loc}}(\mathbb{R}^n)$ satisfies*

$$\lim_{|\boldsymbol{x}|\to\infty} V(\boldsymbol{x}) = \infty.$$

Then the operator H has a pure point spectrum: there exists an orthonormal basis $\{\psi_n\}_{n\in\mathbb{N}}$ for $\mathscr{H}$ consisting of eigenfunctions of H with eigenvalues $\lambda_1 \leq \lambda_2 \leq \cdots \leq \lambda_n \leq \cdots$ of finite multiplicity,

$$H\psi_n = \lambda_n\psi_n,$$

and $\lim_{n\to\infty} \lambda_n = \infty$.

Recall that the essential spectrum $\sigma_{\text{ess}}(A)$ of a self-adjoint operator A consists of all non-isolated points of $\sigma(A)$ and of eigenvalues of infinite multiplicity. The following result gives a sufficient condition for the essential spectrum of the Schrödinger operator to fill $[0, \infty)$.

Theorem 1.7. *Suppose that $V = V_1 + V_2$, where $V_1(\boldsymbol{x}) \in L^q(\mathbb{R}^n)$, $2q \geq n$ for $n > 4$ and $q \geq 2$ for $n \leq 4$[1], and $V_2(\boldsymbol{x}) \in L^\infty(\mathbb{R}^n)$ satisfies*

$$\lim_{|\boldsymbol{x}|\to\infty} V_2(\boldsymbol{x}) = 0.$$

Then $\sigma_{\text{ess}}(H) = [0, \infty)$, so that $\sigma(H) \cap (-\infty, 0)$ consists of isolated eigenvalues of H of finite multiplicity.

---

[1] In the special case $n = 4$ one has $q > 2$.

The next result gives a sufficient condition for the Schrödinger operator with decaying potential to have only negative eigenvalues.

Theorem 1.8 (Kato). *Suppose that* $V(\boldsymbol{x}) \in L^{\infty}(\mathbb{R}^n)$ *and*

$$\lim_{|\boldsymbol{x}| \to \infty} |\boldsymbol{x}| V(\boldsymbol{x}) = 0.$$

Then the Schrödinger operator $H = H_0 + V$ *has no positive eigenvalues.*

Recall that the absolutely continuous spectrum and the singular spectrum of a self-adjoint operator A on $\mathscr{H}$ are defined, respectively, by $\sigma_{\mathrm{ac}}(A) = \sigma(A|_{\mathscr{H}_{\mathrm{ac}}})$ and $\sigma_{\mathrm{sc}}(A) = \sigma(A|_{\mathscr{H}_{\mathrm{sc}}})$, where $\mathscr{H}_{\mathrm{ac}}$ and $\mathscr{H}_{\mathrm{sc}}$ are closed subspaces for A defined as follows. Let P_A be the projection-valued measure for the self-adjoint operator A and let $\nu_\psi = (\mathsf{P}_A \psi, \psi)$ be the finite Borel measure on $\mathbb{R}$ corresponding to $\psi \in \mathscr{H}$, $\psi \neq 0$. Then $\mathscr{H}_{\mathrm{ac}}$ consists of 0 and all $\psi \in \mathscr{H}$ such that the measure ν_ψ is absolutely continuous with respect to the Lebesgue measure on $\mathbb{R}$, and $\mathscr{H}_{\mathrm{sc}}$ consists of 0 and all $\psi \in \mathscr{H}$ such that the measure ν_ψ is continuous singular with respect to the Lebesgue measure on $\mathbb{R}$.

Theorem 1.9. *Suppose that the potential* $V(\boldsymbol{x}) \in L^{\infty}(\mathbb{R}^n)$ *for some* $\varepsilon > 0$ *satisfies*

$$V(\boldsymbol{x}) = O(|\boldsymbol{x}|^{-1-\varepsilon}) \quad as \quad |\boldsymbol{x}| \to \infty.$$

Then the Schrödinger operator $H = H_0 + V$ *has no singular spectrum and* $\sigma_{\mathrm{ac}}(H) = [0, \infty)$. *Moreover,* $\sigma(H) \cap (-\infty, 0)$ *consists of eigenvalues of* H *of finite multiplicity with the only possible accumulation point at* 0.

Finally, for the physically important case $n = 3$ there is a useful estimate for the number of eigenvalues of the Schrödinger operator.

Theorem 1.10 (Birman-Schwinger). *Suppose that* $V(\boldsymbol{x}) \in L^{\infty}(\mathbb{R}^3)$ *and*

$$\int_{\mathbb{R}^3} \int_{\mathbb{R}^3} \frac{|V(\boldsymbol{x})V(\boldsymbol{y})|}{|\boldsymbol{x} - \boldsymbol{y}|^2} d^3\boldsymbol{x}\, d^3\boldsymbol{y} < \infty.$$

Then for the total number N *of eigenvalues of the Schrödinger operator* $H = H_0 + V$, *counted with multiplicities, we have*

$$N \leq \frac{1}{16\pi^2} \int_{\mathbb{R}^3} \int_{\mathbb{R}^3} \frac{|V(\boldsymbol{x})V(\boldsymbol{y})|}{|\boldsymbol{x} - \boldsymbol{y}|^2} d^3\boldsymbol{x}\, d^3\boldsymbol{y}.$$

Problem 1.5. Prove all results stated in this section. (*Hint*: See the list of references to this chapter.)

1.3. The virial theorem. Let $H = H_0 + V$ be the Schrödinger operator whose potential is a homogeneous function on $\mathbb{R}^n$ of degree ρ, i.e., $V(a\boldsymbol{x}) = a^\rho V(\boldsymbol{x})$. The *virial theorem* in quantum mechanics is the relation between the expectation values of the kinetic and potential energy operators H_0 and V in the stationary state.

Theorem 1.11 (The virial theorem). *Let $\psi \in \mathscr{H}$, $\|\psi\| = 1$, be the eigenfunction of the Schrödinger operator with the homogeneous potential of degree ρ, and let $T_0 = (H_0\psi, \psi)$ and $V_0 = (V\psi, \psi)$ be the corresponding expectation values of kinetic and potential energy operators. Then*

$$2T_0 = \rho V_0.$$

Proof. It follows from the Schrödinger equation

$$-\Delta\psi + V(\boldsymbol{x})\psi = \lambda\psi$$

that

$$T_0 + V_0 = -(\Delta\psi, \psi) + (V\psi, \psi) = \lambda.$$

For every $a > 0$ the function $\psi_a(\boldsymbol{x}) = \psi(a\boldsymbol{x})$ satisfies

$$-\Delta\psi_a + a^{\rho+2}V(\boldsymbol{x})\psi_a = a^2\lambda\psi_a,$$

so that by differentiating this equation with respect to a at $a = 1$ and using Euler's homogenous function theorem, we get

$$(1.4) \qquad -\Delta\dot{\psi} + V(\boldsymbol{x})\dot{\psi} = \lambda\dot{\psi} + (2\lambda - (\rho + 2)V(\boldsymbol{x}))\psi.$$

Since $\dot{\psi} \in D(H)$ and H is self-adjoint,

$$((H - \lambda I)\dot{\psi}, \psi) = (\dot{\psi}, (H - \lambda I)\psi) = 0,$$

and we obtain from (1.4) that

$$2\lambda = (\rho + 2)V_0,$$

which completes the proof. $\qquad\qquad\qquad\qquad\qquad\qquad\square$

Remark. Since energy levels of quantum systems are related to the closed orbits of corresponding classical systems, Theorem 1.11 is the quantum analog of the classical virial theorem (see Example 1.2 in Section 1.3 of Chapter 1).

Problem 1.6 (The Raleigh-Ritz principle). Prove that λ is an eigenvalue of a self-adjoint operator H with the eigenfunction ψ if and only if ψ is a critical point of the functional $F(\varphi) = ((H - \lambda I)\varphi, \varphi)$ on $D(H)$.

Problem 1.7. Derive the virial theorem from the Raleigh-Ritz principle.

2. One-dimensional Schrödinger equation

The Schrödinger operator on the real line — the one-dimensional Schrödinger operator — has the form

$$H = -\frac{d^2}{dx^2} + V(x),$$

where the real-valued potential $V(x) \in L^1_{\mathrm{loc}}(\mathbb{R})$, the space of locally integrable functions on $\mathbb{R}$. The corresponding eigenvalue problem

$$H\psi = \lambda\psi$$

reduces to the second order ordinary differential equation

(2.1) $$-y'' + V(x)y = k^2 y, \quad -\infty < x < \infty,$$

where it is convenient to set $\lambda = k^2$. In this section, we will study in detail the one-dimensional Schrödinger equation (2.1) for the case when

(2.2) $$\int_{-\infty}^{\infty} (1 + |x|)|V(x)|dx < \infty.$$

Condition (2.2) is a mathematical formulation of the physical statement that the potential $V(x)$ decays as $|x| \to \infty$. It allows us to compare solutions $y(x, k)$ of the Schrödinger equation (2.1) with the solutions $e^{\pm ikx}$ of the Schrödinger equation for the free quantum particle, which corresponds to the case $V = 0$ in (2.1). All results in this section hold for the real-valued potential $V(x) \in L^1_{\mathrm{loc}}(\mathbb{R})$ satisfying (2.2); to simplify the presentation we will additionally assume that the potential $V(x)$ in (2.2) is continuous on $\mathbb{R}$.

2.1. Jost functions and transition coefficients. Let

$$\sigma(x) = \int_x^{\infty} |V(s)|ds \quad \text{and} \quad \sigma_1(x) = \int_x^{\infty} \sigma(s)ds.$$

Since $V \in L^1(\mathbb{R})$, we have $\lim_{x\to\infty} \sigma(x) = 0$, and using condition (2.2) and the Fubini theorem we get

$$\sigma_1(x) = \int_x^{\infty} \int_s^{\infty} |V(t)|dtds = \int_x^{\infty} \int_x^{t} |V(t)|dsdt = \int_x^{\infty} (t - x)|V(t)|dt < \infty,$$

so that $\sigma \in L^1(x, \infty)$ for all $x \in \mathbb{R}$ and $\lim_{x\to\infty} \sigma_1(x) = 0$. Similarly, the functions

$$\tilde{\sigma}(x) = \int_{-\infty}^{x} |V(s)|ds \quad \text{and} \quad \tilde{\sigma}_1(x) = \int_{-\infty}^{x} \tilde{\sigma}(s)ds$$

satisfy $\lim_{x\to-\infty} \tilde{\sigma}(x) = \lim_{x\to-\infty} \tilde{\sigma}_1(x) = 0$. Condition (2.2) ensures that for real k the differential equation (2.1) has solutions $f_1(x, k)$ and $f_2(x, k)$, uniquely defined by the following asymptotics:

$$f_1(x,k) = e^{ikx} + o(1) \quad \text{as} \quad x \to \infty,$$

$$f_2(x,k) = e^{-ikx} + o(1) \quad \text{as} \quad x \to -\infty.$$

They are called *Jost solutions* and play a fundamental role in the theory of a one-dimensional Schrödinger equation.

Theorem 2.1. *For real k the differential equation (2.1) has solutions $f_1(x,k)$ and $f_2(x,k)$ satisfying the following properties.*

(i) *Estimates for $k \in \mathbb{R}$:*

$$\left| e^{-ikx} f_1(x,k) - 1 \right| \leq \left(\sigma_1(x) - \sigma_1(x + \tfrac{1}{|k|}) \right) e^{\sigma_1(x)},$$

$$\left| e^{ikx} f_2(x,k) - 1 \right| \leq \left(\tilde{\sigma}_1(x) - \tilde{\sigma}_1(x - \tfrac{1}{|k|}) \right) e^{\tilde{\sigma}_1(x)}.$$

(ii) *Asymptotics as $|x| \to \infty$:*

$$\lim_{x \to \infty} e^{-ikx} f_1(x,k) = 1, \quad \lim_{x \to \infty} e^{-ikx} f_1'(x,k) = ik,$$

$$\lim_{x \to -\infty} e^{ikx} f_2(x,k) = 1, \quad \lim_{x \to -\infty} e^{ikx} f_1'(x,k) = -ik.$$

(iii) *Analyticity: $f_1(x,k)$ and $f_2(x,k)$ admit analytic continuation to the upper half-plane $\operatorname{Im} k > 0$, and are continuous functions on $\operatorname{Im} k \geq 0$, uniformly in x on compact subsets of $\mathbb{R}$.*

(iv) *The conjugation property:*

$$\overline{f_1(x,k)} = f_1(x,-\bar{k}), \quad \overline{f_2(x,k)} = f_2(x,-\bar{k}), \quad \operatorname{Im} k \geq 0.$$

(v) *Estimates in part (i) hold for $\operatorname{Im} k \geq 0$. For $k \neq 0$,*

$$\left| f_1(x,k) - e^{ikx} \right| \leq \frac{e^{-\operatorname{Im} kx}}{|k|} \sigma(x) e^{\frac{1}{|k|}\sigma(x)},$$

$$\left| f_2(x,k) - e^{-ikx} \right| \leq \frac{e^{\operatorname{Im} kx}}{|k|} \tilde{\sigma}(x) e^{\frac{1}{|k|}\tilde{\sigma}(x)}.$$

(vi) *Asymptotics as $|k| \to \infty$, $\operatorname{Im} k \geq 0$:*

$$e^{-ikx} f_1(x,k) = 1 + O(|k|^{-1}), \quad e^{ikx} f_2(x,k) = 1 + O(|k|^{-1}).$$

Proof. It follows from the method of variation of parameters that differential equation (2.1) with the boundary condition $\lim_{x \to \infty} e^{-ikx} f_1(x,k) = 1$ is equivalent to the integral equation

$$(2.3) \qquad f_1(x,k) = e^{ikx} - \int_x^\infty \frac{\sin k(x-t)}{k} V(t) f_1(t,k) \, dt.$$

Setting $\varphi(x, k) = e^{-ikx} f_1(x, k)$, we get

$$(2.4) \qquad \varphi(x, k) = 1 - \int_x^\infty \frac{1 - e^{-2ik(x-t)}}{2ik} V(t) \varphi(t, k) dt.$$

For $\operatorname{Im} k \geq 0$ the integral equation (2.4) is of Volterra type and can be solved by the method of successive approximations. Namely, we look for the solution in the form

$$(2.5) \qquad \varphi(x, k) = \sum_{n=0}^\infty \varphi_n(x, k),$$

where $\varphi_0(x, k) = 1$ and

$$\begin{aligned}
\varphi_{n+1}(x, k) &= - \int_x^\infty \frac{1 - e^{-2ik(x-t)}}{2ik} V(t) \varphi_n(t, k) dt \\
&= \int_x^\infty \left(\int_x^t e^{-2ik(x-s)} V(t) \varphi_n(t, k) ds \right) dt.
\end{aligned}$$

From here we conclude that for $\operatorname{Im} k \geq 0$

$$(2.6) \qquad |\varphi_n(x, k)| \leq \frac{\sigma_1(x)^n}{n!}.$$

Indeed, this estimate is true for $n = 0$, and using the Fubini theorem and the induction hypothesis we get

$$\begin{aligned}
|\varphi_{n+1}(x, k)| &\leq \frac{1}{n!} \int_x^\infty \left(\int_x^t |V(t)| \sigma_1(t)^n ds \right) dt \\
&= \frac{1}{n!} \int_x^\infty \left(\int_s^\infty |V(t)| \sigma_1(t)^n dt \right) ds \\
&\leq \frac{1}{n!} \int_x^\infty \sigma(s) \sigma_1(s)^n ds = \frac{\sigma_1(x)^{n+1}}{(n+1)!}.
\end{aligned}$$

Thus

$$(2.7) \qquad |\varphi(x, k)| \leq e^{\sigma_1(x)}.$$

By the Weierstrass M-test the function $\varphi(x, k)$, defined by convergent series (2.5), is analytic for $\operatorname{Im} k > 0$. It is also continuous up to $\operatorname{Im} k = 0$ uniformly in x on compact subsets of $\mathbb{R}$.

The first estimate in part (i) now follows from (2.4) and (2.7). Namely, using $|V(t)| = -\sigma'(t)$ and integration by parts, we obtain

$$|\varphi(x,k) - 1| \leq \int_x^\infty \left| \frac{1 - e^{-2ik(x-t)}}{2ik} \right| |V(t)||\varphi(t,k)|dt$$

$$\leq e^{\sigma_1(x)} \left(\int_x^{x+\frac{1}{|k|}} (t-x)|V(t)|dt + \frac{1}{|k|} \int_{x+\frac{1}{|k|}}^\infty |V(t)|dt \right)$$

$$= e^{\sigma_1(x)} \left(-(t-x)\sigma(t) \Big|_x^{x+\frac{1}{|k|}} + \int_x^{x+\frac{1}{|k|}} \sigma(t)dt + \frac{1}{|k|} \int_{x+\frac{1}{|k|}}^\infty |V(t)|dt \right)$$

$$= e^{\sigma_1(x)} \left(\sigma_1(x) - \sigma_1(x + \tfrac{1}{|k|}) \right).$$

For $k = 0$ this estimate should be understood as $|\varphi(x,0) - 1| \leq \sigma_1(x)e^{\sigma_1(x)}$. The function $f_1(x,k) = e^{ikx}\varphi(x,k)$ satisfies the integral equation (2.3), and (2.7) allows us to differentiate under the integral sign in (2.3). This proves that $f_1(x,k)$ satisfies the differential equation (2.1) and the first estimate in part (i). In particular, $\lim_{x\to\infty} e^{-ikx}f_1(x,k) = 1$ for $\operatorname{Im} k \geq 0$. Differentiating (2.3) and using $e^{-\operatorname{Im} k(t-x)}|\cos k(x-t)| \leq 1$ for $t \geq x$ and $\operatorname{Im} k \geq 0$, we obtain

$$(2.8) \qquad |e^{-ikx}f_1'(x,k) - ik| \leq \int_x^\infty |V(t)||\varphi(t,k)|dt \leq \sigma(x)e^{\sigma_1(x)},$$

so that $\lim_{x\to\infty} e^{-ikx}f_1'(x,k) = ik$ for $\operatorname{Im} k \geq 0$.

The conjugation property follows from the uniqueness of the solution $f_1(x,k)$ satisfying the estimate in part (i). To prove the uniqueness, denote by $\chi(x,k)$ the homogeneous solution of the integral equation (2.4) with the property that $\alpha(x) = \sup_{x \leq t < \infty} |\chi(t,k)| < \infty$ for all $x \in \mathbb{R}$. Consider the inequality

$$|\chi(x,k)| \leq \int_x^\infty \left(\int_x^t |V(t)||\chi(t,k)|ds \right) dt,$$

which follows from the homogeneous form of equation (2.4). Repeating the proof of the estimate (2.6), we get $|\chi(x,k)| \leq \alpha(x)\dfrac{\sigma_1(x)^n}{n!}$, and passing to the limit $n \to \infty$ gives $\chi(x,k) = 0$.

To prove the first estimate in part (v), consider the inequality

$$(2.9) \qquad |\varphi(x,k) - 1| \leq \frac{\sigma(x)}{|k|} - \frac{1}{|k|} \int_x^\infty \sigma'(t)|\varphi(t,k) - 1|dt$$

which follows from (2.4) for $k \neq 0$. Iterating (2.9) we get

$$|\varphi(x,k) - 1| \leq \frac{\sigma(x)}{|k|} e^{\frac{1}{|k|}\sigma(x)}.$$

The asymptotic in part (vi) follows from the estimate in part (v), which completes the proof for the solution $f_1(x, k)$. The existence and analytic properties of the solution $f_2(x, k)$ are proved similarly by considering the integral equation

$$(2.10) \qquad f_2(x, k) = e^{-ikx} + \int_{-\infty}^{x} \frac{\sin k(x - t)}{k} V(t) f_2(t, k) dt. \qquad \square$$

Corollary 2.2. *For* $\operatorname{Im} k \geq 0$ *and* $k \neq 0$,

$$\lim_{x \to \infty} e^{-ikx}(\dot{f}_1(x, k) - ix f_1(x, k)) = 0, \qquad \lim_{x \to -\infty} e^{ikx}(\dot{f}_2(x, k) + ix f_2(x, k)) = 0,$$

where the dot stands for the partial derivative with respect to k.

Proof. Differentiating (2.4) with respect to k we obtain the following integral equation for $\dot{\varphi}(x, k) = e^{-ikx}(\dot{f}_1(x, k) - ix f_1(x, k))$:

$$\dot{\varphi}(x, k) = g(x, k) - \int_{x}^{\infty} \frac{1 - e^{-2ik(x-t)}}{2ik} V(t) \dot{\varphi}(t, k) dt,$$

where

$$g(x, k) = \frac{1}{k} \int_{x}^{\infty} (t - x) e^{-2ik(x-t)} V(t) \varphi(t, k) dt + \frac{1}{k}(1 - \varphi(x, k)).$$

It follows the estimate in part (i) of Theorem 2.1 and (2.7) that $|g(x, k)| \leq \frac{2}{|k|}\sigma_1(x)e^{\sigma_1(x)}$, and repeating the proof of the estimate (2.7), we obtain that $|\dot{\varphi}(x, k)| \leq \frac{2}{|k|}\sigma_1(x)e^{2\sigma_1(x)}$. Since $\lim_{x \to \infty} \sigma_1(x) = 0$, this proves the statement for $f_1(x, k)$. The corresponding result for $f_2(x, k)$ is proved similarly. $\square$

For real $k \neq 0$ the pairs $f_1(x, k)$, $f_1(x, -k) = \overline{f_1(x, k)}$ and $f_2(x, k)$, $f_2(x, -k) = \overline{f_2(x, k)}$ are fundamental solutions of the differential equation (2.1). Indeed, the Wronskian $W(y_1, y_2) = y_1' y_2 - y_1 y_2'$ of two solutions of (2.1) does not depend on x, and we get from part (ii) that

$$W(f_1(x, k), f_1(x, -k)) = \lim_{x \to \infty} W(f_1(x, k), f_1(x, -k)) = 2ik,$$

$$W(f_2(x, k), f_2(x, -k)) = \lim_{x \to -\infty} W(f_2(x, k), f_2(x, -k)) = -2ik.$$

Therefore for such k we have

$$(2.11) \qquad f_2(x, k) = a(k) f_1(x, -k) + b(k) f_1(x, k),$$

where the *transition coefficients* $a(k)$ and $b(k)$ are given by

$$(2.12) \quad a(k) = \frac{1}{2ik} W(f_1(x, k), f_2(x, k)), \quad b(k) = \frac{1}{2ik} W(f_2(x, k), f_1(x, -k)),$$

and satisfy $\overline{a(k)} = a(-k)$, $\overline{b(k)} = b(-k)$. Similarly we find that

$$(2.13) \qquad f_1(x, k) = a(k) f_2(x, -k) - b(-k) f_2(x, k).$$

Back substituting (2.13) for $f_1(x, k)$ and $f_1(x, -k)$ into (2.11), for real $k \neq 0$ we find the so-called *normalization condition,*

$$(2.14) \qquad\qquad |a(k)|^2 = 1 + |b(k)|^2.$$

It follows from parts (iii) and (vi) of Theorem 2.1 that coefficient $a(k)$ admits analytic continuation to $\operatorname{Im} k > 0$ and

$$(2.15) \qquad\qquad a(k) = 1 + O(|k|^{-1}) \quad \text{as} \quad |k| \to \infty.$$

Moreover, the function $ka(k)$ is continuous in $\operatorname{Im} k \geq 0$.

Normalization condition (2.14) implies that all zeros of $a(k)$ are in the upper half-plane $\operatorname{Im} k > 0$. Let k_0 be such a zero, $a(k_0) = 0$. It follows from the first equation in (2.12) that $W(f_1(x, k_0), f_2(x, k_0)) = 0$, so that the Jost solutions $f_1(x, k_0)$, $f_2(x, k_0)$ are linearly dependent,

$$f_1(x, k_0) = c_0 f_2(x, k_0),$$

for some $c_0 \neq 0$. It follows from part (i) of Theorem 2.1 that for $\operatorname{Im} k > 0$ the solution $f_1(x, k)$ is exponentially decaying as $x \to \infty$, and the solution $f_2(x, k)$ is exponentially decaying as $x \to -\infty$. Therefore $f_1(\cdot, k_0) \in L^2(\mathbb{R})$ is an eigenfunction of the Schrödinger operator H with the eigenvalue $\lambda_0 = k_0^2$. Since H is symmetric, its eigenvalues are real:

$$\lambda_0 \| f_1(\cdot, k_0) \|^2 = \int_{-\infty}^{\infty} \left(-f_1''(x, k_0) + V(x) f_1(x, k_0) \right) \overline{f_1(x, k_0)} \, dx$$

$$= \int_{-\infty}^{\infty} f_1(x, k_0) \left(-\overline{f_1''(x, k_0)} + V(x) \overline{f_1(x, k_0)} \right) dx = \bar{\lambda}_0 \| f_1(\cdot, k_0) \|^2,$$

as is easily seen by integrating by parts twice; due to part (i) of Theorem 2.1 and estimate (2.8) the boundary terms vanish as $|x| \to \infty$. Thus $k_0 = i\varkappa_0$ is pure imaginary with $\varkappa_0 > 0$ so that $\lambda_0 = -\varkappa_0^2 < 0$, and the corresponding eigenfunction $f_1(x, i\varkappa_0)$ is real-valued.

Proposition 2.1. *The function $a(k)$ in the upper half-plane $\operatorname{Im} k > 0$ has only finitely many pure imaginary simple zeros $k_l = i\varkappa_l$, and*

$$\dot{a}(i\varkappa_l) = -ic_l \| f_2(\cdot, i\varkappa_l) \|^2, \quad l = 1, \dots, n,$$

where the dot stands for the derivative and $f_1(x, i\varkappa_l) = c_l f_2(x, i\varkappa_l)$. The function $1/a(k)$ is bounded in some neighborhood of $k = 0$ in $\operatorname{Im} k \geq 0$.

Proof. It follows from (2.14) and (2.15) that $k = 0$ is the only possible accumulation point of zeros of $a(k)$ in $\operatorname{Im} k \geq 0$. First suppose that functions $f_1(x, k)$ and $f_2(x, k)$ are linearly independent for $k = 0$. Then the continuous in $\operatorname{Im} k > 0$ function $ka(k)$ satisfies $\lim_{k \to 0} ka(k) = W(f_1(x, 0), f_2(x, 0)) \neq 0$. Therefore $a(k) \neq 0$ in some neighborhood of 0, and $a(k)$ has only finitely many zeros in $\operatorname{Im} k > 0$.

The case when $f_1(x,0) = cf_2(x,0)$, $c \neq 0$, is more subtle. Suppose that there is a converging to 0 subsequence $k_n = i\varkappa_n$ of zeros of $a(k)$ in $\operatorname{Im} k > 0$. It follows from part (i) of Theorem 2.1 that there exists $A > 0$ such that for all $\varkappa \geq 0$ we have $f_1(x, i\varkappa) > \frac{1}{2}e^{-\varkappa x}$ for $x \geq A$ and $f_2(x, i\varkappa) > \frac{1}{2}e^{\varkappa x}$ for $x \leq A$, so that

$$\int_A^\infty f_1(x, i\varkappa_n) f_1(x, 0) dx, \quad \int_{-\infty}^{-A} f_2(x, i\varkappa_n) f_2(x, 0) dx \geq \frac{1}{4\varkappa} e^{-\varkappa A}.$$

Using integration by parts, we get, as before,

$$\varkappa_n^2 \int_{-\infty}^\infty f_1(x, i\varkappa_n) f_1(x, 0) dx = \int_{-\infty}^\infty \left(f_1''(x, i\varkappa_n) - V(x) f_1(x, i\varkappa_n) \right) f_1(x, 0) dx$$

$$= \int_{-\infty}^\infty f_1(x, i\varkappa_n) \left(f_1''(x, 0) - V(x) f_1(x, 0) \right) dx = 0.$$

On other hand, we have

$$(2.16) \qquad 0 = \int_A^\infty f_1(x, i\varkappa_n) f_1(x, 0) dx + \int_{-A}^A f_1(x, i\varkappa_n) f_1(x, 0) dx$$

$$+ c \cdot c_n \int_{-\infty}^{-A} f_2(x, i\varkappa_n) f_2(x, 0) dx.$$

Since $f_1(x, k)$ is continuous on $\operatorname{Im} k \geq 0$, uniformly in x on the compact subsets of $\mathbb{R}$, we have

$$\lim_{n \to \infty} \int_{-A}^A f_1(x, i\varkappa_n) f_1(x, 0) dx = \int_{-A}^A f_1(x, 0)^2 dx \geq 0.$$

Using

$$\lim_{n \to \infty} c \cdot c_n = c \lim_{n \to \infty} \frac{f_1(x, i\varkappa_n)}{f_2(x, i\varkappa_n)} = c \frac{f_1(x, 0)}{f_2(x, 0)} = c^2 > 0,$$

we obtain from (2.16) that for large enough n,

$$0 > \frac{1}{4\varkappa_n} e^{-\varkappa_n A}.$$

This is clearly a contradiction, so $a(k)$ has only finitely many zeros. The proof of local boundness of $1/a(k)$ in this case is left to the reader.

Next, consider the differential equation (2.1) together with the equation obtained from it by differentiating with respect to k:

$$(2.17) \qquad\qquad -y'' + V(x)y = k^2 y,$$

$$(2.18) \qquad\qquad -\dot{y}'' + V(x)\dot{y} = k^2 \dot{y} + 2ky.$$

Set $y = f_1(x, k)$ in equation (2.17), $y = f_2(x, k)$ in equation (2.18), and multiply (2.17) by $\dot{f}_2(x, k)$ and (2.18) by $f_1(x, k)$. Subtracting the resulting equations, we obtain

$$W(f_1(x, k), \dot{f}_2(x, k))' = 2k f_1(x, k) f_2(x, k).$$

Similarly, setting $y = f_2(x, k)$ in (2.17), $y = f_1(x, k)$ in (2.18), cross multiplying and subtracting, we obtain

$$-W(\dot{f}_1(x, k), f_2(x, k))' = 2k f_1(x, k) f_2(x, k).$$

Therefore

(2.19)
$$W(f_1(x, k), \dot{f}_2(x, k))\Big|_{-A}^{x} = 2k \int_{-A}^{x} f_1(x, k) f_2(x, k) dx,$$

(2.20)
$$-W(\dot{f}_1(x, k), f_2(x, k))\Big|_{x}^{A} = 2k \int_{x}^{A} f_1(x, k) f_2(x, k) dx.$$

Now suppose that $a(k_0) = 0$. We have

$$W(f_1(x, k), f_2(x, k)) = 2ik a(k),$$

so that differentiating with respect to k and setting $k = k_0$ gives

$$W(\dot{f}_1(x, k_0), f_2(x, k_0)) + W(f_1(x, k_0), \dot{f}_2(x, k_0)) = 2ik_0 \dot{a}(k_0).$$

Since $f_1(x, k_0) = c_0 f_2(x, k_0)$, it follows from Theorem 2.1 and Corollary 2.2 that boundary terms in (2.19)–(2.20) vanish as $A \to \infty$, and we obtain, using that $f(x, k_0)$ is real-valued:

$$\dot{a}(k_0) = -i \int_{-\infty}^{\infty} f_1(x, k_0) f_2(x, k_0) dx = -i c_0 \| f_2(\,\cdot\,, k_0) \|^2. \qquad \square$$

The eigenvalues of the Schrödinger operator H are simple. Indeed, the function $f_1(x, i\varkappa)$ is a solution of the differential equation (2.1) for $\lambda = -\varkappa^2 < 0$, which decays exponentially as $x \to \infty$. Since the Wronskian of any two solutions of (2.1) is constant, the second solution of (2.1), linearly independent with $f_1(x, i\varkappa)$, is exponentially increasing as $x \to \infty$. Thus every exponentially decaying as $x \to \infty$ solution of (2.1) for $\lambda = -\varkappa^2$ is a constant multiple of $f_1(x, i\varkappa)$. In particular, this proves that the point spectrum of H is simple. This result also follows from the simplicity of zeros of the function $a(k)$ and the eigenfunction expansion theorem, proved in the next section.

Problem 2.1 (The oscillation theorem). Let $\lambda_1 = -\varkappa_1^2 < \cdots < \lambda_n = \varkappa_n^2 < 0$ be the eigenvalues of the one-dimensional Schrödinger operator H. Prove that corresponding eigenfunctions $f_1(x, i\varkappa_l)$ have exactly $l - 1$ simple zeros, $l = 1, \ldots, n$.

Problem 2.2. Prove that for $\operatorname{Im} k \geq 0$ the Jost solution $f_1(x, k)$ admits the representation

$$f_1(x, k) = e^{ikx} + \int_{x}^{\infty} K_1(x, t) e^{ikt} dt,$$

where

$$K(x, x) = \tfrac{1}{2} \int_{x}^{\infty} V(t) dt \quad \text{and} \quad |K_1(x, t)| \leq \tfrac{1}{2} \sigma(\tfrac{x+t}{2}) e^{\sigma_1(x) - \sigma_1(\frac{x+t}{2})}.$$

Moreover, the kernel $K_1(x, t)$ is differentiable and

$$|\tfrac{\partial K_1}{\partial x}(x, t) + \tfrac{1}{4}V(\tfrac{x+t}{2})| \leq \tfrac{1}{2}\sigma_1(x)\sigma(\tfrac{x+t}{2})e^{\sigma_1(x)},$$

with the similar inequality for $\frac{\partial K_1}{\partial t}(x, t)$. Correspondingly, for $\operatorname{Im} k \leq 0$ the Jost solution $f_2(x, k)$ admits the integral representation

$$f_2(x, k) = e^{-ikx} + \int_{-\infty}^{x} K_2(x, t)e^{-ikt}dt,$$

where the kernel $K_2(x, t)$ satisfies similar estimates. (*Hint:* Show that (2.3) is equivalent to the integral equation

$$K_1(x, t) = \tfrac{1}{2}\int_{\frac{x+t}{2}}^{\infty} V(s)ds + \tfrac{1}{2}\int_{x}^{\infty} V(s)\int_{t-(s-x)}^{t+(s-x)} K_1(s, u)duds$$

with the condition $K_1(x, t) = 0$ for $t < x$, which is solved by the method of successive approximations.)

Problem 2.3. Show that transition coefficients $a(k)$ and $b(k)$ have the representations

$$a(k) = 1 - \frac{1}{2ik}\int_{-\infty}^{\infty} V(x)dx - \frac{1}{2ik}\int_{0}^{\infty} A(t)e^{ikt}dt, \quad b(k) = \frac{1}{2ik}\int_{-\infty}^{\infty} B(t)e^{-ikt}dt,$$

where $A(t) \in L^1(0, \infty)$ and $B(t) \in L^1(-\infty, \infty)$.

Problem 2.4. Show that for $\operatorname{Im} k > 0$ the transition coefficient $a(k)$ satisfies the *dispersion relation*

$$a(k) = \exp\left\{\frac{1}{2\pi i}\int_{-\infty}^{\infty}\frac{\log(1 + |b(p)|^2)}{p - k}dp\right\}\prod_{l=1}^{n}\frac{k - i\varkappa_l}{k + i\varkappa_l}.$$

2.2. Eigenfunction expansion. Here we explicitly construct the resolvent kernel for the one-dimensional Schrödinger operator H, and show that it is self-adjoint with the domain consisting of functions $\psi \in L^2(\mathbb{R})$, which are twice differentiable on $\mathbb{R}$, and such that $-\psi'' + V(x)\psi \in L^2(\mathbb{R})$. Using the method of complex integration we derive the eigenfunction expansion theorem for the operator H, which generalizes the corresponding result for the operator H_0 of a free quantum particle, considered in Section 2.3 of Chapter 2.

For $\lambda \in \mathbb{C} \setminus [0, \infty)$ let

$$(2.21) \qquad R_\lambda(x, y) = \begin{cases} -\dfrac{f_1(x, k)f_2(y, k)}{2ika(k)} & \text{if } x \geq y, \\[2ex] -\dfrac{f_1(y, k)f_2(x, k)}{2ika(k)} & \text{if } x \leq y, \end{cases}$$

where the branch of the function $k = \sqrt{\lambda}$ on $\mathbb{C} \setminus [0, \infty)$ is defined by the condition that $\operatorname{Im} k > 0$. For fixed x and y the function $R_\lambda(x, y)$ is meromorphic on $\mathbb{C} \setminus [0, \infty)$ with simple poles at $\lambda_l = -\varkappa_l^2$, $l = 1, \ldots, n$. For fixed

$\lambda \neq \lambda_l$ the function $R_\lambda(x, y) = R_\lambda(y, x)$ is continuous in x and y, and it follows from part (v) of Theorem 2.1 and (2.15) that

$$(2.22) \qquad |R_\lambda(x, y)| \leq C \, \frac{e^{-\operatorname{Im} k |x-y|}}{|k|}, \quad C > 0.$$

The kernel $R_\lambda(x, y)$ for $\lambda \neq \lambda_l$ defines a bounded integral operator R_λ in $L^2(\mathbb{R})$ by the formula

$$(R_\lambda \psi)(x) = \int_{-\infty}^{\infty} R_\lambda(x, y) \psi(y) dy.$$

Indeed, it follows from (2.22) that for $\psi \in L^2(\mathbb{R})$,

$$|k|^2 \, \|R_\lambda \psi\|^2 \leq C^2 \int_{-\infty}^{\infty} \left(\int_{-\infty}^{\infty} e^{-\operatorname{Im} k |x-y|} |\psi(y)| dy \right)^2 dx$$

$$= C^2 \int_{-\infty}^{\infty} \int_{-\infty}^{\infty} e^{-\operatorname{Im} k(|y_1|+|y_2|)} \int_{-\infty}^{\infty} |\psi(x + y_1)\psi(x + y_2)| dx dy_1 dy_2$$

$$\leq 4C^2 (\operatorname{Im} k)^{-2} \|\psi\|^2.$$

In particular,

$$\|R_\lambda\| \leq \frac{2C}{|k| \operatorname{Im} k}.$$

Lemma 2.1. *The operator H is self-adjoint and $R_\lambda = (H - \lambda I)^{-1}$ for $\lambda \in \mathbb{C} \setminus \{[0, \infty) \cup \{\lambda_1, \ldots, \lambda_n\}\}$, where I is the identity operator in $L^2(\mathbb{R})$.*

Proof. Let $g \in L^2(\mathbb{R})$. As in the variation of parameters method, it follows from (2.21) that for $\lambda \in \mathbb{C} \setminus \{[0, \infty) \cup \{\lambda_1, \ldots, \lambda_n\}\}$ the function $y = R_\lambda g \in L^2(\mathbb{R})$ is twice differentiable a.e. on $\mathbb{R}$ and satisfies the differential equation

$$-y'' + V(x)y = \lambda y + g(x).$$

Thus $(H - \lambda I)R_\lambda = I$ and, in particular, $\operatorname{Im}(H \pm iI) = L^2(\mathbb{R})$, so that H is self-adjoint. $\qquad \square$

Let $\mathfrak{H}_0$ be the Hilbert space of $\mathbb{C}^2$-valued functions $\Phi(\lambda) = \begin{pmatrix} \varphi_1(\lambda) \\ \varphi_2(\lambda) \end{pmatrix}$ on $[0, \infty)$ with the norm

$$\|\Phi\|_0^2 = \int_0^{\infty} (|\varphi_1(\lambda)|^2 + |\varphi_2(\lambda)|^2) d\sigma(\lambda),$$

where $d\sigma(\lambda) = \frac{1}{2\sqrt{\lambda}} d\lambda$ and $\sqrt{\lambda} \geq 0$ for $\lambda \geq 0$ (see Section 2.3 of Chapter 2, where one should put $\hbar = 1$ and $m = \frac{1}{2}$). For $\lambda \geq 0$ set

$$(2.23) \qquad u_j(x, \sqrt{\lambda}) = \frac{1}{a(\sqrt{\lambda})} f_j(x, \sqrt{\lambda}), \quad j = 1, 2,$$

and for $\psi \in C_0^2(\mathbb{R})$ define the $\mathbb{C}^2$-valued function $\mathscr{U}\psi$ on $[0, \infty)$ with components $(\mathscr{U}\psi)_1$ and $(\mathscr{U}\psi)_2$ by

$$(2.24) \qquad (\mathscr{U}\psi)_j(\lambda) = \frac{1}{\sqrt{2\pi}} \int_{-\infty}^{\infty} \psi(x)\overline{u_j(x, \sqrt{\lambda})}dx, \quad j = 1, 2.$$

By Proposition 2.1 the function $\mathscr{U}\psi$ is bounded at $\lambda = 0$, and for large λ using differential equation (2.1) and integration by parts we get

$$(\mathscr{U}\psi)_j(\lambda) = \frac{1}{\lambda\sqrt{2\pi}} \int_{-\infty}^{\infty} \left(-\psi''(x) + V(x)\psi(x)\right)\overline{u_j(x, \sqrt{\lambda})}dx,$$

which shows that $\mathscr{U}\psi \in \mathfrak{H}_0$.

Let P be the orthogonal projection on the subspace of $\mathscr{H} = L^2(\mathbb{R})$ spanned by the eigenfunctions $\psi_l(x) = f_1(x, i\varkappa_l)$, $l = 1, \dots, n$. The functions ψ_l are real-valued and orthogonal, so that for $\psi \in \mathscr{H}$

$$(P\psi)(x) = \sum_{l=1}^{n} \frac{1}{\|\psi_l\|^2} \psi_l(x) \int_{-\infty}^{\infty} \psi(y)\psi_l(y)dy.$$

Theorem 2.3. *The operator $\mathscr{U}$ extends to a partial isometry operator $\mathscr{U}$: $\mathscr{H} \to \mathfrak{H}_0$,*

$$\mathscr{U}^*\mathscr{U} = I - P \quad \text{and} \quad \mathscr{U}\mathscr{U}^* = I_0,$$

and establishes the isomorphism $(I - P)\mathscr{H} \simeq \mathfrak{H}_0$. Here I and I_0 are, respectively, identity operators in $\mathscr{H}$ and $\mathfrak{H}_0$. For the spectrum of the corresponding Schrödinger operator H we have $\sigma(H) = \{-\varkappa_1^2, \dots, -\varkappa_n^2\} \cup [0, \infty)$, and

$$\mathscr{H} = \mathscr{H}_{\mathrm{pp}} \oplus \mathscr{H}_{\mathrm{ac}},$$

where $\mathscr{H}_{\mathrm{pp}} = P\mathscr{H}$ and $\mathscr{H}_{\mathrm{ac}} = (I - P)\mathscr{H}$ are, respectively, invariant subspaces associated with pure point and absolutely continuous spectra of H. The operator $\mathscr{U}H\mathscr{U}^*$ is a multiplication by λ operator in $\mathfrak{H}_0$, so that the absolutely continuous spectrum $[0, \infty)$ of H has multiplicity two, and the operator $\mathscr{U}$ "diagonalizes" the restriction of H to the subspace $\mathscr{H}_{\mathrm{ac}}$.

Proof. To proof the relation $\mathscr{U}^*\mathscr{U} = I - P$ — the so-called *completeness relation* — it is sufficient to establish the following classical eigenfunction expansion formula for $\psi \in C_0^2(\mathbb{R})$:

$$\psi(x) = \frac{1}{2\pi} \int_0^{\infty} \left(u_1(x, k)\int_{-\infty}^{\infty}\psi(y)\overline{u_1(y, k)}dy + u_2(x, k)\int_{-\infty}^{\infty}\psi(y)\overline{u_2(y, k)}dy\right)dk$$

$$(2.25) \qquad + \sum_{l=1}^{n} \frac{1}{\|\psi_l\|^2}\psi_l(x) \int_{-\infty}^{\infty} \psi(y)\psi_l(y)dy,$$

where all integrals are absolutely convergent, and we put $\sqrt{\lambda} = k$, $d\sigma(\lambda) = dk$. Indeed, assuming that (2.25) holds, we get for $\psi, \varphi \in C_0^2(\mathbb{R})$

$$(\psi, \varphi) = \int_{-\infty}^{\infty} \psi(x)\overline{\varphi(x)}dx = \frac{1}{2\pi}\int_{-\infty}^{\infty}\overline{\varphi(x)}\left(\int_0^{\infty}\left(u_1(x,k)\int_{-\infty}^{\infty}\psi(y)\overline{u_1(y,k)}dy\right.\right.$$

$$\left.\left. + u_2(x,k)\int_{-\infty}^{\infty}\psi(y)\overline{u_2(y,k)}dy\right)dk\right)dx + \sum_{l=1}^{n}\frac{1}{\|\psi_l\|^2}(\psi,\psi_l)(\psi_l,\varphi)$$

$$= (\mathscr{U}\psi, \mathscr{U}\varphi)_0 + \sum_{l=1}^{n}\frac{1}{\|\psi_l\|^2}(\psi,\psi_l)(\psi_l,\varphi),$$

where $(\ ,\ )_0$ is the inner product in $\mathfrak{H}_0$, and the interchange of integrals over x and k is justified by the Fubini theorem. Thus

$$(2.26) \qquad\qquad (\mathscr{U}\psi, \mathscr{U}\varphi)_0 = ((I - P)\psi, \varphi)$$

for $\psi, \varphi \in C_0^2(\mathbb{R})$, and we conclude that $\mathscr{U}$ extends to a bounded operator from $\mathscr{H}$ and $\mathfrak{H}_0$ satisfying (2.26) for all $\psi, \varphi \in \mathscr{H}$.

To prove the eigenfunction expansion formula (2.25), for $\lambda \in \mathbb{C} \setminus [0, \infty)$ set $g(x, \lambda) = (R_\lambda\psi)(x)$. The function $g(x, \lambda)$ for fixed x is meromorphic in $\lambda \in \mathbb{C} \setminus [0, \infty)$ with simple poles at $\lambda = \lambda_l$, $l = 1, \ldots, n$. It follows from (2.21) and Proposition 2.1 that

$$\operatorname{Res}_{\lambda=\lambda_l} g(x,\lambda) = \frac{ic_l}{\dot{a}(i\varkappa_l)}f_2(x, i\varkappa_l)\int_{-\infty}^{\infty}f_2(y, i\varkappa_l)\psi(y)dy = -\frac{(\psi, \psi_l)}{\|\psi_l\|^2}\psi_l(x).$$

Since $HR_\lambda = R_\lambda H = I + \lambda R_\lambda$, we have

$$-g''(x,\lambda) + V(x)g(x,\lambda) = \psi(x) + \lambda g(x,\lambda),$$

so that

$$g(x,\lambda) = -\frac{1}{\lambda}\psi(x) + \frac{1}{\lambda}(R_\lambda\varphi)(x),$$

where $\varphi = -\psi'' + V(x)\psi \in C_0(\mathbb{R})$. From (2.22) we get

$$|(R_\lambda\varphi)(x)| \le \frac{C}{|\sqrt{\lambda}|} \quad \text{as} \quad |\lambda| \to \infty,$$

so that

$$(2.27) \qquad g(x,\lambda) = -\frac{1}{\lambda}\psi(x) + O(|\lambda|^{-3/2}) \quad \text{as} \quad |\lambda| \to \infty.$$

It follows from Proposition 2.1 and (2.21) that

$$(2.28) \qquad\qquad g(x,\lambda) = O(|\lambda|^{-1/2}) \quad \text{as} \quad \lambda \to 0.$$

For $0 < \varepsilon < 1 < N$ let $\mathcal{C} = \mathcal{C}_{\varepsilon,N}$ be the contour consisting of the following parts: (i) the arc $\mathcal{C}_\varepsilon$ of a circle $\lambda = \varepsilon e^{i\theta}$, $\varepsilon \le |\arg\theta| \le \pi$, oriented clockwise; (ii) the arc $\mathcal{C}_N$ of a circle $\lambda = Ne^{i\theta}$, $\varepsilon_N \le |\arg\theta| \le \pi$, oriented anti-clockwise, where $N\sin\varepsilon_N = \varepsilon\sin\varepsilon$; (iii) the segments $\mathcal{I}_\pm$ of the straight lines $\operatorname{Im} \lambda = \pm \varepsilon \sin \varepsilon$ which connect the boundaries of the arcs. Choose ε and N such that all poles λ_l of $g(x, \lambda)$ are inside $\mathcal{C}$, and consider the integral

$$I = \frac{1}{2\pi i} \int_{\mathcal{C}} g(x, \lambda) d\lambda.$$

On one hand, by the Cauchy residue theorem

$$I = \sum_{l=1}^{n} \operatorname{Res}_{\lambda=\lambda_l} g(x, \lambda) = -\sum_{l=1}^{n} \frac{(\psi, \psi_l)}{\|\psi_l\|^2} \psi_l(x).$$

On the other hand, it follows from (2.27) and (2.28) that

$$\lim_{N \to \infty} \frac{1}{2\pi i} \int_{\mathcal{C}_N} g(x, \lambda) d\lambda = -\psi(x) \quad \text{and} \quad \lim_{\varepsilon \to 0} \frac{1}{2\pi i} \int_{\mathcal{C}_\varepsilon} g(x, \lambda) d\lambda = 0.$$

Thus we obtain

$$\psi(x) - \sum_{l=1}^{n} \frac{1}{\|\psi_l\|^2} (\psi, \psi_l) \psi_l(x) = \lim_{\varepsilon \to 0} \lim_{N \to \infty} \frac{1}{2\pi i} \left(\int_{\mathcal{I}_+} g(x, \lambda) d\lambda - \int_{\mathcal{I}_-} g(x, \lambda) d\lambda \right)$$

$$(2.29) \qquad = \frac{1}{2\pi i} \int_0^\infty \left(\int_{-\infty}^\infty (R_{\lambda+i0}(x, y) - R_{\lambda-i0}(x, y)) \psi(y) dy \right) d\lambda,$$

where

$$R_{\lambda \pm i0}(x, y) = \lim_{\varepsilon \to 0} R_{\lambda \pm i\varepsilon}(x, y).$$

To compute the difference $R_{\lambda+i0}(x, y) - R_{\lambda-i0}(x, y)$, observe that on the cut $\lambda \geq 0$ we have $\sqrt{\lambda + i0} = k \geq 0$ and $\sqrt{\lambda - i0} = -k \leq 0$. It follows from (2.21) that for $x \geq y$

$$R_{\lambda+i0}(x, y) - R_{\lambda-i0}(x, y) = -\frac{1}{2ik} \left(\frac{f_1(x, k) f_2(y, k)}{a(k)} + \frac{f_1(x, -k) f_2(y, -k)}{a(-k)} \right),$$

and using the equations

$$f_1(x, k) = \frac{1}{a(-k)} f_2(x, -k) - \frac{b(-k)}{a(-k)} f_1(x, -k),$$

$$f_2(y, -k) = \frac{1}{a(k)} f_1(y, k) + \frac{b(-k)}{a(k)} f_2(y, k),$$

which follow from (2.11) and (2.13), we obtain

$$R_{\lambda+i0}(x, y) - R_{\lambda-i0}(x, y) = \frac{i}{2k|a(k)|^2} (f_1(x, k) \overline{f_1(y, k)} + f_2(x, k) \overline{f_2(y, k)})$$

$$= \frac{i}{2k} (u_1(x, k) \overline{u_1(y, k)} + u_2(x, k) \overline{u_2(y, k)}),$$

where $\lambda = k^2$. Substituting this into (2.29) and using the symmetry $R_\lambda(x, y) = R_\lambda(y, x)$ of the resolvent kernel, we get the eigenfunction expansion (2.25).

Let $\mathscr{U}_{\mathrm{ac}}$ be the restriction of the operator $\mathscr{U}$ to the subspace $\mathscr{H}_{\mathrm{ac}} = (I - P)\mathscr{H}$. It follows from the completeness relation that the operator $\mathscr{U}_{\mathrm{ac}}$

is an isometry, so that $\operatorname{Im}\mathscr{U} = \operatorname{Im}\mathscr{U}_{\mathrm{ac}}$ is a closed subspace of $\mathfrak{H}_0$. Thus to verify the *orthogonality relation* $\mathscr{U}\mathscr{U}^* = I_0$, it is sufficient to show that $\operatorname{Im}\mathscr{U} = \mathfrak{H}_0$. Using integration by parts we easily get that a self-adjoint operator $\mathscr{U}_{\mathrm{ac}} H \mathscr{U}_{\mathrm{ac}}^{-1}$ is a multiplication by λ operator in $\operatorname{Im}\mathscr{U}$ with the domain $\mathscr{U}D(H)$. Moreover,

$$\mathscr{U}R_\mu = R_\mu^{(0)}\mathscr{U}, \quad \mu \in \mathbb{C}\setminus[0,\infty),$$

where $R_\mu^{(0)}$ is the resolvent of the multiplication by λ operator in $\mathfrak{H}_0$, so that $\operatorname{Im}\mathscr{U}$ is an invariant subspace for $R_\mu^{(0)}$. Now it follows from Lemma 2.2 in Section 2.2 of Chapter 2 that there are Borel subsets $E_1, E_2 \subseteq [0,\infty)$ such that

$$\operatorname{Im}\mathscr{U} = \left\{ \tilde{\Phi} = \begin{pmatrix} \chi_{E_1}\varphi_1 \\ \chi_{E_2}\varphi_2 \end{pmatrix}, \quad \Phi = \begin{pmatrix} \varphi_1 \\ \varphi_2 \end{pmatrix} \in \mathfrak{H}_0 \right\}.$$

If, say, $E_1^{\mathrm{c}} = [0,\infty)\setminus E_1$ has positive Lebesgue measure, then for $\lambda = k^2 \in E_1^{\mathrm{c}}$ we have

$$\int_{-\infty}^{\infty} \overline{u_1(x,k)}\psi(x)dx = 0 \quad \text{for all} \quad \psi \in C_0^2(\mathbb{R}),$$

so that $u(x,k) = 0$ for all $x \in \mathbb{R}$ — a contradiction. Therefore, $\operatorname{Im}\mathscr{U} = \mathfrak{H}_0$. $\qquad\qquad\square$

Let $\mathscr{U}_0 : \mathscr{H} \to \mathfrak{H}_0$ be the corresponding unitary operator for the Schrödinger operator H_0 of a free particle, constructed in Section 2.3 of Chapter 2 (with $\hbar = 1$ and $m = \frac{1}{2}$), and set $U = \mathscr{U}^*\mathscr{U}_0 : \mathscr{H} \to \mathscr{H}_{\mathrm{ac}}$.

Corollary 2.4. *The restriction of H to the absolutely continuous subspace $\mathscr{H}_{\mathrm{ac}}$ is unitarily equivalent to H_0,*

$$H|_{\mathscr{H}_{\mathrm{ac}}} = UH_0U^{-1}.$$

Remark. In physics literature, completeness and orthogonality relations, understood in the distributional sense, are usually written as

$$\frac{1}{2\pi}\int_0^\infty \left(u_1(x,k)\overline{u_1(y,k)} + u_2(x,k)\overline{u_2(y,k)}\right)dk + \sum_{l=1}^{n}\frac{\psi_l(x)\psi_l(y)}{\|\psi_l\|^2} = \delta(x-y)$$

and

$$\frac{1}{2\pi}\int_{-\infty}^\infty u_j(x,k)\overline{u_k(x,p)}dx = \delta_{jk}\delta(k-p),$$

where $i,k = 1,2$ and $k,p > 0$. When transition coefficients $a(k)$ and $b(k)$ are differentiable for $k \neq 0^2$, the orthogonality relation can be derived from the identity

$$(2.30) \qquad \frac{d}{dx}W(u_j(x,k), \overline{u_k(x,p)}) = (k^2 - p^2)u_j(x,k)\overline{u_k(x,p)},$$

---

^{2}This is the case when $V(x)$ satisfies $\int_{-\infty}^{\infty}(1 + |x|^2)|V(x)|dx < \infty$.

which immediately follows from the differential equation (2.1). Namely, consider the case $j = k = 1$ and integrate (2.30) over $[-N, N]$. Using (2.13) and asymptotics of Jost solutions, we get

$$\int_{-N}^{N} u_1(x, k)\overline{u_1(x, p)}dx = \frac{1}{(k^2 - p^2)a(k)\overline{a(p)}}W(f_1(x, k), \overline{f_1(x, p)})\Big|_{-N}^{N}$$

$$= \frac{i}{a(k)\overline{a(p)}}\left(\frac{e^{i(k-p)N} - a(k)\overline{a(p)}e^{-i(k-p)N} + \overline{b(k)}b(p)e^{i(k-p)N}}{k - p}\right.$$

$$\left. + \frac{a(k)b(p)e^{-i(k+p)N} - \overline{b(k)}a(p)e^{i(k+p)N}}{k + p}\right) + o(1) \quad \text{as} \quad N \to \infty.$$

Since $k, p > 0$, by the Riemann-Lebesgue lemma the distributional limit as $N \to \infty$ of the terms in the third line is zero. Since $a(k)$ and $b(k)$ are assumed differentiable, in the second line we can replace $\overline{a(p)}$ and $b(p)$ by $\overline{a(k)}$ and $b(k)$, respectively, since the difference goes to zero as $N \to \infty$ by the Riemann-Lebesgue lemma. Finally, using (2.14) we obtain

$$\lim_{N\to\infty} \int_{-N}^{N} u_1(x, k)\overline{u_1(x, p)}dx = 2 \lim_{N\to\infty} \frac{\sin(k - p)N}{k - p} = 2\pi\delta(k - p),$$

where the last equality is the distributional form of the orthogonality relation for the Fourier transform.

Remark. The eigenfunctions of the continuous spectrum $u_j(x, \lambda)$ also satisfy the normalization condition (2.14) in Section 2.2 of Chapter 2:

$$\lim_{\Delta\to 0} \frac{1}{\Delta}(U_{j,k+\Delta} - U_{j,k}, U_{l,k+\Delta} - U_{l,k}) = \delta_{jl}, \quad j, l = 1, 2,$$

where

$$U_{j,k}(x) = \int_{k_0^2}^{k^2} u_j(x, \sqrt{\lambda})\frac{d\lambda}{2\sqrt{\lambda}} = \int_{k_0}^{k} \frac{f_j(x, p)}{a(p)}dp, \quad j = 1, 2.$$

Remark. The qualitative structure of the spectrum of the Schrödinger operator H is determined by the structure of the level sets $H_c(p, x) = \lambda$ of the classical Hamiltonian function $H_c(p, x) = p^2 + V(x)$. Namely, condition $\lim_{|x|\to\infty} V(x) = 0$ implies that the level sets for $\lambda > 0$ are non-compact and the classical motion is unbounded in both directions, and these values of λ fill the absolutely continuous spectrum $[0, \infty)$ of H of multiplicity two. For $\lambda < 0$ the level sets are compact and classical motion is periodic. According to BWS quantization rules (see Section 2.5 in Chapter 2), the energy levels — eigenvalues of H — correspond to the closed orbits, and condition (2.2) guarantees that H has only finitely many eigenvalues.

Problem 2.5. Find the energy levels for the potential

$$V(x) = -\frac{V_0}{\cosh^2 \varkappa x}, \quad V_0 > 0.$$

In particular, show that when $V_0 = 2\varkappa^2$ there is only one eigenvalue $E = -\varkappa^2$.

Problem 2.6. Show that the operator $R_\lambda - R_\lambda^{(0)}$, where $R_\lambda^{(0)} = (H_0 - \lambda I)^{-1}$ and $\lambda \in \mathbb{C} \setminus [0, \infty)$, is of the trace class, and

$$\operatorname{Tr}(R_\lambda - R_\lambda^{(0)}) = -\frac{d}{d\lambda} \log a(\sqrt{\lambda}).$$

(*Hint:* Use Wronskian identities (2.19)-(2.20) from the previous section.)

2.3. S-matrix and scattering theory. It was shown in the previous section that for potentials $V(x)$ satisfying (2.2) the operators $H = H_0 + V$ and H_0 have the same absolutely continuous spectrum. Among many partial isometry operators in $\mathscr{H}$ which establish the unitary equivalence between $H|_{\mathscr{H}_{\mathrm{ac}}}$ and H_0, there are two operators $W_\pm$ of fundamental physical significance. They are defined as

$$(2.31) \qquad W_\pm = \lim_{t \to \pm\infty} e^{iHt} e^{-itH_0},$$

where the limit is understood in the strong operator topology, and are called *wave operators* (or *Möller operators*). In general, the wave operators exist for a Schrödinger operator $H = H_0 + V$ on $\mathbb{R}^n$ with the potential $V(\boldsymbol{q})$ decaying sufficiently fast as $|\boldsymbol{q}| \to \infty$. In this section, we will show that for potentials $V(x)$ satisfying (2.2) the limits (2.31) exist.

The wave operators satisfy the partial isometry relation

$$(2.32) \qquad W_\pm^* W_\pm = I.$$

Indeed, strong limits of unitary operators preserve the inner products, so that $(W_\pm \psi, W_\pm \varphi) = (\psi, \varphi)$ for all $\psi, \varphi \in \mathscr{H}$. Convergence $A_n \to A$ as $n \to \infty$ in the strong operator topology does not necessarily imply that $A_n^* \to A^*$, so that we cannot conclude that $W_\pm W_\pm^* = I$. Thus in general the wave operators are not unitary. In fact, it can be shown that $\operatorname{Im} W_\pm = \mathscr{H}_{\mathrm{ac}} = (I - P)\mathscr{H}$ or, equivalently,

$$(2.33) \qquad W_\pm W_\pm^* = I - P,$$

where P is the orthogonal projection on the subspace $\mathscr{H}_{\mathrm{pp}}$. Though the proof of (2.33) is rather non-trivial, it is easy to show that $\operatorname{Im} W_\pm \subseteq (I - P)\mathscr{H}$. Indeed, let $\psi \in \mathscr{H}$ be a bound state — an eigenvector for H,

$$H\psi = \lambda\psi.$$

Then for every $\varphi \in \mathscr{H}$ we have

$$(W_\pm \varphi, \psi) = \lim_{t \to \pm\infty} e^{i\lambda t}(e^{-itH_0}\varphi, \psi) = 0$$

by the Riemann-Lebesgue lemma, as we have seen in Section 2.3 of Chapter 2.

Assuming that wave operators exist and satisfy (2.33), it is easy to prove that

$$(2.34) \qquad f(H)W_\pm = W_\pm f(H_0)$$

for any measurable function f on $\mathbb{R}$. Indeed, according to the spectral theorem, it is sufficient to prove this property for the functions $f(\lambda) = e^{i\tau\lambda}$ for all $\tau \in \mathbb{R}$. In this case, (2.34) immediately follows from the identity

$$e^{i\tau H} e^{i(t-\tau)H} e^{-i(t-\tau)H_0} = e^{itH} e^{-itH_0} e^{i\tau H_0}$$

by passing to the limits $t \to \pm\infty$. Using (2.34) and (2.32) we conclude

$$W_\pm^* H W_\pm = H_0,$$

so that restrictions of the wave operators to the subspace $\mathscr{H}_{\mathrm{ac}}$ establish the unitary equivalence between $H|_{\mathscr{H}_{\mathrm{ac}}}$ and H_0.

Remark. The physical meaning of the wave operators is the following. The one-parameter group $U(t) = e^{-itH}$ of unitary operators describes the evolution of the quantum particle moving in a short-ranged potential field. For large $|t|$ the particle with positive energy moves far away from the center and as $|t| \to \infty$ its evolution is described by the one-parameter group $U_0(t) = e^{-itH_0}$ which corresponds to the free motion. Mathematically this is expressed by the fact that for every $\varphi_- \in \mathscr{H}$ there exists a vector $\psi \in \mathscr{H}$ such that

$$\lim_{t \to -\infty} \|e^{-itH}\psi - e^{-itH_0}\varphi_-\| = 0,$$

and such a vector is given by $\psi = W_-\varphi_-$. Similarly, for every $\varphi_+ \in \mathscr{H}$ the vector $\psi = W_+\varphi_+$ satisfies

$$\lim_{t \to \infty} \|e^{-itH}\psi - e^{-itH_0}\varphi_+\| = 0.$$

As a physical interpretation, orthogonality of $\operatorname{Im} W_\pm$ to the subspace $\mathscr{H}_{\mathrm{pp}}$ is explained by the fact that for all times t bound states are localized near the potential center, whereas the free quantum particle goes to infinity as $|t| \to \infty$.

The wave operators are used to describe the scattering of the quantum particle by the potential center. Namely, given $\varphi_- \in \mathscr{H}$, there is $\psi = W_-\varphi_- \in \mathscr{H}$ such that the solution $\psi(t) = U(t)\psi$ of the non-stationary Schrödinger equation with Hamiltonian H and initial condition $\psi(0) = \psi$ as $t \to -\infty$ approaches the solution $\varphi_-(t) = U_0(t)\varphi_-$ of the non-stationary Schrödinger equation with free Hamiltonian H_0 and initial condition $\varphi_-(0) = \varphi_-$. Since $\operatorname{Im} W_- = \operatorname{Im} W_+$, there is $\varphi_+ \in \mathscr{H}$ such that $\psi = W_+\varphi_+$ and the solution $\psi(t)$ of the non-stationary Schrödinger equation with Hamiltonian H and initial condition $\psi(0) = \psi$ as $t \to \infty$ approaches the solution $\varphi_+(t) = U_0(t)\varphi_+$ of the non-stationary Schrödinger equation with free Hamiltonian H_0 and initial condition $\varphi_+(0) = \varphi_+$. The passage from the solution $\varphi_-(t)$ of the free Schrödinger equation describing the motion of quantum particle at $t = -\infty$ to the solution $\varphi_+(t)$ describing its motion at $t = \infty$ is the result of scattering of the quantum particle by a potential center. All information about the scattering is contained in the *scattering operator* S (also called *S-matrix* in physics), which relates initial and final conditions φ_- and φ_+:

$$\varphi_+ = S\varphi_-.$$

From (2.32) and $\psi = W_-\varphi_- = W_+\varphi_+$ we immediately get that

$$(2.35) \qquad S = W_+^* W_-,$$

and from (2.32)–(2.34) we obtain that the scattering operator S is a unitary operator in $\mathscr{H}$ and commutes with the free Hamiltonian H_0:

$$S^*S = SS^* = I \quad \text{and} \quad SH_0 = H_0S.$$

This is the outline of a *non-stationary approach* to scattering theory.

In the *stationary approach* to scattering theory, the wave operators $W_\pm$ and the scattering operator S are constructed as integral operators in the coordinate representation with integral kernels expressed through special solutions of the stationary Schrödinger equation. Here we present this construction for the one-dimensional Schrödinger operator considered in the previous section.

Namely, let $u_1(x, k)$ and $u_2(x, k)$ be solutions of the one-dimensional Schrödinger equation (2.1) given by (2.23), where $\sqrt{\lambda} = k > 0$. It follows from Theorem 2.1 and (2.11), (2.13) that these solutions have the following asymptotics:

$$(2.36) \qquad u_1(x, k) = s_{11}(k)e^{ikx} + o(1) \qquad \text{as} \quad x \to \infty,$$

$$(2.37) \qquad u_1(x, k) = e^{ikx} + s_{21}(k)e^{-ikx} + o(1) \quad \text{as} \quad x \to -\infty,$$

and

$$(2.38) \qquad u_2(x, k) = e^{-ikx} + s_{12}(k)e^{ikx} + o(1) \quad \text{as} \quad x \to \infty,$$

$$(2.39) \qquad u_2(x, k) = s_{22}(k)e^{-ikx} + o(1) \qquad \text{as} \quad x \to -\infty,$$

where

$$(2.40) \qquad s_{11}(k) = s_{22}(k) = \frac{1}{a(k)}, \quad s_{12}(k) = \frac{b(k)}{a(k)}, \quad s_{21}(k) = -\frac{\overline{b(k)}}{a(k)}.$$

Solutions $u_1(x, k)$ and $u_2(x, k)$ are called *scattering solutions*, the function $s_{11}(k)$ is called the *complex transmission coefficient*, and functions $s_{12}(k)$ and

$s_{21}(k)$, respectively, are called *right* and *left complex reflection coefficients*. Correspondingly, $T = |s_{11}(k)|^2$ is called the *transmission coefficient* and $R = |s_{12}(k)|^2 = |s_{21}(k)|^2$ — the *reflection coefficient*. The 2×2 matrix

$$S(k) = \begin{pmatrix} s_{11}(k) & s_{12}(k) \\ s_{21}(k) & s_{22}(k) \end{pmatrix}$$

is called the *scattering matrix*. It follows from (2.14) that the scattering matrix is unitary, $S^*(k)S(k) = I$, where I is the 2×2 identity matrix, and satisfies $\overline{S(k)} = S(-k)$. In particular, unitarity of the scattering matrix implies that $T + R = 1$, which is called *conservation of probability*.

The physical interpretation of the solutions $u_1(x, k)$ and $u_2(x, k)$ is the following. For simplicity, suppose that the potential $V(x)$ is smooth and vanishes for $|x| > A$, so that asymptotics (2.36) and (2.38) become equalities for $x > A$, and asymptotics (2.37) and (2.39) become equalities for $x < -A$. In this case elements $s_{ij}(k)$ of the scattering matrix are smooth functions for $k \neq 0$. As in Section 2.3 of Chapter 2, let $\varphi_1(k)$ and $\varphi_2(k)$ be the wave packets: smooth functions on $(0, \infty)$ supported in some neighborhood U_0 of $k_0 > 0$. The functions, called *scattering waves*,

$$\psi_j(x, t) = \int_0^\infty \varphi_j(k) u_j(x, k) e^{-ik^2 t} dk, \quad j = 1, 2,$$

are solutions of the time-dependent Schrödinger equation

$$i \frac{\partial \psi}{\partial t} = H\psi.$$

In the region $|x| > A$ the solution $\psi_1(x, t)$ can be simplified as follows:

$$\psi_1(x, t) = \int_0^\infty \varphi_1(k) s_{11}(k) e^{ikx - ik^2 t} dk, \qquad \text{when} \quad x > A,$$

$$\psi_1(x, t) = \int_0^\infty \varphi_1(k)(e^{ikx - ik^2 t} + s_{21}(k) e^{-ikx - ik^2 t}) dk, \quad \text{when} \quad x < -A.$$

Using the method of stationary phase (see Section 2.3 in Chapter 2) we get as $t \to -\infty$,

$$\psi_1(x, t) = O(|t|^{-1}), \qquad \text{when} \quad x > A,$$

$$\psi_1(x, t) = \frac{1}{\sqrt{|t|}} \varphi_1\left(\frac{x}{2t}\right) e^{\frac{ix^2}{4t} + \frac{i\pi}{4}} + O(|t|^{-1}), \qquad \text{when} \quad x < -A,$$

and as $t \to \infty$,

$$\psi_1(x, t) = \frac{1}{\sqrt{t}} \varphi_1\left(\frac{x}{2t}\right) s_{11}\left(\frac{x}{2t}\right) e^{\frac{ix^2}{4t} - \frac{i\pi}{4}} + O(t^{-1}), \qquad \text{when} \quad x > A,$$

$$\psi_1(x, t) = \frac{1}{\sqrt{t}} \varphi_1\left(-\frac{x}{2t}\right) s_{21}\left(-\frac{x}{2t}\right) e^{\frac{ix^2}{4t} - \frac{i\pi}{4}} + O(t^{-1}), \quad \text{when} \quad x < -A.$$

Assuming that the neighborhood U_0 is "sufficiently small", we see as in Section 2.3 of Chapter 2 that as $t \to -\infty$ the solution $\psi_1(x,t)$ represents a plane wave with amplitude $|\varphi_1(k_0)|$ moving from $x = -\infty$ to the right, toward the potential center, with velocity $v = 2k_0$. When $t \to \infty$, the solution $\psi_1(x,t)$ is a superposition of two plane waves: the transmitted wave with amplitude $|s_{11}(k_0)|$ times the original amplitude, located to the right of the potential center and moving toward ∞ with velocity v, and the reflected wave with amplitude factor $|s_{21}(k_0)|$, located to the left of the potential center and moving toward $-\infty$ with velocity $-v$. Corresponding transmission and reflection coefficients are $T = |s_{11}(k_0)|^2$ and $R = |s_{21}(k_0)|^2$.

Analogously, we have as $t \to -\infty$,

$$\psi_2(x,t) = \frac{1}{\sqrt{|t|}}\,\varphi_2\left(-\frac{x}{2t}\right) e^{\frac{ix^2}{4t} - \frac{i\pi}{4}} + O(|t|^{-1}), \qquad \text{when} \quad x > A,$$

$$\psi_2(x,t) = O(|t|^{-1}), \qquad \text{when} \quad x < -A,$$

and as $t \to \infty$,

$$\psi_2(x,t) = \frac{1}{\sqrt{t}}\,\varphi_2\left(\frac{x}{2t}\right) s_{12}\left(\frac{x}{2t}\right) e^{\frac{ix^2}{4t} - \frac{i\pi}{4}} + O(t^{-1}), \qquad \text{when} \quad x > A,$$

$$\psi_2(x,t) = \frac{1}{\sqrt{t}}\,\varphi_2\left(-\frac{x}{2t}\right) s_{22}\left(-\frac{x}{2t}\right) e^{\frac{ix^2}{4t} - \frac{i\pi}{4}} + O(t^{-1}), \quad \text{when} \quad x < -A.$$

As $t \to -\infty$, the solution $\psi_2(x,t)$ is a plane wave with amplitude $|\varphi_2(k_0)|$ moving from $x = \infty$ to the left, toward the potential center, with velocity $-v$. When $t \to \infty$, the solution $\psi_2(x,t)$ is a superposition of two plane waves: the transmitted wave with the same amplitude factor $|s_{22}(k_0)|$, located to the left of the potential center and moving toward $-\infty$ with velocity $-v$, and the reflected wave with amplitude factor $|s_{12}(k_0)|$, located to the right of the potential center and moving toward ∞ with velocity v. Formulas (2.40) show that transmission and reflection coefficients for the solution $\psi_2(x,t)$ are the same as for the solution $\psi_1(x,t)$, and are independent of the direction of travel. This is the so-called *reciprocity property of the transmission coefficient.*

The solutions $u_1(x,k)$ and $u_2(x,k)$ of the stationary Schrödinger equation have the property that corresponding solutions $\psi_1(x,t)$ and $\psi_2(x,t)$ of the time-dependent Schrödinger equation become the plane waves as $t \to -\infty$. In accordance with this interpretation, we denote them, respectively, by $u_1^{(-)}(x,k)$ and $u_2^{(-)}(x,k)$. Solutions of the stationary Schrödinger equation,

$$u_1^{(+)}(x,k) = \overline{u_2(x,k)} = u_2(x,-k) \quad \text{and} \quad u_2^{(+)}(x,k) = \overline{u_1(x,k)} = u_1(x,-k),$$

admit a similar interpretation: they correspond to the solutions of the time-dependent Schrödinger equation which become the plane waves as $t \to \infty$.

Introducing

$$U_+(x,k) = \begin{pmatrix} u_1^{(+)}(x,k) \\ u_2^{(+)}(x,k) \end{pmatrix} \quad \text{and} \quad U_-(x,k) = \begin{pmatrix} u_1^{(-)}(x,k) \\ u_2^{(-)}(x,k) \end{pmatrix}$$

we get by a straightforward computation using (2.11) and (2.13) that

$$(2.41) \qquad\qquad U_+(x,k) = \overline{S(k)}U_-(x,k).$$

The following result establishes equivalence between stationary and non-stationary approaches to scattering theory.

Theorem 2.5. *For the one-dimensional Schrödinger operator $H = H_0 + V$ with potential $V(x)$ satisfying condition (2.2) the wave operators exist and are given explicitly by*

$$W_\pm = \mathscr{U}_\pm^* \mathscr{U}_0.$$

Here $\mathscr{U}_\pm : \mathscr{H} \to \mathfrak{H}_0$ are the integral operators given by

$$\mathscr{U}_\pm(\psi) = \frac{1}{\sqrt{2\pi}} \int_{-\infty}^{\infty} \psi(x)\overline{U_\pm(x,\sqrt{\lambda})}dx,$$

and $\mathscr{U}_0 : \mathscr{H} \to \mathfrak{H}_0$ establishes the unitary equivalence between the operator H_0 in $\mathscr{H}$ and the multiplication by λ operator in $\mathfrak{H}_0$,

$$\mathscr{U}_0(\psi) = \frac{1}{\sqrt{2\pi}} \int_{-\infty}^{\infty} \psi(x) \begin{pmatrix} e^{-i\sqrt{\lambda}x} \\ e^{i\sqrt{\lambda}x} \end{pmatrix} dx.$$

The scattering operator S in $\mathscr{H}$ and the multiplication by the scattering matrix $S(\sqrt{\lambda})$ operator in $\mathfrak{H}_0$ are unitarily equivalent:

$$S = \mathscr{U}_0^* S(\sqrt{\lambda})\mathscr{U}_0.$$

Proof. The formula for the scattering operator immediately follows from the definition of the operators $\mathscr{U}_\pm$, formula (2.41), and orthogonality relation $\mathscr{U}_-\mathscr{U}_-^* = I_0$ (see Theorem 2.3). To prove that $W_\pm = \mathscr{U}_\pm^* \mathscr{U}_0$, it is sufficient to show that for all $\Phi = \begin{pmatrix} \varphi_1 \\ \varphi_2 \end{pmatrix} \in \mathfrak{H}_0$, where $\varphi_1(k)$ and $\varphi_2(k)$ are smooth functions on $(0,\infty)$ with compact support,

$$\lim_{t\to\pm\infty} \|(e^{-itH}\mathscr{U}_\pm^* - e^{-itH_0}\mathscr{U}_0^*)\Phi\|_{\mathscr{H}} = 0.$$

Indeed, setting $\chi^{(\pm)}(t) = (e^{-itH}\mathscr{U}_\pm^* - e^{-itH_0}\mathscr{U}_0^*)\Phi$, we have

$$\chi^{(\pm)}(x,t) = \frac{1}{\sqrt{2\pi}} \int_0^{\infty} \left((u_1^{(\pm)}(x,k) - e^{ikx})\varphi_1(k)\right.$$

$$\left. + (u_2^{(\pm)}(x,k) - e^{-ikx})\varphi_2(k)\right)e^{-ik^2t}dk,$$

and it is not difficult to estimate the integral

$$(2.42) \qquad \|\chi^{(\pm)}(t)\|^2 = \int_{-\infty}^{\infty} |\chi^{(\pm)}(x,t)|^2 dx$$

as $t \to \pm\infty$. Namely, it follows from the Riemann-Lebesgue lemma that for any fixed $A > 0$ the contribution to (2.42) from the interval $[-A, A]$ goes to 0 as $t \to \pm\infty$. In integrals over $-\infty \le x \le -A$ and $A \le x \le \infty$ we replace $u_{1,2}^{(\pm)}(x,k)$ by their asymptotics (2.36)–(2.39). Using estimates in Theorem 2.1, part (v), it is easy to show that for sufficiently large A the difference can be made arbitrarily small uniformly in t. To finish the proof it remains to show that for every continuous function $\varphi(k)$ on $(0, \infty)$ with compact support the integrals

$$J_1^{(\pm)}(t) = \frac{1}{2} \int_A^\infty \left| \int_0^\infty \varphi(k) e^{\mp ikx - ik^2 t} dk \right|^2 dx$$

and

$$J_2^{(\pm)}(t) = \frac{1}{2} \int_{-\infty}^A \left| \int_0^\infty \varphi(k) e^{\pm ikx - ik^2 t} dk \right|^2 dx$$

vanish as $t \to \pm\infty$. Consider, for instance, the integral $J_1^{(+)}(t)$. Given $\varepsilon > 0$, there exists a smooth function $\eta(k)$ on $(0, \infty)$ with compact support $[\alpha, \beta]$ such that

$$\|\varphi - \eta\|_{\mathscr{H}_0}^2 < \frac{\varepsilon}{4\pi}.$$

We have by the Plancherel theorem,

$$J_1^{(+)}(t) = \frac{1}{2} \int_A^\infty \left| \int_0^\infty ((\varphi(k) - \eta(k)) + \eta(k)) e^{-ikx - ik^2 t} dk \right|^2 dx$$

$$\le \int_A^\infty \left| \int_0^\infty (\varphi(k) - \eta(k)) e^{-ikx - ik^2 t} dk \right|^2 dx + \int_A^\infty \left| \int_0^\infty \eta(k) e^{-ikx - ik^2 t} dk \right|^2 dx$$

$$\le \int_{-\infty}^\infty \left| \int_0^\infty (\varphi(k) - \eta(k)) e^{-ikx - ik^2 t} dk \right|^2 dx + \int_A^\infty \left| \int_0^\infty \eta(k) e^{-ikx - ik^2 t} dk \right|^2 dx$$

$$\le \frac{\varepsilon}{2} + \int_A^\infty \left| \int_0^\infty \eta(k) e^{-ikx - ik^2 t} dk \right|^2 dx.$$

To estimate the remaining integral we use integration by parts to obtain

$$\int_0^\infty \eta(k) e^{-i(kx + k^2 t)} dk = \int_0^\infty \frac{\eta'(k)}{i(x + 2kt)} e^{-i(kx + k^2 t)} dk,$$

and since $\dfrac{|\eta'(k)|}{x + 2kt} \le \dfrac{C}{x + 2\alpha t}$ for $x, t > 0$, we have

$$\int_A^\infty \left| \int_0^\infty \eta(k) e^{-i(kx + k^2 t)} dk \right|^2 dx \le C^2 \int_A^\infty \frac{dx}{(x + 2\alpha t)^2} = \frac{C^2}{A + 2\alpha t}.$$

Choosing $t > \dfrac{2C^2}{\alpha\varepsilon}$, we obtain $J_1^{(+)}(t) < \varepsilon$. Other integrals are treated similarly. $\qquad\qquad\qquad\qquad\qquad\qquad\qquad\qquad\qquad\qquad\qquad\qquad\qquad\qquad$ $\square$

Corollary 2.6. *The wave operators satisfy orthogonality and completeness relations* $W_\pm^* W_\pm = I$ *and* $W_\pm W_\pm^* = I - P$.

Proof. The result of Theorem 2.3, proved for the operator $\mathscr{U} = \mathscr{U}_-$, obviously holds for the operator $\mathscr{U}_+$. Thus relations (2.32) and (2.33) follow from the corresponding orthogonality and completeness relations for the operators $\mathscr{U}_\pm$. $\qquad\qquad\qquad\qquad\qquad\qquad\qquad\qquad\qquad\qquad\qquad\qquad\qquad$ $\square$

Remark. The orthogonality relation (2.32) is trivial once the existence of the wave operators is established. Thus Theorem 2.5 gives another proof of the orthogonality relation in Theorem 2.3.

Problem 2.7 (The Cook's criterion). In abstract scattering theory, prove that the wave operators $W_\pm$ exist if for all $\varphi \in \mathscr{H}$,

$$\int_0^\infty \|V e^{-itH_0}\varphi\| dt < \infty.$$

Problem 2.8. Find the scattering matrix $S(k)$ for the potential $V(x)$ from Problem 2.5 and show that when $V_0 = 2\varkappa^2$, the scattering matrix is diagonal.

Problem 2.9 (Quantum tunneling). Find the scattering matrix $S(k)$ for the rectangular potential barrier: $V(x) = 0$ for $x < 0$ and $x > 2a$, and $V(x) = V_0 > 0$ for $0 \le x \le 2a$. Show that when E varies from 0 to V_0, T increases from 0 to $(1 + V_0 a^2)^{-1}$ — the penetration of a potential barrier by a quantum particle.

2.4. Other boundary conditions. Here we consider two examples of the Schrödinger operator

$$H = -\frac{d^2}{dx^2} + V(x)$$

with the potential $V(x)$ having different asymptotics as $x \to \pm\infty$.

Example 2.1. Suppose that the potential $V(x)$ satisfies

$$\int_{-\infty}^0 (1 + |x|)|V(x) - c^2| dx < \infty \quad \text{and} \quad \int_0^\infty (1 + |x|)|V(x)| dx < \infty$$

for some $c > 0$. The Schrödinger operator H has negative discrete spectrum, consisting of finitely many simple eigenvalues $\lambda_1 < \cdots < \lambda_n < 0$, and absolutely continuous spectrum $[0, \infty)$, which is simple for $0 < \lambda < c^2$ and is of multiplicity two for $\lambda > c^2$. This qualitative structure of the spectrum is determined by the structure of the level sets $H_c(p, x) = \lambda$ of the classical Hamiltonian function $H_c(p, x) = p^2 + V(x)$: the eigenvalues could only appear for compact level sets, when the classical motion is periodic, while the values of λ with non-compact level sets, where the classical motion is infinite, belong to the absolutely continuous spectrum. The absolutely continuous spectrum has multiplicity one or two depending on whether the corresponding classical motion is unbounded in one or in both directions, i.e., when $0 < \lambda < c^2$ or $\lambda > c^2$.

These results can be proved using methods in Section 2.1. Namely, set $\lambda = k^2$ and define the function $k_1 = \sqrt{k^2 - c^2}$ by the condition that $\operatorname{Im} k_1 \geq 0$ for $\operatorname{Im} k \geq 0$. In particular, $\operatorname{sgn} k_1 = \operatorname{sgn} k$ for k real, $|k| > c$. The differential equation

$$(2.43) \qquad\qquad -y'' + V(x)y = k^2 y$$

for real k has two linear independent solutions $f_1(x,k)$, $f_1(x,-k) = \overline{f_1(x,k)}$, where $f_1(x,k)$ has the asymptotics

$$f_1(x,k) = e^{ikx} + o(1) \quad \text{as} \quad x \to \infty.$$

For real k, $|k| > c$, equation (2.43) also has two linear independent solutions $f_2(x,k)$, $f_2(x,-k) = \overline{f_2(x,-k)}$, where $f_2(x,k)$ has the asymptotics

$$f_2(x,k) = e^{-ik_1 x} + o(1) \quad \text{as} \quad x \to -\infty.$$

(Actually, these solutions satisfy estimates similar to that in Theorem 2.1.) As in Section 2.1, for real k, $|k| > c$, we have

$$f_2(x,k) = a(k)f_1(x,-k) + b(k)f_1(x,k),$$

where

$$(2.44) \quad a(k) = \frac{1}{2ik}W(f_1(x,k), f_2(x,k)), \quad b(k) = \frac{1}{2ik}W(f_2(x,k), f_1(x,-k)),$$

and $a(-k) = \overline{a(k)}$, $b(-k) = \overline{b(k)}$. However, $W(f_2(x,k), f_2(x,-k)) = -2ik_1$, so that for real k, $|k| > c$, we have

$$f_1(x,k) = \frac{k}{k_1}a(k)f_2(x,-k) - \frac{k}{k_1}b(-k)f_2(x,k).$$

From here we obtain the normalization condition

$$|a(k)|^2 - |b(k)|^2 = \frac{k_1}{k}, \quad |k| > c.$$

For fixed x solutions $f_1(x,k)$ and $f_2(x,k)$ can be analytically continued to the upper half-plane $\operatorname{Im} k > 0$. For $-c < k < c$ solution $f_2(x,k)$ is real-valued and satisfies

$$f_2(x,k) = a(k)f_1(x,-k) + \overline{a(k)f_1(x,-k)},$$

where $a(k)$ is still given by the same formula (2.44). The function $a(k)$ does not vanish for real k and admits meromorphic continuation to the upper half-plane $\operatorname{Im} k > 0$ where it has finitely many pure imaginary simple zeros $i\varkappa_1, \dots, i\varkappa_n$ which correspond to the eigenvalues $\lambda_1 = -\varkappa_1^2, \dots, \lambda_n = -\varkappa_n^2$.

The absolutely continuous spectrum of the Schrödinger operator H fills $[0, \infty)$. For $0 < \lambda = k^2 < c^2$ the spectrum is simple and

$$u(x, \lambda) = \frac{1}{a(k)} f_2(x, k), \quad 0 < k < c,$$

are the corresponding normalized eigenfunctions of the continuous spectrum. For $\lambda = k^2 > c^2$ the spectrum has multiplicity two and

$$u_1(x, \lambda) = \frac{1}{a(k)} f_2(x, k), \quad u_2(x, k) = \frac{k_1}{ka(k)} f_1(x, k), \quad k > c,$$

are the corresponding normalized eigenfunctions of the continuous spectrum. Denoting by $\psi_l(x)$ the normalized eigenfunctions of H corresponding to the eigenvalues λ_l, we get the eigenfunction expansion theorem: for $\psi \in L^2(\mathbb{R})$,

$$\psi(x) = \sum_{l=1}^{n} C_l \psi_l(x) + \int_0^c C(k) u(x, k) dk$$

$$+ \int_c^\infty (C_1(k) u_1(x, k) + C_2(k) u_2(x, k)) dk,$$

where $C_l = (\psi, \psi_l)$, $l = 1, \ldots, n$, and

$$C(k) = \int_{-\infty}^\infty \psi(x) \overline{u(x, k)} dx, \quad C_j(k) = \int_{-\infty}^\infty \psi(x) \overline{u_j(x, k)} dx, \quad j = 1, 2.$$

Example 2.2. Suppose that the potential $V(x)$ grows as $x \to -\infty$ and decays as $x \to \infty$:

$$\int_a^\infty (1 + |x|) |V(x)| dx < \infty \quad \text{for all} \quad a,$$

and there exists $x_0 \in \mathbb{R}$ such that the spectrum of the Sturm-Liouville problem

$$-y'' + V(x)y = \lambda y, \quad -\infty < x \le x_0, \quad \text{and} \quad y'(x_0) = 0$$

is bounded below and discrete. The latter condition is a quantitative formulation of the property $\lim_{x \to -\infty} V(x) = \infty$.

Then for every real k there exists a solution $f(x, k)$ of the differential equation (2.43) with the asymptotics

$$f(x, k) = e^{ikx} + o(1) \quad \text{as} \quad x \to \infty,$$

and there is a function $s(k)$ such that

$$u(x, k) = f(x, -k) + s(k) f(x, k)$$

is square integrable over $(-\infty, a)$ for any real a. The Schrödinger operator H has simple absolutely continuous spectrum $[0, \infty)$ and discrete spectrum consisting of finitely many negative eigenvalues. Functions $u(x, k)$ are normalized eigenfunctions of the continuous spectrum and the corresponding eigenfunction expansion theorem has the following form: for every $\psi \in L^2(\mathbb{R})$,

$$\psi(x) = \sum_{l=1}^{n} C_l \psi_l(x) + \int_0^{\infty} C(k) u(x, k) dk,$$

where $C_l = (\psi, \psi_l)$, $l = 1, \dots, n$, and

$$C(k) = \int_{-\infty}^{\infty} \psi(x) \overline{u(x, k)} dx.$$

Problem 2.10. Give an explicit form of the eigenfunction expansion theorem for the potential $V(x) = e^x$.

Problem 2.11. Find energy levels for the *Morse potential* $V(x) = e^{-2\alpha x} - 2e^{-\alpha x}$, $\alpha > 0$.

Problem 2.12. Give an explicit form of the eigenfunction expansion theorem for the potential $V(x) = Fx$, describing the motion of a quantum particle in a homogeneous field; according to Problem 1.4, the corresponding Schrödinger operator is self-adjoint. (*Hint:* Solve the Schrödinger equation explicitly in the momentum representation, and express normalized eigenfunctions of the continuous spectrum in the coordinate representation in terms of Airy-Fock functions, defined in Section 6.2.)

Problem 2.13. Give an explicit form of the eigenfunction expansion theorem for the potential $V(x) = -\frac{1}{2}kx^2$, where $k > 0$ (according to Problem 1.4, the corresponding Schrödinger operator is self-adjoint).

3. Angular momentum and $\mathrm{SO}(3)$

3.1. Angular momentum operators. In Chapter 1 (see Example 1.10 in Section 1.4) for a classical particle in $\mathbb{R}^3$ we introduced the angular momentum vector $\boldsymbol{M}_c = \boldsymbol{x} \times \boldsymbol{p}$ with components

$$M_{c1} = x_2 p_3 - x_3 p_2, \quad M_{c2} = x_3 p_1 - x_1 p_3, \quad M_{c3} = x_1 p_2 - x_2 p_1.$$

According to Example 2.1 in Section 2.6 of Chapter 1, they have the following Poisson brackets with respect to the canonical Poisson structure on $T^*\mathbb{R}^3$:

$$(3.1) \quad \{M_{c1}, M_{c2}\} = -M_{c3}, \quad \{M_{c2}, M_{c3}\} = -M_{c1}, \quad \{M_{c3}, M_{c1}\} = -M_{c2}.$$

The square of the angular momentum $\boldsymbol{M}_c^2 = M_{c1}^2 + M_{c2}^2 + M_{c3}^2$ satisfies

$$\{\boldsymbol{M}_c^2, M_{c1}\} = \{\boldsymbol{M}_c^2, M_{c2}\} = \{\boldsymbol{M}_c^2, M_{c3}\} = 0.$$

If the Hamiltonian function

$$H_{\mathrm{c}}(\boldsymbol{p}, \boldsymbol{x}) = \frac{\boldsymbol{p}^2}{2m} + V(\boldsymbol{x})$$

is invariant under rotations, $V(\boldsymbol{x}) = V(|\boldsymbol{x}|)$, then components of the angular momentum are the integrals of motion

$$\{H_{\mathrm{c}}, M_{\mathrm{c}1}\} = \{H_{\mathrm{c}}, M_{\mathrm{c}2}\} = \{H_{\mathrm{c}}, M_{\mathrm{c}3}\} = 0 \quad \text{and} \quad \{H_{\mathrm{c}}, \boldsymbol{M}_{\mathrm{c}}^2\} = 0.$$

This can also be verified directly using Poisson brackets

$$(3.2) \quad \{M_{\mathrm{c}j}, p_k\} = -\varepsilon_{jkl} p_l \quad \text{and} \quad \{M_{\mathrm{c}j}, x_k\} = -\varepsilon_{jkl} x_l, \quad i, j, k = 1, 2, 3,$$

where ε_{jkl} is a totally anti-symmetric tensor, $\varepsilon_{123} = 1$.

Correspondingly, in quantum mechanics the components of the angular momentum operator $\boldsymbol{M} = \boldsymbol{Q} \times \boldsymbol{P}$ are defined by

$$M_1 = Q_2 P_3 - Q_3 P_2, \quad M_2 = Q_3 P_1 - Q_1 P_3, \quad M_3 = Q_2 P_3 - Q_3 P_2,$$

where $\boldsymbol{Q} = (Q_1, Q_2, Q_3)$ and $\boldsymbol{P} = (P_1, P_2, P_3)$ are, respectively, coordinate and momentum operators. Since operators Q_i and P_k commute for $i \neq k$, there is no ordering problem when defining quantum angular momentum operators. It follows from Heisenberg commutation relations that their quantum brackets are the same as the corresponding Poisson brackets (3.1):

$$(3.3) \quad \{M_1, M_2\}_\hbar = -M_3, \quad \{M_2, M_3\}_\hbar = -M_1, \quad \{M_3, M_1\}_\hbar = -M_2.$$

Equivalently,

$$(3.4) \quad [M_1, M_2] = i\hbar M_3, \quad [M_2, M_3] = i\hbar M_1, \quad [M_3, M_1] = i\hbar M_2.$$

The operator of the square of the total angular momentum $\boldsymbol{M}^2 = M_1^2 + M_2^2 + M_3^2$ satisfies

$$[\boldsymbol{M}^2, M_1] = [\boldsymbol{M}^2, M_2] = [\boldsymbol{M}^2, M_3] = 0.$$

Correspondingly, for the Hamiltonian operator

$$H = \frac{\boldsymbol{P}^2}{2m} + V(\boldsymbol{Q})$$

with spherically symmetric potential $V(\boldsymbol{x}) = V(|\boldsymbol{x}|)$ operators M_1, M_2, M_3 and, therefore, $\boldsymbol{M}^2$, are quantum integrals of motion:

$$[H, M_1] = [H, M_2] = [H, M_3] = 0 \quad \text{and} \quad [H, \boldsymbol{M}^2] = 0.$$

This can be verified directly by using quantum brackets

$$(3.5) \quad \{M_j, P_k\}_\hbar = -\varepsilon_{jkl} P_l, \quad \{M_j, Q_k\}_\hbar = -\varepsilon_{jkl} Q_l, \quad j, k, l = 1, 2, 3,$$

which are the same as Poisson brackets (3.2), and follow from Heisenberg commutation relations.

In the coordinate representation $\mathscr{H} = L^2(\mathbb{R}^3, d^3\boldsymbol{x})$ the operators of angular momentum are given by the following first order self-adjoint differential operators:

$$(3.6) \qquad M_1 = i\hbar\left(x_3\frac{\partial}{\partial x_2} - x_2\frac{\partial}{\partial x_3}\right),$$

$$(3.7) \qquad M_2 = i\hbar\left(x_1\frac{\partial}{\partial x_3} - x_3\frac{\partial}{\partial x_1}\right),$$

$$(3.8) \qquad M_3 = i\hbar\left(x_2\frac{\partial}{\partial x_1} - x_1\frac{\partial}{\partial x_2}\right).$$

They have the property that

$$M_1\psi = M_2\psi = M_3\psi = 0$$

for any spherically symmetric smooth function $\psi(\boldsymbol{x}) = \psi(|\boldsymbol{x}|)$. In other words, angular momentum operators act only on the angle coordinates in $\mathbb{R}^3$. Namely, let

$$x_1 = r\sin\vartheta\cos\varphi, \quad x_2 = r\sin\vartheta\sin\varphi, \quad x_3 = r\cos\vartheta,$$

where $0 \leq \vartheta < \pi$, $0 \leq \varphi < 2\pi$, be the spherical coordinates in $\mathbb{R}^3$. Explicit computation gives

$$M_1 = i\hbar\left(\sin\varphi\frac{\partial}{\partial\vartheta} + \cot\vartheta\cos\varphi\frac{\partial}{\partial\varphi}\right),$$

$$M_2 = -i\hbar\left(\cos\varphi\frac{\partial}{\partial\vartheta} - \cot\vartheta\sin\varphi\frac{\partial}{\partial\varphi}\right),$$

$$M_3 = -i\hbar\frac{\partial}{\partial\varphi}.$$

Thus

$$\boldsymbol{M}^2 = -\hbar^2\left(\frac{1}{\sin\vartheta}\frac{\partial}{\partial\vartheta}\left(\sin\vartheta\frac{\partial}{\partial\vartheta}\right) + \frac{1}{\sin^2\vartheta}\frac{\partial^2}{\partial\varphi^2}\right),$$

so that the operator $-\dfrac{1}{\hbar^2}\boldsymbol{M}^2$ is the spherical part of the Laplace operator in $\mathbb{R}^3$,

$$(3.9) \qquad \Delta = \frac{1}{r^2}\frac{\partial}{\partial r}\left(r^2\frac{\partial}{\partial r}\right) - \frac{1}{\hbar^2 r^2}\boldsymbol{M}^2.$$

3.2. Representation theory of $\mathrm{SO}(3)$. Quantum operators of angular momentum are related with the representation theory of the rotation group $\mathrm{SO}(3)$ — the group of 3×3 orthogonal matrices with determinant 1. $\mathrm{SO}(3)$ is a compact Lie group, isomorphic to the real projective space $\mathbb{RP}^3$ as a smooth manifold. There is a Lie group isomorphism $\mathrm{SO}(3) \simeq \mathrm{SU}(2)/\{\pm I\}$, where $\mathrm{SU}(2)$ is a Lie group of 2×2 unitary matrices with determinant 1.

The Lie algebra $\mathfrak{so}(3)$ of SO(3) is a three-dimensional Lie algebra of 3×3 skew-symmetric matrices, with the basis

$$X_1 = \begin{pmatrix} 0 & 0 & 0 \\ 0 & 0 & -1 \\ 0 & 1 & 0 \end{pmatrix}, \quad X_2 = \begin{pmatrix} 0 & 0 & 1 \\ 0 & 0 & 0 \\ -1 & 0 & 0 \end{pmatrix}, \quad X_3 = \begin{pmatrix} 0 & -1 & 0 \\ 1 & 0 & 0 \\ 0 & 0 & 0 \end{pmatrix}.$$

The matrices X_1, X_2, X_3 correspond, respectively, to the one-parameter subgroups of SO(3) consisting of rotations about coordinate axes in $\mathbb{R}^3$. They satisfy commutation relations

$$[X_1, X_2] = X_3, \quad [X_2, X_3] = X_1, \quad [X_3, X_1] = X_2,$$

which are similar to (3.4). To establish the connection between quantum angular momentum operators and representation theory of SO(3), consider the regular representation R of SO(3) in $\mathscr{H} = L^2(\mathbb{R}^3, d^3\boldsymbol{x})$, defined by

$$(R(g)\psi)(\boldsymbol{x}) = \psi(g^{-1}\boldsymbol{x}), \quad g \in \mathrm{SO}(3), \ \psi \in \mathscr{H}.$$

Lemma 3.1. *We have*

$$R(e^{u_1 X_1 + u_2 X_2 + u_3 X_3}) = e^{-\frac{i}{\hbar}(u_1 M_1 + u_2 M_2 + u_3 M_3)}.$$

Proof. Put $\boldsymbol{uX} = u_1 X_1 + u_2 X_2 + u_3 X_3$ and $\boldsymbol{uM} = u_1 M_1 + u_2 M_2 + u_3 M_3$. It follows from the Stone theorem and direct computation that

$$i\hbar \left.\frac{\partial}{\partial u_j}\right|_{\boldsymbol{u}=0} R(e^{\boldsymbol{uX}}) = M_j, \quad j = 1, 2, 3,$$

where self-adjoint operators M_1, M_2, M_3 are given by (3.6)-(3.8). Next, for fixed u_1, u_2, u_3, consider the following one-parameter group $U(t) = R(e^{t\boldsymbol{uX}})$ of unitary operators in $\mathscr{H}$. By the Stone theorem, $U(t) = e^{-itA}$, where

$$A = i \left.\frac{d}{dt}\right|_{t=0} U(t) = \frac{1}{\hbar}\boldsymbol{uM}. \qquad \square$$

Remark. We have $M_j = i\hbar\rho(X_j)$, where $\rho = dR$ is the corresponding regular representation of the Lie algebra $\mathfrak{so}(3)$ in $\mathscr{H}$.

All irreducible unitary representations R_l of the Lie group SO(3) are finite-dimensional and are parametrized by non-negative integers $l \geq 0$. The corresponding irreducible representation $\rho_l = dR_l$ of the Lie algebra $\mathfrak{so}(3)$ in $2l + 1$-dimensional complex vector space V_l can be explicitly described as follows. Introduce Hermitian operators $T_j = i\rho_l(X_j)$, $j = 1, 2, 3$, which satisfy commutation relations

$$[T_1, T_2] = iT_3, \quad [T_3, T_1] = iT_2, \quad [T_2, T_3] = iT_1.$$

The vector space V_l has an orthonormal basis $\{e_{lm}\}_{m=-l}^{m=l}$ such that

$$(3.10) \qquad (T_1 - iT_2)e_{lm} = -\sqrt{(l+m)(l-m+1)}\,e_{lm-1},$$

$$(3.11) \qquad (T_1 + iT_2)e_{lm} = -\sqrt{(l-m)(l+m+1)}\,e_{lm+1},$$

$$(3.12) \qquad T_3 e_{lm} = m e_{lm}.$$

In particular, $(T_1 + iT_2)e_{ll} = 0$, so that V_l is the highest weight module. The representation ρ_l is irreducible, and we have by Schur's lemma

$$\boldsymbol{T}^2 = l(l+1)I_l,$$

where I_l is the identity operator in V_l. This can also be verified directly by using (3.10)–(3.12).

Remark. When l is a half-integer, i.e., $l \in \frac{1}{2} + \mathbb{Z}_{\geq 0}$, and $m = -l, -l + 1, \ldots, l-1, l$, formulas (3.10)–(3.12) still define an irreducible highest weight representation ρ_l of the Lie algebra $\mathfrak{so}(3)$ of dimension $2l + 1$, and every irreducible n-dimensional representation of $\mathfrak{so}(3)$ is isomorphic to the representation ρ_l with $l = \frac{n-1}{2}$. For half-integer l, representations ρ_l are not integrable: they give rise to two-valued representations of SO(3), the so-called *spinor representations*. However, considered as representations of the Lie algebra $\mathfrak{su}(2) \simeq \mathfrak{so}(3)$, ρ_l correspond to the irreducible unitary representations of the Lie group SU(2). It follows from the representation theory of SU(2) that for $l, l' \in \frac{1}{2}\mathbb{Z}_{\geq 0}$,

$$(3.13) \qquad V_l \otimes V_{l'} = \bigoplus_{j=|l-l'|}^{l+l'} V_j,$$

the so-called *Clebsch-Gordan decomposition*. In physics, it corresponds to the *addition of angular momenta*.

The regular representation R of SO(3) in $\mathscr{H} = L^2(\mathbb{R}^3, d^3\boldsymbol{x})$ is not irreducible. We have

$$(3.14) \qquad \mathscr{H} = L^2(S^2, d\boldsymbol{n}) \otimes L^2(\mathbb{R}_{>0}, r^2 dr),$$

where $d\boldsymbol{n}$ is the measure on S^2 induced by the Lebesgue measure on $\mathbb{R}^3$. The group SO(3) acts by rotations in the first factor of the tensor product (3.14), whereas in the second factor it acts as the identity operator. Thus the problem of decomposing the regular representation R reduces to finding SO(3) invariant subspaces of the Hilbert space $L^2(S^2, d\boldsymbol{n})$. The result is the orthogonal sum decomposition

$$(3.15) \qquad L^2(S^2, d\boldsymbol{n}) = \bigoplus_{l=0}^{\infty} \mathscr{D}_l,$$

where $\mathscr{D}_l \simeq V_l$. Under this isomorphism, the orthonormal basis Y_{lm} in $\mathscr{D}_l$ which corresponds to the basis e_{lm} in V_l is given by the normalized spherical functions

$$Y_{lm}(\vartheta, \varphi) = \frac{1}{\sqrt{2\pi}} e^{im\varphi} P_l^m(\cos\vartheta), \quad m = -l, \ldots, l,$$

where $P_l^m(x)$ are normalized associated Legendre polynomials,

$$P_l^m(x) = (-1)^m \sqrt{\frac{(l+m)!}{(l-m)!}} \sqrt{\frac{2l+1}{2}} \frac{1}{2^l l!} (1-x^2)^{-\frac{m}{2}} \frac{d^{l-m}}{dx^{l-m}} (x^2-1)^l, \quad |x| < 1.$$

The self-adjoint operator $\boldsymbol{M}^2$ in the Hilbert space $L^2(S^2, d\boldsymbol{n})$ has pure point spectrum consisting of eigenvalues $\hbar^2 l(l+1)$ of multiplicities $2l+1$, $l = 0, 1, 2, \ldots$, and the decomposition (3.15) gives its eigenfunction expansion: for every $\psi \in L^2(S^2, d\boldsymbol{n})$,

$$\psi = \sum_{l=0}^{\infty} \sum_{m=-l}^{l} C_{lm} Y_{lm}, \quad \text{where} \quad C_{lm} = \int_0^{2\pi} \int_0^{\pi} \psi(\vartheta, \varphi) \overline{Y_{lm}(\vartheta, \varphi)} \sin\vartheta \, d\vartheta \, d\varphi.$$

Problem 3.1. Prove all results in this section. (*Hint*: See the list of references to this chapter.)

4. Two-body problem

4.1. Separation of the center of mass.

Consider the Schrödinger operator for the two-body problem (see Example 2.2 in Section 2.4 in Chapter 2)

$$H = -\frac{\hbar^2}{2m_1} \Delta_1 - \frac{\hbar^2}{2m_2} \Delta_2 + V(\boldsymbol{x}_1 - \boldsymbol{x}_2).$$

Introducing

$$\boldsymbol{X} = \frac{m_1 \boldsymbol{x}_1 + m_2 \boldsymbol{x}_2}{m_1 + m_2} \quad \text{and} \quad \boldsymbol{x} = \boldsymbol{x}_1 - \boldsymbol{x}_2,$$

the coordinate of the center of mass and the relative coordinate, we get

$$H = -\frac{\hbar^2}{2M} \Delta_{\boldsymbol{X}} - \frac{\hbar^2}{2\mu} \Delta_{\boldsymbol{x}} + V(\boldsymbol{x}),$$

where $M = m_1 + m_2$ is the total mass and $\mu = \dfrac{m_1 m_2}{m_1 + m_2}$ is the reduced mass. The Hamiltonian operator H can be diagonalized by the method of separation of variables. Namely, consider the following decomposition of the two-body Hilbert space $\mathscr{H} = L^2(\mathbb{R}^6)$ into the tensor product of Hilbert spaces:

$$(4.1) \qquad \mathscr{H} = L^2(\mathbb{R}^3, d^3\boldsymbol{X}) \otimes L^2(\mathbb{R}^3, d^3\boldsymbol{x}).$$

The operator

$$(4.2) \qquad H_{\boldsymbol{X}} = -\frac{\hbar^2}{2M}\Delta_{\boldsymbol{X}}$$

acts as the identity operator in the second factor of (4.1), and the operator

$$(4.3) \qquad H_{\boldsymbol{x}} = -\frac{\hbar^2}{2\mu}\Delta_{\boldsymbol{x}} + V(\boldsymbol{x})$$

acts as the identity operator in the first factor of (4.1). For the solutions of the eigenvalue problem

$$(4.4) \qquad H\psi = E\psi$$

in the product form

$$\psi(\boldsymbol{X}, \boldsymbol{x}) = \Psi(\boldsymbol{X})\psi(\boldsymbol{x}),$$

the variables are separated:

$$H_{\boldsymbol{X}}\Psi(\boldsymbol{X}) = E_1\Psi(\boldsymbol{X}) \quad \text{and} \quad H_{\boldsymbol{x}}\psi(\boldsymbol{x}) = E_2\psi(\boldsymbol{x}),$$

where $E = E_1 + E_2$. Since

$$\Psi(\boldsymbol{X}) = \left(\frac{1}{2\pi\hbar}\right)^{\frac{3}{2}} e^{\frac{i}{\hbar}\boldsymbol{k}\boldsymbol{X}}, \quad \boldsymbol{k}^2 = 2ME_1,$$

the quantum two-body problem reduces to the problem of a quantum particle moving in the potential field, and is described by the Hamiltonian operator (4.3) in the Hilbert space $L^2(\mathbb{R}^3, d^3\boldsymbol{x})$.

4.2. Three-dimensional scattering theory. Here we outline the scattering theory for the Schrödinger operator $H = -\Delta + V(\boldsymbol{x})$ (where we put $\hbar = 1$ and $\mu = \frac{1}{2}$) with rapidly decreasing potential. Specifically, assume that the bounded, real-valued function $V(\boldsymbol{x})$ on $\mathbb{R}^3$ satisfies

$$(4.5) \qquad V(\boldsymbol{x}) = O(|\boldsymbol{x}|^{-3-\varepsilon}) \quad \text{as} \quad |\boldsymbol{x}| \to \infty$$

for some $\varepsilon > 0$. Then for every $\boldsymbol{k} \in \mathbb{R}^3$ the Schrödinger equation

$$(4.6) \qquad -\Delta\psi(\boldsymbol{x}) + V(\boldsymbol{x})\psi(\boldsymbol{x}) = k^2\psi(\boldsymbol{x}), \quad k = |\boldsymbol{k}|,$$

has two solutions $u^{(\pm)}(\boldsymbol{x}, \boldsymbol{k})$ satisfying the following asymptotics as $|\boldsymbol{x}| \to \infty$:

$$(4.7) \qquad u^{(\pm)}(\boldsymbol{x}, \boldsymbol{k}) = e^{i\boldsymbol{k}\boldsymbol{x}} + f^{(\pm)}(k, \boldsymbol{\omega}, \boldsymbol{n})\frac{e^{\pm ikr}}{r} + o\left(\frac{1}{r}\right),$$

where $\boldsymbol{k} = k\boldsymbol{\omega}$, $\boldsymbol{x} = r\boldsymbol{n}$. Asymptotics (4.7) are called *Sommerfeld's radiation conditions*. To prove the existence of solutions $u^{(\pm)}(\boldsymbol{x}, \boldsymbol{k})$, one should consider the following integral equations:

$$(4.8) \qquad u^{(\pm)}(\boldsymbol{x}, \boldsymbol{k}) = e^{i\boldsymbol{k}\boldsymbol{x}} + \int_{\mathbb{R}^3} G^{(\pm)}(\boldsymbol{x} - \boldsymbol{y}, k)V(\boldsymbol{y})u^{(\pm)}(\boldsymbol{y}, \boldsymbol{k})d^3\boldsymbol{y},$$

where

$$G^{(\pm)}(\boldsymbol{x}, k) = -\frac{1}{4\pi}\frac{e^{\pm ikr}}{r}.$$

Integral equations (4.8) are equivalent to the Schrödinger equation (4.6) with Sommerfeld's radiation conditions (4.7), and are called *Lippman-Schwinger equations*. Their analysis uses Fredholm theory and Kato's Theorem 1.8.

Solutions $u^{(\pm)}(\boldsymbol{x}, \boldsymbol{k})$ are called *stationary scattering waves*. They are analogous to the solutions $u_j^{(\pm)}(x, k)$ for the one-dimensional case, where the vector $\boldsymbol{\omega} \in S^2$ replaces the index $j = 1, 2$. The absolutely continuous spectrum of H fills $[0, \infty)$ and has a uniform infinite multiplicity, parametrized by the two-dimensional sphere S^2. Solutions $u^{(\pm)}(\boldsymbol{x}, \boldsymbol{k})$ are normalized eigenfunctions of the continuous spectrum. In general, the operator H has finitely many negative eigenvalues $\lambda_l < 0$ of finite multiplicities m_l, $l = 1, \ldots, n$.

In terms of the operators $\mathscr{U}_\pm : \mathscr{H} \to \mathscr{H}$ given by

$$\mathscr{U}_\pm(\psi)(\boldsymbol{k}) = (2\pi)^{-\frac{3}{2}} \int_{\mathbb{R}^3} \psi(\boldsymbol{x}) u^{(\pm)}(\boldsymbol{x}, \boldsymbol{k}) d^3\boldsymbol{x},$$

completeness and orthogonality relations take the form

$$\mathscr{U}_\pm^* \mathscr{U}_\pm = I - P, \quad \mathscr{U}_\pm \mathscr{U}_\pm^* = I,$$

where P is the orthogonal projection on the invariant subspace for the operator H associated with the pure point spectrum. Now let

$$W_\pm = \mathscr{U}_\pm^* \mathscr{U}_0,$$

where $\mathscr{U}_0 = \mathscr{F}^{-1}$ is the inverse Fourier transform. As in Section 2.3, one can prove the following result.

Theorem 4.1. *The operators $W_\pm$ are wave operators for the Schrödinger operator H. The corresponding scattering operator $S = W_+^* W_-$ has the form*

$$S = \mathscr{F} \hat{S} \mathscr{F}^{-1},$$

where $\hat{S}$ is the integral operator,

$$(\hat{S}\psi)(k, \boldsymbol{\omega}) = \psi(k, \boldsymbol{\omega}) + \frac{ik}{2\pi} \int_{S^2} f(k, \boldsymbol{\omega}, \boldsymbol{\omega}')\psi(k, \boldsymbol{\omega}')d\boldsymbol{\omega}',$$

and $f(k, \boldsymbol{\omega}, \boldsymbol{\omega}') = f^{(+)}(k, \boldsymbol{\omega}, \boldsymbol{\omega}')$.

The function $f(k, \boldsymbol{\omega}, \boldsymbol{\omega}')$ is called the *scattering amplitude*.

Problem 4.1. Prove that if

$$\sup_{\boldsymbol{x} \in \mathbb{R}^3} \int_{\mathbb{R}^3} |V(\boldsymbol{y})| \frac{1}{|\boldsymbol{x} - \boldsymbol{y}|} d^3\boldsymbol{y} < 4\pi,$$

then the Lippman-Schwinger integral equations can be solved by Neumann series and for fixed $\boldsymbol{x}$ and $\boldsymbol{\omega}$ solutions $u^{(\pm)}(\boldsymbol{x}, \boldsymbol{k})$, $\boldsymbol{k} = k\boldsymbol{\omega}$, admit analytic continuation to the upper half-plane $\operatorname{Im} k > 0$. Show also that in this case the Schrödinger operator H has no eigenvalues.

Problem 4.2 (Born's approximation). Show that $u^{(+)}(\boldsymbol{x}, \boldsymbol{k}) = e^{i\boldsymbol{k}\boldsymbol{x}} + o(1)$ as $k \to \infty$, and deduce from this that

$$f(k, \boldsymbol{n}, \boldsymbol{\omega}) + \frac{1}{4\pi} \int_{\mathbb{R}^3} e^{ik(\boldsymbol{n}-\boldsymbol{\omega})\boldsymbol{x}} V(\boldsymbol{x}) d^3\boldsymbol{x} = o(1) \quad \text{as} \quad k \to \infty,$$

uniformly on $\boldsymbol{n}, \boldsymbol{\omega} \in S^2$.

Problem 4.3. Prove unitarity of the operator $\hat{S}$ using Schrödinger equation (4.6), radiation conditions (4.7), and the Green's formula.

Problem 4.4 (The optical theorem). Show that

$$\int_{S^2} |f(k, \boldsymbol{n}, \boldsymbol{\omega})|^2 d\boldsymbol{n} = \frac{4\pi}{k} \operatorname{Im} f(k, \boldsymbol{\omega}, \boldsymbol{\omega}).$$

(Here the left-hand side is the *total cross-section* in direction $\boldsymbol{\omega}$ at the energy $E = k^2$.)

Problem 4.5. Let $\lambda_l = -\varkappa_l^2 < 0$ be the eigenvalues of H with multiplicities m_l, $l = 1, \ldots, n$. Prove that for fixed $\boldsymbol{x}$ and $\boldsymbol{\omega}$ solutions $u^{(\pm)}(\boldsymbol{x}, \boldsymbol{k})$ admit a meromorphic continuation to the upper half-plane $\operatorname{Im} k > 0$ with poles of orders m_l at $i\varkappa_l$, $l = 1, \ldots, n$.

Problem 4.6. Prove that the wave operators $W_\pm$ exist using the non-stationary approach. (*Hint*: Show that when $V(\boldsymbol{x})$ satisfies (4.5), Cook's criterion, formulated in Problem 2.7, is applicable.)

4.3. Particle in a central potential. The eigenvalue problem for the Schrödinger operator

$$H = -\frac{\hbar^2}{2\mu} \Delta + V(\boldsymbol{x})$$

simplifies, when H commutes with the SO(3)-action in $\mathcal{H} = L^2(\mathbb{R}^3, d^3\boldsymbol{x})$:

$$[H, T(g)] = 0 \quad \text{for all} \quad g \in \text{SO}(3).$$

This reduces to the conditions

$$(4.9) \qquad\qquad [H, M_i] = 0, \quad i = 1, 2, 3,$$

and is equivalent to the property that the potential V is spherically symmetric, $V(\boldsymbol{x}) = V(r)$, $r = |\boldsymbol{x}|$. In particular

$$[H, M_3] = [H, \boldsymbol{M}^2] = 0,$$

where $\boldsymbol{M}^2 = M_1^2 + M_2^2 + M_3^2$, and operators M_3 and $\boldsymbol{M}^2$ are commuting quantum integrals of motion for the Hamiltonian H. As follows from results in Section 3.2, one can look for the solutions of the eigenvalue problem

$$H\psi = E\psi$$

satisfying

$$M_3\psi = m\psi \quad \text{and} \quad \boldsymbol{M}^2\psi = \hbar^2 l(l+1)\psi, \quad m = -l, \ldots, l.$$

Using (3.9), we get

$$H = -\frac{\hbar^2}{2\mu r^2}\frac{\partial}{\partial r}\left(r^2\frac{\partial}{\partial r}\right) + \frac{\boldsymbol{M}^2}{\mu r^2} + V(r),$$

so that in accordance with the decompositions (3.14) and (3.15), we look for the solutions in the form

$$\psi(\boldsymbol{x}) = R_l(r)Y_{lm}(\boldsymbol{n}), \quad \boldsymbol{x} = r\boldsymbol{n},$$

where Y_{lm} are normalized spherical functions. Separating the variables, one gets the following ordinary differential equation for the function $R_l(r)$:

$$\frac{\hbar^2}{2\mu r^2}\frac{d}{dr}\left(r^2\frac{dR_l}{dr}\right) + \frac{\hbar^2 l(l+1)}{\mu r^2}R_l + V(r)R_l = ER_l.$$

Introducing

$$f_l(r) = rR_l(r),$$

we obtain the so-called *radial Schrödinger equation*

$$(4.10) \qquad -\frac{\hbar^2}{2\mu}\frac{d^2 f_l}{dr^2} + \frac{\hbar^2 l(l+1)}{2\mu r^2}f_l + V(r)f_l = Ef_l.$$

Since for the continuous potential $V(\boldsymbol{x})$ the solution $\psi(\boldsymbol{x})$ is also continuous, equation (4.10) should be supplemented by the boundary condition $f_l(0) = 0$.

The radial Schrödinger equation looks similar to the Schrödinger equation for a one-dimensional particle, if one introduces the so-called *effective potential*

$$V_{\text{eff}}(r) = V(r) + \frac{\hbar^2 l(l+1)}{2\mu r^2},$$

where the second term is called the centrifugal energy. However, since f_l is defined only for $r > 0$, equation (4.10) is equivalent to equation (2.1) with the potential satisfying $V(x) = \infty$ for $x < 0$, which describes the infinite potential barrier at $x = 0$.

Since the radial Schrödinger operators

$$H_l = -\frac{\hbar^2}{2\mu}\frac{d^2}{dr^2} + V(r) + \frac{\hbar^2 l(l+1)}{2\mu r^2}$$

are obtained from the three-dimensional Schrödinger operator

$$H = -\frac{\hbar^2}{2\mu}\Delta + V(r)$$

by separation of variables, the operators H_l with boundary condition $f_l(0) = 0$ are self-adjoint in $L^2(0,\infty)$ whenever H is a self-adjoint operator in $L^2(\mathbb{R}^3)$.

In particular, it follows from Theorem 1.9 in Section 1.2 that when the bounded potential V satisfies

$$V(r) = O(r^{-1-\varepsilon}) \quad \text{as} \quad r \to \infty$$

for some $\varepsilon > 0$, then H_l are self-adjoint operators with simple absolutely continuous spectrum filling $[0, \infty)$, and negative eigenvalues with possible accumulation point at 0. If

$$V(r) = O(r^{-2-\varepsilon}) \quad \text{as} \quad r \to \infty,$$

then the operators H_l have only finitely many negative eigenvalues, and for l large enough have no eigenvalues at all. The same conclusion holds if

$$(4.11) \qquad \int_0^\infty r|V(r)|dr < \infty.$$

Such potentials $V(r)$ are called *short-range potentials*. Decaying at infinity potentials which do not satisfy (4.11) are called *long-range potentials*.

For a short-range potential $V(r)$ the differential equation (4.10) for $E \neq 0$ has two linearly independent solutions $f_l^{\pm}(r)$ satisfying the following asymptotics as $r \to \infty$:

$$(4.12) \qquad f_l^{\pm}(r) = e^{\pm \varkappa r}(1 + o(1)),$$

where $\varkappa = \sqrt{-2\mu E}$ is pure imaginary, $\operatorname{Im} \varkappa > 0$ for $E > 0$, and $\varkappa > 0$ for $E < 0$. When $r \to 0$, the most singular term in the radial Schrödinger equation is given by the centrifugal energy. Since the elementary differential equation

$$f'' = \frac{l(l+1)}{r^2} f$$

has two linearly independent solutions r^{-l} and r^{l+1}, the solution $f_l(r)$ satisfying the boundary condition $f_l(0) = 0$ has the asymptotics

$$(4.13) \qquad f_l(r) = Cr^{l+1} + o(1) \quad \text{as} \quad r \to 0,$$

and is uniquely determined (up to a constant). Since

$$f_l(r) = C_1 f_l^{+}(r) + C_2 f_l^{-}(r),$$

for some constants C_1 and C_2 depending on E, the differential equation (4.10) has no square-integrable solutions for $E > 0$. The corresponding solution $f_l(r)$ is bounded on $[0, \infty)$, and is an eigenfunction of the continuous spectrum. This agrees with the description in Section 1.6 of Chapter 1, since for $E > 0$ the classical particle in the central potential $V_{\text{eff}}(r)$ goes to infinity with finite velocity. For $E < 0$ the equation $C_1(E) = 0$ determines the eigenvalues of H_l, which are simple. This also agrees with the classical picture, since classical motion is finite for $E < 0$. For a short range potential the equation $C_1(E) = 0$ has only finitely many solutions. When

$V_{\text{eff}}(r) > 0$ for $r > 0$ — the case of repulsive potential — the operator H_l has no eigenvalues.

Remark. When $V(r) = 0$, the radial Schrödinger operator H_l has only simple absolutely continuous spectrum filling $[0, \infty)$. The substitution

$$f_l(r) = \sqrt{\xi}\, J(\xi), \quad \xi = \frac{kr}{\hbar} \quad \text{and} \quad k = |\varkappa| = \sqrt{2\mu E} > 0,$$

reduces differential equation (4.10) to the Bessel equation

$$\xi^2 \frac{d^2 J}{d\xi^2} + \xi \frac{dJ}{d\xi} + (\xi^2 - \nu^2) J = 0$$

of the half-integer order $\nu = l + \frac{1}{2}$. The corresponding solution regular at 0 — the Bessel function of the first kind $J_{l+\frac{1}{2}}(\xi)$ — is given explicitly by

$$J_{l+\frac{1}{2}}(\xi) = (-1)^l \sqrt{\frac{2}{\pi\xi}}\, \xi^{l+1} \left(\frac{1}{\xi} \frac{d}{d\xi} \right)^l \frac{\sin\xi}{\xi}$$

and

$$(4.14) \qquad J_{l+\frac{1}{2}}(\xi) = \sqrt{\frac{2}{\pi\xi}} \sin\left(\xi - \frac{l\pi}{2} \right) \quad \text{as} \quad \xi \to \infty.$$

It can be shown that

$$f_{El}(r) = \sqrt{\frac{kr}{\hbar}}\, J_{l+\frac{1}{2}}\left(\frac{kr}{\hbar} \right)$$

satisfy the normalization condition (2.14) in Section 2.2 of Chapter 2:

$$(4.15) \qquad \lim_{\Delta \to 0} \frac{1}{\Delta} \int_0^\infty \left(\int_k^{k+\Delta} f_{El}(r)\, d\sigma(E) \right)^2 dr = 1,$$

where $k = \sqrt{2\mu E}$ and $d\sigma(E) = \sqrt{\frac{\mu}{2E}}\, dE = dk$. The corresponding eigenfunction expansion theorem is the special case of the classical Fourier-Bessel transform for integer l, which generalizes the Fourier sine transform: for every $f \in L^2(0, \infty)$,

$$f(r) = \int_0^\infty c_l(E) f_{El}(r)\, d\sigma(E), \quad c_l(E) = \int_0^\infty f(r) f_{El}(r)\, dr.$$

In general, for every $l = 0, 1, \ldots$, denote by $f_{El}(r)$ the solution of the radial Schrödinger equation (4.10) with the following asymptotics:

$$(4.16) \qquad f_{El}(r) = \sqrt{\frac{2}{\pi\hbar}} \sin\left(\frac{kr}{\hbar} - \frac{l\pi}{2} + \delta_l \right) + o(1) \quad \text{as} \quad r \to \infty.$$

The function $\delta_l(k)$, $k = \sqrt{2\mu E}$, is called the *phase shift*. It follows from (4.14) that for the free case $\delta_l(k) = 0$. The eigenfunctions of the continuous spectrum $f_{El}(r)$ satisfy the same normalization condition (4.15) as the eigenfunctions for the free case. The function $S_l(k) = e^{2i\delta_l(k)}$ plays the role of the scattering matrix for the radial Schrödinger operator H_l. It admits a meromorphic continuation to the upper half-plane $\operatorname{Im} k > 0$ with simple poles $k = i\varkappa_{kl} = i\sqrt{-2\mu E_{kl}}$, where E_{kl} are the eigenvalues of H_l.

Let $E_{0l} < E_{2l} < \cdots < E_{N_l-1\,l} < 0$ be the eigenvalues of H_l, and let $f_{jl}(r)$, $j = 0, \ldots, N_l - 1$, be the corresponding normalized eigenfunctions. By the oscillation theorem, the eigenfunctions $f_{jl}(r)$ have j simple zeros in $(0, \infty)$. The functions

$$\psi_{jlm}(\boldsymbol{x}) = \frac{f_{jl}(r)}{r}Y_{lm}(\boldsymbol{n}), \quad m = -l, \ldots, l,$$

are normalized eigenfunctions of the Schrödinger operator H with eigenvalues E_{jl}, and the functions

$$\psi_{Elm}(\boldsymbol{x}) = \frac{f_{El}(r)}{r}Y_{lm}(\boldsymbol{n}), \quad m = -l, \ldots, l,$$

are normalized eigenfunctions of the continuous spectrum. The corresponding eigenfunction expansion theorem for the Schrödinger operator H with spherically symmetric potential states that for every $\psi \in L^2(\mathbb{R}^3, d^3\boldsymbol{x})$,

$$\psi(\boldsymbol{x}) = \sum_{l=0}^{\infty}\sum_{m=-l}^{l}\int_0^{\infty} c_{lm}(E)\psi_{Elm}(\boldsymbol{x})d\sigma(E) + \sum_{l=0}^{\infty}\sum_{m=-l}^{l}\sum_{j=0}^{N_l-1} c_{jlm}\psi_{jlm}(\boldsymbol{x}),$$

where

$$c_{lm}(E) = \int_{\mathbb{R}^3}\psi(\boldsymbol{x})\overline{\psi_{Elm}(\boldsymbol{x})}d^3\boldsymbol{x}, \quad c_{jlm} = \int_{\mathbb{R}^3}\psi(\boldsymbol{x})\overline{\psi_{jlm}(\boldsymbol{x})}d^3\boldsymbol{x}.$$

Remark. In physics, it is traditional to call the parameter j in the eigenfunction $\psi_{jlm}(\boldsymbol{x})$ the *radial quantum number* and denote it by n_r. The parameter l is called the *azimuthal quantum number*, and the parameter m — the *magnetic quantum number*. This terminology originated from the old quantum theory, where to each value E_{kl} there corresponds a classical orbit. The parameter $n = n_r + l + 1$ is called the *principal quantum number*, so that $n_r = n - l - 1$ is always the number of zeros of the corresponding radial eigenfunction $f_{jl}(r)$.

Remark. In general, the eigenvalues E_{jl} of a Schrödinger operator H with spherically symmetric potential have multiplicity $2l + 1$. For special potentials, due to the extra symmetry of a problem, there may be "accidental degeneracy" with respect to the azimuthal quantum number l. This is the case for the Schrödinger operator of the hydrogen atom, considered in the next section.

Problem 4.7. Prove all results stated in this section. (*Hint*: See the list of references to this chapter.)

Problem 4.8. Find the energy levels of a particle with angular momentum $l = 0$ in the centrally symmetric potential well $V(r) = -V_0 < 0$ when $0 < r < a$ and $V(r) = 0$ when $r > a$.

Problem 4.9. Find the spectrum of the Schrödinger operator with the potential $V(r) = ar^{-2} + br^2$, $a, b > 0$.

Problem 4.10. Prove that the scattering wave $u(\boldsymbol{x}, \boldsymbol{k}) = u^{(+)}(\boldsymbol{x}, \boldsymbol{k})$ in the central potential is given by

$$u(\boldsymbol{x}, \boldsymbol{k}) = \frac{1}{kr}\sqrt{\frac{\pi}{2}} \sum_{l=0}^{\infty}(2l + 1)i^l e^{i\delta_l(k)} f_{El}(r) P_l(\cos\vartheta),$$

where $\boldsymbol{x} \cdot \boldsymbol{k} = kr\cos\vartheta$ and, as in Section 4.2, $\hbar = 1$ and $\mu = \frac{1}{2}$.

Problem 4.11. Using the result of the previous problem, show that

$$f(k, \boldsymbol{n}, \boldsymbol{\omega}) = f(k, \cos\vartheta) = \frac{1}{2ik} \sum_{l=0}^{\infty}(2l + 1)(S_l(k) - 1)P_l(\cos\vartheta)$$

— the *partial wave decomposition* of the scattering amplitude — and get the following formula for the total cross-section

$$\sigma_{\text{tot}}(k) = \frac{4\pi}{k^2} \sum_{l=0}^{\infty}(2l + 1)\sin^2\delta_l(k).$$

5. Hydrogen atom and SO(4)

The hydrogen atom is described by the long-range potential

$$V(\boldsymbol{r}) = -\frac{\alpha}{r}$$

— the *Coulomb potential,* where $\alpha > 0$[3]. The corresponding eigenvalue problem for the Schrödinger operator with Coulomb potential is called the *Coulomb problem.* Here we present its exact solution using the so-called *Coulomb units* $\hbar = 1$, $\mu = 1$, and $\alpha = 1$[4] .

5.1. Discrete spectrum.
To determine the discrete spectrum, it is convenient to put $2E = -\varkappa^2 < 0$, so that the eigenvalue equation (4.10) becomes

(5.1) $$f_l'' + \left(\frac{2}{r} - \frac{l(l + 1)}{r^2} - \varkappa^2\right)f_l = 0.$$

Asymptotics (4.12)–(4.13) for the short-range potentials suggest that for the long-range Coulomb potential one can look for a square-integrable solution in the following form:

$$f_l(r) = r^{l+1}e^{-\varkappa r}\Lambda(r).$$

---

[3]For the hydrogen atom $\alpha = e^2$, where e is the electron charge. The case $\alpha = Ze^2$, where $Z = 2, 3, \ldots$, corresponds to the hydrogen ions.

[4]When $\alpha = e^2$, Coulomb units coincide with *atomic units.*

Substituting it into (5.1), we get the equation

$$(5.2) \qquad \Lambda_l'' + \left(\frac{2(l+1)}{r} - 2\varkappa \right) \Lambda_l' + \left(\frac{2}{r} - \frac{2\varkappa(l+1)}{r} \right) \Lambda_l = 0,$$

which can be solved by the power series

$$(5.3) \qquad \Lambda_l(r) = \sum_{k=0}^{\infty} a_k r^k.$$

Substituting these power series into (5.2) yields the following recurrence relation for the coefficients a_k:

$$a_{k+1} = \frac{2\varkappa(k+l+1) - 2}{(k+1)(k+2l+2)} a_k, \quad k = 1, 2, \ldots,$$

where $a_0 \neq 0$. The power series (5.3) converges for all $r > 0$ by the ratio test. When $\varkappa > 0$ is such that $a_k \neq 0$ for all k, we have

$$\lim_{k \to \infty} \frac{(k+1)a_{k+1}}{a_k} = 2\varkappa.$$

Thus for every $\varepsilon > 0$ there is N such that for $k > N$ all a_k are of the same sign, say $a_k > 0$, and

$$\frac{a_{k+1}}{a_k} \geq \frac{2\varkappa - \varepsilon}{k+1}.$$

Then for $k > N$ we get

$$a_k \geq C \frac{(2\varkappa - \varepsilon)^k}{k!}$$

with some constant $C > 0$ (depending on ε), and since for fixed N the sum of the first N terms in (5.3) grows like r^N as $r \to \infty$, for r large enough we obtain[5]

$$\Lambda_l(r) \geq Ce^{(2\varkappa - \varepsilon)r} - C_1 r^N > C_2 e^{(2\varkappa - \varepsilon)r}.$$

This proves that for such values of $\varkappa$ the function $f_l(r)$ is not square-integrable on $(0, \infty)$. However, for the special values

$$\varkappa = \varkappa_{kl} = \frac{1}{k+l+1}$$

the power series (5.3) terminates: $\Lambda_l(r)$ becomes a polynomial $\Lambda_{kl}(r)$ of order k and $f_{kl}(r) = r^{l+1} e^{-\varkappa_{kl} r} \Lambda_{kl}(r) \in L^2(0, \infty)$. Setting $n = k + l + 1$, we get an explicit formula for the eigenvalues

$$E_n = -\frac{1}{2n^2}, \quad n = 1, 2, \ldots.$$

---

[5]It can be shown that $\Lambda_l(r) = Ce^{2\varkappa r}(1 + o(1))$ as $r \to \infty$.

For fixed n the integer l varies from 0 to $n-1$, and for each l there are $2l+1$ orthogonal eigenfunctions $\dfrac{f_{kl}(r)}{r} Y_{lm}(\boldsymbol{n})$ with the eigenvalue E_n. Thus the overall multiplicity of the eigenvalue E_n is

$$\sum_{l=0}^{n-1} (2l+1) = n^2.$$

The eigenfunctions $f_{kl}(r)$ of the radial Schrödinger equation (5.1) corresponding to the eigenvalue E_n can be expressed in terms of classical Laguerre polynomials. Namely, for $\varkappa = \dfrac{1}{n}$ the substitution $x = \dfrac{2r}{n}$ reduces the differential equation (5.2) to

$$xQ'' + (p+1-x)Q' + kQ = 0,$$

where $p = 2l+1$ and $Q(x) = \Lambda_l(r)$. This is the differential equation satisfied by the associated Laguerre polynomials $Q_k^p(x)$, defined by[6]

$$Q_k^p(x) = e^x x^{-p} \frac{d^k}{dx^k}(e^{-x} x^{k+p}),$$

which are polynomials of order k with leading coefficient $(-1)^k$, having k zeros in $(0, \infty)$. For every p the associated Laguerre polynomials $\{Q_k^p(x)\}_{k=0}^{\infty}$ are orthogonal polynomials on $(0, \infty)$ with respect to the measure $e^{-x} x^p dx$, and have the property

$$\int_0^{\infty} e^{-x} x^{p+1} Q_k^p(x)^2 dx = k!(k+p)!(2k+p+1).$$

The orthonormal eigenfunctions of the Schrödinger operator H for the hydrogen atom corresponding to the eigenvalue E_n have the form

$$(5.4) \qquad \psi_{klm}(\boldsymbol{q}) = \frac{2}{n^2} \frac{1}{\sqrt{k!(n+l)!}} \left(\frac{2r}{n}\right)^l e^{-\frac{r}{n}} Q_k^{2l+1}\left(\frac{2r}{n}\right) Y_{lm}(\boldsymbol{n}),$$

where $n = k+l+1$, $l = 0, \ldots, n-1$, and $m = -l, \ldots, l$.

Remark. Returning to the physical units $\alpha = e^2$, where e is the electron charge, we get for the energy levels of the hydrogen atom

$$E_n = -\frac{\mu e^4}{2n^2\hbar^2},$$

where μ is the reduced mass of the electron and the nucleus. In particular, the ground state energy is $E_1 = -13.6$ eV; its absolute value is the *ionization energy* — the energy required to remove the electron from the hydrogen atom. Using Bohr's formula for the frequencies of the spectral lines

$$\hbar\omega_{mn} = E_n - E_m,$$

---

[6]Polynomials $L_n^m(x) = (-1)^m \frac{n!}{(n-m)!} Q_{n-m}^m(x)$ are also being used.

we get for the hydrogen atom

$$\omega_{mn} = \frac{\mu e^4}{2\hbar^3}\left(\frac{1}{n^2} - \frac{1}{m^2}\right), \quad m < n.$$

When $n = 1$ and $m = 2, 3, \dots$, we get classical Lyman series, $n = 2$ and $m = 3, 4, \dots$ give Balmer series, and $n = 3$ and $m = 4, 5, \dots$ give Paschen series. These series of spectral lines were discovered experimentally long before the formulation of quantum mechanics.

Remark. The energy levels of the hydrogen atom can also be determined from Bohr-Wilson-Sommerfeld quantization rules. Namely, in spherical coordinates (r, ϑ, φ) in $\mathbb{R}^3$ the Lagrangian of Kepler's problem takes the form (see Section 1.6 in Chapter 1)

$$L = \tfrac{1}{2}\mu(\dot{r}^2 + r^2\dot{\vartheta}^2 + r^2\sin^2\vartheta\,\dot{\varphi}^2) + \frac{\alpha}{r},$$

and the corresponding generalized momenta are

$$p_r = \mu\dot{r}, \quad p_\vartheta = \mu r^2\dot{\vartheta}, \quad p_\varphi = \mu r^2\sin^2\vartheta\,\dot{\varphi}.$$

The BWS quantization conditions

$$(5.5) \qquad \oint p_r\,dr = 2\pi\hbar(k + \tfrac{1}{2}),$$

$$(5.6) \qquad \oint p_\vartheta\,d\vartheta = 2\pi(l - m + \tfrac{1}{2}),$$

$$(5.7) \qquad \oint p_\varphi\,d\varphi = 2\pi\hbar m,$$

where integration goes over the closed orbit of the Kepler problem with the energy $E < 0$, exactly determine the energy levels E_n, $n = k + l + 1$.

Indeed, we have $p_\varphi = M_{c3}$, where $\boldsymbol{M}_c = (M_{c1}, M_{c2}, M_{c3})$ is the classical angular momentum. It is constant along the orbit, so that (5.7) determines the eigenvalues of operator M_3 (see Section 3.2). To evaluate $\oint p_\vartheta\,d\vartheta$, we use polar coordinates (r, χ) in the orbit plane P (see Section 1.6 in Chapter 1). Since $\dot{\chi}^2 = \dot{\vartheta}^2 + \sin^2\vartheta\,\dot{\varphi}^2$, along the orbit we get $p_\chi d\chi = p_\vartheta d\vartheta + p_\varphi d\varphi$, where $p_\chi = \mu r^2\dot{\chi} = |\boldsymbol{M}_c|$. Condition (5.6) now follows from

$$(5.8) \qquad \oint p_\chi\,d\chi = 2\pi|\boldsymbol{M}_c| = 2\pi\hbar(l + \tfrac{1}{2}),$$

which is the quantization rule for the square of total angular momentum[7]. Finally, to evaluate $\oint p_r\,dr$, we use equations (1.7) and (1.11) in Section 1.6

---

[7]It gives the quantized values $\hbar^2(l + \tfrac{1}{2})^2$, which for large l agree well with the eigenvalues $\hbar^2 l(l + 1)$ of the operator $\boldsymbol{M}^2$.

of Chapter 1 and get

$$p_r dr = \mu \dot{r} dr = \mu \frac{dr}{d\chi} \dot{\chi}\, dr = \frac{|\boldsymbol{M}_{\mathrm{c}}|}{r^2} \left(\frac{dr}{d\chi}\right)^2 d\chi = p_\chi \frac{e^2 \sin^2 \chi}{(1 + e\cos\chi)^2} d\chi,$$

where $0 < e < 1$ is the eccentricity of the orbit. We have

$$\oint p_r dr = p_\chi \int_0^{2\pi} \frac{e^2 \sin^2 \chi}{(1 + e\cos\chi)^2} d\chi = 2\pi p_\chi \left(\frac{1}{\sqrt{1 - e^2}} - 1\right),$$

as can be shown by using the substitution $z = e^{i\chi}$ and the Cauchy residue theorem[8]. It follows from equation (1.12) in Section 1.6 of Chapter 1 that

$$\sqrt{1 - e^2} = \frac{|\boldsymbol{M}_{\mathrm{c}}|\sqrt{2|E|}}{\alpha\sqrt{\mu}},$$

and we obtain from (5.5) and (5.8),

$$|E| = -E_n = \frac{\mu\alpha^2}{2n^2\hbar^2}, \quad n = k + l + 1.$$

5.2. Continuous spectrum. Set $2E = k^2$, where $k > 0$, and consider the equation

$$(5.9) \qquad f_l'' + \left(\frac{2}{r} - \frac{l(l+1)}{r^2} + k^2\right) f_l = 0.$$

Substitution

$$f_l(r) = r^{l+1} e^{-ikr} F_l(r)$$

reduces (5.9) to

$$(5.10) \qquad F_l'' + \left(\frac{2(l+1)}{r} - 2ik\right) F_l' + \left(\frac{2}{r} - \frac{2ik(l+1)}{r}\right) F_l = 0.$$

The solution of equation (5.10) satisfying $F_l(0) = 1$ can be explicitly written as

$$F_l(r) = F\left(l + 1 + i\lambda, 2l + 2, 2ikr\right), \quad \lambda = \frac{1}{k},$$

where $F(\alpha, \gamma, z)$ is a confluent hypergeometric function, defined by the absolutely convergent series

$$F(\alpha, \gamma, z) = \sum_{n=0}^{\infty} \frac{\Gamma(\alpha + n)\Gamma(\gamma)}{\Gamma(\alpha)\Gamma(\gamma + n)} \frac{z^n}{n!}$$

for all α and $\gamma \neq 0, -1, -2, \ldots$. The confluent hypergeometric function is an entire function of the variable z and satisfies the differential equation

$$zF'' + (\gamma - z)F' - \alpha F = 0.$$

---

[8]This integral was evaluated by Sommerfeld in 1916.

For $\alpha = -k$ and $\gamma = p + 1$, where $k, p = 0, 1, 2, \ldots$, the confluent hypergeometric function reduces to associated Laguerre polynomials

$$Q_k^p(x) = \frac{(k+p)!}{p!} F(-k, p+1, x),$$

considered in the previous section. For $\operatorname{Re}\gamma > \operatorname{Re}\alpha > 0$ the function $F(\alpha, \gamma, z)$ admits the integral representation

$$(5.11) \qquad F(\alpha, \gamma, z) = \frac{\Gamma(\gamma)}{\Gamma(\alpha)\Gamma(\gamma - \alpha)} \int_0^1 t^{\alpha-1}(1 - t)^{\gamma-\alpha-1} e^{zt} dt,$$

obtained by the Laplace method. The confluent hypergeometric function satisfies the functional equation

$$(5.12) \qquad F(\alpha, \gamma, z) = e^z F(\gamma - \alpha, \gamma, -z)$$

and has the following asymptotics as $z \to \infty$:

$$(5.13) \qquad F(\alpha, \gamma, z) = \frac{\Gamma(\gamma)}{\Gamma(\gamma - \alpha)}(-z)^{-\alpha}\left(1 + O(z^{-1})\right)$$

$$+ \frac{\Gamma(\gamma)}{\Gamma(\alpha)} e^z z^{\alpha-\gamma}\left(1 + O(z^{-1})\right).$$

It follows from (5.12) that $f_l(r) = r^{l+1} e^{-ikr} F(l + 1 + i\lambda, 2l + 2, 2ikr)$ is real-valued and it follows from (5.13) that the function

$$(5.14) \qquad f_{El}(r) = \frac{(2k)^{l+1}}{\sqrt{2\pi}(2l + 1)!} e^{\frac{\pi}{2k}} |\Gamma(l + 1 - i\lambda)| f_l(r)$$

has the following asymptotics as $r \to \infty$:

$$(5.15) \qquad f_{El}(r) = \sqrt{\frac{2}{\pi}} \sin\left(kr + \lambda \log 2kr - \frac{l\pi}{2} + \delta_l\right),$$

where

$$(5.16) \qquad \delta_l(k) = \arg\Gamma(l + 1 - i\lambda), \quad \lambda = \frac{1}{k} = \frac{1}{\sqrt{2E}},$$

is the phase shift. The partial S-matrix for the Coulomb problem is

$$S_l(k) = e^{2i\delta_l(k)} = \frac{\Gamma(l + 1 - i\lambda)}{\Gamma(l + 1 + i\lambda)}.$$

It admits meromorphic continuation to the upper half-plane $\operatorname{Im} k > 0$ and has simple poles at $k = i\varkappa_{jl}$, which correspond to the eigenvalues E_{kl} of the radial Schrödinger operator H_l. The radial eigenfunctions of the continuous spectrum (5.14) satisfy normalization condition (4.15).

Remark. It is instructive to compare asymptotics (5.15) for the Coulomb potential, which is long-range, with the corresponding formula (4.16) for the general short-range potential. The long range nature of the Coulomb interaction manifests itself by the extra logarithmic term $\lambda \log 2kr$ in (5.15).

The eigenfunction expansion theorem for the Schrödinger operator H of the hydrogen atom has the same form as in Section 4.3: the eigenfunctions $\psi_{klm}(\boldsymbol{x})$ are given by (5.4) where $k = 0, 1, \ldots, N_l = \infty$, and the eigenfunctions of the continuous spectrum are given by

$$\psi_{Elm}(\boldsymbol{x}) = \frac{f_{El}(r)}{r} Y_{lm}(\boldsymbol{n}), \quad l = 0, 1, \ldots, \quad m = -l, \ldots, l.$$

5.3. Hidden SO(4) **symmetry.** As we have seen in Section 1.6 of Chapter 1, classical system with the Hamiltonian function

$$H_{\mathrm{c}}(\boldsymbol{p}, \boldsymbol{x}) = \frac{\boldsymbol{p}^2}{2m} - \frac{\alpha}{r},$$

in addition to the angular momentum $\boldsymbol{M}_{\mathrm{c}}$, has three extra integrals of motion given by the Laplace-Runge-Lenz vector

$$\boldsymbol{W}_{\mathrm{c}} = \frac{\boldsymbol{p}}{m} \times \boldsymbol{M}_{\mathrm{c}} - \frac{\alpha \boldsymbol{x}}{r}.$$

According to Example 2.2 in Section 2.6 of Chapter 1, the integrals $\boldsymbol{M}_{\mathrm{c}}$ and $\boldsymbol{W}_{\mathrm{c}}$ for the Kepler problem have the Poisson brackets

$$\{M_{\mathrm{c}j}, M_{\mathrm{c}k}\} = -\varepsilon_{jkl} M_{\mathrm{c}l}, \quad \{W_{\mathrm{c}j}, M_{\mathrm{c}k}\} = -\varepsilon_{jkl} W_{\mathrm{c}l},$$
$$\{W_{\mathrm{c}j}, W_{\mathrm{c}k}\} = 2H_{\mathrm{c}}\varepsilon_{jkl} M_{\mathrm{c}l},$$

where $j, k, l = 1, 2, 3$ and $\varepsilon_{123} = 1$.

The quantum Kepler problem is the Coulomb problem. In the coordinate representation $\mathscr{H} = L^2(\mathbb{R}^3, d^3\boldsymbol{x})$, its Hamiltonian is

$$H = \frac{\boldsymbol{P}^2}{2} - \frac{\alpha}{r},$$

where $r = |\boldsymbol{x}|$ and we put $m = 1$. The quantum Laplace-Runge-Lenz vector — the Laplace-Runge-Lenz operator $\boldsymbol{W}$ — is defined by

$$\boldsymbol{W} = \frac{1}{2}(\boldsymbol{P} \times \boldsymbol{M} - \boldsymbol{M} \times \boldsymbol{P}) - \frac{\alpha \boldsymbol{Q}}{r},$$

or in components

$$W_j = \frac{1}{2}\varepsilon_{jkl}(P_k M_l + M_l P_k) - \frac{\alpha Q_j}{r}, \quad j = 1, 2, 3.$$

Here Q_i are multiplication by x_i operators, and it is always understood that there is a summation over repeated indices $1, 2, 3$. Using commutation relations (3.5), we also have

$$(5.17) \qquad \boldsymbol{W} = \boldsymbol{P} \times \boldsymbol{M} - \frac{\alpha \boldsymbol{Q}}{r} - i\hbar \boldsymbol{P} = -\boldsymbol{M} \times \boldsymbol{P} - \frac{\alpha \boldsymbol{Q}}{r} + i\hbar \boldsymbol{P},$$

where the term $i\hbar \boldsymbol{P}$ plays the role of a "quantum correction". The following result reveals the hidden symmetry of the Coulomb problem.

Proposition 5.1. *The Schrödinger operator H of the hydrogen atom has quantum integrals of motion $\boldsymbol{M}$ and $\boldsymbol{W}$,*

$$[H, M_i] = [H, W_i] = 0, \quad i = 1, 2, 3,$$

satisfying $\boldsymbol{W} \cdot \boldsymbol{M} = \boldsymbol{M} \cdot \boldsymbol{W} = 0$ and

$$(5.18) \qquad \boldsymbol{W}^2 = \alpha^2 + 2H\boldsymbol{M}^2 + 2\hbar^2 H.$$

Moreover, self-adjoint operators $\boldsymbol{M}$ and $\boldsymbol{W}$ have the following commutation relations:

$$[M_j, M_k] = i\hbar\varepsilon_{jkl}M_l, \quad [W_j, M_k] = i\hbar\varepsilon_{jkl}W_l, \quad [W_j, W_k] = -2i\hbar\varepsilon_{jkl}M_l H.$$

Proof. We know that $[H, M_j] = 0$. To prove that W_j are quantum integrals of motion, we first compute $\left[\boldsymbol{P}^2, \dfrac{Q_j}{r}\right]$. It follows from Heisenberg's commutation relations that

$$(5.19) \qquad \left[P_j, \frac{1}{r}\right] = i\hbar\frac{Q_j}{r^3},$$

so that using Leibniz rule and relations $r^2 = x_1^2 + x_2^3 + x_3^2$, $[M_l, r] = 0$, and $Q_j P_k - Q_k P_j = \varepsilon_{jkl}M_l$, we obtain

$$\left[\boldsymbol{P}^2, \frac{Q_j}{r}\right] = 2i\hbar\frac{Q_j}{r^3} + 2i\hbar\varepsilon_{jkl}\frac{Q_k}{r^3}M_l.$$

Now using the first equation in (5.17) and (5.19), we get

$$[H, W_j] = \left[\frac{\boldsymbol{P}^2}{2} - \frac{\alpha}{r}, \varepsilon_{jkl}P_k M_l - \frac{\alpha Q_j}{r} - i\hbar P_j\right]$$

$$= -i\alpha\hbar\frac{Q_j}{r^3} - i\alpha\hbar\varepsilon_{jkl}\frac{Q_k}{r^3}M_l + i\alpha\hbar\varepsilon_{jkl}\frac{Q_k}{r^3}M_l + i\alpha\hbar\frac{Q_j}{q^3} = 0.$$

It is easy to prove the relation $\boldsymbol{W} \cdot \boldsymbol{M} = W_1 M_1 + W_2 M_2 + W_3 M_3 = 0$. Indeed, it follows from the definition of $\boldsymbol{M}$ and commutativity of P_k and Q_l for $k \neq l$ that

$$\boldsymbol{M} \cdot \boldsymbol{P} = \boldsymbol{P} \cdot \boldsymbol{M} = \boldsymbol{M} \cdot \boldsymbol{Q} = \boldsymbol{Q} \cdot \boldsymbol{M} = 0,$$

and using the first equation in (5.17) and commutation relations for the components of the angular momentum, we immediately get $\boldsymbol{W} \cdot \boldsymbol{M} = 0$. To prove that $\boldsymbol{M} \cdot \boldsymbol{W} = 0$, one should use the second equation in (5.17).

Verification of (5.18) is more involved. We have

$$\boldsymbol{W}^2 = W_j W_j = \left(\varepsilon_{jkl} M_l P_k - \frac{\alpha Q_j}{r} + i\hbar P_j\right)\left(\varepsilon_{jmn} P_m M_n - \frac{\alpha Q_j}{r} - i\hbar P_j\right)$$

$$= \varepsilon_{jkl}(\varepsilon_{jmn} M_l P_k P_m M_n - \alpha M_l P_k \frac{Q_j}{r} - i\hbar M_l P_k P_j) - \alpha \varepsilon_{jmn} \frac{Q_j}{q} P_m M_n$$

$$+ \alpha^2 \frac{Q_j Q_j}{r^2} + i\alpha\hbar \frac{Q_j}{r} P_j + i\hbar\varepsilon_{jmn} P_j P_m M_n - i\alpha\hbar P_j \frac{Q_j}{r} + \hbar^2 P_j P_j$$

$$= \alpha^2 I + \hbar^2 \boldsymbol{P}^2 + \varepsilon_{jkl}\varepsilon_{jmn} M_l P_k P_m M_n - \alpha\varepsilon_{jkl}\left(M_l P_k \frac{Q_j}{q} + \frac{Q_j}{r} P_k M_l\right)$$

$$- i\alpha\hbar \left[P_j, \frac{Q_j}{r}\right].$$

Since operators M_j and P_j commute, the identity $(\boldsymbol{a} \times \boldsymbol{b})^2 = \boldsymbol{a}^2 \boldsymbol{b}^2 - (\boldsymbol{a} \cdot \boldsymbol{b})^2$ is still applicable and we get

$$\varepsilon_{jkl}\varepsilon_{jmn} M_l P_k P_m M_n = -(\boldsymbol{M} \times \boldsymbol{P})\cdot(\boldsymbol{P} \times \boldsymbol{M}) = \boldsymbol{M}^2 \boldsymbol{P}^2 - (\boldsymbol{M} \cdot \boldsymbol{P})^2 = \boldsymbol{M}^2 \boldsymbol{P}^2.$$

Using (5.19) we easily obtain (summation over repeated indices)

$$\left[P_j, \frac{Q_j}{r}\right] = i\hbar \left(-\frac{3}{r} + \frac{r^2}{r^3}\right) = -\frac{2i\hbar}{r}.$$

Since $[\boldsymbol{M}^2, r] = 0$ it follows from $\boldsymbol{M} = \boldsymbol{Q} \times \boldsymbol{P}$ that

$$\varepsilon_{jkl}\left(M_l P_k \frac{Q_j}{r} + \frac{Q_j}{r} P_k M_l\right) = \boldsymbol{M}^2 \frac{1}{r} + \frac{1}{r}\boldsymbol{M}^2 = \frac{2\boldsymbol{M}^2}{r}.$$

Putting everything together, we get (5.18).

It is also not difficult to establish commutation relations between M_j and W_k. Using (3.5) and properties of ε_{jkl}, we get

$$[M_j, W_k] = \left[M_j, \varepsilon_{kmn} P_m M_n - \frac{\alpha Q_k}{r} - i\hbar P_k\right]$$

$$= i\hbar(\varepsilon_{jmp}\varepsilon_{kmn} P_p M_n + \varepsilon_{jnp}\varepsilon_{kmn} P_m M_p) - i\hbar\varepsilon_{jkl}\left(\frac{\alpha Q_l}{r} + i\hbar P_l\right)$$

$$= i\hbar(P_j M_k - M_k P_j) - i\hbar\varepsilon_{jkl}\left(\frac{\alpha r_l}{r} + i\hbar P_l\right) = i\hbar\varepsilon_{jkl} W_l.$$

Finally, to establish commutation relations between components of $\boldsymbol{W}$, we use the representations

$$(5.20) \qquad \boldsymbol{W} = \boldsymbol{Q} \boldsymbol{P}^2 - \boldsymbol{P}(\boldsymbol{Q} \cdot \boldsymbol{P}) - \frac{\alpha \boldsymbol{Q}}{r} = \boldsymbol{P}^2 \boldsymbol{Q} - (\boldsymbol{P} \cdot \boldsymbol{Q})\boldsymbol{P} - \frac{\alpha \boldsymbol{Q}}{r}.$$

The first formula in (5.20) follows from the first formula in (5.17) by using

$$\varepsilon_{jkl} P_k M_l = \varepsilon_{jkl}\varepsilon_{lmn} P_k Q_m P_l = \varepsilon_{jkl}(\varepsilon_{ljk} P_k Q_j P_k + \varepsilon_{lkj} P_k Q_k P_j)$$

$$= P_k(P_k Q_j - P_j Q_k) = Q_j \boldsymbol{P}^2 - P_j(\boldsymbol{P} \cdot \boldsymbol{Q}) - 2i\hbar P_j,$$

and relation $\boldsymbol{P} \cdot \boldsymbol{Q} = \boldsymbol{Q} \cdot \boldsymbol{P} - 3i\hbar I$. The second formula in (5.20) follows from the first by using Heisenberg's commutation relations. Now we have

$$\varepsilon_{jkl}[W_k, W_l] = 2\varepsilon_{jmn} W_m W_n$$

$$= 2\varepsilon_{jmn}\left(\boldsymbol{P}^2 Q_m - (\boldsymbol{P} \cdot \boldsymbol{Q})P_m - \frac{\alpha Q_m}{r}\right)\left(Q_n \boldsymbol{P}^2 - P_n(\boldsymbol{Q} \cdot \boldsymbol{P}) - \frac{\alpha Q_n}{r}\right)$$

$$= 2M_j\left(-\boldsymbol{P}^2(\boldsymbol{Q} \cdot \boldsymbol{P}) + (\boldsymbol{P} \cdot \boldsymbol{Q})\boldsymbol{P}^2 + \frac{\alpha}{r}(\boldsymbol{Q} \cdot \boldsymbol{P}) - (\boldsymbol{P} \cdot \boldsymbol{Q})\frac{\alpha}{r}\right)$$

$$= -2i\hbar M_j\left(\boldsymbol{P}^2 - \frac{2\alpha}{r}\right) = -4i\hbar M_j H,$$

where we have used $[\boldsymbol{M}, \boldsymbol{P} \cdot \boldsymbol{Q}] = 0$ and $\left[\boldsymbol{P} \cdot \boldsymbol{Q}, \dfrac{1}{r}\right] = \dfrac{i\hbar}{r}$. This proves that $[W_k, W_l] = -2i\hbar\varepsilon_{jkl}M_j H$. $\qquad\square$

Let $\mathcal{H}_0 = \mathsf{P}_H(-\infty, 0)\mathcal{H}$, where P_H is the projection-valued measure for the Schrödinger operator H. Since self-adjoint operators $\boldsymbol{M}$ and $\boldsymbol{W}$ commute with H, the subspace $\mathcal{H}_0$ is an invariant subspace for these operators. The Schrödinger operator H is non-negative on $\mathcal{H}_0$, so that on this subspace the operator $(-2H)^{-1/2}$ is well defined. Now on $\mathcal{H}_0$ we set

$$\boldsymbol{J}^{(\pm)} = \frac{1}{2}(\boldsymbol{M} \pm (-2H)^{-1/2}\boldsymbol{W}).$$

It follows from Proposition 5.1 that self-adjoint operators $J_j^{(\pm)}$ on $\mathcal{H}_0$ satisfy the commutation relations

$$(5.21) \qquad \left[J_j^{(\pm)}, J_k^{(\pm)}\right] = i\hbar\varepsilon_{jkl}J_l^{(\pm)}, \quad \left[J_j^{(+)}, J_k^{(-)}\right] = 0$$

and

$$(5.22) \qquad (\boldsymbol{J}^{(+)})^2 = (\boldsymbol{J}^{(-)})^2 = -\frac{1}{4}\left(\hbar^2 I + \frac{\alpha^2}{2H}\right).$$

Equations (5.21) are commutation relations for the generators of the Lie algebra $\mathfrak{so}(4)$, which correspond to the Lie isomorphism $\mathfrak{so}(4) \simeq \mathfrak{so}(3) \oplus \mathfrak{so}(3)$, and exhibit the hidden SO(4) symmetry of the Coulomb problem! Together with (5.22), they allow us to find the energy levels pure algebraically.

Namely, the eigenvalues of the operators $(\boldsymbol{J}^{(+)})^2$ and $(\boldsymbol{J}^{(-)})^2$ are, respectively, $\hbar^2 l_1(l_1 + 1)$ and $\hbar^2 l_2(l_2 + 1)$, and it follows from (5.22) that $l_1 = l_2 = l$ so that the corresponding eigenvalue of H is

$$E_n = -\frac{\alpha^2}{2\hbar^2 n^2}, \quad n = 2l + 1.$$

Assuming that

$$(5.23) \qquad \mathcal{H}_0 \simeq \bigoplus_{l \in \frac{1}{2}\mathbb{Z}_{\geq 0}} V_l \otimes V_l,$$

where summation goes over all non-negative integer and half-integer values of l, we get from the Clebsch-Gordan decomposition (3.13) that the multiplicity of the eigenvalue E_n is

$$\dim V_l \otimes V_l = \sum_{j=0}^{2l}(2j+1) = (2l+1)^2 = n^2.$$

To prove the orthogonal sum decomposition (5.23), one should consider the Schrödinger equation for the hydrogen atom in the momentum representation. Using spherical coordinates in $\mathbb{R}^3$ and the elementary integral

$$\int_0^\infty \frac{\sin r}{r}dr = \frac{\pi}{2},$$

it is easy to show that for every $\psi \in \mathscr{S}(\mathbb{R}^3)$

$$(5.24) \qquad \frac{1}{2\pi^2\hbar}\int_{\mathbb{R}^3}\frac{\hat{\psi}(\boldsymbol{q})}{|\boldsymbol{p}-\boldsymbol{q}|^2}d^3\boldsymbol{q} = \frac{1}{(\sqrt{2\pi\hbar})^3}\int_{\mathbb{R}^3}\frac{1}{r}\psi(\boldsymbol{x})e^{-\frac{i}{\hbar}\boldsymbol{px}}d^3\boldsymbol{x},$$

where $\hat{\psi} = \mathscr{F}_\hbar(\psi)$ is the $\hbar$-dependent Fourier transform[9]. Let ψ be the eigenfunction of H with the the eigenvalue $E < 0$. It follows from (5.24) that in the momentum representation the corresponding Schrödinger equation — the eigenvalue equation $H\psi = E\psi$ — takes the form

$$(5.25) \qquad (\boldsymbol{p}^2 + p_0^2)\hat{\psi}(\boldsymbol{p}) = \lambda \int_{\mathbb{R}^3}\frac{\hat{\psi}(\boldsymbol{q})}{|\boldsymbol{p}-\boldsymbol{q}|^2}d^3\boldsymbol{q},$$

where $p_0 = \sqrt{-2\mu E}$ and $\lambda = \dfrac{\alpha\mu}{p_0\hbar}$. Next, consider the homogeneous coordinates $\dfrac{\boldsymbol{p}}{p_0}$ in $\mathbb{R}^3$ as coordinates of the stereographic projection of the unit sphere S^3 in $\mathbb{R}^4$. Namely, denoting by $\boldsymbol{n}$ the unit vector from the origin to the north pole of S^3, we get for the point $\boldsymbol{u} \in S^3$ corresponding to $\dfrac{\boldsymbol{p}}{p_0} \in \mathbb{R}^3$:

$$\boldsymbol{u} = \frac{\boldsymbol{p}^2 - p_0^2}{\boldsymbol{p}^2 + p_0^2}\boldsymbol{n} + \frac{2p_0}{\boldsymbol{p}^2 + p_0^2}\boldsymbol{p}.$$

For the volume form on S^3 we have

$$d\Omega = \frac{(2p_0)^3}{(\boldsymbol{p}^2 + p_0^2)^3}d^3\boldsymbol{p}, \qquad \int_{S^3}d\Omega = 2\pi^2.$$

Using

$$|\boldsymbol{p}-\boldsymbol{q}|^2 = \frac{(\boldsymbol{p}^2 + p_0^2)(\boldsymbol{q}^2 + p_0^2)}{(2p_0)^2}|\boldsymbol{u}-\boldsymbol{v}|^2,$$

---

[9]Since $\boldsymbol{x}$ is the variable in the coordinate representation, here $\boldsymbol{q}$ is another variable in the momentum representation.

where $\boldsymbol{v} \in S^3$ corresponds to $\dfrac{\boldsymbol{q}}{p_0} \in \mathbb{R}^3$, and introducing

$$\Psi(\boldsymbol{u}) = \frac{1}{\sqrt{p_0}} \frac{(\boldsymbol{p}^2 + p_0^2)^2}{(2p_0)^2} \, \hat{\psi}(\boldsymbol{p}),$$

we can rewrite equation (5.25) as

$$(5.26) \qquad\qquad \Psi(\boldsymbol{u}) = \frac{\lambda}{2\pi^2} \int_{S^3} \frac{\Psi(\boldsymbol{v})}{|\boldsymbol{u} - \boldsymbol{v}|^2} d\Omega_{\boldsymbol{v}}.$$

We also have

$$(5.27) \qquad \int_{S^3} |\Psi(\boldsymbol{u})|^2 d\Omega_{\boldsymbol{u}} = \int_{\mathbb{R}^3} \frac{\boldsymbol{p}^2 + p_0^2}{2p_0^2} |\hat{\psi}(\boldsymbol{p})|^2 d^3\boldsymbol{p} = \int_{\mathbb{R}^3} |\hat{\psi}(\boldsymbol{p})|^2 d^3\boldsymbol{p}.$$

Indeed, it follows from the Schrödinger equation that

$$\frac{1}{2\mu} \int_{\mathbb{R}^3} \boldsymbol{p}^2 |\hat{\psi}(\boldsymbol{p})|^2 d^3\boldsymbol{p} = \frac{1}{2\mu} \int_{\mathbb{R}^3} \left| \frac{\partial \psi(\boldsymbol{x})}{\partial \boldsymbol{x}} \right|^2 d^3\boldsymbol{x} = \int_{\mathbb{R}^3} (E - V(r)) |\psi(\boldsymbol{x})|^2 d^3\boldsymbol{x},$$

and by the virial theorem (see Section 1.3) we get

$$\frac{1}{2\mu} \int_{\mathbb{R}^3} \left| \frac{\partial \psi(\boldsymbol{x})}{\partial \boldsymbol{x}} \right|^2 d^3\boldsymbol{x} = -\frac{1}{2\mu} \int_{\mathbb{R}^3} \Delta\psi(\boldsymbol{x}) \overline{\psi(\boldsymbol{x})} d^3\boldsymbol{x} = -\frac{1}{2} \int_{\mathbb{R}^3} V(r) |\psi(\boldsymbol{x})|^2 d^3\boldsymbol{x},$$

where $V(r) = -\dfrac{\alpha}{r}$, so that

$$\int_{\mathbb{R}^3} V(r) |\psi(\boldsymbol{x})|^2 d^3\boldsymbol{x} = 2E \int_{\mathbb{R}^3} |\psi(\boldsymbol{x})|^2 d^3\boldsymbol{x} = 2E \int_{\mathbb{R}^3} |\hat{\psi}(\boldsymbol{p})|^2 d^3\boldsymbol{p}.$$

This shows that the eigenvalue problem for the Schrödinger equation is equivalent to the eigenvalue problem for the integral equation (5.26) in $L^2(S^3, d\Omega)$. The latter is a classical problem in the theory of spherical harmonics. Namely, the function $G(\boldsymbol{u}) = -\dfrac{1}{4\pi^2 u^2}$, where $u = |\boldsymbol{u}|$, $\boldsymbol{u} \in \mathbb{R}^4$, is the fundamental solution for the Laplace operator Δ in $\mathbb{R}^4$,

$$\Delta = \frac{\partial^2}{\partial u_1^2} + \frac{\partial^2}{\partial u_2^2} + \frac{\partial^2}{\partial u_3^2} + \frac{\partial^2}{\partial u_4^2},$$

and equation (5.26) is the equation satisfied by the spherical harmonics on S^3: the eigenfunctions of the spherical part Δ_0 of the Laplace operator on $\mathbb{R}^4$, defined by

$$\Delta = \frac{1}{u^3} \frac{\partial}{\partial u} \left(u^3 \frac{\partial}{\partial u} \right) + \frac{1}{u^2} \Delta_0.$$

It is well known from the representation theory of the Lie group SO(4) that spherical harmonics are restrictions to S^3 of homogeneous harmonic polynomials in $\mathbb{R}^4$ of degree $n - 1$, $n \in \mathbb{N}$. Correspondingly, equation (5.26) is obtained as a limit $|\boldsymbol{u}| \to 1$ of the Green's formula for the homogeneous harmonic function of degree $n - 1$ in the unit ball in $\mathbb{R}^4$, by using the property of a double layer potential. This gives $\lambda = n$, which once again establishes the exact formula for the energy levels of the hydrogen atom. The isomorphism $\mathscr{H}_0 \simeq L^2(S^3, d\Omega)$ follows from (5.27), and (5.23) — from the decomposition

$$L^2(S^3, d\Omega) \simeq \bigoplus_{l \in \frac{1}{2}\mathbb{Z}_{\geq 0}} V_l \otimes V_l$$

of the regular representation of $SO(4)$ into the direct sum of irreducible components. One can also obtain the explicit form (5.4) of the eigenfunctions by using the representation theory of $SO(4)$. We leave all these details to the interested reader.

Remark. For $E > 0$ one should consider the subspace $\mathscr{H}_1 = \mathsf{P}_H(0, \infty)\mathscr{H}$ and define $\boldsymbol{J}^{(\pm)} = \frac{1}{2}(\boldsymbol{M} \pm (2H)^{-1/2}\boldsymbol{W})$. Instead of (5.21), these operators satisfy commutation relations of the Lie algebra $\mathfrak{so}(3, 1)$ of the Lorentz group $SO(3, 1)$. The problem of finding eigenfunctions of the continuous spectrum for the Coulomb problem can be solved by using harmonic analysis on the three-dimensional Lobachevsky space.

Problem 5.1. Show that components of angular momentum $\boldsymbol{M}$ are generators of $\mathfrak{so}(4)$ that correspond to the infinitesimal rotations in the subspace of $\mathbb{R}^4$ with coordinates $(0, \boldsymbol{p})$, and components of the Laplace-Runge-Lenz vector $\boldsymbol{W}$ correspond to the infinitesimal rotations in the (p_0p_1), (p_0, p_2), and (p_0p_3) planes in $\mathbb{R}^4$.

6. Semi-classical asymptotics – I

Here we describe the relation between classical and quantum mechanics by considering the behavior of the wave function $\psi(\boldsymbol{q}, t)$ — the solution of the time dependent Schrödinger equation[10]

$$(6.1) \qquad i\hbar\frac{\partial\psi}{\partial t} = -\frac{\hbar^2}{2m}\Delta\psi + V(\boldsymbol{q})\psi$$

as $\hbar \to 0$. The substitution

$$(6.2) \qquad \psi(\boldsymbol{q}, t) = e^{-\frac{i}{\hbar}S(\boldsymbol{q}, t; \hbar)}$$

reduces (6.1) to the following non-linear partial differential equation:

$$(6.3) \qquad \frac{\partial S}{\partial t} + \frac{1}{2m}\left(\frac{\partial S}{\partial \boldsymbol{q}}\right)^2 + V(\boldsymbol{q}) = \frac{i\hbar}{2m}\Delta S.$$

It is remarkable that (6.3) differs from the Hamilton-Jacobi equation (2.4) for the Hamiltonian function $H_c(\boldsymbol{p}, \boldsymbol{q}) = \dfrac{\boldsymbol{p}^2}{2m} + V(\boldsymbol{q})$, considered in Section 2.3 of Chapter 1, only by the term in the right-hand side which is proportional to $\hbar$. Thus as $\hbar \to 0$, equations of motion in quantum mechanics turn into classical equations of motion.

---

[10]Here it is convenient to denote Cartesian coordinates on $\mathbb{R}^n$ by $\boldsymbol{q} = (q_1, \ldots, q_n)$.

For the stationary state $\psi(\boldsymbol{q},t) = \psi(\boldsymbol{q})e^{-\frac{i}{\hbar}Et}$ the substitution (6.2) becomes

$$(6.4) \qquad \psi(\boldsymbol{q},t) = e^{-\frac{i}{\hbar}(\sigma(\boldsymbol{q};\hbar)-Et)}.$$

The corresponding non-linear partial differential equation

$$(6.5) \qquad \frac{1}{2m}\left(\frac{\partial\sigma}{\partial\boldsymbol{q}}\right)^2 + V(\boldsymbol{q}) = E + \frac{i\hbar}{2m}\Delta\sigma$$

differs from the corresponding Hamilton-Jacobi equation for the abbreviated action, considered in Section 2.5 of Chapter 1, by the term proportional to $\hbar$ in the right-hand side.

In this section we consider *semi-classical asymptotics*: asymptotics of the partial differential equations (6.3) and (6.5) as $\hbar \to 0$. They describe the precise relation between quantum mechanics and classical mechanics, and provide a quantitative form of the correspondence principle, discussed in Section 2 of Chapter 2. In particular, using semi-classical asymptotics for the stationary Schrödinger equation, we derive Bohr-Wilson-Sommerfeld quantization rules, postulated in Section 2.5 of Chapter 2.

6.1. Time-dependent asymptotics. Here we consider the problem of finding *short-wave asymptotics* — asymptotics as $\hbar \to 0$ of the solution $\psi_\hbar(q,t)$ of the Cauchy problem for the Schrödinger equation (6.1) in one dimension,

$$i\hbar\frac{\partial\psi}{\partial t} = -\frac{\hbar^2}{2m}\frac{\partial^2\psi}{\partial q^2} + V(q)\psi,$$

with the initial condition

$$\psi_\hbar(q,t)|_{t=0} = \varphi(q)e^{\frac{i}{\hbar}s(q)}.$$

It is assumed that the real-valued functions $\varphi(q)$ and $s(q)$ are smooth, $s(q), \varphi(q) \in C^\infty(\mathbb{R})$, and that the "amplitude" $\varphi(q)$ has compact support. The substitution (6.2) is $\psi_\hbar(q,t) = e^{\frac{i}{\hbar}S(q,t,\hbar)}$, and differential equation (6.3) takes the form

$$(6.6) \qquad \frac{\partial S}{\partial t} + \frac{1}{2m}\left(\frac{\partial S}{\partial q}\right)^2 + V(q) = \frac{i\hbar}{2m}\frac{\partial^2 S}{\partial q^2}.$$

To determine the asymptotic behavior of $S(q,t,\hbar)$ as $\hbar \to 0$, we assume that as $\hbar \to 0$

$$S(q,t,\hbar) = \sum_{n=0}^{\infty}(-i\hbar)^n S_n(q,t),$$

and substitute this expansion into (6.6). Comparing terms with the same powers of $\hbar$ we obtain that $S_0(q,t)$ satisfies the initial value problem

$$(6.7) \qquad \frac{\partial S_0}{\partial t} + \frac{1}{2m}\left(\frac{\partial S_0}{\partial q}\right)^2 + V(q) = 0,$$

and

$$(6.8) \qquad S_0(q,t)|_{t=0} = s(q),$$

whereas $S_1(q,t)$ satisfies the differential equation

$$(6.9) \qquad \frac{\partial S_1}{\partial t} + \frac{1}{m}\frac{\partial S_0}{\partial q}\frac{\partial S_1}{\partial q} = -\frac{1}{2m}\frac{\partial^2 S_0}{\partial q^2},$$

the so-called *transport equation*, and

$$(6.10) \qquad S_1(q,t)|_{t=0} = \varphi(q).$$

The functions $S_n(q,t)$ for $n > 1$ satisfy non-homogeneous differential equations similar to (6.9).

The initial value problem (6.7)–(6.8) is the Cauchy problem for the Hamilton-Jacobi equation with the Hamiltonian function

$$H_c(p,q) = \frac{p^2}{2m} + V(q),$$

considered in Section 2.3 of Chapter 1. According to Proposition 2.1 in Section 2.3 of Chapter 1, the solution of (6.7)–(6.8) is given by the method of characteristics,

$$(6.11) \qquad S_0(q,t) = s(q_0) + \int_0^t L(\gamma'(\tau))d\tau.$$

Here $L(q,\dot{q}) = \frac{1}{2}m\dot{q}^2 - V(q)$ is the Lagrangian function, and $\gamma(\tau)$ is the characteristic — the classical trajectory which starts at q_0 at time $\tau = 0$ with the momentum $p_0 = \dfrac{\partial s}{\partial q}(q_0)$, and ends at q at time $\tau = t$, where q_0 is uniquely determined by q. (We are assuming that the Hamiltonian phase flow g_t satisfies the assumptions made in Section 2.3 of Chapter 1.) It follows from Theorem 2.7 in Section 2.3 of Chapter 1 that along the characteristic,

$$\frac{\partial S_0}{\partial q}(q,t) = m\frac{d\gamma}{dt}(t),$$

so that

$$(6.12) \qquad \left(\frac{\partial}{\partial t} + \frac{\partial S_0}{\partial q}\right)S_1(\gamma(t),t) = \frac{d}{dt}S_1(\gamma(t),t).$$

Now we can solve the Cauchy problem (6.9)–(6.10) for the transport equation explicitly. Consider the flow $\pi_t : \mathbb{R} \to \mathbb{R}$, defined in Section 2.3 of

Chapter 1, and denote[11] by $\gamma(Q, q; \tau)$ the characteristic connecting points q at $\tau = 0$ and $Q = \pi^t(q)$ at $\tau = t$ (under our assumptions the flow π_t is a diffeomorphism and the mapping $q \mapsto Q$ is one to one). Differentiating the equation

$$\frac{\partial S_0}{\partial Q}(Q, t) = m\frac{\partial \gamma}{\partial t}(Q, q; t)$$

with respect to q we obtain

$$\frac{\partial^2 S_0}{\partial Q^2}(Q, t)\frac{\partial Q}{\partial q} = m\frac{\partial^2 \gamma}{\partial q \partial t}(Q, q; t) = m\frac{d}{dt}\left(\frac{\partial Q}{\partial q}\right),$$

so that (6.9) can be rewritten as

$$\frac{d}{dt}S_1(Q, t) = -\frac{1}{2}\frac{d}{dt}\log\frac{\partial Q}{\partial q},$$

and using (6.10) we obtain

$$S_1(Q, t) = \varphi(q)\left|\frac{\partial Q}{\partial q}(q)\right|^{-\frac{1}{2}}.$$

Therefore,

$$(6.13)\qquad \psi_\hbar(Q, t) = \varphi(q)\left|\frac{\partial Q}{\partial q}(q)\right|^{-\frac{1}{2}} e^{\frac{i}{\hbar}(S(Q, q; t) + s(q))}(1 + O(\hbar)),$$

where $S(Q, q; t)$ is the classical action along the characteristic that starts at q at time $\tau = 0$ and ends at Q at time $\tau = t$.

The rigorous proof that (6.13) is an asymptotic expansion as $\hbar \to 0$ uses the assumptions made in Section 2.3 of Chapter 1, and is left to the interested reader. Here we just mention that asymptotics (6.13) is consistent with the *conservation of probability*: for any Borel subset $E \subset \mathbb{R}$,

$$\int_{E_t}|\psi_\hbar(Q, t)|^2 dQ = \int_E |\varphi(q)|^2 dq + O(\hbar)$$

as $\hbar \to 0$, where $E_t = \pi_t(E)$.

Remark. When assumptions in Section 2.3 of Chapter 1 are not satisfied, the situation becomes more complicated. Namely, in this case there may be several characteristics $\gamma_j(\tau)$ which end at Q at $\tau = t$ having q_j as their corresponding initial points. In this case,

$$\psi_\hbar(Q, t) = \sum_j \varphi(q_j)\left|\frac{\partial Q}{\partial q}(q_j)\right|^{-\frac{1}{2}} e^{\frac{i}{\hbar}(S(Q, q_j; t) + s(q_j)) - \frac{\pi i}{2}\mu_j}(1 + O(\hbar)),$$

where $\mu_j \in \mathbb{Z}$ is the *Morse index* of the characteristic γ_j. It is defined as the number of focal points of the phase curve $(q(\tau), p(\tau))$ with initial data

---

[11]There should not be any confusion with the quantum coordinate operator Q.

q_j and $p_j = \dfrac{\partial s}{\partial q}(q_j)$ with respect to the configuration space $\mathbb{R}$. It is a special case of a more general Maslov index.

The case of n degrees of freedom is considered similarly. Under the assumptions in Section 2.3 of Chapter 1, the solution $\psi_\hbar(q, t)$ of the Schrödinger equation (6.1) with the initial condition

$$\psi_\hbar(q, t)|_{t=0} = \varphi(q)e^{\frac{i}{\hbar}s(q)},$$

where real-valued functions $s(q)$ and $\varphi(q)$ are smooth and $\varphi(q)$ is compactly supported, has the following asymptotics as $\hbar \to 0$:

$$(6.14) \qquad \psi_\hbar(Q, t) = \varphi(q) \left| \det\left(\frac{\partial Q}{\partial q}(q) \right) \right|^{-\frac{1}{2}} e^{\frac{i}{\hbar}(S(Q,q;t)+s(q))}(1 + O(\hbar)).$$

6.2. Time independent asymptotics. Here we consider semi-classical asymptotics for the one-dimensional Schrödinger equation

$$(6.15) \qquad -\frac{\hbar^2}{2m}\frac{d^2\psi}{dx^2} + V(x)\psi = E\psi.$$

This asymptotic method is also known as the *WKB method* (after G. Wentzel, H. Kramers, and L. Brillouin). The substitution (6.4) is $\psi_\hbar(x) = e^{\frac{i}{\hbar}\sigma(x;\hbar)}$ and equation (6.5) takes the form

$$\frac{1}{2m}\left(\frac{d\sigma}{dx}\right)^2 + V(x) = E + \frac{i\hbar}{2m}\frac{d^2\sigma}{dx^2}.$$

As in the previous section, we use the expansion

$$\sigma(x, \hbar) = \sum_{n=0}^{\infty} (-i\hbar)^n \sigma_n(x),$$

and for the first two terms obtain

$$(6.16) \qquad \frac{1}{2m}\sigma_0'^2 = E - V(x) \quad \text{and} \quad \sigma_0'\sigma_1' = -\tfrac{1}{2}\sigma_0''.$$

Let $p(x) = \sqrt{2m(E - V(x))}$ be the classical momentum of a particle moving in a potential $V(x)$ with the energy E. The solution of the first equation in (6.16) is given by

$$\sigma_0 = \pm \int p(x)dx,$$

and from the second equation in (6.16) we get

$$\sigma_1 = -\tfrac{1}{2}\log p.$$

The wave function in the WKB approximation has the form

$$(6.17) \qquad \psi_\hbar(x) = \frac{1}{\sqrt{|p(x)|}}(C_1 e^{\frac{i}{\hbar}\int p(x)dx} + C_2 e^{-\frac{i}{\hbar}\int p(x)dx})(1 + O(\hbar)),$$

where $p(x)$ is real in the classical region $V(x) < E$, and is pure imaginary in the classically forbidden region $V(x) > E$ (see Section 1.5 of Chapter 1).

Note that the asymptotic (6.17) is valid only if $\sigma'^2 \ll \hbar\sigma''$, which in the first approximation gives

$$\left| \frac{d}{dx}\left(\frac{\hbar}{p(x)} \right) \right| \ll 1.$$

Introducing the classical force $F = -\dfrac{dV}{dx}$, this condition can be written as

$$(6.18) \qquad \frac{m\hbar F}{p^3} \ll 1.$$

Thus the WKB approximation does not work when the classical momentum $p(x)$ is small. In particular, it is not applicable near the turning points where $E = V(x)$ and, therefore, $p(x) = 0$. Let $x = a$ be a turning point. If $F_0 = F(a) \neq 0$, using the approximation $E - V(x) \simeq F_0(x - a)$ near $x = a$, one can replace (6.15) by

$$\frac{\hbar^2}{2m}\,\psi'' = F_0(x - a)\psi.$$

This differential equation can be explicitly solved by the Laplace method. Its bounded solution is

$$\psi(x) = \Phi(\xi), \quad \xi = \left(\frac{2mF_0}{\hbar^2} \right)^{\frac{1}{3}}(a - x),$$

where $\Phi(\xi)$ is the Airy-Fock function, defined by the improper integral

$$\Phi(\xi) = \frac{1}{\sqrt{\pi}} \int_0^\infty \cos\left(\tfrac{1}{3}t^3 + \xi t \right) dt.$$

The large ξ asymptotics of the Airy-Fock function, obtained by the steepest descent method, are the following:

$$(6.19) \qquad \Phi(\xi) = \frac{1}{2\sqrt[4]{\xi}}e^{-\frac{2}{3}\xi^{\frac{3}{2}}}\left(1 + O(\xi^{-\frac{3}{2}}) \right) \quad \text{as} \quad \xi \to \infty,$$

$$(6.20) \qquad \Phi(\xi) = \frac{1}{\sqrt[4]{|\xi|}} \sin\left(\tfrac{2}{3}|\xi|^{\frac{3}{2}} + \tfrac{\pi}{4} \right)\left(1 + O(|\xi|^{-\frac{3}{2}}) \right) \quad \text{as} \quad \xi \to -\infty.$$

Using asymptotics (6.19)–(6.20) one can obtain connection formulas relating WKB approximations for classical and forbidden regions. The rigorous justification of this approach is technically rather involved and we will not present it here. Instead we highlight the simplest case when there is only one turning point $x = a$ with $F_0 > 0$, so that $x < a$ is a classically forbidden region. From (6.17) we obtain that the WKB wave function in the forbidden region $x < a$ is exponentially decaying,

$$\psi_{\mathrm{WKB}}(x) = \frac{A}{\sqrt{|p(x)|}}e^{\frac{1}{\hbar}\int_a^x |p(s)|\,ds},$$

whereas in the classical region $x < a$ it is oscillating,

$$\psi_{\mathrm{WKB}}(x) = \frac{1}{\sqrt{p(x)}}\left(C_1 e^{\frac{i}{\hbar}\int_a^x p(s)ds} + C_2 e^{-\frac{i}{\hbar}\int_a^x p(s)ds}\right) = \frac{B}{\sqrt{p(x)}}\sin\left(\tfrac{1}{\hbar}\int_a^x p(s)ds + \alpha\right).$$

As follows from (6.18), the WKB approximation near the turning point $x = a$ remains valid whenever $|x - a| \gg \frac{1}{2}\left(\frac{\hbar^2}{mF_0}\right)^{\frac{1}{3}}$, which is equivalent to the condition $|\xi| \gg 1$. On the other hand, when $|x - a| \ll 1$ the wave function has the asymptotic

$$\psi_\hbar(x) = C\Phi(\xi)(1 + O(\hbar)).$$

To determine the unknown phase α and to find the relation between coefficients A, B, and C, we consider the domain $\frac{1}{2}\left(\frac{\hbar^2}{mF_0}\right)^{\frac{1}{3}} \ll |x - a| \ll 1$ and compare the WKB approximation for the wave function with the large ξ asymptotics of the Airy-Fock function. Namely, for $x < a$ we have $\xi \gg 1$ and

$$\frac{2}{3}\xi^{\frac{3}{2}} = \frac{2}{3\hbar}\sqrt{2mF_0}(a - x)^{\frac{3}{2}} \simeq \frac{1}{\hbar}\int_x^a |p(s)|ds, \quad \sqrt[4]{\xi} = \frac{\sqrt{|p(x)|}}{\sqrt[6]{2m\hbar F_0}}.$$

Comparing (6.19) with the WKB wave function for $x < a$ we obtain that $2A = \sqrt[6]{2m\hbar F_0}\,C$. For $x > a$ we have $\xi \ll -1$ and

$$\frac{2}{3}|\xi|^{\frac{3}{2}} = \frac{2}{3\hbar}\sqrt{2mF_0}(x - a)^{\frac{3}{2}} \simeq \frac{1}{\hbar}\int_a^x p(s)ds, \quad \sqrt[4]{|\xi|} = \frac{\sqrt{p(x)}}{\sqrt[6]{2m\hbar F_0}}.$$

Comparing (6.20) with the WKB wave function for $x < a$ we obtain that $\alpha = \frac{\pi}{4}$ and $B = \sqrt[6]{2m\hbar F_0}\,C$, so that $B = 2A$. Thus for this example the WKB wave function is

$$(6.21)\qquad \psi_{\mathrm{WKB}}(x) = \begin{cases} \dfrac{A}{\sqrt{|p(x)|}}e^{\frac{1}{\hbar}\int_a^x |p(s)|ds} & \text{when } x < a, \\[2ex] \dfrac{2A}{\sqrt{p(x)}}\sin\left(\tfrac{1}{\hbar}\int_a^x p(s)ds + \tfrac{\pi}{4}\right) & \text{when } x > a. \end{cases}$$

The case when there is only one turning point $x = b$ with $F_0 < 0$, so that $x > b$ is a classically forbidden region, reduces to the previous example by reversing the orientation $x \mapsto -x$. As the result, we obtain

$$(6.22)\qquad \psi_{\mathrm{WKB}}(x) = \begin{cases} \dfrac{2B}{\sqrt{p(x)}}\sin\left(\tfrac{1}{\hbar}\int_x^b p(s)ds + \tfrac{\pi}{4}\right) & \text{when } x < b, \\[2ex] \dfrac{B}{\sqrt{|p(x)|}}e^{-\frac{1}{\hbar}\int_b^x |p(s)|ds} & \text{when } x > b. \end{cases}$$

Problem 6.1 (Penetration through a potential barrier). Consider a potential barrier for a given energy E — a potential $V(x)$ such that $\{x \in \mathbb{R} : V(x) > E\} = (a, b)$. Show that

$$T_{\mathrm{WKB}} = c^{-\frac{2}{\hbar}\int_a^b |p(x)|dx}.$$

Verify directly that transmission coefficient for the potential in Problem 2.9 in the semi-classical approximation is given by this formula.

Problem 6.2 (Above barrier reflection). Suppose that a potential $V(x)$ admits analytic continuation into the upper half-plane, and let E be the energy such that $E > V(x)$ for all real x. Suppose that there is only one complex x_0 such that $V(x_0) = E$. Show that in the semi-classical approximation

$$R_{\mathrm{WKB}} = e^{-\frac{4}{\hbar}\,\mathrm{Im}\,\int_a^{x_0} p(x)dx},$$

where $p(x) = \sqrt{2m(E - V(x))}$ and $a \in \mathbb{R}$ (the choice of a does not change the imaginary part of the integral in the exponent). Verify directly that the reflection coefficient for the potential in Problem 2.9 in the semi-classical approximation is given by this formula.

Problem 6.3. Find transmission and reflection coefficients for the parabolic barrier — a potential $V(x) = -\frac{1}{2}kx^2$, where $k > 0$ — and verify directly that in the semi-classical approximation they satisfy, for $E < 0$ and $E > 0$, respectively, formulas in Problems 6.1 and 6.2.

6.3. Bohr-Wilson-Sommerfeld quantization rules. The WKB method allows us to determine energy levels in the semi-classical approximation. Consider, for simplicity, the finite motion of a one-dimensional particle in a potential well: in a potential $V(x)$ at the energy E such that there are two turning points a and b. The classical region is $a \leq x \leq b$, and the motion is periodic with the period

$$T = 2\int_a^b \frac{dx}{\dot{x}} = 2m\int_a^b \frac{dx}{p} = \sqrt{2m}\int_a^b \frac{dx}{\sqrt{E - V(x)}}.$$

The regions $x < a$ and $x > b$ are forbidden (see Section 1.5 of Chapter 1).

It follows from (6.21)–(6.22) that the WKB wave function exponentially decays in the forbidden regions $x < a$ and $x > b$. Using (6.21) we see that in the classical region the WKB wave function has the form

$$\psi_{\mathrm{WKB}}(x) = \frac{2A}{\sqrt{p(x)}} \sin\left(\frac{1}{\hbar}\int_a^x p(s)ds + \frac{\pi}{4}\right),$$

while using (6.22) we obtain

$$\psi_{\mathrm{WKB}}(x) = \frac{2B}{\sqrt{p(x)}} \sin\left(\frac{1}{\hbar}\int_x^b p(s)ds + \frac{\pi}{4}\right)$$

$$= \frac{2B}{\sqrt{p(x)}} \sin\left(\frac{1}{\hbar}\int_a^b p(s)ds + \frac{\pi}{2} - \left(\frac{1}{\hbar}\int_a^x p(s)ds + \frac{\pi}{4}\right)\right).$$

These two expressions define the same function in $a \leq x \leq b$ if and only if there is a positive integer n such that

$$(6.23) \qquad\qquad \frac{1}{\hbar}\int_a^b p(x)dx + \frac{\pi}{2} = (n + 1)\pi,$$

in which case $A = (-1)^{n+1}B$. Equation (6.23), written in the form

$$\int_a^b \sqrt{2m(E - V(x))}\,dx = \pi\hbar(n + \tfrac{1}{2}),$$

determines the semi-classical energy levels. It follows from (6.21)–(6.22) that the integer n is equal to the number of zeros of the WKB wave function. In accordance with the oscillation theorem, the case $n = 0$ corresponds to the ground state, case $n = 1$ — to the next energy level state, etc.

Denoting, as in Chapters 1 and 2, coordinate x by q, we can rewrite (6.23) as

$$(6.24) \qquad \oint pdq = 2\pi\hbar(n + \tfrac{1}{2}),$$

where integration goes over the closed classical orbit in the phase plane $\mathbb{R}^2$ with canonical coordinates p, q. Condition (6.24) is the famous Bohr-Wilson-Sommerfeld quantization rule for the case of one degree of freedom (see Section 2.5 of Chapter 2). We emphasize that, in general, the BWS quantization rule is applicable only for large n and gives the semi-classical asymptotic of energy levels E_n. However, as was shown in Section 2.6 of Chapter 2, the energy levels of the harmonic oscillator obtained by the BWS quantization rule are exact. The BWS rules are also exact for the energy levels of the hydrogen atom (see Section 5.1).

This semi-classical analysis can be generalized for quantum systems with n degrees of freedom which correspond to completely integrable classical Hamiltonian systems (see Section 2.6 of Chapter 1). Bohr-Wilson-Sommerfeld quantization rules take the form

$$(6.25) \qquad \oint_\gamma \boldsymbol{p}d\boldsymbol{q} = 2\pi\hbar(n_\gamma + \tfrac{1}{4}\operatorname{ind}\gamma).$$

Here integration goes over all 1-cycles γ in the Lagrangian submanifold $\Lambda = \{(\boldsymbol{p}, \boldsymbol{q}) \in \mathbb{R}^{2n} : H_c(\boldsymbol{p}, \boldsymbol{q}) = E, F_2(\boldsymbol{p}, \boldsymbol{q}) = E_2, \ldots, F_n(\boldsymbol{p}, \boldsymbol{q}) = E_n\}$, and $\operatorname{ind}\gamma$ is the Maslov index of a cycle γ in Λ. For the action-angle variables $(\boldsymbol{I}, \boldsymbol{\varphi})$ the integration in (6.25) goes over the basic 1-cycles on the n-torus T^n, and we obtain quantization conditions $I_i = (n_i + \tfrac{1}{2})\hbar$, $i = 1, \ldots, n$.

7. Notes and references

The classical text [**LL58**] is the source of many basic facts on the Schrödinger equation, written from the physics perspective. The textbook [**Foc78**], another classic, shows meticulous attention to detail. There are numerous mathematics papers and monographs devoted to the different aspects of the Schrödinger equation, and here we mention only the sources being used. For the self-adjointness criteria in Section 1.1 we refer the reader to the encyclopaedia survey [**RSS94**] and to the treatise

[**RS75**], and references therein; in particular, in [**RS75**] one can find the proof that the Schrödinger operator for a complex atom is essentially self-adjoint. For the proof of Sears theorem, see [**BS91**]. Theorems 1.6 and 1.7 (the latter in the special case $V_1 = 0$) in Section 1.2 are proved in [**BS91**] and [**RSS94**]; see [**RS78**] for the proof of Theorem 1.7 in the general case, and for other generalizations. The proof of the Kato theorem can be found in [**RS78**]; see also [**RSS94**]. For the proof of the Birman-Schwinger bound, see [**RS78**]. The monograph [**HS96**], besides the introduction to spectral theory, contains the proofs of many results in Sections 1.1-1.2; see also the monograph [**CFKS08**] for these and other results.

Section 2 is based on the fundamental paper [**Fad64**], classical surveys [**Fad59, Fad74**], and the monograph [**Mar86**], devoted to the inverse scattering problems. For more information and details on the material in Section 2.1 and Problems 2.2, 2.3, 2.4, see [**Fad64**] and [**Mar86**]. The complex integration of the resolvent kernel is a powerful method for proving the eigenfunction expansion theorem, especially in the presence of the absolutely continuous spectrum, and in Section 2.2 we followed the elegant approach from [**Fad59**]; Problem 2.6 is taken from [**Fad74**]. Section 2.3 is based on [**Fad59, Fad64, Fad74**]; see also the monograph [**New02**] for a comprehensive exposition of the scattering theory from the physics perspective, and the recent book [**Yaf92**] for its abstract mathematical formulation. We refer the reader to [**Fad74**] and references therein for more details on the material in Section 2.4.

Section 3 is fairly standard; from a physics point of view, its material can be found in almost any textbook on quantum mechanics; see [**LL58, Foc78**] for a clear exposition. It was H. Weyl who introduced Lie groups into quantum mechanics [**Wey50**], and nowadays representation theory of Lie groups is a part of the theoretical physics curriculum; see, e.g., [**BR86**]. There are also many mathematics textbooks and monographs on representation theory of compact Lie groups; see, e.g., [**Kir76, FH91**], as well as [**Vil68**]. Section 4 contains the standard material presented in any textbook on quantum mechanics; see [**LL58, Foc78, Mes99**] for the physics presentation. Our goal here was precise mathematical formulation of the basic facts; for the proofs of all results in this section (and of the problems as well), and for their generalizations, we refer the reader to the treatise [**RS79**].

The solution of the eigenvalue problem for the hydrogen atom, given by E. Schrödinger in 1926, was the major triumph of quantum mechanics. Before that, the energy levels of the hydrogen atom were obtained by N. Bohr in 1913 using the Bohr model (later replaced by the Bohr-Sommerfeld model and the BWS quantization rules), and in 1926 by W. Pauli by using the quantum Laplace-Runge-Lenz vector. Sections 5.1 and 5.2 are standard and our exposition follows [**LL58, Foc78**]. We started Section 5.3 by presenting Pauli's approach (see [**BR86**] for more details on representation theory). The hidden SO(4) symmetry of the hydrogen atom was discovered by V.A. Fock, and the end of Section 5.3 is based on [**Foc78**] (see

[**BI66a, BI66b**] for more details and for the treatment of the absolutely continuous spectrum).

There exists an extensive literature on semi-classical asymptotics and the WKB method, which we just barely touched on in Section 6. We refer to [**LL58**] and [**Dav76**] for a physics discussion and to [**GS77, BW97**] for a mathematical introduction. Detailed exposition of the one-dimensional WKB method can be found in [**Olv97**] — a reference book on special functions used in this chapter. We refer to the monographs [**MF81**] and [**Ler81**] for the detailed exposition of the much more complicated multi-dimensional WKB method.

Chapter 4

Spin and Identical Particles

1. Spin

1.1. Spin operators. So far we have been tacitly assuming that the Hilbert space for a quantum particle is $\mathscr{H} = L^2(\mathbb{R}^3, d^3\boldsymbol{x})$. Based on this assumption, in Section 5.1 of Chapter 3 we have found the energy levels of the hydrogen atom. In particular, in the ground state the third component M_3 of the quantum angular momentum operator $\boldsymbol{M}$ has eigenvalue 0. However, the famous Stern-Gerlach experiment has shown that the electron in the hydrogen atom also has a "magnetic angular momentum", whose third component in the ground state may take two values which differ by a sign. Therefore in addition to the "mechanical angular momentum" operator $\boldsymbol{M}$ with components M_1, M_2, M_3, the electron also has the "internal angular momentum" operator $\boldsymbol{S}$ with components S_1, S_2, S_3, called *spin*. The spin describes internal degrees of freedom of the electron and is independent of its position in the space[1]. The Hilbert space of states of the electron is $\mathscr{H}_S = \mathscr{H} \otimes \mathbb{C}^2$, i.e., it has twice as many states than a spinless particle. The Hilbert space $\mathscr{H}_S$ consists of two-component vector-valued functions on $\mathbb{R}^3$,

$$\Psi = \begin{pmatrix} \psi_1(\boldsymbol{x}) \\ \psi_2(\boldsymbol{x}) \end{pmatrix} \in \mathscr{H}_S,$$

where

$$\|\Psi\|^2 = \int_{\mathbb{R}^3} |\psi_1(\boldsymbol{x})|^2 d^3\boldsymbol{x} + \int_{\mathbb{R}^3} |\psi_2(\boldsymbol{x})|^2 d^3\boldsymbol{x} < \infty.$$

---

[1] The spin is a pure quantum effect which is absent in classical mechanics.

To every observable A in $\mathcal{H}$ there corresponds an observable $A \otimes I_2$ in $\mathcal{H}_S$, given by the 2×2 block-diagonal matrix $\begin{pmatrix} A & 0 \\ 0 & A \end{pmatrix}$. Observables of the form $I \otimes S$, where I is the identity operator in $\mathcal{H}$ and S is a self-adjoint operator in $\mathbb{C}^2$, describe the inner degrees of freedom and commute with all observables $A \otimes I_2$. The complete set of the observables in $\mathcal{H}_S$ consists of the operators $Q_1 \otimes I_2, Q_2 \otimes I_2, Q_3 \otimes I_2$ and $I \otimes S_1, I \otimes S_2, I \otimes S_3$, where Q_j are coordinate operators, and S_j are *spin operators* — traceless self-adjoint operators in $\mathbb{C}^2$, satisfying the same commutation relations as the quantum operators of the angular momentum,

$$[S_1, S_2] = i\hbar S_3, \quad [S_2, S_3] = i\hbar S_1, \quad [S_3, S_1] = i\hbar S_2.$$

In terms of the standard basis $e_1 = \begin{pmatrix} 1 \\ 0 \end{pmatrix}$, $e_2 = \begin{pmatrix} 0 \\ 1 \end{pmatrix}$ of $\mathbb{C}^2$, $S_j = \frac{\hbar}{2}\sigma_j$, $j = 1, 2, 3$, where σ_j are the so-called *Pauli matrices*

$$\sigma_1 = \begin{pmatrix} 0 & 1 \\ 1 & 0 \end{pmatrix}, \quad \sigma_2 = \begin{pmatrix} 0 & -i \\ i & 0 \end{pmatrix}, \quad \sigma_3 = \begin{pmatrix} 1 & 0 \\ 0 & -1 \end{pmatrix},$$

which are commonly used in physics. It is also convenient to represent $\Psi \in \mathcal{H}_S$ as a function $\psi(\boldsymbol{x}, \sigma)$ of two variables, where $\boldsymbol{x} \in \mathbb{R}^3$ and σ takes two values $\frac{1}{2}$ and $-\frac{1}{2}$, by setting $\psi(\boldsymbol{x}, \frac{1}{2}) = \psi_1(\boldsymbol{x})$ and $\psi(\boldsymbol{x}, -\frac{1}{2}) = \psi_2(\boldsymbol{x})$. Every function $\psi(\boldsymbol{x}, \sigma)$ is a finite linear combination of the functions $\psi(\boldsymbol{x})\chi(\sigma)$ — the decomposable elements of the tensor product $L^2(\mathbb{R}^3) \otimes \mathbb{C}^2$. Here we are identifying $\mathbb{C}^2$ with the complex vector space of functions $\chi : \{-\frac{1}{2}, \frac{1}{2}\} \to \mathbb{C}$ by assigning to the standard basis e_1, e_2 the functions χ_1, χ_2, defined by $\chi_1(\frac{1}{2}) = 1, \chi_1(-\frac{1}{2}) = 0$ and $\chi_2(\frac{1}{2}) = 0, \chi_2(-\frac{1}{2}) = 1$. This identification is also commonly used in physics.

In this notation the operator S_3 becomes a multiplication by $\hbar\sigma$ operator,

$$S_3\psi(\boldsymbol{x}, \sigma) = \hbar\sigma\psi(\boldsymbol{x}, \sigma),$$

and

$$S_1\psi(\boldsymbol{x}, \sigma) = \hbar|\sigma|\psi(\boldsymbol{x}, -\sigma), \quad S_2\psi(\boldsymbol{x}, \sigma) = -i\hbar\sigma\psi(\boldsymbol{x}, -\sigma).$$

Here and in what follows we will always assume that spin operators act only on the variable σ, and will often write S_j instead of $I \otimes S_j$. The operator $\boldsymbol{S}^2 = S_1^2 + S_2^2 + S_3^2$ is called the square of the total spin operator, and $\boldsymbol{S}^2 = \hbar^2 s(s+1)I_2$, where $s = |\sigma| = \frac{1}{2}$ is a total spin. In physics terminology, the electron has spin $\frac{1}{2}$.

1.2. Spin and representation theory of $\mathrm{SU}(2)$**.** Mathematical interpretation of spin is provided by the representation theory of the Lie algebra $\mathfrak{su}(2)$, discussed in Section 3.2 of Chapter 3. Namely, let

$$A_1 = -\frac{i}{2}\sigma_1, \quad A_2 = -\frac{i}{2}\sigma_2, \quad A_3 = -\frac{i}{2}\sigma_3$$

be the standard basis of the Lie algebra $\mathfrak{su}(2)$ — 2×2 traceless skew-Hermitian matrices satisfying commutation relations

$$[A_1, A_2] = A_3, \quad [A_2, A_3] = A_1, \quad [A_3, A_1] = A_2,$$

and let $\rho_{\frac{1}{2}}$ be the two-dimensional (fundamental) representation of $\mathfrak{su}(2)$ in $V_{\frac{1}{2}} \simeq \mathbb{C}^2$. The spin operators are given by

$$S_j = i\hbar\rho_{\frac{1}{2}}(A_j), \quad j = 1, 2, 3,$$

and $\rho_{\frac{1}{2}}$ is called the representation of spin $\frac{1}{2}$. In general, representation ρ_s in the $2s + 1$-dimensional complex vector space V_s is called the representation of spin $s \in \frac{1}{2}\mathbb{Z}_{\geq 0}$. The action of operators $T_j = i\rho_s(A_j)$ in V_s is described by formulas (3.10)–(3.12) in Section 3.2 of Chapter 3 (where l is replaced by s). There is another explicit realization of the representation ρ_s in the vector space $\mathscr{P}_s$ of polynomials $f(z)$ of degree not greater than $2s$. Namely, the mapping

$$e_{sm} \mapsto \frac{z^{s+m}}{\sqrt{(s+m)!(s-m)!}}, \quad m = -s, -s+1, \ldots, s-1, s,$$

establishes the isomorphism between the vector space V_s with the inner product determined by the orthonormal basis e_{sm}, and the vector space $\mathscr{P}_s$ with the inner product

$$(1.1) \qquad (f, g) = \frac{(2s+1)!}{\pi} \int_{\mathbb{C}} \frac{f(z)\overline{g(z)}}{(1+|z|^2)^{2s+2}} d^2z, \quad f, g \in \mathscr{P}_s.$$

Corresponding operators $T_j = i\rho_s(A_j)$ in $\mathscr{P}_s$ are given by

$$(1.2) \qquad T_1 - iT_2 = \frac{d}{dz}, \quad T_1 + iT_2 = z^2\frac{d}{dz} - 2sz, \quad T_3 = z\frac{d}{dz} - s.$$

It is easy to verify that T_1, T_2, T_3 are Hermitian operators with respect to the inner product (1.1). Representation ρ_s of the Lie algebra $\mathfrak{su}(2)$ is integrable, $\rho_s = dR_s$, where R_s is an irreducible unitary representation of the Lie group $\mathrm{SU}(2)$, defined by

$$R_s(g)(f)(z) = (\bar{\beta}z + \alpha)^{2s} f\left(\frac{\bar{\alpha}z - \beta}{\bar{\beta}z + \alpha}\right), \quad g = \begin{pmatrix} \alpha & \beta \\ -\bar{\beta} & \bar{\alpha} \end{pmatrix} \in \mathrm{SU}(2), \quad f \in \mathscr{P}_s.$$

Remark. The spin $\frac{1}{2}$ representation $R_{\frac{1}{2}}$ in $\mathbb{C}^2$ can also be explicitly described as

$$R_{\frac{1}{2}}(g)\boldsymbol{z} = \begin{pmatrix} \alpha & \beta \\ -\bar{\beta} & \bar{\alpha} \end{pmatrix}\begin{pmatrix} z_1 \\ z_2 \end{pmatrix}, \quad \boldsymbol{z} = \begin{pmatrix} z_1 \\ z_2 \end{pmatrix} \in \mathbb{C}^2,$$

and is called the fundamental representation.

Remark. Representation R_s of SU(2) can be lifted to a family of representations of the general unitary group U(2) acting in the same vector space $\mathscr{P}_s$. Namely, every $g \in$ U(2) can be written as $g = \alpha g_0$, where $\alpha^2 = \det g$ and is determined up to a sign, and $g_0 \in$ SU(2). Then for every integer l with the property that $2s + l$ is even, the formula

$$R_{l,s}(g) = \alpha^l R_s(g_0)$$

defines an irreducible unitary representation of U(2) in $\mathscr{P}_s$. The condition that $\frac{l}{2} + s$ is an integer ensures that $R_{l,s}$ does not depend on the choice of a sign in the definition of α. The representation $R_{l,s}$ is the highest weight representation of U(2) with the dominant weights $\lambda_1 = \frac{l}{2} + s \geq \lambda_2 = \frac{l}{2} - s$, $\lambda_1 + \lambda_2 = l$. We will denote the corresponding U(2)-module as $V_{\lambda,l}$, where $\lambda = (\lambda_1, \lambda_2)$ is a dominant weight with $\lambda_1 + \lambda_2 = l$. As an SU(2)-module, $V_{\lambda,l} = V_s$, where $s = \frac{1}{2}(\lambda_1 - \lambda_2)$.

Returning to the basic principles of quantum mechanics, formulated in Section 1 of Chapter 2, we now extend the first postulate **A1** to the case of particles with spin.

A10 (Particles with Spin). The Hilbert space of states for a quantum particle of a spin $s \in \{0, \frac{1}{2}, 1, \frac{3}{2}, \dots\}$ is

$$\mathscr{H}_S = L^2(\mathbb{R}^3) \otimes V_s,$$

where V_s is a $2s + 1$-dimensional complex vector space of the irreducible representation R_s of the Lie group SU(2) of spin s. The spin operators $\boldsymbol{S} = (S_1, S_2, S_3)$ are given by

$$S_j = i\hbar \rho_s(A_j), \quad j = 1, 2, 3,$$

where A_1, A_2, A_3 are standard generators of the Lie algebra $\mathfrak{su}(2)$ and $\rho_s = dR_s$ is the corresponding representation of $\mathfrak{su}(2)$ in V_s. Particles with integer spin are called *bosons*, and particles with half-integer spin are called *fermions*.

As in the case of spin $\frac{1}{2}$, it will be convenient to represent $\Psi \in \mathscr{H}_S$ as a function $\psi(\boldsymbol{x}, \sigma)$ of two variables, where $\boldsymbol{x} \in \mathbb{R}^3$ and the variable σ takes $2s+1$ values from the set $\{-s, -s+1, \dots, s-1, s\}$. For a given σ, the function $\psi(\boldsymbol{x}, \sigma)$ is the k-th component, $k = s - \sigma + 1$, of the $2s + 1$-dimensional vector $\Psi(\boldsymbol{x})$. Every function $\psi(\boldsymbol{x}, \sigma)$ is a finite linear combination of the functions $\psi(\boldsymbol{x})\chi(\sigma)$ — the decomposable elements of the tensor product $L^2(\mathbb{R}^3) \otimes \mathbb{C}^{2s+1}$, where we are identifying $\mathbb{C}^{2s+1}$ with the complex vector space of functions $\chi : \{-s, -s + 1, \dots, s - 1, s\} \to \mathbb{C}$ by assigning to the elements e_k of the standard basis $e_1, \dots, e_{2s+1}$ the functions χ_k, defined by $\chi_k(s + 1 - k) = 1$ and being 0 otherwise.

The complete set of observables for a quantum particle of spin s consists of position operators $Q_1 \otimes I_{2s+1}, Q_2 \otimes I_{2s+1}, Q_3 \otimes I_{2s+1}$, where I_{2s+1} is the identity operator in V_s, and spin operators $I \otimes S_1, I \otimes S_2, I \otimes S_3$. The total angular momentum operators $\boldsymbol{J} = (J_1, J_2, J_3)$ in $\mathscr{H}_S$ are self-adjoint operators in $\mathscr{H}_S$ given by

$$\boldsymbol{J} = \boldsymbol{M} \otimes I_{2s+1} + I \otimes \boldsymbol{S}.$$

They satisfy the same commutation relations as the spin operators,

$$[J_1, J_2] = i\hbar J_3, \quad [J_2, J_3] = i\hbar J_1, \quad [J_3, J_1] = i\hbar J_2.$$

As angular momentum operators $\boldsymbol{M}$ (see Lemma 3.1 in Section 3.2 of Chapter 3), the total angular momentum operators $\boldsymbol{J}$ are associated with the unitary representation of the Lie group $\mathrm{SU}(2)$ in the Hilbert space $\mathscr{H}_S = L^2(\mathbb{R}^3, d^3\boldsymbol{x}) \otimes V_s$. Namely, let R be the unitary representation of $\mathrm{SU}(2)$ in $\mathscr{H} = L^2(\mathbb{R}^3, d^3\boldsymbol{x})$ corresponding to the adjoint representation of $\mathrm{SU}(2)$ in $\mathfrak{su}(2) \simeq \mathbb{R}^3$,

$$(R(g)\psi)(\boldsymbol{x}) = \psi(\mathrm{Ad}\,g^{-1}\,\boldsymbol{x}), \quad g \in \mathrm{SU}(2), \quad \psi \in \mathscr{H},$$

where

$$(\mathrm{Ad}\,g\,\boldsymbol{x}) \cdot \boldsymbol{\sigma} = g(\boldsymbol{x} \cdot \boldsymbol{\sigma})g^{-1}, \quad \boldsymbol{x} \cdot \boldsymbol{\sigma} = x_1\sigma_1 + x_2\sigma_2 + x_3\sigma_3.$$

Lemma 1.1. *We have*

$$(R \otimes R_s)(e^{a_1 A_1 + a_2 A_2 + a_3 A_3}) = e^{-\frac{i}{\hbar}(a_1 J_1 + a_2 J_2 + a_3 J_3)},$$

where $\boldsymbol{a} = (a_1, a_2, a_3) \in \mathbb{R}^3$.

Problem 1.1. Derive formulas (1.2) and show that operators T_1, T_2, T_3 are Hermitian with respect to the inner product (1.1).

Problem 1.2. Prove Lemma 1.1.

2. Charged spin particle in the magnetic field

2.1. Pauli Hamiltonian. The classical particle of charge e moving in the electromagnetic field with the vector potential $\boldsymbol{A}(\boldsymbol{x})$ and the scalar potential $\varphi(\boldsymbol{x})$ is described by the Hamiltonian function

$$H_c(\boldsymbol{p}, \boldsymbol{x}) = \frac{1}{2m}\left(\boldsymbol{p} - \frac{e}{c}\boldsymbol{A}\right)^2 + e\varphi(\boldsymbol{x}), \quad \boldsymbol{x} = (x, y, z) \in \mathbb{R}^3$$

(see Problem 1.27 in Section 1.7 of Chapter 1). As we have seen in Example 2.3 in Section 2.4 of Chapter 2, the corresponding Hamiltonian operator of the spinless particle is given by

$$\tag{2.1} H^0 = \frac{1}{2m}\left(\boldsymbol{P} - \frac{e}{c}\boldsymbol{A}\right)^2 + e\varphi(\boldsymbol{x}).$$

However, quantum particles with spin interact with the magnetic field, so that the Hamiltonian H^0, which is an operator in the Hilbert space $\mathscr{H} =

$L^2(\mathbb{R}^3, d^3\boldsymbol{x})$, is not adequate when $s \neq 0$. The correct prescription is to include the spin degrees of freedom and consider the Hamiltonian operator in the Hilbert space $\mathscr{H}_S = \mathscr{H} \otimes V_s$. For the case of an electron — particle with spin $\frac{1}{2}$ — the corresponding Hamiltonian is given by

$$(2.2) \qquad H = \frac{1}{2m}(\boldsymbol{P}_A \cdot \boldsymbol{\sigma})^2 + e\varphi(\boldsymbol{x}), \quad \boldsymbol{P}_A = \boldsymbol{P} - \frac{e}{c}\boldsymbol{A}.$$

Using the basic facts that Pauli matrices anti-commute and satisfy $\sigma_1^2 = \sigma_2^2 = \sigma_3^2 = I_2$ and

$$\sigma_1\sigma_2 = i\sigma_3, \quad \sigma_2\sigma_3 = i\sigma_1, \quad \sigma_3\sigma_1 = i\sigma_2,$$

we readily obtain

$$(\boldsymbol{P}_A \cdot \boldsymbol{\sigma})^2 = \left(P_1 - \frac{e}{c}A_1\right)^2 + \left(P_2 - \frac{e}{c}A_2\right)^2 + \left(P_3 - \frac{e}{c}A_3\right)^2$$
$$+ i\left[P_1 - \frac{e}{c}A_1, P_2 - \frac{e}{c}A_2\right]\sigma_3 + i\left[P_2 - \frac{e}{c}A_2, P_3 - \frac{e}{c}A_3\right]\sigma_1$$
$$+ i\left[P_3 - \frac{e}{c}A_3, P_1 - \frac{e}{c}A_1\right]\sigma_2$$
$$= \left(\boldsymbol{P} - \frac{e}{c}\boldsymbol{A}\right)^2 - \frac{\hbar e}{c}(B_1\sigma_1 + B_2\sigma_2 + B_3\sigma_3).$$

Here $\boldsymbol{B} = (B_1, B_2, B_3)$ is the magnetic field, $\boldsymbol{B} = \operatorname{curl} \boldsymbol{A}$, and in the last line we have used Heisenberg commutation relations (cf. a similar computation in Example 2.3 in Section 2.4 of Chapter 2). The Hamiltonian (2.2) takes the form

$$(2.3) \qquad H = H^0 - \mu\,\boldsymbol{B}\cdot\boldsymbol{\sigma}, \quad \mu = \frac{\hbar e}{2mc},$$

and is known as the *Pauli Hamiltonian*; the quantity μ is the total *magnetic moment* of an electron. It is customary to put $\mu = -\mu_{\mathrm{B}}$, where

$$\mu_{\mathrm{B}} = \frac{\hbar|e|}{2mc} = 0.927 \times 10^{-20}\,\frac{\text{erg}}{\text{gauss}}$$

is the so-called *Bohr magneton*. The corresponding time-dependent Schrödinger equation

$$i\hbar\frac{\partial\psi}{\partial t} = H\psi$$

in this context is called the *Pauli wave equation*.

Remark. The Pauli wave equation is obtained from the *relativistic Dirac equation* for the electron in the limit $c \to \infty$ by neglecting the terms of order c^{-2}.

The corresponding Hamiltonian for a quantum particle of spin s and charge e has a similar form

$$H = H^0 - \boldsymbol{B}\cdot\boldsymbol{\mu}, \quad \boldsymbol{\mu} = \frac{\mu}{\hbar s}\boldsymbol{S},$$

where $\boldsymbol{S} = (S_1, S_2, S_3)$ are spin operators of spin s, quantity μ is the total magnetic moment of a particle, and the operator $\boldsymbol{\mu}$ is its "intrinsic" magnetic moment.

2.2. Particle in a uniform magnetic field. Here we consider the problem of finding the energy levels of a charged spin $\frac{1}{2}$ particle moving in a constant uniform magnetic field. Choosing the z-axis along the direction of the field, $\boldsymbol{B} = (0, 0, B)$, we get $A_1(\boldsymbol{x}) = -By, A_2(\boldsymbol{x}) = A_3(\boldsymbol{x}) = 0$, and $\varphi(\boldsymbol{x}) = 0$. The Pauli Hamiltonian takes the form

$$(2.4) \qquad H = \frac{1}{2m}\left[\left(P_1 + \frac{e}{c}By\right)^2 + P_2^2 + P_3^2\right] - \mu B\sigma_3.$$

The eigenvalue equation

$$H\psi(\boldsymbol{x}, \sigma) = E\psi(\boldsymbol{x}, \sigma)$$

in $\mathscr{H}_S$ reduces to two (for each value of $\sigma = \pm\frac{1}{2}$) eigenvalue equations

$$\frac{1}{2m}\left[\left(P_1 + \frac{e}{c}By\right)^2 + P_2^2 + P_3^2\right]\psi - \mu B\sigma\psi = E\psi$$

in $\mathscr{H}$. Separating the variables

$$\psi(\boldsymbol{x}) = e^{\frac{i}{\hbar}(k_1 x + k_3 z)}\chi(y),$$

we obtain

$$(2.5) \qquad -\frac{\hbar^2}{2m}\chi'' + \frac{m\omega_{\mathrm{B}}^2}{2}(y - y_0)^2\chi = \left(E + \mu B\sigma - \frac{k_3^2}{2m}\right)\chi,$$

where

$$y_0 = -\frac{ck_1}{eB} \quad \text{and} \quad \omega_{\mathrm{B}} = \frac{|e|B}{mc}.$$

Equation (2.5) coincides with the eigenvalue problem for the one-dimensional harmonic oscillator with the frequency ω_{B}, solved in Section 2.6 of Chapter 2. Thus we obtain

$$(2.6) \qquad E = (n + \tfrac{1}{2})\hbar\omega_{\mathrm{B}} + \frac{k_3^2}{2m} - \mu B\sigma, \quad n = 0, 1, \ldots; \quad \sigma = \pm\frac{1}{2}.$$

The first term in this formula gives discrete energy levels, called *Landau levels*, for the motion in the plane perpendicular to the magnetic field.

The spectrum of Pauli Hamiltonian (2.4) is absolutely continuous and fills $[\frac{1}{2}(\hbar\omega_{\mathrm{B}} - \mu B), \infty)$ with variable infinite multiplicity. It can be visualized as a countable union of branches $[(n + \frac{1}{2})\hbar\omega_{\mathrm{B}} - \mu B\sigma, \infty)$, each with a multiplicity parametrized by $\mathbb{R} \times \{\pm 1\}$ (all $k_1 \in \mathbb{R}$ and $\pm k_3$ for fixed k_3

correspond to the same E). Corresponding normalized eigenfunctions of the continuous spectrum are

$$\psi_n(\boldsymbol{x}; k_1, k_3) = \frac{1}{2\pi\hbar} \sqrt[4]{\frac{m\omega_{\mathrm{B}}}{\pi\hbar}} \frac{1}{\sqrt{2^n n!}} e^{\frac{i}{\hbar}(k_1 x + k_3 z) - \frac{m\omega_{\mathrm{B}}}{2\hbar}(y - y_0)^2} H_n\left(\sqrt{\frac{m\omega_{\mathrm{B}}}{\hbar}}\, y\right),$$

where $-\infty < k_1, k_3 < \infty$.

For the electron, $\hbar\omega_{\mathrm{B}} = \frac{1}{2}\mu_{\mathrm{B}}$, and formula (2.6) takes the form

$$E = (n + \tfrac{1}{2} + \sigma)\mu_{\mathrm{B}} B + \frac{k_3^2}{2m}.$$

In this case there is an extra degeneration of the spectrum: energy levels with n and $\sigma = \frac{1}{2}$ and those with $n + 1$ and $\sigma = -\frac{1}{2}$ coincide.

Problem 2.1. Derive the eigenfunction expansion theorem for the Hamiltonian (2.4).

Problem 2.2. Show that the operators

$$X_0 = Q_1 + \frac{c}{eB} P_2 \quad \text{and} \quad Y_0 = -\frac{c}{eB} P_1,$$

which correspond to the coordinates of a center of a circle (see Problem 1.5 in Section 1.3 of Chapter 1), are quantum integrals of motion, i.e., commute with the Hamiltonian H, but

$$\{X_0, Y_0\}_\hbar = \frac{c}{eB} I.$$

3. System of identical particles

Consider the quantum system of N particles of spins $s_1, \ldots, s_N$. According to the basic principles of quantum mechanics, the Hilbert space of states of the system is given by

$$\mathscr{H}_N = \mathscr{H}_{S_1} \otimes \cdots \otimes \mathscr{H}_{S_N} = \underbrace{L^2(\mathbb{R}^3) \otimes \cdots \otimes L^2(\mathbb{R}^3)}_{N} \otimes V_{s_1} \otimes \cdots \otimes V_{s_N}.$$

In terms of the variables $\boldsymbol{\xi}_i = (\boldsymbol{x}_i, \sigma_i)$, where $\boldsymbol{x}_i \in \mathbb{R}^3$ and $\sigma_i \in \{-s_i, -s_i + 1, \ldots, s_i - 1, s_i\}$, the *total wave function* of the system can be written as $\Phi(\boldsymbol{\xi}_1, \ldots, \boldsymbol{\xi}_N)$. The total wave function is a finite linear combination of the product functions

$$\Phi(\boldsymbol{\xi}_1, \ldots, \boldsymbol{\xi}_N) = \Psi(\boldsymbol{x}_1, \ldots, \boldsymbol{x}_N)\chi(\sigma_1, \ldots, \sigma_N),$$

where $\Psi(\boldsymbol{x}_1, \ldots, \boldsymbol{x}_N) \in L^2(\mathbb{R}^{3N})$ is a coordinate part, and $\chi(\sigma_1, \ldots, \sigma_N)$ is a spin part of the total wave function. The spin part, in accordance with the tensor product structure $V_{s_1} \otimes \cdots \otimes V_{s_N}$, is a finite linear combination of the products $\chi_{k_1}(\sigma_1) \ldots \chi_{k_N}(\sigma_N)$ of functions χ_k, defined in Section 1.

The Hamiltonian of the system of N-particles (without the spin interaction) in the coordinate representation has the form

$$H_N = -\sum_{i=1}^{N} \frac{\hbar^2}{2m_i}\Delta_i + \sum_{i=1}^{N} V_i(\boldsymbol{x}_i) + \sum_{1\leq i<k\leq N} V_{ik}(\boldsymbol{x}_i - \boldsymbol{x}_k),$$

where the first term is the operator of kinetic energy of the system of N particles, the second term describes the interaction of the particles with the external field, and the last term describes their pair-wise interaction. Note that though the Hamiltonian H_N acts only on the coordinate part $\Psi(\boldsymbol{x}_1,\ldots,\boldsymbol{x}_N)$ of the total wave function, the physical properties of the system do depend on the spins of the particles.

3.1. The symmetrization postulate. For the system of N identical particles of mass m and spin s the corresponding Hilbert space is

$$(3.1) \qquad \mathscr{H}_N = L^2(\mathbb{R}^{3N}) \otimes V_s^{\otimes N}.$$

The irreducible representation R_s of the Lie group $\mathrm{SU}(2)$ in the vector space V_s naturally defines a representation $R_s^{\otimes N}$ in $V_s^{\otimes N}$,

$$R_s^{\otimes N}(g)(v_1 \otimes \cdots \otimes v_N) = R_s(g)v_1 \otimes \cdots \otimes R_s(g)v_N, \quad g \in \mathrm{SU}(2).$$

The corresponding representation $\rho_s^{\otimes N}$ of the Lie algebra $\mathfrak{su}(2)$ is given by

$$\rho_s^{\otimes N}(x)(v_1 \otimes \cdots \otimes v_N) = \sum_{k=1}^{N} v_1 \otimes \cdots \otimes \rho_s(x)v_k \otimes \cdots \otimes v_N, \quad x \in \mathfrak{su}(2).$$

The total spin operators $\boldsymbol{S} = (S_1, S_2, S_3)$ of the system of N identical particles are given by

$$(3.2) \qquad S_j = i\hbar\rho_s^{\otimes N}(A_j) = \sum_{k=1}^{N} S_j^{(k)}, \quad j = 1, 2, 3.$$

Here $S_j^{(k)} = I_{2s+1} \otimes \cdots \otimes i\hbar\rho_s(A_j) \otimes \cdots \otimes I_{2s+1}$ are the spin operators of the k-th particle; they act non-trivially only in the k-th factor of the tensor product $V_s^{\otimes N}$.

Representations $R_s^{\otimes N}$ and $\rho_s^{\otimes N}$ are reducible and contain all representations V_l with spins $l \in \frac{1}{2}\mathbb{Z}_{\geq 0}$ such that $sN - l$ are non-negative integers. Indeed, the square $\boldsymbol{S}^2$ of the total spin operator has the form

$$\boldsymbol{S}^2 = S_1^2 + S_2^2 + S_3^2 = S_-S_+ + S_3(S_3 + I),$$

where $S_\pm = S_1 \pm iS_2$ and I here is the identity operator in $V_s^{\otimes N}$. It is easy to see from (3.2) that the eigenvalues of the operator S_3 are $-\hbar sN$, $\hbar(-sN + 1),\ldots,\hbar(sN - 1), \hbar sN$. Now by Schur's lemma, restriction of $\boldsymbol{S}^2$ to the irreducible representation V_l is $\hbar^2 l(l + 1)$ times the identity operator in V_l. Since V_l contains a highest weight vector which is annihilated by the operator S_+ and is an eigenvector of the operator S_3 with eigenvalue $\hbar l$, we conclude that $V_s^{\otimes N}$ contains all irreducible representations with spins $l \in \frac{1}{2}\mathbb{Z}_{\geq 0}$ such that $sN - l$ are non-negative integers. The problem of determining the multiplicities with which representations V_l appear in the decomposition $V_s^{\otimes N}$ into the direct sum of irreducible representations will be discussed below. It plays a fundamental role in classifying the energy levels of the system of N identical particles of spin s.

There is also a natural action of the symmetric group Sym_N on N elements on $\mathscr{H}_N$, given by

$$P_\pi \Phi(\boldsymbol{\xi}_1, \ldots, \boldsymbol{\xi}_N) = \Phi(\boldsymbol{\xi}_{\pi^{-1}(1)}, \ldots, \boldsymbol{\xi}_{\pi^{-1}(N)}), \quad \pi \in \mathrm{Sym}_N.$$

The corresponding Hamiltonian in this case has the form

$$(3.3) \qquad H_N = -\frac{\hbar^2}{2m} \sum_{i=1}^{N} \Delta_i + \sum_{i=1}^{N} V_{\mathrm{ext}}(\boldsymbol{x}_i) + \sum_{1 \leq i < k \leq N} V_{\mathrm{int}}(\boldsymbol{x}_i - \boldsymbol{x}_k),$$

and commutes with the Sym_N-action:

$$[H_N, P_\pi] = 0 \quad \text{for all} \quad \pi \in \mathrm{Sym}_N.$$

Thus the operator H_N can be restricted to the Sym_N-invariant subspaces of $\mathscr{H}$, in particular to the subspace $\mathscr{H}_N^{(\mathrm{S})}$ of totally symmetric functions $\Phi(\boldsymbol{\xi}_1, \ldots, \boldsymbol{\xi}_N)$,

$$P_\pi \Phi = \Phi \quad \text{for all} \quad \pi \in \mathrm{Sym}_N,$$

and to the subspace $\mathscr{H}_N^{(\mathrm{A})}$ of totally anti-symmetric functions $\Phi(\boldsymbol{\xi}_1, \ldots, \boldsymbol{\xi}_N)$,

$$P_\pi \Phi = (-1)^{\varepsilon(\pi)} \Phi \quad \text{for all} \quad \pi \in \mathrm{Sym}_N,$$

where $\varepsilon(\pi)$ is the parity of a permutation π.

In classical mechanics, it is possible to label the identical particles (say by the integers $1, 2, \ldots, N$), and to trace the trajectory of each labeled particle separately. It follows from the Heisenberg uncertainty relations that in quantum mechanics it is impossible to trace the time evolution of labeled particles separately. In other words, in quantum mechanics identical particles are truly "indistinguishable", so that the Hilbert space (3.1) contains "too many states". It is another postulate of quantum mechanics which specifies "correct" Hilbert space of states for the system of identical particles.

A11 (The Symmetrization Postulate). The Hilbert space of states for the system of N identical particles of spin s is the totally symmetric subspace $\mathscr{H}_N^{(\mathrm{S})}$ of the N-fold tensor product $(L^2(\mathbb{R}^3) \otimes V_s)^{\otimes N}$ for the case of bosons (particles with integer spin), and is the totally anti-symmetric subspace $\mathscr{H}_N^{(\mathrm{A})}$ for the case of fermions (particles with half-integer spin).

Remark. The symmetrization postulate implies the *Pauli exclusion principle*, which states that no two fermions can occupy the same state. The Pauli exclusion principle is fundamental for atomic and molecular physics, and for the whole body of chemistry.

For the case of identical particles of spin 0 it follows from the symmetrization postulate that the corresponding Hilbert space $\mathscr{H}_N^{(S)}$ is the subspace of all totally symmetric functions $\Psi(\boldsymbol{x}_1, \dots, \boldsymbol{x}_N)$ in $L^2(\mathbb{R}^{3N})$, and in order to find the energy levels of the system, one needs to solve the Schrödinger equation

$$(3.4) \qquad\qquad H_N \Psi = E_N \Psi$$

only in the totally symmetric subspace $\mathscr{H}_N^{(S)}$ of $\mathscr{H}_N$. Solutions of the eigenvalue problem (3.4) that do not belong to this subspace have no physical interpretation.

For the spin $\frac{1}{2}$ particles the situation is different: the symmetry properties of the coordinate part of the total wave function depend on the symmetry properties of the corresponding spin part. Consider first the simple example of a system of two electrons. In this case we have the direct sum decomposition of Hilbert spaces,

$$\mathscr{H}_2 = \mathscr{H}_2^{(S)} \oplus \mathscr{H}_2^{(A)},$$

which immediately follows from the representation

$$\Phi(\boldsymbol{\xi}_1, \boldsymbol{\xi}_2) = \tfrac{1}{2}(\Phi(\boldsymbol{\xi}_1, \boldsymbol{\xi}_2) + \Phi(\boldsymbol{\xi}_2, \boldsymbol{\xi}_1)) + \tfrac{1}{2}(\Phi(\boldsymbol{\xi}_1, \boldsymbol{\xi}_2) - \Phi(\boldsymbol{\xi}_2, \boldsymbol{\xi}_1)).$$

Using the identity

$$\Psi(\boldsymbol{x}_1, \boldsymbol{x}_2)\chi(\sigma_1, \sigma_2) - \Psi(\boldsymbol{x}_2, \boldsymbol{x}_1)\chi(\sigma_2, \sigma_1)$$
$$= \tfrac{1}{2}(\Psi(\boldsymbol{x}_1, \boldsymbol{x}_2) + \Psi(\boldsymbol{x}_2, \boldsymbol{x}_1))(\chi(\sigma_1, \sigma_2) - \chi(\sigma_2, \sigma_1))$$
$$+ \tfrac{1}{2}(\Psi(\boldsymbol{x}_1, \boldsymbol{x}_2) - \Psi(\boldsymbol{x}_2, \boldsymbol{x}_1))(\chi(\sigma_1, \sigma_2) + \chi(\sigma_2, \sigma_1)),$$

we can represent $\mathscr{H}_2^{(A)}$ as a direct sum of Hilbert spaces

$$\mathscr{H}_2^{(A)} = \mathfrak{S}_0 \oplus \mathfrak{S}_1,$$

where

$$\mathfrak{S}_0 = \left(L^2(\mathbb{R}^3) \otimes L^2(\mathbb{R}^3)\right)^{(S)} \otimes \wedge^2 \mathbb{C}^2$$

consists of finite linear combinations of the product functions $\Phi(\boldsymbol{\xi}_1, \boldsymbol{\xi}_2) = \Psi(\boldsymbol{x}_1, \boldsymbol{x}_2)\chi(\sigma_1, \sigma_2)$, where $\Psi(\boldsymbol{x}_1, \boldsymbol{x}_2) \in L^2(\mathbb{R}^3) \otimes L^2(\mathbb{R}^3)$ is symmetric and $\chi(\sigma_1, \sigma_2) \in \wedge^2 \mathbb{C}^2$ is anti-symmetric, and

$$\mathfrak{S}_1 = \left(L^2(\mathbb{R}^3) \otimes L^2(\mathbb{R}^3)\right)^{(A)} \otimes \mathrm{Sym}^2 \mathbb{C}^2$$

consists of finite linear combinations of the product functions where $\Psi(\boldsymbol{x}_1, \boldsymbol{x}_2)$ is anti-symmetric and $\chi(\sigma_1, \sigma_2)$ is symmetric. The vector space $\wedge^2 \mathbb{C}^2$ is one-dimensional and is spanned by the function χ_{00}, which corresponds to the vector

$$\tfrac{1}{\sqrt{2}}(e_1 \otimes e_2 - e_2 \otimes e_1) \in \mathbb{C}^2 \otimes \mathbb{C}^2,$$

and the vector space $\mathrm{Sym}^2 \mathbb{C}^2$ is three-dimensional and is spanned by the functions $\chi_{11}, \chi_{10}, \chi_{1\text{-}1}$, which correspond to the vectors

$$e_1 \otimes e_1, \quad \tfrac{1}{\sqrt{2}}(e_1 \otimes e_2 + e_2 \otimes e_1), \quad e_2 \otimes e_2 \in \mathbb{C}^2 \otimes \mathbb{C}^2.$$

Comparing the decomposition

$$\mathbb{C}^2 \otimes \mathbb{C}^2 = \wedge^2 \mathbb{C}^2 \oplus \mathrm{Sym}^2 \mathbb{C}^2$$

with the Clebsch-Gordan decomposition

$$V_{\frac{1}{2}} \otimes V_{\frac{1}{2}} = V_0 \oplus V_1,$$

discussed in Section 3.2 of Chapter 3, we see that[2] $\wedge^2 \mathbb{C}^2 = V_0$ and $\mathrm{Sym}^2 \mathbb{C}^2 = V_1$. In physics terminology, V_0 is the *spin singlet space*, and V_1 is the *spin triplet space*. We summarize the obtained results in the following statement.

Proposition 3.1. *For a system of two identical particles of spin $\frac{1}{2}$ the total wave function of spin 0 has the form $\Psi(\boldsymbol{x}_1, \boldsymbol{x}_2)\chi_{00}(\sigma_1, \sigma_2)$, where $\Psi(\boldsymbol{x}_1, \boldsymbol{x}_2)$ is symmetric and satisfies the Schrödinger equation (3.4). The total wave function of spin 1 has the form $\Psi(\boldsymbol{x}_1, \boldsymbol{x}_2)\chi_{1m}(\sigma_1, \sigma_2)$, where $m = -1, 0, 1$, and $\Psi(\boldsymbol{x}_1, \boldsymbol{x}_2)$ is anti-symmetric and satisfies the Schrödinger equation (3.4).*

Remark. The coordinate part $\Psi(\boldsymbol{x}_1, \boldsymbol{x}_2)$ of the total wave function has the usual probabilistic interpretation. Thus for a simple example of a system of two electrons with $V_{\mathrm{int}}(\boldsymbol{x}) = 0$, the coordinate part of the total wave function has the form

$$\Psi(\boldsymbol{x}_1, \boldsymbol{x}_2) = \tfrac{1}{\sqrt{2}}(\psi_1(\boldsymbol{x}_1)\psi_2(\boldsymbol{x}_2) \pm \psi_1(\boldsymbol{x}_2)\psi_2(\boldsymbol{x}_1)),$$

where $\psi_1(\boldsymbol{x})$ and $\psi_2(\boldsymbol{x})$ are two one-electron coordinate wave functions, the plus sign refers to the spin singlet state, and the minus sign refers to the spin triplet state. The corresponding probability density is

$$|\Psi(\boldsymbol{x}_1, \boldsymbol{x}_2)|^2 d^3\boldsymbol{x}_1 d^3\boldsymbol{x}_2 = \tfrac{1}{2}\big(|\psi_1(\boldsymbol{x}_1)|^2|\psi_2(\boldsymbol{x}_2)|^2 + |\psi_1(\boldsymbol{x}_2)|^2|\psi_2(\boldsymbol{x}_1)|^2$$

$$\pm 2\,\mathrm{Re}(\psi_1(\boldsymbol{x}_1)\psi_2(\boldsymbol{x}_2)\overline{\psi_1(\boldsymbol{x}_2)\psi_2(\boldsymbol{x}_1)}))d^3\boldsymbol{x}_1 d^3\boldsymbol{x}_2,$$

where the last term is called the *exchange energy*. It follows from this formula that when two electrons are in a spin triplet state, there is a zero probability of finding them at the same point $\boldsymbol{x} \in \mathbb{R}^3$. However, when electrons are in a spin singlet state, there is a non-zero probability of finding them at the same point in space, which is due to the exchange energy.

---

[2]This can also be verified directly by checking the action of total spin operators.

Consider now the system of N particles of spin $\frac{1}{2}$. In this case the total wave function $\Phi(\boldsymbol{\xi}_1,\ldots,\boldsymbol{\xi}_N) \in \mathscr{H}_N^{(\mathrm{A})}$ satisfies

$$(3.5) \qquad H_N\Phi = E_N\Phi \quad \text{and} \quad \boldsymbol{S}^2\Phi = s(s+1)\Phi,$$

and describes the bound state with total spin s, where $\frac{1}{2}N - s$ is a non-negative integer. It seems natural to assume that if one finds a coordinate wave function $\Psi(\boldsymbol{x}_1,\ldots,\boldsymbol{x}_N)$ satisfying (3.4), and a spin wave function $\chi(\sigma_1,\ldots,\sigma_N)$ which is the eigenfunction of $\boldsymbol{S}^2$, then the total wave function $\Phi(\boldsymbol{\xi}_1,\ldots,\boldsymbol{\xi}_N)$ can be obtained by the anti-symmetrization:

$$\Phi(\boldsymbol{\xi}_1,\ldots,\boldsymbol{\xi}_N) = \sum_{\pi\in\mathrm{Sym}_N} (-1)^{\varepsilon(\pi)}\Psi(\boldsymbol{x}_{\pi(1)},\ldots,\boldsymbol{x}_{\pi(N)})\chi(\sigma_{\pi(1)},\ldots,\sigma_{\pi(N)}).$$

However, it may happen that for a coordinate wave function $\Psi(\boldsymbol{x}_1,\ldots,\boldsymbol{x}_N)$ satisfying (3.4), the corresponding total wave function $\Phi(\boldsymbol{\xi}_1,\ldots,\boldsymbol{\xi}_N)$ is identically zero! The requirement that the total wave function $\Phi(\boldsymbol{\xi}_1,\ldots,\boldsymbol{\xi}_N)$ is not identically zero determines admissible symmetry properties of the coordinate wave function $\Psi(\boldsymbol{x}_1,\ldots,\boldsymbol{x}_N)$, which are described by the irreducible representations of the symmetric group Sym_N.

3.2. Young diagrams and representation theory of Sym_N**.** The number of irreducible representations of the symmetric group Sym_N is the number of its conjugacy classes, which in turn is the number of partitions $\lambda = (\lambda_1,\ldots,\lambda_n)$ of N: $N = \lambda_1 + \cdots + \lambda_n$, where $\lambda_1 \geq \lambda_2 \geq \cdots \geq \lambda_n \geq 1$. Denote by $\mathrm{Par}(N)$ the set of all partitions of N. To each partition $\lambda \in \mathrm{Par}(N)$ there is an irreducible representation G_λ of Sym_N, constructed as follows. Let Y_λ be the *Young diagram* associated with $\lambda \in \mathrm{Par}(N)$ — a collection of N boxes arranged into n left-justified rows, with the first row containing λ_1 boxes, the second row containing λ_2 boxes, etc. Thus the diagram

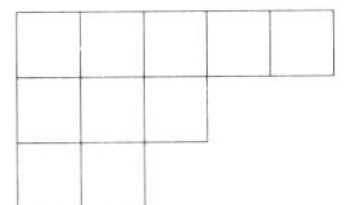

corresponds to the partition $10 = 5 + 3 + 2$. A *Young tableau A* associated with the Young diagram Y_λ is an assignment of N integers $1, 2, \ldots, N$ to the N boxes such that each box gets a different integer; denote by $\mathfrak{T}_\lambda$ the set of all Young tableaux associated with the partition $\lambda \in \mathrm{Par}(N)$. A canonical Young tableau $A_\lambda \in \mathfrak{T}_\lambda$ is obtained by numbering the boxes consecutively along the rows from left to right. For a Young tableau $A \in \mathfrak{T}_\lambda$, say the canonical one, define two subgroups of Sym_N:

$$\mathrm{Row}(A) = \{\pi \in \mathrm{Sym}_N : \pi \text{ preserves the rows of } A\},$$
$$\mathrm{Col}(A) = \{\pi \in \mathrm{Sym}_N : \pi \text{ preserves the columns of } A\}.$$

Now let $\mathfrak{A} = \mathbb{C}\mathrm{Sym}_N$ be the group algebra of Sym_N — a complex vector space with the basis $\{e_\pi\}_{\pi\in\mathrm{Sym}_N}$ and the multiplication map defined by $e_{\pi_1} \cdot e_{\pi_2} = e_{\pi_1\pi_2}$. Denote by R the regular representation of Sym_N in $\mathfrak{A}$, defined by

$$R(\pi)x = e_\pi \cdot x, \quad x \in \mathfrak{A}.$$

With a given Young tableau A one can associate two elements in the group algebra $\mathfrak{A}$: the row symmetrizer

$$\mathrm{R}_A = \sum_{\pi\in\mathrm{Row}(A)} \pi,$$

and the column anti-symmetrizer

$$\mathrm{C}_A = \sum_{\pi\in\mathrm{Col}(A)} (-1)^{\varepsilon(\pi)}\pi.$$

Define the *Young symmetrizer* by

$$\Pi_A = \mathrm{C}_A\mathrm{R}_A \in \mathfrak{A},$$

and consider the left Sym_N-module[3] $G_\lambda = \mathfrak{A} \cdot \Pi_A$, and the representation $T_\lambda : \mathrm{Sym}_N \to \mathrm{End}\,G_\lambda$, where Sym_N acts by left multiplication. The following result is fundamental.

Theorem 3.1. *Each T_λ is an irreducible representation of* Sym_N, *and representations T_λ and T_μ are not isomorphic if $\lambda \neq \mu$. The dimension of the representation T_λ is given by the Frobenius formula*

$$d_\lambda = \frac{N!}{l_1!\ldots l_n!} \prod_{i<j}(l_i - l_j), \quad l_i = \lambda_i + n - i, \quad i = 1,\ldots,n,$$

and $\Pi_A^2 = d_\lambda\Pi_A$. Every irreducible representation of Sym_N *is isomorphic to the representation T_λ for some partition $\lambda \in \mathrm{Par}(N)$.*

Remark. It follows from the representation theory of finite groups that

$$R = \bigoplus_{\lambda\in\mathrm{Par}(N)} d_\lambda G_\lambda \quad \left(\mathfrak{A} = \bigoplus_{\lambda\in\mathrm{Par}(N)} \mathrm{End}\,G_\lambda \text{ as algebras}\right),$$

so that

$$N! = \sum_{\lambda\in\mathrm{Par}(N)} d_\lambda^2.$$

There is another expression for dimensions d_λ as a product over the boxes of the Young diagram Y_λ, given by the so-called hook-length formula.

---

[3]Another choice of a Young tableau $A \in \mathfrak{T}_\lambda$ gives a Sym_N-module equivalent to G_λ.

Remark. One can explicitly construct the bases in the subspaces G_λ by the following procedure. Let A be the standard Young tableau for a partition $\lambda \in \mathrm{Par}(N)$: a Young tableau with the condition that in each row of Y_λ, the numbers in the boxes are strictly increasing from left to right, and in each column — strictly increasing from top to bottom. Let $\pi \in \mathrm{Sym}_N$ be the unique permutation such that $A = \pi(A_\lambda)$ and put $e_A = \Pi_A e_\pi$. Then the vectors e_A, where A runs through all standard Young tableaux for the partition λ, form a basis of G_λ. In other words, the basis elements are obtained by successive symmetrization on the rows of Y_λ with a subsequent anti-symmetrization on the columns.

The partition $\lambda = (N)$ corresponds to the Young diagram with only one row and produces the trivial representation of Sym_N. The partition $\lambda = (1, \ldots, 1)$ corresponds to the transposed Young diagram — a diagram with only one column — and produces another one-dimensional representation, the alternating representation $\pi \mapsto (-1)^{\varepsilon(\pi)}$. In general, let $\lambda' = (\lambda_1', \ldots, \lambda_m')$ be the partition conjugated to the partition λ, which is defined by transposing the Young diagram Y_λ by interchanging the rows and columns (λ_i' is the number of terms in the partition λ which are greater than or equal to i). Then

$$(3.6) \qquad\qquad T_{\lambda'} = T_\lambda \otimes T_{(1,\ldots,1)}.$$

Other partitions correspond to the symmetries of mixed type.

The next result will play a crucial role in determining the symmetry properties of coordinate wave functions.

Proposition 3.2. *The tensor product $T_\lambda \otimes T_\mu$ contains the trivial representation of Sym_N if and only if $\mu = \lambda$, in which case it has multiplicity one. In terms of a basis $\{e_i\}_{i=1}^{d_\lambda}$ in G_λ such that the representation T_λ is given by orthogonal $d_\lambda \times d_\lambda$ matrices $t_\lambda(\pi)_{ij}$,*

$$T_\lambda(\pi)e_i = \sum_{j=1}^{d_\lambda} t_\lambda(\pi)_{ji} e_j, \quad i = 1, \ldots, d_\lambda,$$

the one-dimensional subspace of the trivial representation is spanned by the vector $\sum_{i=1}^{d_\lambda} e_i \otimes e_i \in G_\lambda \otimes G_\lambda$. Correspondingly, the tensor product $T_\lambda \otimes T_\mu$ contains the alternating representation of Sym_N if and only if $\mu = \lambda'$, in which case it has multiplicity one. In terms of a basis $\{e_i'\}_{i=1}^{d_\lambda}$ in $G_{\lambda'}$ such that the representation $T_{\lambda'}$ is given by the matrices $t_{\lambda'}(\pi)_{ij} = (-1)^{\varepsilon(\pi)} t_\lambda(\pi)_{ij}$, the one-dimensional subspace of the alternating representation is spanned by the vector $\sum_{i=1}^{d_\lambda} e_i \otimes e_i' \in G_\lambda \otimes G_{\lambda'}$.

Problem 3.1. Prove all the statements in this section. (*Hint*: See the list of references to this chapter).

3.3. Schur-Weyl duality and symmetry of the wave functions. In addition to the diagonal action of the Lie group $\mathrm{SU}(2)$ in the tensor product $(\mathbb{C}^2)^{\otimes N}$, discussed in Section 3.1, there is also a left action of a symmetric group Sym_N by permuting factors[4]:

$$T(\pi)(v_1 \otimes \cdots \otimes v_N) = v_{\pi^{-1}(1)} \otimes \cdots \otimes v_{\pi^{-1}(N)}, \quad \pi \in \mathrm{Sym}_N.$$

For every partition $\lambda \in \mathrm{Par}(N)$ denote by Π_λ the Young symmetrizer corresponding to the canonical Young tableau A_λ, and by $S_\lambda \mathbb{C}^2$ the image of the operator $T(\Pi_\lambda)$ in $(\mathbb{C}^2)^{\otimes N}$. The subspace $S_\lambda \mathbb{C}^2$ is zero-dimensional when the Young diagram Y_λ has more than two rows, and is an irreducible $\mathrm{U}(2)$-module, called the *Weyl module*, when the Young diagram Y_λ has one or two rows. For the partition $\lambda = (N)$ the Weyl module $S_\lambda \mathbb{C}^2 = \mathrm{Sym}^N \mathbb{C}^2$ and is isomorphic to the highest weight module $V_{\lambda,N}$ with the dominant weight $(N, 0)$, defined in Section 1.2. For the partition $\lambda = (\lambda_1, \lambda_2)$ the Weyl module $S_\lambda \mathbb{C}^2$ is isomorphic to the highest weight module $V_{\lambda,N}$ with the dominant weight λ.

Remark. This construction is a special case of the general construction given by H. Weyl, which applies to the symmetric group action in the tensor product $V^{\otimes N}$, where V is a finite-dimensional vector space. The Weyl modules $S_\lambda V$ associated with the Young diagram Y_λ are zero-dimensional when the number of rows is greater than $n = \dim V$, and are isomorphic to the highest weight modules of $\mathrm{GL}(V)$ with the dominant weight λ when the number of rows is less than or equal to n. The association $V \mapsto S_\lambda V$ is called the *Schur functor*.

The following result, the *Schur-Weyl duality*, establishes explicit pairing between irreducible representations of $\mathrm{U}(2)$ and Sym_N that appear in the decomposition of the representation of $\mathrm{SU}(2) \times \mathrm{Sym}_N$ in $(\mathbb{C}^2)^{\otimes N}$ into the irreducible components.

Theorem 3.2 (Schur-Weyl duality). *There is a direct sum decomposition into irreducible components of the representation $R_{\frac{1}{2}}^{\otimes N} \times T$:*

$$(\mathbb{C}^2)^{\otimes N} = \bigoplus_{\lambda \in \mathrm{Par}(N,2)} S_\lambda \mathbb{C}^2 \otimes G_\lambda,$$

where $\mathrm{Par}(N, 2)$ is the set of partitions of N with number of terms less than or equal to 2.

Setting $\lambda_1 = \frac{N}{2} + s, \lambda_2 = \frac{N}{2} - s$ (cf. Section 1.2), from the Frobenius formula we obtain

$$(3.7) \qquad d_\lambda = \dim G_\lambda = \binom{N}{\frac{N}{2} - s} - \binom{N}{\frac{N}{2} - s - 1}.$$

---

[4]It moves the vector in the i-th place of the tensor product to the $\pi(i)$-th place.

Now return to the main problem of finding total wave functions satisfying (3.5). It follows from the Schur-Weyl duality that there are $(2s+1)d_\lambda$ linear independent spin wave functions $\chi_{mj}(\sigma_1, \ldots, \sigma_N)$ of spin s satisfying

$$(3.8) \qquad \boldsymbol{S}^2 \chi_{mj} = s(s+1)\chi_{mj} \quad \text{and} \quad S_3\chi_{mj} = m\chi_{mj},$$

where $m = -s, -s+1, \ldots, s-1, s$ and $j = 1, \ldots, d_\lambda$. Here d_λ corresponds to the partition $\lambda = (\frac{N}{2} + s, \frac{N}{2} - s)$ (or $\lambda = (N)$ for $s = \frac{N}{2}$) and is given by (3.7). The spin s representation ρ_s of SU(2) acts on the index m of the spin wave function $\chi_{mj}(\sigma_1, \ldots, \sigma_N)$, and the representation T_λ of Sym_N acts on the index j. The total wave function has the form

$$(3.9) \qquad \Phi_m(\boldsymbol{\xi}_1, \ldots, \boldsymbol{\xi}_N) = \sum_{j=1}^{d_\lambda} \Psi_j(\boldsymbol{x}_1, \ldots, \boldsymbol{x}_N)\chi_{mj}(\sigma_1, \ldots, \sigma_N),$$

where $m = -s, \ldots, s$. It follows from Proposition 3.2 that the total wave function $\Phi_m(\boldsymbol{\xi}_1, \ldots, \boldsymbol{\xi}_N)$ is totally anti-symmetric if and only if the coordinate wave functions $\Psi_j(\boldsymbol{x}_1, \ldots, \boldsymbol{x}_N)$ transform according to the conjugated representation $T_{\lambda'}$ of Sym_N,

$$(3.10) \quad (P_\pi \Psi_i)(\boldsymbol{x}_1, \ldots, \boldsymbol{x}_N) = \sum_{j=1}^{d_\lambda} t_{\lambda'}(\pi)_{ji}\Psi_j(\boldsymbol{x}_1, \ldots, \boldsymbol{x}_N), \quad i = 1, \ldots, d_\lambda.$$

We summarize these results as follows.

Theorem 3.3. *Coordinate wave functions* $\Psi_j(\boldsymbol{x}_1, \ldots, \boldsymbol{x}_N)$ *associated with the total wave function of spin s of a system of N identical particles of spin $\frac{1}{2}$ are those solutions of the N-particle Schrödinger equation (3.4) that transform according to the representation $T_{\lambda'}$ of the symmetric group Sym_N, where λ' is a conjugated partition to $\lambda = (\frac{N}{2} + s, \frac{N}{2} - s)$. The spin wave functions $\chi_{mj}(\sigma_1, \ldots, \sigma_N)$ satisfy (3.8), and corresponding total wave functions $\Phi_m(\boldsymbol{\xi}_1, \ldots, \boldsymbol{\xi}_N)$ are given by (3.9). In general, for a given energy level E there are $2s + 1$ linearly independent total wave functions of spin s, $0 \le s \le \frac{N}{2}$, and $\frac{N}{2} - s$ is an integer.*

Remark. For the highest value of total spin $s = \frac{N}{2}$ the coordinate wave function $\Psi(\boldsymbol{x}_1, \ldots, \boldsymbol{x}_N)$ is totally anti-symmetric. Since $(N)' = (1, \ldots, 1) \notin \mathrm{Par}(N, 2)$ when $N > 2$, coordinate wave function for the case $N > 2$ is never totally symmetric.

Using explicit description of the Young symmetrizer in the previous section, we obtain the following result.

Corollary 3.4 (V.A. Fock). *The coordinate wave function* $\Psi(\boldsymbol{x}_1, \ldots, \boldsymbol{x}_N)$ *associated with the total wave function of spin s is characterized by the following symmetry properties.*

(i) $\Psi(\boldsymbol{x}_1, \ldots, \boldsymbol{x}_N)$ *is anti-symmetric with respect to the group of arguments* $\boldsymbol{x}_1, \ldots, \boldsymbol{x}_k$, $k = \frac{N}{2} + s$.

(ii) $\Psi(\boldsymbol{x}_1, \ldots, \boldsymbol{x}_N)$ *is anti-symmetric with respect to the group of arguments* $\boldsymbol{x}_{k+1}, \ldots, \boldsymbol{x}_N$.

(iii) $\Psi(\boldsymbol{x}_1, \ldots, \boldsymbol{x}_N)$ *satisfies*

$$\Psi(\boldsymbol{x}_1, \ldots, \boldsymbol{x}_N) = \sum_{i=1}^{k} \Psi(\boldsymbol{x}_1, \ldots, \boldsymbol{x}_{i-1}, \boldsymbol{x}_{k+1}, \boldsymbol{x}_{i+1}, \ldots, \boldsymbol{x}_k, \boldsymbol{x}_i, \boldsymbol{x}_{k+2}, \ldots, \boldsymbol{x}_N).$$

The case of N identical particles of arbitrary spin s is treated similarly. The symmetry properties of coordinate wave functions can be obtained by using Schur-Weyl duality for the action of $U(l) \times \mathrm{Sym}_N$ in $\mathbb{C}^l$, where $l = 2s + 1$. It has the form

$$(\mathbb{C}^l)^{\otimes N} = \bigoplus_{\lambda \in \mathrm{Par}(N,l)} S_\lambda \mathbb{C}^l \otimes G_\lambda,$$

where $\mathrm{Par}(N, l)$ is the set of partitions of N with number of terms less than or equal to l, and $S_\lambda \mathbb{C}^l$ is the Weyl module corresponding to the highest weight representation of $U(l)$ with the dominant weight λ.

Problem 3.2. Prove Corollary 3.4.

Problem 3.3 (One-electron approximation). Suppose that $V_{\mathrm{int}}(\boldsymbol{x}) = 0$, and let $\psi_1(\boldsymbol{x}), \ldots, \psi_k(\boldsymbol{x})$, $k \geq \frac{N}{2}$, be the one-electron wave functions: eigenfunctions of the one-electron Schrödinger operator H_1 (see (3.4)). Show that the coordinate wave function

$$\Psi(\boldsymbol{x}_1, \ldots, \boldsymbol{x}_N) = \begin{vmatrix} \psi_1(\boldsymbol{x}_1) & \ldots & \psi_1(\boldsymbol{x}_k) \\ \ldots & \ldots & \ldots \\ \psi_k(\boldsymbol{x}_1) & \ldots & \psi_k(\boldsymbol{x}_k) \end{vmatrix} \begin{vmatrix} \psi_1(\boldsymbol{x}_{k+1}) & \ldots & \psi_1(\boldsymbol{x}_N) \\ \ldots & \ldots & \ldots \\ \psi_{N-k}(\boldsymbol{x}_{k+1}) & \ldots & \psi_{N-k}(\boldsymbol{x}_N) \end{vmatrix},$$

where $|\cdots|$ stands for the matrix determinant, satisfy the symmetry properties in Corollary 3.4.

4. Notes and references

The notion of electron spin is fundamental for the agreement of quantum mechanics with experiments. See [**Mes99**] for a thorough discussion of the principal evidence supporting the hypothesis of electron spin (Zeeman effect, Stern-Gerlach experiment). Classic texts [**LL58, Foc78**] also introduce the concept of spin from a physics perspective. Representation theory of the Lie group $SU(2)$ — the simplest compact non-abelian Lie group — is discussed in detail in [**Vil68**]; for the description of all irreducible representations of $U(n)$ see [**FH91**]. Our discussion of the Pauli Hamiltonian in Section 2.1 follows [**Foc78**]; the case of a particle in a uniform magnetic field in Section 2.2 was first considered by L.D. Landau in 1930 (see [**LL58**]). Problem 2.2 shows that in the presence of a magnetic field the operators

X_0 and Y_0, associated with coordinates of classical trajectories of a charged particle, do not commute and cannot be measured simultaneously.

The Pauli wave equation in Section 2.1 is non-relativistic, i.e., it does not take into account the postulates of special relativity. The correct quantum description of the electron in an external electromagnetic field was proposed by Dirac and is given by the Dirac equation; the Pauli wave equation is obtained as its non-relativistic limit [**Foc78**]. The Dirac equation is a single particle equation which describes an electron and a particle with charge $-e$ — a positron. In order to describe quantum processes of creation and annihilation of particles, one needs to pass from quantum mechanics to quantum field theory, which deals with systems with infinitely many degrees of freedom, and we refer to [**IZ80**] for an introduction.

The symmetrization postulate, formulated in Section 3.1, is another fundamental principle of quantum mechanics, and we refer to [**Mes99**] for its physical motivation. The N-particle Schrödinger equation describes all the properties of atoms. However, it is almost impossible to solve this equation analytically (for the potentials in atomic physics), and even numerically when $N > 2$; the case of the helium atom ($N = 3$) illustrates the difficulties of the problem. For this purpose various approximate methods like the one-electron approximation, the Hartree-Fock method and its modifications, have been developed (see [**LL58, Foc78**] for the details and [**FY80**] for a clear introduction). A correct description of the properties of atoms and an explanation of D.I. Mendeleev's periodic table of elements, based on these methods, has been a major triumph of quantum mechanics. For a detailed account we refer the interested reader to [**LL58, Foc78**] and [**FY80**]. The N-particle Schrödinger equation, in principle, also describes[5] all properties of molecules studied in chemistry. However, this "reductionism" does not work due to the complexity of the quantum N-body problem, and approximate methods like Hartree-Fock approximation play a central role in quantum chemistry.

The vector space $\mathbb{C}^2 \otimes \mathbb{C}^2$ describes the spin degrees of freedom of two electrons, and the spin states which are not represented as a tensor product $u \otimes v$ are called *entangled*. Entangled states are used to describe the Einstein-Podolski-Rozen paradox, a quantum experiment, which apparently violates the principle of locality (see its description in [**Bel87, Sak94**]). The fact that quantum mechanics is correct and allows for an accurate description of the microworld is beyond any doubt, and is confirmed by numerous physical experiments. The Einstein-Podolski-Rosen paradox is only a "paradox" because classical intuition does not always correspond to the physical reality. Moreover, it follows from the *Bell inequalities* [**Bel87**] that any attempt to turn quantum mechanics into a deterministic theory by introducing the so-called *hidden variables* is fallacious; we refer to the web site `http://en.wikipedia.org/wiki/EPR_paradox` for further references. The vector space $\mathbb{C}^2 \otimes \cdots \otimes \mathbb{C}^2$, describing the spin degrees of freedom of several

---

[5] In this sense, quantum mechanics is a "theory of everything" for chemistry.

identical particles of spin $\frac{1}{2}$, also plays a fundamental role in *quantum computation theory*. This is a rapidly developing field, and we refer the interested reader to the monograph [**KSV02**] for the mathematical introduction, and to the web site `http://en.wikipedia.org/wiki/Quantum_computer` for updates on the specialized literature. Another rapidly developing and exciting new application of quantum mechanics is *quantum cryptography*, reviewed at `http://en.wikipedia.org/wiki /Quantum_cryptography`.

Section 3.2 is a crash course on the representation theory of the symmetric group Sym_N — a beautiful piece of classical mathematics which has many interesting applications in modern combinatorics and representation theory. Our goal here was to give a concise and clear presentation of the basic necessary facts, and we refer to the classic text [**Wey50**] and to [**FH91**] for detailed accounts of representation theory, and to [**GW98**] for the Schur-Weyl duality. In Section 3.3 we emphasize the role of Schur-Weyl duality in studying the symmetry properties of coordinate wave functions. Physics textbooks like [**LL58**], [**Dav76**] and [**Sak94**] carefully treat the basic examples, but are somewhat vague in describing the form of the total wave function of N particles. The monograph [**Foc78**] is very precise in this regard, but in his presentation V.A. Fock is trying to reduce the usage of group theory to a minimum. Our goal in Section 3.3 was to fill this gap and to give a clear description of the symmetry properties of coordinate and spin parts of the total wave function by using the language of representation theory. For the case of N particles of spin $\frac{1}{2}$ this problem was solved by V.A. Fock in 1940; see [**Foc78**], as well as in yet another classic [**Wig59**]. For more applications of group theory to quantum mechanics we refer to the latter text, and to [**Wey50**].

Part 2

Functional Methods and Supersymmetry

Chapter 5

Path Integral Formulation of Quantum Mechanics

1. Feynman path integral

Here we present Feynman's path integral approach to quantum mechanics, which expresses the propagator of a quantum system with the Hamiltonian H — the kernel of the evolution operator $U(t) = e^{-\frac{i}{\hbar}tH}$ — as a "sum over trajectories" of the corresponding classical system.

1.1. The fundamental solution of the Schrödinger equation. Recall (see Section 1.3 of Chapter 2) that in terms of the evolution operator $U(t) = e^{-\frac{i}{\hbar}tH}$, the solution $\psi(t)$ of the initial value problem for the time-dependent Schrödinger equation

$$\tag{1.1} i\hbar\frac{d\psi}{dt}(t) = H\psi(t),$$

$$\tag{1.2} \psi(t)|_{t=0} = \psi$$

is given by $\psi(t) = U(t)\psi$. For the Hamiltonian operator

$$H = \frac{\boldsymbol{P}^2}{2m} + V(\boldsymbol{Q})$$

of a quantum particle in $\mathbb{R}^n$ moving in the potential field $V(\boldsymbol{q})$, the initial value problem (1.1)–(1.2) in coordinate representation becomes the following

Cauchy problem:

$$(1.3) \qquad i\hbar\frac{\partial\psi}{\partial t} = -\frac{\hbar^2}{2m}\Delta\psi + V(\boldsymbol{q})\psi,$$

$$(1.4) \qquad \psi(\boldsymbol{q},t)|_{t=0} = \psi(\boldsymbol{q}),$$

where Δ is the Laplace operator on $\mathbb{R}^n$. Under rather general conditions on the potential $V(\boldsymbol{q})$ (i.e., when $V \in L^\infty_{\mathrm{loc}}(\mathbb{R}^n)$ is bounded from below) the Cauchy problem (1.3)–(1.4) has a fundamental solution: a function $K(\boldsymbol{q},\boldsymbol{q}',t)$ which satisfies partial differential equation (1.3) with respect to the variable $\boldsymbol{q}$ in the distributional sense (i.e., $K(\boldsymbol{q},\boldsymbol{q}',t)$ is a weak solution of the Schrödinger equation), and the initial condition

$$(1.5) \qquad K(\boldsymbol{q},\boldsymbol{q}',t)\big|_{t=0} = \delta(\boldsymbol{q}-\boldsymbol{q}').$$

The solution of the Cauchy problem (1.3)–(1.4) can be formally written as

$$(1.6) \qquad \psi(\boldsymbol{q}',t) = \int_{\mathbb{R}^n} K(\boldsymbol{q}',\boldsymbol{q},t)\psi(\boldsymbol{q})d^n\boldsymbol{q},$$

where the integral is understood in the distributional sense.

For $\psi \in L^2(\mathbb{R}^n)$ formula (1.6) should be understood as

$$\psi(\boldsymbol{q}',t) = \underset{R\to\infty}{\mathrm{l.i.m.}}\int_{|\boldsymbol{q}|\le R} K(\boldsymbol{q}',\boldsymbol{q},t)\psi(\boldsymbol{q})d^n\boldsymbol{q} = \mathrm{l.i.m.}\int_{\mathbb{R}^n} K(\boldsymbol{q}',\boldsymbol{q},t)\psi(\boldsymbol{q})d^n\boldsymbol{q},$$

where l.i.m. stands for the limit in the L^2-norm. In general, $\psi(\boldsymbol{q},t)$ is only a weak solution of the Schrödinger equation (1.3), and it is a regular solution when ψ belongs to the Gårding domain D_0 (see Problem 1.7 in Section 1.3 of Chapter 2). The fundamental solution $K(\boldsymbol{q},\boldsymbol{q}',t)$ is the distributional kernel (in the sense of the Schwartz kernel theorem) of the evolution operator $U(t)$.

Using the group property $U(t+t') = U(t)U(t')$, we can rewrite (1.6) in the form

$$(1.7) \qquad \psi(\boldsymbol{q}',t') = \int_{\mathbb{R}^n} K(\boldsymbol{q}',t';\boldsymbol{q},t)\psi(\boldsymbol{q},t)d^n\boldsymbol{q},$$

where $K(\boldsymbol{q}',t';\boldsymbol{q},t) = K(\boldsymbol{q}',\boldsymbol{q},t'-t)$. The function $|K(\boldsymbol{q}',t';\boldsymbol{q},t)|^2$ has a physical meaning of conditional probability distribution of finding a quantum particle at a point $\boldsymbol{q}' \in \mathbb{R}^n$ at time t' provided that it was at the point $\boldsymbol{q} \in \mathbb{R}^n$ at time t.

Remark. In physics terminology, the distributional kernel $K(\boldsymbol{q}',t';\boldsymbol{q},t)$ of the evolution operator $U(t'-t)$ is called the *complex probability amplitude*, or simply the *amplitude* or *propagator*. In Dirac's notation, $K(\boldsymbol{q}',t';\boldsymbol{q},t) = \langle \boldsymbol{q}',t' \,|\, \boldsymbol{q},t\rangle$.

Finding the propagator of a given quantum system is a fundamental problem of quantum mechanics. When the spectral decomposition of the Hamiltonian operator H is known, the propagator can be obtained in a closed form. Namely, suppose for simplicity that H has a pure point spectrum, i.e., there is an orthonormal basis of the Hilbert space $\mathscr{H} = L^2(\mathbb{R}^n, d^n q)$ consisting of the eigenfunctions $\{\psi_n(q)\}_{n=0}^\infty$ of H with the eigenvalues E_n. We have

$$\psi(q) = \sum_{n=0}^\infty c_n \psi_n(q), \quad c_n = \int_{\mathbb{R}^n} \psi(q)\overline{\psi_n(q)}d^n q,$$

so that

$$(U(t)\psi)(q) = \sum_{n=0}^\infty e^{-\frac{i}{\hbar}E_n t} c_n \psi_n(q)$$

and

$$\psi(q', t') = \sum_{n=0}^\infty e^{-\frac{i}{\hbar}E_n t'} \psi_n(q') \int_{\mathbb{R}^n} \overline{\psi_n(q)}\psi(q)d^n q,$$

where the series and integrals converge in the L^2-sense. If the change of orders of summation and integration was justified, we could write

$$(1.8) \qquad K(q', t; q, t) = \sum_{n=0}^\infty e^{-\frac{i}{\hbar}E_n T} \psi_n(q')\overline{\psi_n(q)}, \quad T = t' - t.$$

The series (1.8) converges in the distributional sense, and gives a representation of the propogator in terms of the spectral decomposition of H.

A similar representation of the propagator exists when the Hamiltonian H has absolutely continuous spectrum. Consider the simplest case of a free quantum particle with the Hamiltonian operator

$$H_0 = \frac{\boldsymbol{P}^2}{2m}.$$

The corresponding Schrödinger equation can be solved by the Fourier method (see Section 2.3 in Chapter 2), and we obtain

$$\psi(q', t') = \text{l.i.m.}(2\pi\hbar)^{-\frac{n}{2}} \int_{\mathbb{R}^n} e^{\frac{i}{\hbar}(q'\boldsymbol{p} - \frac{\boldsymbol{p}^2}{2m}T)} \hat{\psi}(\boldsymbol{p}, t)d^n \boldsymbol{p}$$

$$= \text{l.i.m.} \int_{\mathbb{R}^n} K(q', t'; q, t)\psi(q, t)d^n q,$$

where

$$(1.9) \qquad K(q', t'; q, t) = \frac{1}{(2\pi\hbar)^n} \int_{\mathbb{R}^n} e^{\frac{i}{\hbar}(\boldsymbol{p}(q'-q) - \frac{\boldsymbol{p}^2}{2m}T)}d^n \boldsymbol{p}, \quad T = t' - t.$$

Using the classical Fresnel integral formula

$$(1.10) \qquad \int_{-\infty}^\infty e^{iax^2}dx = e^{\frac{\pi i\, \text{sgn}(a)}{4}}\sqrt{\frac{\pi}{|a|}},$$

and completing the square in (1.9), we obtain the following expression for the propagator of a free quantum particle:

$$(1.11) \qquad K(\boldsymbol{q}', t'; \boldsymbol{q}, t) = \left(\frac{m}{2\pi i \hbar T}\right)^{\frac{n}{2}} e^{\frac{im}{2\hbar T}(\boldsymbol{q} - \boldsymbol{q}')^2},$$

where $i^{\frac{n}{2}} = e^{-\frac{\pi i n}{4}}$ and $T > 0$.

Remark. Formally replacing the actual "physical" time t in the Schrödinger equation

$$i\hbar \frac{\partial \psi}{\partial t} = -\frac{\hbar^2}{2m} \Delta \psi$$

for a free quantum particle of mass m by the *Euclidean time* (or *pure imaginary time*) $-it$, we obtain the heat (diffusion) equation

$$\frac{1}{D} \frac{\partial u}{\partial t} = \Delta u$$

on $\mathbb{R}^n$ with the diffusion coefficient $D = \dfrac{\hbar}{2m}$. It is quite remarkable that in agreement with this formal procedure the propagator (1.11) for a free quantum particle is obtained from the heat kernel on $\mathbb{R}^n$ by the analytic continuation $T \mapsto -iT$.

Problem 1.1. Show that

$$\lim_{R \to \infty} \int_0^R \sin x^2 \, dx = \lim_{R \to \infty} \int_0^R \cos x^2 \, dx = \frac{1}{2}\sqrt{\frac{\pi}{2}}$$

by integrating the function e^{-z^2} over an appropriate "pizza-sliced" contour.

Problem 1.2. Show directly that

$$\psi(\boldsymbol{q}', t) = \left(e^{-\frac{i}{\hbar}tH_0}\psi\right)(\boldsymbol{q}') = \text{l.i.m.} \left(\frac{m}{2\pi i \hbar t}\right)^{\frac{n}{2}} \int_{\mathbb{R}^n} e^{\frac{im}{2\hbar t}(\boldsymbol{q} - \boldsymbol{q}')^2} \psi(\boldsymbol{q}) d^n \boldsymbol{q}.$$

Problem 1.3. Give an expression for the propagator when H is a Schrödinger operator with rapidly decreasing potential, considered in Section 2 in Chapter 3.

1.2. Feynman path integral in the phase space. For a general Hamiltonian operator $H = H_0 + V$, where $V = V(\boldsymbol{Q})$, there is no simple formula for the propagator $K(\boldsymbol{q}', t'; \boldsymbol{q}, t)$ like (1.11). This is because operators H_0 and V do not commute, so that $e^{-\frac{i}{\hbar}tH} \neq e^{-\frac{i}{\hbar}tH_0} e^{-\frac{i}{\hbar}tV}$.

It was Feynman's fundamental discovery that there is another representation of the propagator for a quantum system in terms of the corresponding classical system. We start by describing Feynman's approach for the case of a quantum particle with one degree of freedom. It is based on the so-called Lie-Kato-Trotter product formula, which allows us to express the exponential $e^{i(A+B)}$ of two non-commuting self-adjoint operators in terms of the individual exponentials e^{iA} and e^{iB}.

Theorem 1.1 (Lie-Kato-Trotter product formula). *Let A and B be self-adjoint operators on $\mathscr{H}$ such that $A+B$ is essentially self-adjoint on $D(A)\cap D(B)$. Then for $\psi \in \mathscr{H}$,*

$$e^{i(A+B)}\psi = \lim_{n\to\infty}(e^{\frac{i}{n}A}e^{\frac{i}{n}B})^n\psi.$$

Proof. We consider only the special case when A and B are bounded operators, which was already proved by Sophus Lie. Set $C_n = e^{i(A+B)/n}$ and $D_n = e^{iA/n}e^{iB/n}$. We have the telescopic sum

$$C_n^n - D_n^n = C_n^n - C_n^{n-1}D_n + C_n^{n-1}D_n - C_n^{n-2}D_n^2 + \cdots + C_n D_n^{n-1} - D_n^n$$

$$= \sum_{k=0}^{n-1} C_n^{n-k-1}(C_n - D_n)D_n^k,$$

and since by the Taylor formula $\|C_n - D_n\| \leq \dfrac{c}{n^2}$ for some constant $c > 0$, we obtain

$$\|C_n^n - D_n^n\| \leq \frac{c}{n}.$$

This proves the result with convergence in the uniform topology. $\square$

We will always assume that the Hamiltonian $H = H_0 + V$ is essentially self-adjoint on $D(H_0) \cap D(V)$, so that the Lie-Kato-Trotter formula is applicable. (According to Theorem 1.2 in Section 1.1 of Chapter 3, for the case $n = 3$ we can assume that $V = V_1 + V_2$, where $V_1 \in L^2(\mathbb{R}^3)$ and $V_2 \in L^\infty(\mathbb{R}^3)$.) Applying the Lie-Kato-Trotter formula to $A = -\frac{T}{\hbar}H_0$ and $B = -\frac{T}{\hbar}V$, we obtain that in the strong operator topology

$$(1.12) \qquad e^{-\frac{i}{\hbar}TH} = \lim_{n\to\infty}(e^{-\frac{i\Delta t}{\hbar}H_0}\,e^{-\frac{i\Delta t}{\hbar}V})^n, \quad \Delta t = \frac{T}{n},$$

where $T = t' - t$. In the coordinate representation the operator $e^{-\frac{i\Delta t}{\hbar}V}$ is a multiplication by $e^{-\frac{i}{\hbar}V(q)\Delta t}$ operator, and the distributional kernel of the operator $e^{-\frac{i\Delta t}{\hbar}H_0}$ is given by (1.9), where T is replaced by Δt. Thus the distributional kernel $K(q', q; \Delta t)$ of the operator $e^{-\frac{i\Delta t}{\hbar}H_0}\,e^{-\frac{i\Delta t}{\hbar}V}$ is given by

$$(1.13) \qquad K(q', q; \Delta t) = \frac{1}{2\pi\hbar}\int_{-\infty}^{\infty} e^{\frac{i}{\hbar}(p(q'-q)-(\frac{p^2}{2m}+V(q))\Delta t)}dp.$$

Since the kernel of a product of two operators is a composition of corresponding individual kernels, for the distributional kernel $K_n(q', t'; q, t)$ of the operator $(e^{-\frac{i\Delta t}{\hbar}H_0}\,e^{-\frac{i\Delta t}{\hbar}V})^n$ we obtain the following integral representation:

$$(1.14) \qquad K_n(q', t'; q, t) = \int_{\mathbb{R}^{n-1}}\cdots\int \prod_{k=0}^{n-1} K(q_{k+1}, q_k; \Delta t)\prod_{k=1}^{n-1} dq_k,$$

where $q_0 = q$, $q_n = q'$. Now replace each factor $K(q_{k+1}, q_k; \Delta t)$ in (1.14) by its integral representation (1.13), where the corresponding variable of integration is denoted by p_k, $k = 0, \ldots, n-1$. Changing the order of integrations in the resulting $(2n-1)$-fold integral and using (1.12), we obtain the following remarkable representation of the propagator of a quantum particle as a limit of multiple integrals when the number of integrations goes to infinity:

$$(1.15) \qquad K(q', t'; q, t) = \lim_{n \to \infty} K_n(q', t'; q, t)$$

$$= \lim_{n \to \infty} \int \cdots \int_{\mathbb{R}^{2n-1}} \exp\left\{ \frac{i}{\hbar} \sum_{k=0}^{n-1} (p_k(q_{k+1} - q_k) - H_c(p_k, q_k)\Delta t) \right\} \frac{dp_0}{2\pi\hbar} \prod_{k=1}^{n-1} \frac{dp_k dq_k}{2\pi\hbar}.$$

Here $H_c(p, q) = \dfrac{p^2}{2m} + V(q)$ is the classical Hamiltonian function, and $q_0 = q$, $q_n = q'$.

Heuristically, formula (1.15) admits the following interpretation. To every point $(p_0, p_1, \ldots, p_{n-1}, q_1, \ldots, q_{n-1}) \in \mathbb{R}^{2n-1}$ assign a piece-wise linear path σ in the extended phase space $\mathbb{R}^2 \times \mathbb{R}$ of the classical particle, defined by the following *time slicing* procedure. Let $t_k = t + k\Delta t$ and $\sigma(\tau) = (p(\tau), q(\tau), \tau)$, where

$$p(\tau) = p_k, \quad q(\tau) = q_k + (\tau - t_k)\frac{q_{k+1} - q_k}{t_{k+1} - t_k},$$

and $\tau \in [t_k, t_{k+1}]$, $k = 0, \ldots, n - 1$. Then for the Riemann integrable potentials $V(q)$ we have

$$(1.16) \quad \sum_{k=0}^{n-1} (p_k(q_{k+1} - q_k) - H_c(p_k, q_k)\Delta t) = S(\sigma) + o(1), \quad \text{as} \quad n \to \infty,$$

where

$$S(\sigma) = \int_\sigma (pdq - H_c d\tau) = \int_t^{t'} (p(\tau)\dot{q}(\tau) - H_c(p(\tau), q(\tau)))d\tau$$

is the action functional of a classical system with the Hamiltonian function $H_c(p, q)$ (see Section 2.2 in Chapter 1). This suggests to interpret (1.15) as a kind of "integral" over the space $\tilde{P}(\mathbb{R}^2)_{q,t}^{q',t'}$ of all paths $\sigma(\tau) = (p(\tau), q(\tau), \tau)$ in the extended phase space $\mathbb{R}^2 \times \mathbb{R}$ such that[1] $q(t) = q$ and $q(t') = q'$, which was used in the formulation of the principle of the least action in the phase space (see Section 2.2 in Chapter 1). Thus we put

$$(1.17) \qquad K(q', t'; q, t) = \int_{\tilde{P}(\mathbb{R}^2)_{q,t}^{q',t'}} e^{\frac{i}{\hbar} S(\sigma)} \mathscr{D}p \mathscr{D}q,$$

---

[1]Note that there is no condition on the values of $p(\tau)$ at the endpoints.

where the "measure" $\mathscr{D}p\mathscr{D}q$ on $\tilde{P}(\mathbb{R}^2)_{q,t}^{q',t'}$ is given by

$$(1.18) \qquad \mathscr{D}p\mathscr{D}q = \lim_{n\to\infty} \frac{dp_0}{2\pi\hbar} \prod_{k=1}^{n-1} \frac{dp_k dq_k}{2\pi\hbar}.$$

Representation (1.17)–(1.18) is the famous *Feynman path integral in the phase space* for the propagator of a quantum particle. It expresses the propagator $K(q',t';q,t)$ as the "weighted sum" over all possible "histories" of the classical particle, the paths $\sigma \in \tilde{P}(\mathbb{R}^2)_{q,t}^{q',t'}$, where each path σ has a complex weight $\exp\{\frac{i}{\hbar}S(\sigma)\}$. On one hand, this representation clearly shows the fundamental difference between classical mechanics and quantum mechanics. Thus in classical mechanics, the particle moves along classical trajectories, which are the critical points (extremals) of the action functional $S(\sigma)$, whereas in quantum mechanics all possible paths contribute to the complex probability amplitude $K(q',t';q,t)$. On the other hand, representation (1.17)–(1.18) clearly points at the relation between quantum mechanics and classical mechanics in the semi-classical limit $\hbar \to 0$. Namely, as $\hbar \to 0$ the weights $\exp\{\frac{i}{\hbar}S(\sigma)\}$ become rapidly oscillating and their contributions to (1.17) will mutually cancel, unless the action is almost constant. The latter happens near the critical points, which give the major contribution to (1.17). This is how classical trajectories of a particle emerge from its quantum description, which will be discussed in detail in Section 6.1.

Remark. The Feynman path integral in the phase space in not an integral in the sense of abstract integration theory, since the formal expression $\mathscr{D}p\mathscr{D}q$ does not define a measure[2] on the path space $\tilde{P}(\mathbb{R}^2)_{q,t}^{q',t'}$. Besides, the functional $\exp\{\frac{i}{\hbar}S(\sigma)\}$ has absolute value 1 and cannot be integrable with respect to any measure $d\mu$ with the property $\mu(\tilde{P}(\mathbb{R}^2)_{q,t}^{q',t'}) = \infty$. The rigorous mathematical meaning of (1.17)–(1.18) is the original formula (1.15), which expresses the propagator as a limit of multiple integrals as the number of integrations goes to infinity. This explains the apparent "paradox" that quantum mechanics is defined entirely in classical terms by (1.17). It is not so, since there is no "intrinsic" definition of (1.17) in classical terms, besides the time slicing procedure (1.15), which also requires the special choice of the approximation (1.16) of $S(\sigma)$ by the Riemann sums.

1.3. Feynman path integral in the configuration space. Using the Fresnel integral (1.10) for the integration over p in (1.13), we obtain the following formula for the distributional kernel of the operator $e^{-\frac{i\Delta t}{\hbar}H_0}\,e^{-\frac{i\Delta t}{\hbar}V}$:

$$\sqrt{\frac{m}{2\pi i\hbar\Delta t}}\; e^{\frac{i}{\hbar}\{\frac{m}{2}\frac{(q-q')^2}{\Delta t} - V(q)\Delta t\}}.$$

---

[2]In abstract measure theory the measure is non-negative and countably additive.

Repeating the time slicing procedure from the previous section, instead of
(1.15) we now get

$$
(1.19) \qquad K(q',t';q,t) = \lim_{n\to\infty} \left(\frac{m}{2\pi\hbar i \Delta t} \right)^{\frac{n}{2}}
$$

$$
\times \int_{\mathbb{R}^{n-1}} \cdots \int \exp\left\{ \frac{i}{\hbar} \sum_{k=0}^{n-1} \left(\frac{m}{2} \left(\frac{q_{k+1}-q_k}{\Delta t} \right)^2 - V(q_k) \right) \Delta t \right\} \prod_{k=1}^{n-1} dq_k.
$$

Assuming that $q_k = q(t_k)$, $k = 0, \dots, n$, for some smooth path $\gamma(\tau) = q(\tau)$
in $\mathbb{R}$, where $q_0 = q$ and $q_n = q'$, we obtain as $n \to \infty$,

$$
\sum_{k=0}^{n-1} \left(\frac{m}{2} \left(\frac{q_{k+1}-q_k}{\Delta t} \right)^2 - V(q_k) \right) \Delta t = S(\gamma) + o(1),
$$

where

$$
S(\gamma) = \int_t^{t'} L(\gamma'(\tau))d\tau = \int_t^{t'} L(q(\tau),\dot{q}(\tau))d\tau, \quad L(q,\dot{q}) = \tfrac{1}{2}m\dot{q}^2 - V(q),
$$

is the action functional of a classical particle of mass m moving in the
potential field $V(q)$ (see Section 1.3 in Chapter 1). This suggests to interpret
the limit of multiple integrals (1.19) as the *Feynman path integral in the
configuration space*

$$
(1.20) \qquad K(q',t';q,t) = \int_{P(\mathbb{R})_{q,t}^{q',t'}} e^{\frac{i}{\hbar}S(\gamma)} \mathscr{D}q.
$$

Here $P(\mathbb{R})_{q,t}^{q',t'}$ is the space of smooth parametrized paths γ in the configuration space $\mathbb{R}$ connecting points q and q', and the "measure" $\mathscr{D}q$ on $P(\mathbb{R})_{q,t}^{q',t'}$
is given by

$$
(1.21) \qquad \mathscr{D}q = \lim_{n\to\infty} \left(\frac{m}{2\pi\hbar i \Delta t} \right)^{\frac{n}{2}} \prod_{k=1}^{n-1} dq_k.
$$

As the Feynman path integral in the phase space, the Feynman path integral in the configuration space is not actually an integral in the sense of
integration theory, and the correct mathematical meaning of (1.20)-(1.21) is
given by (1.19). Formula (1.20) expresses the propagator $K(q',t';q,t)$ as the
sum over all histories in the configuration space of classical particle — paths
$\gamma \in P(\mathbb{R})_{q,t}^{q',t'}$, by assigning to each path γ a complex weight $\exp\{\frac{i}{\hbar}S(\gamma)\}$.

Remark. Convergence in (1.19), as well as in (1.15), is understood in the distributional sense. In particular, for every $\psi \in L^2(\mathbb{R})$,

$$\left(e^{-\frac{i}{\hbar}TH}\psi\right)(q) = \lim_{n\to\infty} \left(\frac{m}{2\pi\hbar i \Delta t}\right)^{\frac{n}{2}}$$

$$\times \int_{\mathbb{R}^n} \cdots \int \exp\left\{\frac{i}{\hbar}\sum_{k=0}^{n-1}\left(\frac{m}{2}\left(\frac{q_{k+1}-q_k}{\Delta t}\right)^2 - V(q_k)\right)\Delta t\right\}\psi(q_n)\prod_{k=1}^{n}dq_k,$$

where all integrals are understood as $\int_{-\infty}^{\infty} = \lim_{R\to\infty}\int_{|q|\leq R}$, and all limits — in the L^2-sense.

Remark. It is said in physics textbooks that the Feynman path integral in the configuration space is obtained from the Feynman path integral in the phase space by evaluating the Fresnel integral over $\mathscr{D}p$,

$$(1.22) \qquad \int_{P(\mathbb{R})_{q,t}^{q',t'}} e^{\frac{i}{\hbar}\int_t^{t'}\left(\frac{m}{2}\dot{q}^2 - V(q)\right)d\tau}\mathscr{D}q = \int_{\tilde{P}(\mathbb{R}^2)_{q,t}^{q',t'}} e^{\frac{i}{\hbar}\int_t^{t'}(p\dot{q}-H_c(p,q))d\tau}\mathscr{D}p\mathscr{D}q.$$

Note that the symbol $\mathscr{D}q$ has two different meanings: in the left-hand side of (1.22) it is defined by (1.21), whereas in the right-hand side it is defined as a part of (1.18).

1.4. Several degrees of freedom. Let

$$H = \frac{\boldsymbol{P}^2}{2m} + V(\boldsymbol{Q})$$

be the Hamiltonian operator of a quantum particle in $\mathbb{R}^n$ moving in the potential field $V(\boldsymbol{q})$. As in the case of one degree of freedom, by using the Lie-Kato-Trotter product formula, the propagator $K(\boldsymbol{q}',t';\boldsymbol{q},t)$ is expressed as a limit of multiple integrals when the number of integrations goes to infinity,

$$(1.23) \qquad K(\boldsymbol{q}',t';\boldsymbol{q},t) = \lim_{N\to\infty}\int_{\mathbb{R}^{(2N-1)n}}\cdots\int$$

$$\exp\left\{\frac{i}{\hbar}\sum_{k=0}^{N-1}\left(\boldsymbol{p}_k(\boldsymbol{q}_{k+1}-\boldsymbol{q}_k) - H_c(\boldsymbol{p}_k,\boldsymbol{q}_k)\Delta t\right)\right\}\frac{d^n\boldsymbol{p}_0}{(2\pi\hbar)^n}\prod_{k=1}^{N-1}\frac{d^n\boldsymbol{p}_k d^n\boldsymbol{q}_k}{(2\pi\hbar)^n}.$$

Here $H_c(\boldsymbol{p},\boldsymbol{q}) = \dfrac{\boldsymbol{p}^2}{2m} + V(\boldsymbol{q})$ is the classical Hamiltonian function, and $\boldsymbol{q}_0 = \boldsymbol{q}$, $\boldsymbol{q}_N = \boldsymbol{q}'$. This representation is symbolized by the Feynman path integral in the phase space $\mathscr{M} = T^*\mathbb{R}^n$,

$$(1.24) \qquad K(\boldsymbol{q}',t';\boldsymbol{q},t) = \int_{\tilde{P}(\mathscr{M})_{q,t}^{q',t'}} e^{\frac{i}{\hbar}\int_\sigma(\boldsymbol{p}\dot{\boldsymbol{q}}-H_c(\boldsymbol{p},\boldsymbol{q}))d\tau}\mathscr{D}\boldsymbol{p}\mathscr{D}\boldsymbol{q},$$

where

$$\mathscr{D}p\mathscr{D}q = \lim_{N\to\infty} \frac{d^n p_0}{(2\pi\hbar)^n} \prod_{k=1}^{N-1} \frac{d^n p_k\, d^n q_k}{(2\pi\hbar)^n}$$

and $\tilde{P}(\mathscr{M})_{q,t}^{q',t'}$ is the space of all admissible paths σ in the extended phase space $\mathscr{M}\times\mathbb{R}$ connecting points (q,t) and (q',t') (see Section 2.2 in Chapter 1).

Equivalently, the propagator $K(q',t';q,t)$ can be written as

$$(1.25) \qquad K(q',t';q,t) = \lim_{N\to\infty} \left(\frac{m}{2\pi\hbar i\Delta t}\right)^{\frac{nN}{2}}$$

$$\times \int_{\mathbb{R}^{(n-1)N}} \cdots \int \exp\left\{ \frac{i}{\hbar} \sum_{k=0}^{N-1} \left(\frac{m}{2}\left(\frac{q_{k+1}-q_k}{\Delta t}\right)^2 - V(q_k) \right) \Delta t \right\} \prod_{k=1}^{N-1} d^n q_k.$$

This formula is symbolized by the Feynman path integral in the configuration space,

$$(1.26) \qquad K(q',t';q,t) = \int_{P(M)_{q,t}^{q',t'}} e^{\frac{i}{\hbar}\int_t^{t'} L(q,\dot{q})d\tau}\, \mathscr{D}q,$$

where $L(q,\dot{q}) = \frac{1}{2}m\dot{q}^2 - V(q)$ is the corresponding Lagrangian,

$$\mathscr{D}q = \lim_{N\to\infty} \left(\frac{m}{2\pi i\hbar\Delta t}\right)^{\frac{nN}{2}} \prod_{k=1}^{N-1} d^n q_k,$$

and $P(M)_{q,t}^{q',t'}$ is the space of smooth parametrized paths in the configuration space M connecting points q and q'. The precise mathematical meaning of formula (1.26) is the same as of (1.20).

Remark. Formally, definition (1.24) of the Feynman path integral in the phase space can be extended to the case $(\mathscr{M},\omega,H_c)$, where $\mathscr{M} = T^*M$ and $\omega = d\theta$, the canonical Liouville 1-form on $\mathscr{M}$. Namely, one can obtain a distributional kernel $K(q',t';q,t)$ by using the same time slicing as in (1.24) and replacing the 1-form pdq by θ, and the volume form $d^n p_k d^n q_k$ by an appropriate multiple of the Liouville volume form on $\mathscr{M}$. Similarly, representation (1.26) can be formally extended to the case of general Lagrangian system (M,L), where the configuration space M is a Riemannian manifold, by using the same time slicing and replacing $d^n q_k$ by the Riemannian volume form on M. However, in general it is not clear for which unitary operators these Feynman integrals are actual distributional kernels. Moreover, even in the case $M = \mathbb{R}^n$ with the standard Euclidean metric, representations (1.26) and (1.24), where H_c is a Legendre transform of L, do not necessarily agree if L is not of the form "kinetic energy minus potential energy". However, when $L = \frac{1}{2}g_{\mu\nu}(\boldsymbol{x})\dot{x}^{\mu}\dot{x}^{\nu} - V(\boldsymbol{x})$ (cf. Example 1.7 in Section 1.3 of Chapter 1), then with appropriately defined time slicing, Feynman path integrals (1.26) and (1.24) agree, and are equal to the propagator for the Hamiltonian operator

$$H = \frac{\hbar^2}{2}\Delta_g + V.$$

Here Δ_g is the Laplace operator of the Riemannian metric on M, introduced in Example 2.4 in Section 2.4 of Chapter 2.

Remark. Formula (1.24) could serve as a heuristic tool which enables us, in some cases, to quantize the classical Hamiltonian system $(\mathscr{M},\omega,H_{\mathrm{c}})$. In general this is a rather non-trivial problem, especially when the phase space $\mathscr{M}$ is a compact manifold.

2. Symbols of the evolution operator and path integrals

In Sections 2.7 and 3.3 of Chapter 2 we introduced Wick, $\boldsymbol{pq}$-, $\boldsymbol{qp}$-, and Weyl symbols of operators. For the evolution operator $U(T) = e^{-\frac{i}{\hbar}TH}$ these symbols can also be represented by Feynman path integrals. Here we consider only the case of one degree of freedom, since generalization to several degrees of freedom is straightforward.

2.1. The pq-symbol. Let $F_1(p,q,T)$ be the pq-symbol of the evolution operator $U(T)$. Using (1.15) and formula (3.18) in Section 3.3 of Chapter 2 — the relation between the pq-symbol of an operator and its distributional kernel — we get

$$(2.1) \qquad F_1(p,q,T) = \int_{-\infty}^{\infty} K(q-v,T;q,0)e^{\frac{i}{\hbar}pv}\,dv$$

$$= \lim_{n\to\infty}\int_{\mathbb{R}^{2n-2}}\cdots\int \exp\left\{\frac{i}{\hbar}\sum_{k=0}^{n-1}(p_k(q_{k+1}-q_k) - H_{\mathrm{c}}(p_k,q_k)\Delta t)\right\}\prod_{k=1}^{n-1}\frac{dp_{k-1}dq_k}{2\pi\hbar},$$

where $p_{n-1} = p$ and $q_0 = q_n = q$. Formula (2.1) can also be obtained from the composition formula for pq-symbols (see Problem 3.9 in Section 3.4 of Chapter 2). Indeed, it is sufficient to observe that the pq-symbol of the operator $e^{-\frac{i\Delta t}{\hbar}H_0}e^{-\frac{i\Delta t}{\hbar}V}$ is $f(p,q) = e^{-\frac{i\Delta t}{\hbar}H_{\mathrm{c}}(p,q)}$, and to represent the pq-symbol of the operator $(e^{-\frac{i\Delta t}{\hbar}H_0}e^{-\frac{i\Delta t}{\hbar}V})^n$ as the multiple integral

$$\int_{\mathbb{R}^{2n-2}}\cdots\int f(p,q_{n-1})f(p_{n-2},q_{n-2})\cdots f(p_2,q_2)f(p_0,q)$$

$$\times \exp\left\{\frac{i}{\hbar}\sum_{k=0}^{n-2}(p-p_k)(q_k-q_{k+1})\right\}\prod_{k=1}^{n-1}\frac{dp_{k-1}dq_k}{2\pi\hbar},$$

where $q_0 = q$. In the same spirit as (1.15), formula (2.1) can be written as the following Feynman path integral for the pq-symbol:

$$(2.2) \qquad F_1(p, q, T) = \int_{\Omega_{(p,q)}(\mathbb{R}^2)} e^{\frac{i}{\hbar} S(\sigma)} \mathscr{D}p \mathscr{D}q, \quad S(\sigma) = \int_\sigma (pdq - H_c dt),$$

where $\Omega_{(p,q)}(\mathbb{R}^2) = \{\sigma : [0, T] \to \mathbb{R}^2 : \sigma(0) = \sigma(T) = (p, q) \in \mathbb{R}^2\}$ is the space of parametrized loops in the phase space $\mathbb{R}^2$ which start and end at the point (p, q).

Remark. The mathematical meaning of (2.2) is the original formula (2.1) with a special choice of the approximation of $S(\sigma)$ by Riemann sums.

2.2. The qp-symbol. Let $F_2(p, q, T)$ be the qp-symbol of the evolution operator $U(T)$. In this case instead of formula (1.12), convenient for pq-symbols, one should consider the equivalent representation

$$(2.3) \qquad e^{-\frac{i}{\hbar} TH} = \lim_{n \to \infty} (e^{-\frac{i\Delta t}{\hbar} V} e^{-\frac{i\Delta t}{\hbar} H_0})^n.$$

The distributional kernel $\tilde{K}(q', q; \Delta t)$ of the operator $e^{-\frac{i\Delta t}{\hbar} V} e^{-\frac{i\Delta t}{\hbar} H_0}$ is given by

$$(2.4) \qquad \tilde{K}(q', q; \Delta t) = \frac{1}{2\pi\hbar} \int_{-\infty}^{\infty} e^{\frac{i}{\hbar}(p(q'-q) - (\frac{p^2}{2m} + V(q'))\Delta t)} dp,$$

and as in Section 1.2 we obtain

$$(2.5) \qquad K(q', t'; q, t) = \lim_{n \to \infty} \int_{\mathbb{R}^{n-1}} \cdots \int \prod_{k=0}^{n-1} \tilde{K}(q_{k+1}, q_k; \Delta t) \prod_{k=1}^{n-1} dq_k$$

$$= \lim_{n \to \infty} \int_{\mathbb{R}^{2n-1}} \cdots \int \exp\left\{\frac{i}{\hbar} \sum_{k=0}^{n-1} (p_k(q_{k+1} - q_k) - H_c(p_k, q_{k+1})\Delta t)\right\} \frac{dp_0}{2\pi\hbar} \prod_{k=1}^{n-1} \frac{dp_k dq_k}{2\pi\hbar}.$$

Using formula (3.19) in Section 3.3 of Chapter 2, we get from (2.5),

$$(2.6) \qquad F_2(p, q, T) = \int_{-\infty}^{\infty} K(q, T; q + v, 0) e^{\frac{i}{\hbar} pv} dv$$

$$= \lim_{n \to \infty} \int_{\mathbb{R}^{2n-2}} \cdots \int \exp\left\{\frac{i}{\hbar} \sum_{k=0}^{n-1} (p_k(q_{k+1} - q_k) - H_c(p_k, q_{k+1})\Delta t)\right\} \prod_{k=1}^{n-1} \frac{dp_{k-1} dq_k}{2\pi\hbar},$$

where $p_0 = p$ and $q_0 = q_n = q$. Formula (2.6) can also be obtained from the composition formula for qp-symbols (see Problem 3.10 in Section 3.4 of Chapter 2) by observing that the qp-symbol of the operator $e^{-\frac{i\Delta t}{\hbar} V} e^{-\frac{i\Delta t}{\hbar} H_0}$ is again $e^{-\frac{i\Delta t}{\hbar} H_c(p,q)}$. In the same spirit as (1.15), formula (2.6) can be written as the following Feynman path integral for the qp-symbol:

$$(2.7) \qquad F_2(p, q, T) = \int_{\Omega_{(p,q)}(\mathbb{R}^2)} e^{\frac{i}{\hbar} S(\sigma)} \mathscr{D}p\mathscr{D}q.$$

This expression looks exactly like (2.2), so that the formal path integral representation does not distinguish between pq- and qp-symbols! Of course, the mathematical meaning of (2.7) is the original formula (2.6), which requires a special choice of the approximation of $S(\sigma)$ by Riemann sums, which is different from the one used for the pq-symbol.

Problem 2.1. Deduce formula (2.6) from the composition formula for qp-symbols.

2.3. The Weyl symbol. Let $F(p, q; T)$ be the Weyl symbol of the evolution operator $U(T)$. It follows from the Weyl inversion formula — formula (3.12) in Section 3.3 of Chapter 2 — that

$$F(p, q, T) = \int_{-\infty}^{\infty} K(q - \tfrac{1}{2}v, T; q + \tfrac{1}{2}v, 0)e^{\frac{i}{\hbar}pv}dv.$$

However, for general potentials $V(q)$ one cannot perform integration over v by using either (1.15) or (2.5), and we choose another approach. Namely, denote by $\tilde{U}(\Delta t)$ an operator[3] with the Weyl symbol $e^{-\frac{i}{\hbar}H_c(p,q)\Delta t}$, and suppose that

$$(2.8) \qquad U(T) = \lim_{n \to \infty} \tilde{U}(\Delta t)^n.$$

Using the composition formula for the Weyl symbols, formula (3.24) in Section 3.3 of Chapter 2, we can express the Weyl symbol as

$$(2.9) \qquad F(p, q, T) = \lim_{n \to \infty} \int_{\mathbb{R}^{4(n-1)}} \cdots \int \exp\left\{ \frac{i}{\hbar}\left(2\sum_{k=0}^{n-2} \left((p_k - \xi_k)(\eta_{k+1} - \eta_k) \right.\right.\right.$$

$$\left.\left.\left. -(q_k - \eta_k)(\xi_{k+1} - \xi_k) \right) - \sum_{k=0}^{n-1} H_c(p_k, q_k)\Delta t \right)\right\} \prod_{k=1}^{n-1} \frac{dp_k dq_k d\xi_{k-1} d\eta_{k-1}}{\pi\hbar},$$

where $\xi_0 = p, \eta_0 = q$ and $\xi_{n-1} = p_{n-1}, \eta_{n-1} = q_{n-1}$. As for formula (1.15), formula (2.9) can be written as the following Feynman path integral for the Weyl symbol:

$$(2.10) \qquad\qquad F(p, q, T)$$

$$= \int e^{\frac{i}{\hbar}\int_0^T \{2(p(t)-\xi(t))\dot{\eta}(t)-2(q(t)-\eta(t))\dot{\xi}(t)-H_c(p(t),q(t))\}dt} \mathscr{D}p\mathscr{D}q\mathscr{D}\xi\mathscr{D}\eta,$$

---

[3] There is no easy way to express it in terms of $e^{-\frac{i\Delta t}{\hbar}H_0}$ and $e^{-\frac{i\Delta t}{\hbar}V}$.

where

$$\mathscr{D}p\,\mathscr{D}q\,\mathscr{D}\xi\,\mathscr{D}\eta = \lim_{n\to\infty} \prod_{k=1}^{n-1} \frac{dp_k\,dq_k\,d\xi_{k-1}\,d\eta_{k-1}}{\pi\hbar},$$

and "integration" goes over the space of real-valued functions $p(t)$, $q(t)$, $\xi(t)$, and $\eta(t)$ on $[0,T]$ satisfying $\xi(0) = p, \eta(0) = q$ and $\xi(T) = p(T), \eta(T) = q(T)$.

Remark. Expression (2.10) looks very different from (2.2) and (2.7). However, the variables $\xi(t)$ and $\eta(t)$ appear only linearly in the exponential in (2.10), and the integration over them can be performed explicitly. Indeed, using integration by parts (and ignoring subtleties with the boundary terms) we can transform the integral over $\mathscr{D}\xi$ to

$$\int e^{\frac{2i}{\hbar}\int_0^T \xi(t)(\dot{q}(t)-2\dot{\eta}(t))dt}\,\mathscr{D}\xi,$$

which equals the product of delta-functions $\delta(2\dot{\eta}(t) - \dot{q}(t))$ over $[0,T]$ and allows us to replace $2\dot{\eta}(t)$ by $\dot{q}(t)$ in (2.10). Thus we obtain the formula

$$F(p,q,T) = \int e^{\frac{i}{\hbar}\int_0^T (p\dot{q}-H_c(p,q))dt}\,\mathscr{D}p\,\mathscr{D}q,$$

which looks exactly like (2.2) and (2.7)! This heuristic argument shows that the Feynman path integral, understood naively, "erases" the difference between various symbols of the evolution operator. Of course, one should always specify particular finite-dimensional approximation of $S(\sigma)$ in order to get correct results.

Problem 2.2. Find the integral kernel of the operator $\tilde{U}(\Delta t)$.

Problem 2.3. Derive formula (2.9).

2.4. The Wick symbol. Instead of the variables $\bar{z}$ and z, used in Section 2.7 of Chapter 2, it is convenient to use variables $\bar{a} = \sqrt{\hbar}\,\bar{z}$ and $a = \sqrt{\hbar}\,z$, which introduce explicit $\hbar$-dependence in the calculus of Wick symbols. Let $H(\bar{a},a)$ be the Wick symbol of a Hamiltonian operator H. Here we derive a formula for the Wick symbol $U(\bar{a},a;T)$ of the evolution operator $U(T) = e^{-\frac{i}{\hbar}TH}$. Denote by $\tilde{U}(\Delta t)$ an operator with the Wick symbol $e^{-\frac{i}{\hbar}H(\bar{a},a)\Delta t}$, and as in the previous section, suppose that

(2.11) $$U(T) = \lim_{n\to\infty} \tilde{U}(\Delta t)^n.$$

Using the formula for the composition of Wick symbols (see Theorem 2.2 and Problem 2.18 in Section 2.7 of Chapter 2), we get the following expression for the Wick symbol $U_n(\bar{a}, a; T)$ of the operator $\tilde{U}(\Delta t)^n$:

$$(2.12) \qquad U_n(\bar{a}, a; T) = \int \cdots \int_{\mathbb{C}^{n-1}} \exp\left\{ \frac{1}{\hbar}\left(\bar{a}(a_{n-1} - a) - iH(\bar{a}, a_{n-1})\Delta t \right.\right.$$

$$\left.\left. + \sum_{k=1}^{n-1} (\bar{a}_k(a_{k-1} - a_k) - iH(\bar{a}_k, a_{k-1})\Delta t) \right) \right\} \prod_{k=1}^{n-1} \frac{d^2 a_k}{\pi \hbar},$$

where $a_0 = a$. Setting $\bar{a}_n = \bar{a}$, we can rewrite the sum in (2.12) as

$$\sum_{k=1}^{n} (\bar{a}_k(a_{k-1} - a_k) + \bar{a}(a_n - a) - iH(\bar{a}_k, a_{k-1})\Delta t),$$

and obtain

$$(2.13) \qquad U(\bar{a}, a; T) = \lim_{n\to\infty} U_n(\bar{a}, a; T) = \lim_{n\to\infty} \int \cdots \int_{\mathbb{C}^{n-1}}$$

$$\exp\left\{ \frac{1}{\hbar}\left(\sum_{k=1}^{n} (\bar{a}_k(a_{k-1} - a_k) + \bar{a}(a_n - a) - iH(\bar{a}_k, a_{k-1})\Delta t) \right) \right\} \prod_{k=1}^{n-1} \frac{d^2 a_k}{\pi \hbar}.$$

As in the previous examples, we can represent formula (2.13) by the following Feynman path integral for the Wick symbol:

$$(2.14) \qquad U(\bar{a}, a; T) = \int_{\left\{ \begin{array}{l} \bar{a}(T)=\bar{a} \\ a(0)=a \end{array} \right\}} e^{\frac{i}{\hbar}\int_0^T (i\bar{a}\dot{a} - H(\bar{a}, a))dt + \frac{1}{\hbar}\bar{a}(a(T) - a)} \mathscr{D}\bar{a}\mathscr{D}a,$$

where

$$\mathscr{D}\bar{a}\mathscr{D}a = \lim_{n\to\infty} \prod_{k=1}^{n-1} \frac{d^2 a_k}{\pi \hbar}.$$

Here integration goes over all complex-valued functions $a(t)$ and $\bar{a}(t)$ satisfying boundary conditions $a(0) = a$ and $\bar{a}(T) = \bar{a}$, and such that $\bar{a}(t)$ is complex-conjugated to $a(t)$ for $0 < t < T$.

Remark. It should be emphasized that in (2.14) the values $\bar{a}(0)$ and $a(T)$ are not complex-conjugated to the fixed boundary values $a(0) = a$ and $\bar{a}(T) = \bar{a}$, but rather are "variables of integration". This should be compared with formula (1.17), where the boundary values for the function $q(\tau)$ are fixed, and the boundary values for the function $p(\tau)$ are free. The difference is that at $t = 0$ we fix the boundary value of the function $a(t)$, while at $t = T$ we fix the boundary value of another function $\bar{a}(t)$.

Remark. The Wick symbol $U(\bar{a}, a; -iT)$ of the operator $e^{-\frac{1}{\hbar}TH} = U(-iT)$ can also be represented by

$$U(\bar{a}, a; -iT) = \lim_{n\to\infty} U_n(\bar{a}, a; -iT) = \lim_{n\to\infty} \int \cdots \int_{\mathbb{C}^{n-1}}$$

$$\exp\left\{\frac{1}{\hbar}\left(\sum_{k=1}^{n}(\bar{a}_k(a_{k-1} - a_k) - H(\bar{a}_k, a_{k-1})\Delta t) + \bar{a}(a_n - a)\right)\right\} \prod_{k=1}^{n-1} \frac{d^2 a_k}{\pi\hbar}$$

$$(2.15) \qquad = \int_{\left\{\substack{\bar{a}(T)=\bar{a} \\ a(0)=a}\right\}} e^{-\frac{1}{\hbar}\int_0^T (\bar{a}\dot{a} + H(\bar{a},a))dt + \frac{1}{\hbar}\bar{a}(a(T)-a)} \mathscr{D}\bar{a}\mathscr{D}a,$$

which is the so-called *Euclidean path integral* — the Feynman path integral with respect to the Euclidean time, which will be discussed in Section 2.3 of Chapter 6. In particular, if the Hamiltonian H has a pure point spectrum and $e^{-\frac{1}{\hbar}TH}$ is of trace class, then using the relation between the trace and Wick symbols (see Problem 2.17 in Section 2.7 of Chapter 2), we obtain from (2.15)

$$(2.16) \qquad \operatorname{Tr} e^{-\frac{1}{\hbar}TH} = \int_{\left\{\substack{\bar{a}(0)=\bar{a}(T) \\ a(0)=a(T)}\right\}} e^{-\frac{1}{\hbar}\int_0^T (\bar{a}\dot{a} + H(\bar{a},a))dt} \mathscr{D}\bar{a}\mathscr{D}a,$$

where now integration goes over all complex-conjugate functions $a(t)$ and $\bar{a}(t)$ satisfying periodic boundary conditions $a(0) = a(T)$ and $\bar{a}(0) = \bar{a}(T)$, and

$$\mathscr{D}\bar{a}\mathscr{D}a = \lim_{n\to\infty} e^{-\frac{1}{\hbar}\bar{a}_n a_n} \prod_{k=1}^{n} \frac{d^2 a_k}{\pi\hbar}.$$

Indeed, it follows from the identity

$$(2.17) \qquad \sum_{k=1}^{n-1} \bar{a}_k(a_{k-1} - a_k) + \bar{a}(a_{n-1} - a)$$

$$= \frac{1}{2}\sum_{k=1}^{n-1}\left\{\bar{a}_k(a_{k-1} - a_k) + (\bar{a}_{k+1} - \bar{a}_k)a_k\right\} + \frac{1}{2}\left\{\bar{a}(a_{n-1} - a) + (\bar{a}_1 - \bar{a})a\right\}$$

that when $\bar{a}$ and a are variables of integration, one can put $a_n = a$ and $\bar{a}_0 = \bar{a}$, which together with $\bar{a}_n = \bar{a}$ and $a_0 = a$ imply periodic boundary conditions. Thus

$$\sum_{k=1}^{n-1} \bar{a}_k(a_{k-1} - a_k) + \bar{a}(a_{n-1} - a) = \frac{1}{2}\sum_{k=1}^{n}\left\{\bar{a}_k(a_{k-1} - a_k) + (\bar{a}_{k+1} - \bar{a}_k)a_k\right\},$$

which in the limit $n \to \infty$ becomes

$$\frac{1}{2} \int_0^T (-\bar{a}\dot{a} + \dot{\bar{a}}a)dt = - \int_0^T \bar{a}\dot{a}\,dt$$

due to the periodic boundary conditions.

In the holomorphic representation, the matrix symbol $K(\bar{a}, a; T)$ of the evolution operator plays the role of a propagator $K(q', T; q, 0)$ in the coordinate representation. Using the relation between matrix and Wick symbols (see Lemma 2.4 in Section 2.7 of Chapter 2), we obtain

$$(2.18) \qquad K(\bar{a}, a; T) = \int_{\left\{\begin{smallmatrix} \bar{a}(T)=\bar{a} \\ a(0)=a \end{smallmatrix}\right\}} e^{\frac{i}{\hbar} \int_0^T (i\bar{a}\dot{a} - H(\bar{a},a))dt + \frac{1}{\hbar}\bar{a}\,a(T)} \mathscr{D}\bar{a}\mathscr{D}a.$$

Problem 2.4. Find the relation between the Wick symbol of the evolution operator and the propagator. (*Hint:* Use the analog of formula (2.45) in Section 2.7 of Chapter 2.)

3. Feynman path integral for the harmonic oscillator

As a first impression, one may think that Feynman path integrals are not very practical. Indeed, they are defined as limits of multiple integrals when the number of integrations goes to infinity, and it seems difficult to evaluate them. In fact, this is not so: Feynman path integrals turned out to be very useful for various computations, and for several important cases they can be evaluated exactly. Here we consider the basic example[4], the Feynman path integral for the harmonic oscillator.

3.1. Gaussian integration. Evaluation of Feynman path integrals simplifies when corresponding finite-dimensional integrals can be computed exactly for every n (or for large enough n). This is the case for the important class of Gaussian integrals.

Lemma 3.1 (Gaussian integration). *Let A be a positive-definite, real symmetric $n \times n$ matrix. We have*

$$(3.1) \qquad \int_{\mathbb{R}^n} e^{-\frac{1}{2}(A\boldsymbol{q},\boldsymbol{q})+(\boldsymbol{p},\boldsymbol{q})} d^n\boldsymbol{q} = \frac{\sqrt{(2\pi)^n}}{\sqrt{\det A}} \, e^{\frac{1}{2}(A^{-1}\boldsymbol{p},\boldsymbol{p})},$$

where $(\ ,\ )$ stands for the standard Euclidian inner product in $\mathbb{R}^n$.

---

[4]The simplest case of a free particle provides a trivial example, since the Lie-Kato-Trotter product formula reduces to $e^{iA} = (e^{\frac{i}{n}A})^n$, and formula (1.14) gives the same answer as (1.11) for every n.

Proof. Completing the square we obtain

$$-\tfrac{1}{2}(A\boldsymbol{q},\boldsymbol{q}) + (\boldsymbol{p},\boldsymbol{q}) = -\tfrac{1}{2}(A(\boldsymbol{q}-A^{-1}\boldsymbol{p}),(\boldsymbol{q}-A^{-1}\boldsymbol{p})) + \tfrac{1}{2}(A^{-1}\boldsymbol{p},\boldsymbol{p}),$$

so by the change of variables $\boldsymbol{q} = \boldsymbol{x} + A^{-1}\boldsymbol{p}$ the integral reduces to the standard Gaussian integral $\int_{\mathbb{R}^n} e^{-\frac{1}{2}(A\boldsymbol{x},\boldsymbol{x})} d^n\boldsymbol{x}$, which is evaluated by diagonalizing the matrix A and using the one-dimensional Gaussian integral

$$\int_{-\infty}^{\infty} e^{-\frac{1}{2}ax^2} dx = \sqrt{\frac{2\pi}{a}}, \quad a > 0. \qquad \square$$

Formula (3.1) is one of the fundamental mathematical facts which is used in many disciplines, from probability theory to number theory. For applications to quantum mechanics, one needs a version of Lemma 3.1 when the decaying exponential factor is replaced by the oscillating exponential factor, as in (1.10).

Corollary 3.1. *Let A be a real, non-degenerate symmetric $n \times n$ matrix. Then*

$$(3.2) \qquad \int_{\mathbb{R}^n} e^{\frac{i}{2}(A\boldsymbol{q},\boldsymbol{q})+i(\boldsymbol{p},\boldsymbol{q})} d^n\boldsymbol{q} = e^{\frac{\pi i n}{4} - \frac{\pi i \nu}{2}} \frac{\sqrt{(2\pi)^n}}{\sqrt{|\det A|}} e^{-\frac{i}{2}(A^{-1}\boldsymbol{p},\boldsymbol{p})},$$

where the integral is understood in the distributional sense as $\lim_{R\to\infty}\int_{|\boldsymbol{q}|\leq R}$ and ν is the number of negative eigenvalues of A.

Proof. Formula (3.2) follows from (3.1) by analytic continuation. It also can be proved directly by completing the square, diagonalizing the matrix A, and using the Fresnel integral formula (1.10). $\qquad \square$

There is also a complex version of Gaussian integration, given by the following analog of Lemma 3.1.

Lemma 3.2 (Gaussian integration in complex domain). *Let C be a complex $n \times n$ matrix such that its Hermitian part $\tfrac{1}{2}(C+C^*)$ is positive-definite. We have*

$$(3.3) \qquad \int_{\mathbb{C}^n} e^{-(C\boldsymbol{z},\boldsymbol{z})+(\boldsymbol{a},\boldsymbol{z})+(\boldsymbol{z},\boldsymbol{b})} d^{2n}\boldsymbol{z} = \frac{\pi^n}{\det C} e^{(C^{-1}\boldsymbol{a},\boldsymbol{b})},$$

where $(\ ,\ )$ stands for the standard Hermitian inner product in $\mathbb{C}^n$, $\boldsymbol{a},\boldsymbol{b} \in \mathbb{C}^n$, and $d^{2n}\boldsymbol{z} = d^2 z_1 \ldots d^2 z_n$.

Problem 3.1. Prove Lemma 3.2.

3.2. Propagator of the harmonic oscillator. The classical harmonic oscillator with one degree of freedom is described by the Lagrangian function $L(q, \dot{q}) = \frac{1}{2}m(\dot{q}^2 - \omega^2 q^2)$. The corresponding Hamiltonian operator for a quantum harmonic oscillator is given by

$$H = \frac{P^2}{2m} + \frac{m\omega^2 Q^2}{2}.$$

The following result is an exact computation of the propagator $K(q', t'; q, t)$ for the harmonic oscillator, using the Feynman path integral in configuration space.

Proposition 3.1. *The Feynman path integral for the harmonic oscillator is explicitly evaluated as follows:*

$$\int_{P(\mathbb{R})_{q,t}^{q',t'}} e^{\frac{im}{2\hbar}\int_t^{t'}(\dot{q}^2-\omega^2 q^2)d\tau}\,\mathscr{D}q = \sqrt{\frac{m\omega}{2\pi i\hbar \sin\omega T}}\, e^{\frac{im\omega}{2\hbar\sin\omega T}\{(q^2+q'\,^2)\cos\omega T - 2qq'\}},$$

where for $\dfrac{\pi\nu}{\omega} = T_\nu < T < T_{\nu+1} = \dfrac{\pi(\nu+1)}{\omega}$, $\nu \in \mathbb{N}$, *we have*

$$\sqrt{\frac{m\omega}{2\pi i\hbar \sin\omega T}} = e^{-\frac{\pi i}{4} - \frac{\pi i\nu}{2}}\sqrt{\frac{m\omega}{2\pi\hbar|\sin\omega T|}}.$$

When $T \to T_\nu$, *the right-hand side converges in the distributional sense to* $e^{-\frac{\pi i\nu}{2}}\delta(q - q')$ *for even* ν, *and to* $e^{-\frac{\pi i\nu}{2}}\delta(q + q')$ *for odd* ν.

Proof. In this case the $(n - 1)$-fold integral in (1.19) is Gaussian and can be computed exactly using (3.2). Namely, we have

$$\sum_{k=0}^{n-1}\left((q_{k+1} - q_k)^2 - \varepsilon^2 q_k^2\right) = (A_{n-1}\boldsymbol{q}, \boldsymbol{q}) - 2(\boldsymbol{p}, \boldsymbol{q}) + q^2 + q'\,^2,$$

where $\varepsilon = \omega\Delta t$, $\boldsymbol{q} = (q_1, \ldots, q_{n-1})$, $\boldsymbol{p} = (q, 0, \ldots, 0, q')$ are vectors in $\mathbb{R}^{n-1}$, and A_{n-1} is the following three-diagonal $(n - 1) \times (n - 1)$ matrix:

$$A_{n-1} = \begin{pmatrix} 2 - \varepsilon^2 & -1 & 0 & \cdots & 0 & 0 \\ -1 & 2 - \varepsilon^2 & -1 & \cdots & 0 & 0 \\ 0 & -1 & 2 - \varepsilon^2 & \cdots & 0 & 0 \\ \vdots & \vdots & \vdots & \ddots & \vdots & \vdots \\ 0 & 0 & 0 & \cdots & 2 - \varepsilon^2 & -1 \\ 0 & 0 & 0 & \cdots & -1 & 2 - \varepsilon^2 \end{pmatrix}.$$

It follows from (3.2) that

$$\left(\frac{m}{2\pi i\hbar\Delta t}\right)^{\frac{n}{2}}\int_{\mathbb{R}^{n-1}} e^{\frac{im}{2\hbar\Delta t}\sum_{k=0}^{n-1}((q_{k+1}-q_k)^2-\varepsilon^2 q_k^2)}\prod_{k=1}^{n-1}dq_k$$

$$(3.4)\qquad = e^{-\frac{\pi i\nu}{2}}\sqrt{\frac{m}{2\pi i\hbar\Delta t|\det A_{n-1}|}}\, e^{\frac{im}{2\hbar\Delta t}\{q^2+q'^{\,2}-(A_{n-1}^{-1}\boldsymbol{p},\boldsymbol{p})\}},$$

where $\nu = \nu_{n-1}$ is the number of negative eigenvalues of the matrix A_{n-1}. It is easy to find $\det A_{n-1}$ and $(A_{n-1}\boldsymbol{p},\boldsymbol{p})$. Namely, put $a_n = \det A_n$. Expanding $\det A_n$ with respect to the last row, we obtain the three-term recurrence relation

$$(3.5)\qquad a_{n+1} = (2-\varepsilon^2)a_n - a_{n-1}, \quad n = 0,1,2,\dots,$$

with the initial conditions $a_{-1} = 0$ and $a_0 = 1$. Recurrence relation (3.5) has two linearly independent solutions z^n and z^{-n}, where $2-\varepsilon^2 = z+z^{-1}$, and a solution a_n satisfying these initial conditions is given by

$$a_n = \frac{z^{n+1}-z^{-n-1}}{z-z^{-1}}.$$

From here it is easy to obtain that the eigenvalues of the matrix A_n are given by

$$\lambda_k = z+z^{-1}-2\cos\frac{\pi k}{n+1}, \quad k = 1,\dots,n.$$

Since $\varepsilon = \frac{\omega T}{n}$ we have $z = e^{i\theta}$ where $\theta = \varepsilon + O(n^{-2})$ and

$$\det A_{n-1} = \frac{\sin n\theta}{\sin\theta} = \frac{\sin\omega T}{\omega\Delta t}(1+O(n^{-1})) \quad \text{as} \quad n\to\infty,$$

and for n large enough the matrix A_{n-1} has exactly ν negative eigenvalues whenever $T_\nu < T < T_{\nu+1}$. In order to evaluate the inner product $(A_{n-1}\boldsymbol{p},\boldsymbol{p})$ we only need to know the corner elements of the inverse matrix $B = A_{n-1}^{-1}$, which are given by

$$B_{11} = B_{n-1\,n-1} = \frac{a_{n-2}}{a_{n-1}} = \frac{\sin(n-1)\theta}{\sin n\theta}, \quad B_{1\,n-1} = B_{n-1\,1} = \frac{1}{a_n} = \frac{\sin n\theta}{\sin\theta}.$$

Thus we get

$$q^2+q'^{\,2}-(A_{n-1}^{-1}\boldsymbol{p},\boldsymbol{p}) = \frac{1}{\sin n\theta}\left(2\sin\frac{\theta}{2}\cos\frac{(2n-1)\theta}{2}(q^2+q'^{\,2})-2\sin\theta\, qq'\right).$$

Using these formulas and passing to the limit $n\to\infty$ in (3.4), we obtain the expression for the propagator for the case $T \neq T_\nu$. The limit $T\to T_\nu$ is evaluated by using the following standard formula in the theory of distributions:

$$\lim_{t\to 0}\frac{1}{\sqrt{2\pi t}}\, e^{\frac{i(x-y)^2}{2t}} = e^{\frac{\pi i}{4}}\delta(x-y). \qquad \square$$

Remark. It follows from Proposition 3.1 that in the limit $\omega \to 0$ the propagator of the harmonic oscillator turns into the propagator (1.11) of a free particle.

Remark. For even ν, special values T_ν are integer multiples of the period $\dfrac{2\pi}{\omega}$ of the harmonic oscillator (see Section 1.5 in Chapter 1), so when $t'-t = T_\nu$, the extremal connecting q at time t and q' at time t' exists if and only if $q' = q$. Correspondingly, when $t'-t = T_\nu$ for odd ν, the extremal connecting q and q' exists if and only if $q' = -q$. For the general case $t' - t \neq T_\nu$, the extremal connecting q at time t and q' at time t' exists for all q and q'. The integer ν is the Morse index of the trajectory $q(\tau)$ — the number of negative eigenvalues of the corresponding Jacobi operator $\mathcal{J}$,

$$\mathcal{J} = -m\frac{d^2}{d\tau^2} - m\omega^2, \quad t \leq \tau \leq t',$$

with Dirichlet boundary conditions (see Problem 1.7 in Section 1.3 in Chapter 1).

Problem 3.2. Compute Weyl, pq, and qp-symbols of the evolution operator for the harmonic oscillator by using: (a) formula (3.2) and path integral representations from Section 2; (b) formula (3.6) and formulas (3.12), (3.18) and (3.19) in Section 3.3 of Chapter 2.

Problem 3.3. Show that the matrix symbol of the evolution operator for the harmonic oscillator is $K(\bar{a}, a; T) = \exp\{\bar{a}ae^{-i\omega T} - \frac{i}{2}\omega T\}$ by using: (a) series (1.8) in holomorphic representation; (b) Lemma 3.2 and path integral representation from Section 2; (c) formula (3.6) and result of Problem 2.4.

3.3. Mehler identity. It is instructive to compare the closed expression

$$(3.6) \qquad K(q', t'; q, t) = \sqrt{\frac{m\omega}{2\pi i\hbar \sin \omega T}}\, e^{\frac{im\omega}{2\hbar \sin \omega T}\{(q^2+q'^{\,2})\cos \omega T - 2qq'\}}$$

for the propagator of the harmonic oscillator with the series (1.8). Putting $x = \sqrt{\dfrac{m\omega}{\hbar}}\,q$, $y = \sqrt{\dfrac{m\omega}{\hbar}}\,q'$ and using the explicit formula for the normalized eigenfunctions

$$\psi_n(q) = \sqrt[4]{\frac{m\omega}{\pi\hbar}}\,\frac{1}{\sqrt{2^n n!}}\, e^{-\frac{m\omega}{2\hbar}q^2}\, H_n(x)$$

corresponding to the eigenvalues $E_n = \hbar\omega(n + \frac{1}{2})$, where $H_n(q)$ are classical Hermite-Tchebyscheff polynomials (see Section 2.6 in Chapter 2), we obtain the series

$$K(q', t'; q, t) = \sqrt{\frac{m\omega}{\pi\hbar}} \sum_{n=0}^{\infty} \frac{e^{-\frac{m\omega}{2\hbar}(q^2+q'^{\,2})-i\omega T(n+\frac{1}{2})}}{2^n n!} H_n(x)H_n(y),$$

which converges in the distributional sense. Setting $z = e^{-i\omega T}$ and comparing with (3.6), we get the formula

$$(3.7) \qquad \sum_{n=0}^{\infty} \frac{z^n}{2^n n!} H_n(x) H_n(y) = \frac{1}{\sqrt{1-z^2}} \exp\left\{ \frac{2xyz - (x^2+y^2)z^2}{1-z^2} \right\},$$

where $z \neq \pm 1$, and the square root in the right-hand side is understood as in Proposition 3.1. When $|z| < 1$, formula (3.7) is the classical Mehler identity from the theory of Hermite-Tchebyscheff polynomials. Thus we obtained a distributional form of the Mehler identity for $|z| = 1$ by computing the propagator of the harmonic oscillator in two different ways.

Formula (3.6) shows that the propagator $K(q', t'; q, t)$ is a smooth function of q, q' and $T = t' - t$ whenever $T \neq T_\nu$, and it is singular at $T = T_\nu$. Corresponding eigenvalues of the evolution operator $U(T_\nu)$ are $e^{-\pi i \nu (n + \frac{1}{2})}$, so that when ν is even we have $U(T_\nu) = e^{-\frac{\pi i \nu}{2}} I$, and therefore

$$K(q', t + T_\nu; q, t) = e^{-\frac{\pi i \nu}{2}} \delta(q - q'),$$

in perfect agreement with Proposition 3.1. For odd ν using the series (1.8) we obtain

$$e^{-\frac{i}{\hbar} H T_\nu} = e^{-\frac{\pi i \nu}{2}} \sum_{n=0}^{\infty} (-1)^n P_n,$$

where P_n are projection operators on the eigenspaces $\mathbb{C}\psi_n$ of H. Since $H_n(-q) = (-1)^n H_n(q)$, in this case we have

$$K(q', t + T_\nu; q, t) = e^{-\frac{\pi i \nu}{2}} \delta(q + q'),$$

which again agrees with Proposition 3.1.

4. Gaussian path integrals

It was already mentioned in Section 1 that in the semi-classical limit $\hbar \to 0$ the leading contribution to the propagator $K(q, t'; q, t)$ is given by the classical trajectory $q_{\mathrm{cl}}(\tau)$. This suggests to represent the paths $\gamma = q(\tau) \in P(\mathbb{R})_{q,t}^{q',t'}$ as $q(\tau) = q_{\mathrm{cl}}(\tau) + y(\tau)$, where $y(\tau)$ — the *quantum fluctuation* part — satisfies Dirichlet boundary conditions $y(t) = y(t') = 0$. It follows from the principle of the least action (see Section 1.2 in Chapter 1) that

$$(4.1)$$

$$S(q_{\mathrm{cl}} + y) = S_{\mathrm{cl}} + \frac{1}{2} \int_t^{t'} (m\dot{y}^2 - V''(q_{\mathrm{cl}}(\tau))y^2) d\tau + \text{higher order terms in } y,$$

where $S_{\mathrm{cl}} = S(q_{\mathrm{cl}})$ is the classical action. Similarly, for the case of several degrees of freedom,

$$(4.2) \qquad S(\boldsymbol{q}_{\mathrm{cl}} + \boldsymbol{y}) = S_{\mathrm{cl}} + \frac{1}{2} \int_t^{t'} \mathcal{J}(\boldsymbol{y})\boldsymbol{y}\, d\tau + \text{higher order terms in } \boldsymbol{y},$$

where $\mathcal{J}$ is the corresponding Jacobi operator (see Problem 1.7 in Section 1.3 of Chapter 1). It is remarkable that the Gaussian path integral

$$
(4.3) \qquad \int_{\left\{\begin{smallmatrix} y(t')=0 \\ y(t)=0 \end{smallmatrix}\right\}} e^{\frac{im}{2\hbar} \int_t^{t'} \mathcal{J}(y) y \, d\tau} \mathscr{D}\boldsymbol{y}
$$

over the fluctuating part can be evaluated explicitly in terms of the regularized determinant of the second order differential operator $\mathcal{J}$. Here we do this calculation for the case of the free particle and harmonic oscillator, and give another interpretation of formulas (1.11) and (3.6) for the propagators. The Gaussian path integral (4.3) also plays a fundamental role in the semi-classical asymptotics which we discuss in Section 6.1. In general, formula (4.2), with higher order terms in $\boldsymbol{y}$, and Gaussian integration over $\mathscr{D}\boldsymbol{y}$, constitute a basis for the perturbative expansion of the propagator.

4.1. Gaussian path integral for a free particle. The propagator of a free quantum particle is

$$
(4.4) \qquad K(q',t';q,t) = \sqrt{\frac{m}{2\pi i\hbar T}}\, e^{\frac{im(q-q')^2}{2\hbar T}} = \int_{P(\mathbb{R})_{q,t}^{q',t'}} e^{\frac{im}{2\hbar} \int_t^{t'} \dot{q}^2 d\tau} \mathscr{D}q,
$$

and the corresponding classical trajectory is

$$
q_{\mathrm{cl}}(\tau) = q + (\tau - t)\frac{q' - q}{T}, \quad T = t' - t.
$$

Using the decomposition $q(\tau) = q_{\mathrm{cl}}(\tau) + y(\tau)$, where $y(t) = y(t') = 0$, we obtain

$$
S(q) = \tfrac{1}{2} \int_t^{t'} m\dot{q}^2 d\tau = S_{\mathrm{cl}} + S(y),
$$

where

$$
S_{\mathrm{cl}} = \frac{1}{2} \int_t^{t'} m\dot{q}_{\mathrm{cl}}^2 \, d\tau = \frac{m(q - q')^2}{2T}.
$$

Assuming that $\mathscr{D}q = \mathscr{D}y$ under the "change of variable" $q = q_{\mathrm{cl}} + y$, we can rewrite the Feynman path integral for a free particle as

$$
K(q',t';q,t) = e^{\frac{i}{\hbar}S_{\mathrm{cl}}} \int_{\left\{\begin{smallmatrix} y(t')=0 \\ y(t)=0 \end{smallmatrix}\right\}} e^{\frac{im}{2\hbar} \int_t^{t'} \dot{y}^2 d\tau} \mathscr{D}y.
$$

Remarkably, the classical contribution $e^{\frac{i}{\hbar}S_{\mathrm{cl}}} = e^{\frac{im(q-q')^2}{2\hbar T}}$ exactly reproduces the exponential factor in the propagator for a free particle. The integral over the fluctuating part — the Gaussian path integral for a free particle — does not depend on q and q' and, as we know, coincides with the prefactor in (4.4).

A more conceptual way to interpret this result is the following. Let

$$A = -D^2, \quad D = \frac{d}{d\tau},$$

be the second order differential operator on the interval $[t, t']$ with Dirichlet boundary conditions $y(t) = y(t') = 0$. The operator A is self-adjoint on $L^2(t, t')$. For any real-valued, absolutely continuous function $y(\tau)$ satisfying Dirichlet boundary conditions and $y, \dot{y} \in L^2(t, t')$, we have by using integration by parts

$$(Ay, y) = -\int_t^{t'} \ddot{y} y \, d\tau = \int_t^{t'} \dot{y}^2 d\tau.$$

The "integrand" in the fluctuation factor

$$\int_{\left\{ \substack{y(t')=0 \\ y(t)=0} \right\}} e^{\frac{im}{2\hbar} \int_t^{t'} \dot{y}^2 d\tau} \mathscr{D}y$$

is the exponent of the quadratic form of the operator A, and in accordance with the finite-dimensional formula (3.1) it is natural to expect that this Gaussian path integral is proportional to $(\det A)^{-\frac{1}{2}}$. Of course, the problem here is to understand what we mean by a determinant of a differential operator. Clearly, it should be defined by some regularization of the divergent infinite product $\prod_{n=1}^{\infty} \lambda_n$, where λ_n are non-zero eigenvalues of A.

The most natural and useful regularization is given by the so-called operator zeta-function. Namely, let A be a non-negative self-adjoint operator in the Hilbert space $\mathscr{H}$ with pure point spectrum $0 \leq \lambda_1 \leq \lambda_2 \leq \ldots$, such that for some $\alpha > 0$ the operator $(A + I)^{-\alpha}$ is of trace class. Then the zeta-function $\zeta_A(s)$ of the operator A is defined for $\operatorname{Re} s > \alpha$ by the following absolutely convergent series:

$$\zeta_A(s) = \sum_{\lambda_n > 0} \frac{1}{\lambda_n^s}.$$

If $\zeta_A(s)$ admits a meromorphic continuation to a larger domain containing the point $s = 0$ and is regular at $s = 0$, then we define a *regularized determinant* of A by

$$(4.5) \qquad \det{}' A = \exp\left\{ -\frac{d\zeta_A}{ds}(0) \right\}.$$

Here the prime on the symbol det indicates that zero eigenvalues are excluded from the definition of an operator zeta-function. In the special case when 0 is not the eigenvalue of A, it is customary to denote the regularized determinant of A by $\det A$. We will also write

$$\det' A = \prod_{\lambda_n > 0}' \lambda_n,$$

where the prime indicates that the infinite product is regularized by the operator zeta-function. We have $\zeta_{cA}(s) = c^{-s}\zeta_A(s)$ for $c > 0$, so that

$$\det' cA = c^{\zeta_A(0)} \det' A,$$

which shows that $\zeta_A(0)$ plays the role of "regularized scaling dimension" of the Hilbert space $\mathscr{H}$ (with respect to the operator A). When $\dim_{\mathbb{C}} \mathscr{H} = n < \infty$ and $A > 0$, then $\zeta_A(0) = n$ and $\zeta_A'(0) = \log \lambda_1 + \cdots + \log \lambda_n$, and we recover the usual definition of $\det A$.

This basic outline works for the general case of elliptic operators on a compact manifold M. In quantum mechanics, only determinants of differential operators on one-dimensional[5] manifolds $M = [t, t']$ or $M = S^1$ appear. Their systematic study will be presented in Section 5, while here we consider the simplest case of the second derivative operator $A = -D^2$ associated with a free particle. The corresponding eigenvalues of A are $\lambda_n = \left(\dfrac{\pi n}{T}\right)^2$, $n = 1, 2, \ldots$, and we have for the operator zeta-function

$$\zeta_A(s) = \left(\frac{T}{\pi}\right)^{2s} \zeta(2s),$$

where $\zeta(s)$ is the Riemann zeta-function. Using classical formulas $\zeta(0) = -\frac{1}{2}$ and $\zeta'(0) = -\frac{1}{2}\log 2\pi$, we obtain

$$(4.6) \qquad \zeta_A(0) = -\frac{1}{2} \quad \text{and} \quad \zeta_A'(0) = -\log\frac{T}{\pi} - \log 2\pi = -\log 2T.$$

Thus for the operator $A = -D^2$ on the interval $[t, t']$ with Dirichlet boundary conditions we have

$$(4.7) \qquad\qquad\qquad \det A = 2T.$$

The formula

$$\int_{\left\{\substack{y(t')=0 \\ y(t)=0}\right\}} e^{\frac{im}{2\hbar}\int_t^{t'} \dot{y}^2 d\tau} \mathscr{D}y = \sqrt{\frac{m}{2\pi i\hbar T}}$$

agrees with our interpretation that the Gaussian path integral is proportional to $(\det A)^{-\frac{1}{2}}$. The coefficient of proportionality $c_{m,\hbar} = \sqrt{\frac{m}{\pi i\hbar}}$ is determined by comparison with the actual propagator for a free particle.

Remark. In Section 3.1 of Chapter 6 we will prove that for Gaussian Wiener integrals the corresponding constant is $\sqrt{\frac{m}{\pi\hbar}}$.

---

[5]Higher-dimensional manifolds are used in quantum field theory.

Problem 4.1. Show that the propagator of a particle in a constant uniform field f is given by the following formula:

$$K(q', t'; q, t) = \sqrt{\frac{m}{2\pi i \hbar T}}\; e^{\frac{i}{2\hbar}\left\{\frac{m(q-q')^2}{T} + fT(q+q') - \frac{f^2 T^3}{12m}\right\}}.$$

(*Hint:* Use Lagrangian function $L = \frac{1}{2}m\dot{q}^2 + fq$.)

4.2. Gaussian path integral for the harmonic oscillator. It is remarkable that the same interpretation (with the same constant $c_{m,\hbar}$) holds for the harmonic oscillator. Namely, according to Proposition 3.1 we have

$$K(q', t'; q, t) = \int_{P(\mathbb{R})_{q,t}^{q',t'}} e^{\frac{im}{2\hbar}\int_t^{t'}(\dot{q}^2 - \omega^2 q^2)d\tau}\, \mathscr{D}q$$

$$= \sqrt{\frac{m\omega}{2\pi i \hbar \sin \omega T}}\, \exp\left\{\frac{im\omega}{2\hbar \sin \omega T}\left((q^2 + q'^{\,2})\cos \omega T - 2qq'\right)\right\}.$$

Now solving classical equations of motion with the boundary conditions $q(t) = q$ and $q(t') = q'$, we readily compute that for $T \neq T_\nu$,

$$S_{\mathrm{cl}} = \frac{m}{2}\int_t^{t'}(\dot{q}_{\mathrm{cl}}^2 - \omega^2 q_{\mathrm{cl}}^2)d\tau = \frac{m\omega}{2\sin \omega T}\left((q^2 + q'^{\,2})\cos \omega T - 2qq'\right),$$

so that $e^{\frac{i}{\hbar}S_{\mathrm{cl}}}$ is exactly the exponential factor in the propagator. For the fluctuating factor — the Gaussian path integral for the harmonic oscillator — we have

$$\int_{\left\{\substack{y(t')=0 \\ y(t)=0}\right\}} e^{\frac{im}{2\hbar}\int_t^{t'}(\dot{y}^2 - \omega^2 y^2)d\tau}\, \mathscr{D}y = \sqrt{\frac{m}{\pi i \hbar\, \det A_\omega}},$$

where $A_\omega = -D^2 - \omega^2$ is the second order differential operator on the interval $[t, t']$ with Dirichlet boundary conditions. Thus to justify, in this case, the interpretation of the fluctuating factor in terms of the regularized determinant, we need to show that $\det A_\omega = \dfrac{2\sin \omega T}{\omega}$ when $T \neq T_\nu$.

At a heuristic level, this can be done by the following beautiful computation, which goes back to Euler. Namely, the eigenvalues of A_ω are $\lambda_n(\omega) = \left(\dfrac{\pi n}{T}\right)^2 - \omega^2$ (0 is not an eigenvalue for A_ω since $T \neq T_\nu$), and we have

$$\frac{\det A_\omega}{\det A_0} = \prod_{n=1}^\infty \frac{\lambda_n(\omega)}{\lambda_n(0)} = \prod_{n=1}^\infty \left(1 - \frac{\omega^2 T^2}{\pi^2 n^2}\right) = \frac{\sin \omega T}{\omega T}.$$

Since $\det A_0 = 2T$, we get the result! For the rigorous derivation, it is more convenient to consider the positive-definite operator $A_{i\omega}$, where $\omega > 0$, instead of the operator A_ω.

Lemma 4.1. *Let $A_{i\omega} = -D^2 + \omega^2$ be the second order differential operator on the interval $[t, t']$ with Dirichlet boundary conditions. Then*

$$\det A_{i\omega} = \frac{2\sinh\omega T}{\omega}.$$

Proof. Denoting $\zeta_{i\omega}(s) = \zeta_{A_{i\omega}}(s)$, we have for $\operatorname{Re} s > \frac{1}{2}$,

$$\zeta_{i\omega}(s) = \frac{1}{\Gamma(s)}\int_0^\infty e^{-\omega^2 x}\sum_{n=1}^\infty e^{-\frac{\pi^2 n^2}{T^2}x}x^s\frac{dx}{x}$$

$$= \frac{1}{2\Gamma(s)}\int_0^\infty e^{-\omega^2 x}\vartheta\left(\frac{\pi x}{T^2}\right)x^s\frac{dx}{x} - \frac{1}{2\omega^{2s}},$$

where $\vartheta(x) = \sum_{n\in\mathbb{Z}}e^{-\pi n^2 x}$ is the Jacobi theta series. Using the Jacobi inversion formula

$$\vartheta\left(\frac{1}{x}\right) = \sqrt{x}\,\vartheta(x), \quad x > 0,$$

we get the following representation:

$$\zeta_{i\omega}(s) = -\frac{1}{2\omega^{2s}} + \frac{T}{2\sqrt{\pi}\Gamma(s)}\int_0^\infty e^{-\omega^2 x}\vartheta\left(\frac{T^2}{\pi x}\right)x^{s-\frac{1}{2}}\frac{dx}{x}$$

$$= -\frac{1}{2\omega^{2s}} + \frac{T\Gamma(s-\frac{1}{2})}{2\sqrt{\pi}\omega^{2s-1}\Gamma(s)} + \frac{T}{\sqrt{\pi}\Gamma(s)}\int_0^\infty e^{-\omega^2 x}\sum_{n=1}^\infty e^{-\frac{n^2 T^2}{x}}x^{s-\frac{1}{2}}\frac{dx}{x}$$

$$(4.8) \quad = -\frac{1}{2\omega^{2s}} + \frac{T\Gamma(s-\frac{1}{2})}{2\sqrt{\pi}\omega^{2s-1}\Gamma(s)} + \frac{2T}{\sqrt{\pi}\Gamma(s)}\sum_{n=1}^\infty\left(\frac{nT}{\omega}\right)^{s-\frac{1}{2}}K_{s-\frac{1}{2}}(2\omega nT),$$

where

$$K_s(x) = \frac{1}{2}\int_0^\infty e^{-\frac{x}{2}(u+u^{-1})}u^s\frac{du}{u}, \quad x > 0,$$

is Macdonald's K-function (modified Bessel function of the second kind). Since $K_s(x) = O(e^{-x})$ as $x \to \infty$, uniformly in $s \in \mathbb{C}$ on compact subsets, it follows from representation (4.8) and the Weierstrass M-test that $\zeta_{i\omega}(s)$ admits meromorphic continuation to the entire s-plane with simple poles at $s \in -\frac{1}{2} + \mathbb{Z}_{\geq 0}$. Since $\lim_{s\to 0}s\Gamma(s) = 1$, we obtain $\zeta_{i\omega}(0) = -\frac{1}{2}$. Using classical formulas

$$(4.9) \qquad\qquad K_{\frac{1}{2}}(x) = K_{-\frac{1}{2}}(x) = \sqrt{\frac{\pi}{2x}}e^{-x}$$

and $\Gamma(\frac{1}{2}) = \sqrt{\pi}$, we obtain from (4.8),

$$\frac{d\zeta_{i\omega}}{ds}(0) = \log\omega - \omega T + \sum_{n=1}^\infty\frac{1}{n}e^{-2n\omega T}$$

$$= \log\omega - \omega T - \log(1 - e^{-2\omega T}),$$

so that

$$\det A_{i\omega} = \exp\left\{-\frac{d\zeta_{i\omega}}{ds}(0)\right\} = \frac{2\sinh\omega T}{\omega}. \qquad \Box$$

Corollary 4.1. *Let $B_\omega = -D^2 + i\omega D$ be the second order differential operator on the interval $[t, t']$ with Dirichlet boundary conditions. Then $\det B_{2\omega} = \det A_\omega$.*

Proof. The substitution $y(\tau) = e^{i\omega\tau} f(\tau)$ transforms the eigenvalue problem $B_{2\omega} y = \lambda y$ to the eigenvalue problem $A_\omega f = \lambda f$. $\qquad \Box$

Let I be the identity operator in $L^2(t, t')$. In addition to the regularized determinant $\det' A$ of the differential operator $A = -D^2$, it is possible to define the so-called *characteristic determinant* of A — an entire function $\det(A - \lambda I)$ on the complex λ-plane, whose only simple zeros are the eigenvalues $\lambda_n = \left(\dfrac{\pi n}{T}\right)^2$.

Indeed, for $\operatorname{Re}(\lambda_1 - \lambda) > 0$ consider the zeta-function

$$(4.10) \qquad \zeta_{A-\lambda I}(s) = \sum_{n=1}^{\infty} \frac{1}{(\lambda_n - \lambda)^s} = \sum_{n=1}^{\infty} e^{-s\log(\lambda_n - \lambda)},$$

where we are using the principal branch of the logarithm. As before, $\zeta_{A-\lambda I}(s)$ is absolutely convergent for $\operatorname{Re} s > \frac{1}{2}$, admits a meromorphic continuation to the complex s-plane, and is regular at $s = 0$. Thus for $\operatorname{Re}(\lambda_1 - \lambda) > 0$ we define

$$\det(A - \lambda I) = \prod_{n=1}^{\infty}{}'(\lambda_n - \lambda) = \exp\left\{-\frac{\partial\zeta_{A-\lambda I}}{\partial s}(0)\right\},$$

which is a holomorphic function of λ. Now suppose that $\det(A - \lambda I)$ is already defined for $\operatorname{Re}(\lambda_N - \lambda) > 0$ and is holomorphic. To extend it to the domain $\operatorname{Re}(\lambda_{N+1} - \lambda) > 0$, we set

$$\det(A - \lambda I) = \prod_{k=1}^{N}(\lambda_k - \lambda) \prod_{n=N+1}^{\infty}{}'(\lambda_n - \lambda),$$

where the regularized product is defined by the "truncated" zeta-function

$$\zeta_{A-\lambda I}^{(N+1)}(s) = \sum_{n=N+1}^{\infty} \frac{1}{(\lambda_n - \lambda)^s}$$

as

$$(4.11) \qquad \prod_{n=N+1}^{\infty}{}'(\lambda_n - \lambda) = \exp\left\{-\frac{\partial\zeta_{A-\lambda I}^{(N+1)}}{\partial s}(0)\right\}.$$

It requires to prove that $\zeta_{A-\lambda I}^{(N+1)}(s)$ admits a meromorphic continuation to the complex s-plane and is regular at $s = 0$. This is done by using the representation

$$\zeta_{A-\lambda I}^{(N+1)}(s) = \frac{1}{2\Gamma(s)} \int_1^\infty e^{-\lambda x} \vartheta_{N+1}(x) x^s \frac{dx}{x} + \frac{1}{2\Gamma(s)} \int_0^1 e^{-\lambda x} \vartheta_{N+1}(x) x^s \frac{dx}{x},$$

where

$$\vartheta_{N+1}(x) = \vartheta(x) - 2 \sum_{n=1}^N e^{-\lambda_n x} - 1.$$

Since $\mathrm{Re}(\lambda_n - \lambda) > 0$ for all $n > N$, the first integral in this formula is absolutely convergent for all $s \in \mathbb{C}$ and represents a holomorphic function. Using the Jacobi inversion formula and expanding $e^{-\lambda x}, e^{-\lambda_1 x}, \ldots, e^{-\lambda_N x}$ into power series in x, we conclude, just as before, that the second integral admits a meromorphic continuation to the s-plane with simple poles at $s \in -\frac{1}{2} + \mathbb{Z}_{\geq 0}$. Since for $M > N$

$$\prod_{n=N+1}^{\infty}{}' (\lambda_n - \lambda) = \prod_{k=N+1}^{M} (\lambda_k - \lambda) \prod_{n=M+1}^{\infty}{}' (\lambda_n - \lambda),$$

$\det(A - \lambda I)$ is well defined, and is an entire function of λ with simple zeros at $\lambda = \lambda_n$.

To obtain a closed (and simple) formula for $\det(A - \lambda I)$, observe that for $\mathrm{Re}(\lambda_1 - \lambda) > 0$ it follows from (4.10) that

$$\frac{\partial \zeta_{A-\lambda I}}{\partial \lambda}(s) = s \sum_{n=1}^{\infty} \frac{1}{(\lambda_n - \lambda)^{s+1}}.$$

This series is absolutely convergent for $\mathrm{Re}\, s > -\frac{1}{2}$, so that for $\mathrm{Re}(\lambda_1 - \lambda) > 0$ we have

$$(4.12) \qquad \frac{d}{d\lambda} \log \det(A - \lambda I) = -\frac{\partial^2}{\partial s \partial \lambda} \zeta_{A-\lambda I}(0) = \sum_{n=1}^{\infty} \frac{1}{\lambda - \lambda_n}.$$

By analytic continuation, formula (4.12) is valid for all $\lambda \neq \lambda_n$. Since $\lambda_n = \left(\frac{\pi n}{T}\right)^2$, we have

$$\sum_{n=1}^{\infty} \frac{1}{\lambda - \lambda_n} = \frac{T}{2\sqrt{\lambda}} \cot \sqrt{\lambda} T - \frac{1}{2\lambda},$$

so that

$$\det(A - \lambda I) = c \frac{\sin \sqrt{\lambda}\, T}{\sqrt{\lambda}},$$

and comparison between formula $\det(A + \omega^2 I) = \det A_{i\omega}$ and Lemma 4.1 gives $c = 2$. Thus we have proved the following result.

Lemma 4.2. *The characteristic determinant* $\det(A - \lambda I)$ *of the operator* $A = -D^2$ *on the interval* $[t, t']$ *with Dirichlet boundary conditions is well defined, and is an entire function of* λ. *It is given explicitly by*

$$\det(A - \lambda I) = \det A \prod_{n=1}^{\infty} \left(1 - \frac{\lambda}{\lambda_n}\right) = \frac{2 \sin \sqrt{\lambda}\, T}{\sqrt{\lambda}}.$$

Problem 4.2. Prove the Jacobi inversion formula. (*Hint:* Use the Poisson summation formula

$$\sum_{n=-\infty}^{\infty} f(n) = \sqrt{2\pi} \sum_{n=-\infty}^{\infty} \hat{f}(2\pi n),$$

where $f \in \mathscr{S}(\mathbb{R})$ and $\hat{f}$ is the Fourier transform of f.)

Problem 4.3. Prove formula (4.9).

Problem 4.4. Let

$$L = \frac{m}{2}(\dot{x}^2 + \dot{y}^2 + \dot{z}^2) + \frac{eB}{2c}(x\dot{y} - y\dot{x})$$

be the Lagrangian of a classical particle moving in the constant uniform magnetic field $\boldsymbol{B} = (0, 0, B)$. Show that the propagator of the corresponding quantum particle (see Example 2.3 in Section 2.4 of Chapter 2) is given by

$$K(\boldsymbol{r}', t'; \boldsymbol{r}, t) = \left(\frac{m}{2\pi i \hbar T}\right)^{\frac{3}{2}} \frac{\omega T}{\sin \omega T} \exp \frac{im}{2\hbar} \Big\{ \frac{(z - z')^2}{T} + \omega \cot \omega T [(x - x')^2$$
$$+ (y - y')^2] + 2\omega(xy' - yx') \Big\}.$$

(*Hint:* Use Corollary 4.1.)

5. Regularized determinants of differential operators

Here we study the characteristic determinant $\det(A - \lambda I)$ of the Sturm-Liouville operator

$$A = -D^2 + u(x), \quad D = \frac{d}{dx},$$

on the interval $[0, T]$ with Dirichlet or periodic boundary conditions, and its generalizations to the matrix case.

5.1. Dirichlet boundary conditions. Suppose that $u(x) \in C^1([0, T], \mathbb{R})$. The operator A is self-adjoint on $L^2(0, T)$ with domain

$$D(A) = \{y(x) \in W^{2,2}(0, T) : y(0) = y(T) = 0\},$$

where $W^{2,2}(0, T)$ is the Sobolev space. It has a pure point spectrum with simple eigenvalues $\lambda_1 < \lambda_2 < \cdots < \lambda_n < \cdots$, accumulating to ∞. Moreover, as $n \to \infty$,

$$(5.1) \qquad \lambda_n = \frac{\pi^2 n^2}{T^2} + c + O(n^{-2}), \quad \text{where} \quad c = \frac{1}{T} \int_0^T u(x)dx.$$

Consider first the case $A > 0$. Putting

$$\vartheta_A(t) = \sum_{n=1}^{\infty} e^{-\lambda_n t},$$

we have for $\operatorname{Re} s > \frac{1}{2}$,

$$(5.2) \qquad \zeta_A(s) = \frac{1}{\Gamma(s)} \int_0^{\infty} \vartheta_A(t) t^s \frac{dt}{t}$$

$$(5.3) \qquad = \frac{1}{\Gamma(s)} \int_1^{\infty} \vartheta_A(t) t^s \frac{dt}{t} + \frac{1}{\Gamma(s)} \int_0^1 \vartheta_A(t) t^s \frac{dt}{t}.$$

Since $\vartheta_A(t) = O(e^{-\lambda_1 t})$ as $t \to \infty$, the first integral in (5.3) converges absolutely for all $s \in \mathbb{C}$ and represents an entire function. Using asymptotics (5.1) and the Jacobi inversion formula we get as $t \to 0$,

$$\vartheta_A(t) = \frac{1}{2} e^{-ct} \left(\vartheta\left(\frac{\pi t}{T^2} \right) - 1 \right) (1 + O(t)) = \frac{a_{-\frac{1}{2}}}{\sqrt{t}} + a_0 + \widetilde{\vartheta}_A(t),$$

where

$$(5.4) \qquad a_{-\frac{1}{2}} = \frac{T}{2\sqrt{\pi}}, \quad a_0 = -\frac{1}{2}$$

and $\widetilde{\vartheta}_A(t) = O(\sqrt{t})$. Thus for the second integral in (5.3) we have

$$\frac{1}{\Gamma(s)} \int_0^1 \vartheta_A(t) t^s \frac{dt}{t} = \frac{a_{-\frac{1}{2}}}{(s - \frac{1}{2})\Gamma(s)} + \frac{a_0}{s\Gamma(s)} + \frac{1}{\Gamma(s)} \int_0^1 \widetilde{\vartheta}_A(t) t^s \frac{dt}{t},$$

so that it admits a meromorphic continuation to the half-plane $\operatorname{Re} s > -\frac{1}{2}$ and is regular at $s = 0$. Therefore we can define

$$\det{}' A = \prod_{n=1}^{\infty}{}' \lambda_n = \exp\left\{ -\frac{d\zeta_A}{ds}(0) \right\}.$$

Now repeating verbatim the arguments in the proof of Lemma 4.2, we see that for $\operatorname{Re}(\lambda_N - \lambda) > 0$ the truncated zeta-function $\zeta_{A-\lambda I}^{(N+1)}(s)$ admits a meromorphic continuation to $\operatorname{Re} s > -\frac{1}{2}$, and is regular at $s = 0$. Defining the regularized product $\prod_{n=N+1}^{\infty} (\lambda_n - \lambda)$ by the same formula (4.11), we see that

$$\det(A - \lambda I) = \prod_{k=1}^{N} (\lambda_k - \lambda) \prod_{n=N+1}^{\infty}{}' (\lambda_n - \lambda)$$

is an entire function of λ with simple zeros at λ_n. To remove the assumption $A > 0$, replace A by $\tilde{A} = A + (a - \lambda_1)I > 0$, where $a > 0$. Then[6]

$$\det(A - \lambda I) = \det(\tilde{A} - (\lambda + \lambda_1 - a)I),$$

---

[6]The definition of $\det' A$ does not depend on the choice of $a > 0$.

and

$$\det{}'A = \begin{cases} \det A & \text{if } 0 \text{ is not an eigenvalue of } A, \\ \lim_{\lambda \to 0} \lambda^{-1}\det(A + \lambda I) & \text{if } 0 \text{ is an eigenvalue of } A. \end{cases}$$

Let $y_1(x, \lambda)$ be the solution of the second order differential equation

$$(5.5) \qquad\qquad -y'' + u(x)y = \lambda y$$

on $[0, T]$, satisfying initial conditions

$$(5.6) \qquad\qquad y_1(0, \lambda) = 0, \quad y_1'(0, \lambda) = 1.$$

It is known that for every $0 \le x \le T$, the solution $y_1(x, \lambda)$ is an entire function of λ of order $\frac{1}{2}$ and as $\lambda \to \infty$,

$$(5.7) \qquad\qquad y_1(x, \lambda) = \frac{\sin \sqrt{\lambda}x}{\sqrt{\lambda}} + O\left(|\lambda|^{-1}e^{|\operatorname{Re}\sqrt{\lambda}|x}\right).$$

The entire function $d(\lambda) = y_1(T, \lambda)$ has simple zeros at the eigenvalues λ_n of the operator A, and has the following Hadamard product representation:

$$(5.8) \qquad\qquad d(\lambda) = c\,\lambda^{\delta} \prod_{\lambda_n \neq 0} \left(1 - \frac{\lambda}{\lambda_n}\right).$$

Here c is a constant, $\delta = 1$ if 0 is an eigenvalue of A, and $\delta = 0$ otherwise.

Theorem 5.1. *The characteristic determinant* $\det(A - \lambda I)$ *is given by the simple formula*

$$\det(A - \lambda I) = 2d(\lambda).$$

Moreover,

$$\frac{\det(A - \lambda I)}{\det{}'A} = (-\lambda)^{\delta} \prod_{\lambda_n \neq 0} \left(1 - \frac{\lambda}{\lambda_n}\right),$$

which fixes the constant in (5.8) *as* $c = \frac{1}{2}(-1)^{\delta}\det{}'A$.

Proof. Since both functions are entire, it is sufficient to prove the equality $\det(A - \lambda I) = 2d(\lambda)$ for $\operatorname{Re}(\lambda_1 - \lambda) > 0$. In this case, using

$$\zeta_{A - \lambda I}(s) = \frac{1}{\Gamma(s)} \int_0^{\infty} \operatorname{Tr} e^{-(A - \lambda I)t} t^s \frac{dt}{t}$$

for $\operatorname{Re} s > \frac{1}{2}$, and differentiating under the integral sign we get

$$\frac{\partial}{\partial \lambda} \zeta_{A - \lambda I}(s) = \frac{1}{\Gamma(s)} \int_0^{\infty} \operatorname{Tr} e^{-(A - \lambda I)t} t^s dt,$$

which is now absolutely convergent for $\mathrm{Re}\, s > -\frac{1}{2}$. Differentiating with respect to s at $s = 0$, we obtain

$$\frac{\partial^2}{\partial s \partial \lambda} \zeta_{A-\lambda I}(0) = \int_0^\infty \mathrm{Tr}\, e^{-(A-\lambda I)t} dt = \mathrm{Tr}(A - \lambda I)^{-1}.$$

It follows from (5.1) that the operator $R_\lambda = (A - \lambda I)^{-1}$ — the resolvent of A — is of trace class. Thus all our manipulations are justified and we arrive at the following very useful formula:

$$(5.9) \qquad \frac{d}{d\lambda} \log \det(A - \lambda I) = -\,\mathrm{Tr}\, R_\lambda,$$

which generalizes the familiar property of finite-dimensional determinants.

To compute the trace in (5.9), we use the representation of R_λ for $\lambda \neq \lambda_n$ as an integral operator with the continuous kernel $R_\lambda(x, \xi)$ (cf. formula (2.21) in Section 2.2 of Chapter 3). Namely, let $y_2(x, \lambda)$ be another solution of (5.5) with boundary conditions $y_2(T, \lambda) = 0$ and $y_2'(T, \lambda) = 1$, so that

$$W(y_1, y_2)(\lambda) = y_1'(x, \lambda)y_2(x, \lambda) - y_1(x, \lambda)y_2'(x, \lambda) = -d(\lambda).$$

Using the method of variation of parameters, for the solution of the inhomogeneous equation

$$-y'' + u(x)y = \lambda y + f(x), \quad \lambda \neq \lambda_n,$$

satisfying Dirichlet boundary conditions we get

$$y(x) = \int_0^T R_\lambda(x, \xi) f(\xi) d\xi,$$

where

$$(5.10) \qquad R_\lambda(x, \xi) = \begin{cases} -\dfrac{y_1(x, \lambda)y_2(\xi, \lambda)}{d(\lambda)} & \text{if } x \leq \xi, \\[2ex] -\dfrac{y_1(\xi, \lambda)y_2(x, \lambda)}{d(\lambda)} & \text{if } x \geq \xi. \end{cases}$$

Since R_λ is a trace class operator on $L^2(0, T)$ with the integral kernel $R_\lambda(x, \xi)$, which is a continuous function on $[0, T] \times [0, T]$, its operator trace equals the "matrix trace",

$$\mathrm{Tr}\, R_\lambda = \int_0^T R_\lambda(x, x) dx = -\frac{1}{d(\lambda)} \int_0^T y_1(x, \lambda)y_2(x, \lambda) dx.$$

We evaluate the last integral by the same computation used in the proof of Proposition 2.1 in Section 2.1 of Chapter 3. Namely, put $\dot{y}(x, \lambda) = \dfrac{\partial y}{\partial \lambda}(x, \lambda)$, and consider the following pair of equations:

$$-\ddot{y}_1'' + u(x)\dot{y}_1 = \lambda \dot{y}_1 + y_1,$$
$$-y_2'' + u(x)y_2 = \lambda y_2.$$

Multiplying the first equation by $y_2(x, \lambda)$, the second equation by $\dot{y}_1(x, \lambda)$ and subtracting, we obtain

$$y_1 y_2 = \dot{y}_1\, y_2'' - \dot{y}_1''\, y_2 = -W(\dot{y}_1, y_2)',$$

so that

$$(5.11) \qquad \int_0^T y_1(x, \lambda) y_2(x, \lambda)\, dx = -\, W(\dot{y}_1, y_2)\big|_0^T .$$

This formula is valid for any two solutions of the differential equation (5.5). Using boundary conditions for the solutions y_1 and y_2, we finally get

$$(5.12) \qquad \int_0^T y_1(x, \lambda) y_2(x, \lambda)\, dx = \dot{y}_1(T, \lambda).$$

Thus we have proved that for $\mathrm{Re}(\lambda_1 - \lambda) > 0$,

$$(5.13) \qquad \mathrm{Tr}\, R_\lambda = -\frac{d}{d\lambda} \log d(\lambda),$$

which implies that

$$(5.14) \qquad \det(A - \lambda I) = C\, d(\lambda)$$

for all $\lambda \in \mathbb{C}$ and some constant C.

It follows from (5.7) that

$$(5.15) \qquad d(-\mu) = \frac{e^{\sqrt{\mu}\, T}}{2\sqrt{\mu}} \left(1 + O\big(\mu^{-\frac{1}{2}}\big) \right) \qquad \text{as} \quad \mu \to +\infty.$$

Thus in order to determine the constant C in (5.14), it is sufficient to compute the asymptotics of $\det(A + \mu I)$ as $\mu \to +\infty$. We have

$$\zeta_{A+\mu I}(s) = \frac{1}{\Gamma(s)} \int_1^\infty \vartheta_A(t) e^{-\mu t}\, t^s\, \frac{dt}{t} + \frac{1}{\Gamma(s)} \int_0^1 \vartheta_A(t) e^{-\mu t}\, t^s\, \frac{dt}{t}.$$

The first integral is an entire function of s whose derivative at $s = 0$ exponentially decays as $\mu \to +\infty$. For the second integral we have

$$\frac{1}{\Gamma(s)} \int_0^1 \vartheta_A(t) e^{-\mu t}\, t^s\, \frac{dt}{t} = \frac{1}{\Gamma(s)} \int_0^1 \widetilde{\vartheta}_A(t) e^{-\mu t}\, t^s\, \frac{dt}{t}$$

$$+ \frac{1}{\Gamma(s)} \int_0^1 \left(\frac{a_{-\frac{1}{2}}}{\sqrt{t}} + a_0 \right) e^{-\mu t}\, t^s\, \frac{dt}{t}.$$

Since $\widetilde{\vartheta}_A(t) = O(\sqrt{t})$ as $t \to 0$, the first integral is absolutely convergent for $\mathrm{Re}\, s > -\frac{1}{2}$ and its derivative at $s = 0$ is $O(\mu^{-\frac{1}{2}})$ as $\mu \to +\infty$. For the remaining integral we have

$$\frac{1}{\Gamma(s)}\int_0^1\left(\frac{a_{-\frac{1}{2}}}{\sqrt{t}}+a_0\right)e^{-\mu t}\,t^s\,\frac{dt}{t}=\frac{a_{-\frac{1}{2}}\,\mu^{\frac{1}{2}-s}}{\Gamma(s)}\left(\Gamma(s-\tfrac{1}{2})-\int_\mu^\infty e^{-t}t^{s-\frac{1}{2}}\,\frac{dt}{t}\right)$$
$$+\frac{a_0\,\mu^{-s}}{\Gamma(s)}\left(\Gamma(s)-\int_\mu^\infty e^{-t}t^s\,\frac{dt}{t}\right).$$

It is elementary to show that the s-derivative of this integral at $s=0$ has asymptotics $-2\sqrt{\pi}a_{-\frac{1}{2}}\sqrt{\mu}-a_0\log\mu+O(e^{-\mu/2})$ as $\mu\to+\infty$. Using (5.4) we finally obtain

$$\det(A+\mu I)=\frac{e^{\sqrt{\mu}T}}{\sqrt{\mu}}\left(1+O\big(\mu^{-\frac{1}{2}}\big)\right)$$

as $\mu\to+\infty$, and comparison with (5.15) gives $C=2$. $\qquad\square$

Remark. When zero is not an eigenvalue of A, its inverse A^{-1} is a trace class operator and

$$\frac{\det(A-\lambda I)}{\det A}=\det_F(I-\lambda A^{-1}),$$

where $\det_F$ is the *Fredholm determinant.*

Remark. In Section 6.1 we will use Theorem 5.1 for evaluating the fluctuating factor in the semi-classical asymptotics of the propagator, and in Section 3.1 of Chapter 6 — for evaluating Gaussian Wiener integrals.

A similar result holds for the matrix-valued Sturm-Liouville operator with Dirichlet boundary conditions. Namely, let $U(x)=\{u_{ij}(x)\}_{i,j=1}^n$ be a C^1-function on $[0,T]$ which takes values in real, symmetric $n\times n$ matrices, and consider

$$\boldsymbol{A}=-D^2I_n+U(x),$$

where I_n is the $n\times n$ identity matrix. The differential operator $\boldsymbol{A}$ with Dirichlet boundary conditions is self-adjoint on the Hilbert space $L^2([0,T],\mathbb{C}^n)$ of $\mathbb{C}^n$-valued functions, and has a pure point spectrum accumulating to ∞. Its regularized determinant $\det'\boldsymbol{A}$ and characteristic determinant $\det(\boldsymbol{A}-\lambda\boldsymbol{I})$, where $\boldsymbol{I}$ is the identity operator in $L^2([0,T],\mathbb{C}^n)$, are defined as in the $n=1$ case. Let $\boldsymbol{Y}(x,\lambda)$ be the solution of the differential equation

$$-\boldsymbol{Y}''+U(x)\boldsymbol{Y}=\lambda\boldsymbol{Y}$$

satisfying initial conditions

$$\boldsymbol{Y}(0,\lambda)=0,\quad \boldsymbol{Y}'(0,\lambda)=I_n,$$

and put $\boldsymbol{D}(\lambda)=\det\boldsymbol{Y}(T,\lambda)$. The entire function $\boldsymbol{D}(\lambda)$ has properties similar to that of $d(\lambda)$, and the following analog of Theorem 5.1 holds.

Proposition 5.1. *The characteristic determinant* $\det(\boldsymbol{A} - \lambda\boldsymbol{I})$ *is given by the formula*

$$\det(\boldsymbol{A} - \lambda\boldsymbol{I}) = 2^n \boldsymbol{D}(\lambda).$$

Moreover,

$$\frac{\det(\boldsymbol{A} - \lambda\boldsymbol{I})}{\det' \boldsymbol{A}} = (-\lambda)^\delta \prod_{\lambda_n \neq 0} \left(1 - \frac{\lambda}{\lambda_n}\right),$$

where δ is the multiplicity of the eigenvalue $\lambda = 0$.

Problem 5.1 (Gelfand-Levitan trace identity). Prove that

$$\sum_{n=1}^{\infty}(\lambda_n - \lambda_n^{(0)} - c) = \frac{1}{2T}\int_0^T u(x)dx - \frac{u(0) + u(T)}{4},$$

where $\lambda_n^{(0)} = (\frac{\pi n}{T})^2$ and $c = \frac{1}{T}\int_0^T u(x)dx$.

5.2. Periodic boundary conditions. As in the previous section, we assume that $u(x) \in C^1([0,T],\mathbb{R})$. The Sturm-Liouville operator $A = -D^2 + u(x)$ with periodic boundary conditions is self-adjoint on $L^2(0,T)$ with the domain

$$D(A) = \{y(x) \in W^{2,2}(0,T) : y(0) = y(T) \;\text{ and }\; y'(0) = y'(T)\}.$$

It has a pure point spectrum with the eigenvalues

$$\lambda_0 < \lambda_1 \leq \lambda_2 < \cdots < \lambda_{2n-1} \leq \lambda_{2n} < \cdots$$

accumulating to ∞. Moreover, as $n \to \infty$,

$$(5.16) \qquad \lambda_{2n-1} = \frac{4\pi^2 n^2}{T^2} + c + O(n^{-2}), \quad \lambda_{2n} = \frac{4\pi^2 n^2}{T^2} + c + O(n^{-2}),$$

where c is the same as in (5.1). Replacing, if necessary, A by $A - (\lambda_0 + a)I$ with $a > 0$, we can always assume that $\lambda_0 > 0$, and define

$$\vartheta_A(t) = \sum_{n=0}^{\infty} e^{-\lambda_n t}, \;\; t > 0.$$

Using asymptotics (5.16), we get that as $t \to 0$,

$$\vartheta_A(t) = e^{-ct}\vartheta\left(\frac{4\pi t}{T^2}\right)(1 + O(t)) = \frac{a_{-\frac{1}{2}}}{\sqrt{t}} + O(\sqrt{t}),$$

where $a_{-\frac{1}{2}} = \dfrac{T}{2\sqrt{\pi}}$ as in (5.4), but $a_0 = 0$. This allows us to define the regularized determinant $\det' A$ and the characteristic determinant $\det(A - \lambda I)$ exactly as in the previous section. Since $a_0 = 0$, we now get as $\mu \to +\infty$

$$(5.17) \qquad \det(A + \mu I) = e^{\sqrt{\mu}T}\left(1 + O(\mu^{-\frac{1}{2}})\right)$$

(see the end of the proof of Theorem 5.1).

Here we denote[7] by $y_1(x, \lambda)$ and $y_2(x, \lambda)$ solutions of the Sturm-Liouville equation (5.5) satisfying initial conditions $y_1(0, \lambda) = 1$, $y_1'(0, \lambda) = 0$ and $y_2(0, \lambda) = 0$, $y_2'(0, \lambda) = 1$. Solutions y_1 and y_2 are linearly independent for all λ and the matrix

$$Y(x, \lambda) = \begin{pmatrix} y_1(x, \lambda) & y_2(x, \lambda) \\ y_1'(x, \lambda) & y_2'(x, \lambda) \end{pmatrix}$$

satisfies the initial condition $Y(0, \lambda) = I_2$, where I_2 is the 2×2 identity matrix, and has the property $\det Y(x, \lambda) = 1$. For fixed x the matrix $Y(x, \lambda)$ is an entire matrix-valued function of λ having the following asymptotics as $\lambda \to \infty$:

$$(5.18) \quad Y(x, \lambda) = \begin{pmatrix} \cos\sqrt{\lambda}x & \dfrac{1}{\sqrt{\lambda}}\sin\sqrt{\lambda}x \\ \sqrt{\lambda}\sin\sqrt{\lambda}x & \cos\sqrt{\lambda}x \end{pmatrix} \left(I_2 + O\big(|\lambda|^{-1}e^{|\operatorname{Re}\sqrt{\lambda}|x}\big) \right).$$

By definition, the *monodromy matrix* of the periodic Sturm-Liouville problem is the matrix

$$T(\lambda) = Y(T, \lambda).$$

The monodromy matrix satisfies $\det T(\lambda) = 1$ and is an entire matrix-valued function. The following result is the analog of Theorem 5.1 for the periodic boundary conditions.

Theorem 5.2. *One has*

$$\det(A - \lambda I) = -\det_2(T(\lambda) - I_2) = y_1(T, \lambda) + y_2'(T, \lambda) - 2,$$

where $\det_2$ *is the determinant of a* 2×2 *matrix. Moreover,*

$$\frac{\det(A - \lambda I)}{\det' A} = (-\lambda)^\delta \prod_{\lambda_n \neq 0} \left(1 - \frac{\lambda}{\lambda_n} \right),$$

where $\{\lambda_n\}_{n=0}^\infty$ *are the eigenvalues of* A, *and* $0 \leq \delta \leq 2$ *is a multiplicity of the eigenvalue* $\lambda = 0$.

Proof. The proof follows closely the proof of Theorem 5.1, and we will assume that $A > 0$. First, in exact analogy with (5.9) we obtain that for $\lambda \neq \lambda_n$,

$$\frac{d}{d\lambda} \log \det(A - \lambda I) = -\operatorname{Tr} R_\lambda,$$

where $R_\lambda = (A - \lambda I)^{-1}$. To get a closed expression for the integral kernel $R(x, \xi)$ of the operator R_λ, we use the same variation of parameters method as in the previous section, but now for periodic boundary conditions. As a

---

[7]There should be no confusion with the notation in the previous section.

result, we obtain that the symmetric, continuous kernel $R_\lambda(x,\xi)$ is given for $x \leq \xi$ by the formula

$$R_\lambda(x,\xi) = -\big(y_1(x,\lambda), y_2(x,\lambda)\big)\,(T(\lambda)-I_2)^{-1}T(\lambda)\begin{pmatrix} y_2(\xi,\lambda) \\ -y_1(\xi,\lambda) \end{pmatrix}$$

$$= -\mathrm{Tr}_2\left\{(T(\lambda)-I_2)^{-1}T(\lambda)Z(x,\xi;\lambda)\right\},$$

where Tr_2 in the last formula is the matrix trace and

$$Z(x,\xi;\lambda) = \begin{pmatrix} y_1(x,\lambda)y_2(\xi,\lambda) & y_2(x,\lambda)y_2(\xi,\lambda) \\ -y_1(x,\lambda)y_1(\xi,\lambda) & -y_2(x,\lambda)y_1(\xi,\lambda) \end{pmatrix}.$$

As in the proof of Theorem 5.1, we need to compute $\int_0^T Z(x,x;\lambda)dx$. It readily follows from formula (5.11) and the definition of the monodromy matrix that

$$\int_0^T Z(x,x;\lambda)dx = T^{-1}(\lambda)\frac{d}{d\lambda}T(\lambda).$$

Therefore

$$\frac{d}{d\lambda}\log\det(A-\lambda I) = \mathrm{Tr}_2\left((T(\lambda)-I_2)^{-1}\frac{d}{d\lambda}T(\lambda)\right) = \frac{d}{d\lambda}\log\det_2(T(\lambda)-I_2),$$

and $\det(A-\lambda I) = C\det(T(\lambda)-I_2)$. To determine the constant C we set $\lambda = -\mu \to +\infty$ and compare asymptotics (5.17) with the asymptotics

$$\det_2(T(-\mu)-I_2) = 2 - \mathrm{Tr}\,T(-\mu) = 2 - 2\cosh\sqrt{\mu}T(1+O(\mu^{-\frac{1}{2}}))$$

$$= -e^{\sqrt{\mu}T}(1+O(\mu^{-\frac{1}{2}})),$$

which follows from (5.18). Thus $C = -1$. $\square$

Remark. In Section 3.2 of Chapter 6 we will use Theorem 5.2 for calculating Gaussian Wiener integrals over the loop spaces.

In the special case $u(x) = 0$, we have

$$y_1(x,\lambda) = \cos\sqrt{\lambda}x \quad\text{and}\quad y_2(x,\lambda) = \frac{\sin\sqrt{\lambda}x}{\sqrt{\lambda}},$$

so that

$$(5.19) \qquad \det(-D^2 - \lambda I) = 2(\cos\sqrt{\lambda}T - 1) = -4\sin^2\frac{\sqrt{\lambda}T}{2}.$$

Setting $\lambda = -\omega^2 < 0$, for the operator $A_{i\omega} = -D^2 + \omega^2$ we get

$$(5.20) \qquad\qquad\qquad \det A_{i\omega} = 4\sinh^2\frac{\omega T}{2},$$

and also

$$(5.21) \qquad \det{}' A_0 = -\lim_{\lambda\to 0^-}\frac{\det(A-\lambda I)}{\lambda} = \lim_{\omega\to 0}\frac{\det{}' A_{i\omega}}{\omega^2} = T^2.$$

Remark. Using that the spectrum of $A_{i\omega}$ consists of double eigenvalues $\lambda_n(\omega) = \left(\dfrac{2\pi n}{T}\right)^2$, $n = 1, 2, \ldots$, and of the simple eigenvalue ω^2, we can derive formulas (5.20)–(5.21) directly, as was done in Section 4.2 for Dirichlet boundary conditions. There is also an analog of the heuristic computation in Section 4.2:

$$\frac{\det A_{i\omega}}{\det' A_0} = \omega^2 \prod_{n=1}^{\infty} \frac{\lambda_n(\omega)^2}{\lambda_n(0)^2} = \omega^2 \prod_{n=1}^{\infty} \left(1 + \frac{\omega^2 T^2}{4\pi^2 n^2}\right)^2 = \frac{4}{T^2} \sinh^2 \frac{\omega T}{2}.$$

A similar result holds for the general second order differential operator $A = -D^2 + v(x)D + u(x)$ on the interval $[0, T]$ with periodic boundary conditions. Namely, repeating the proof of Theorem 5.2, we have the following

Theorem 5.3. *For the differential operator $A = -D^2 + v(x)D + u(x)$ one has*

$$\det(A - \lambda I) = -e^{-\frac{1}{2} \int_0^T v(x)dx} \det_2(T(\lambda) - I_2)$$

$$= e^{-\frac{1}{2} \int_0^T v(x)dx}\left(y_1(T, \lambda) + y_2'(T, \lambda) - 1 - e^{\int_0^T v(x)dx}\right),$$

where solutions $y_{1,2}(x, \lambda)$ and the monodromy matrix $T(\lambda)$ are defined by the same formulas as for the case $v(x) = 0$.

In particular, the following result will be used in Section 2.2 of Chapter 8.

Corollary 5.4.

$$\det'(-D^2 + \omega D) = \frac{2T}{\omega} \sinh \frac{\omega T}{2}.$$

Proof. The proof is an elementary computation, using Theorem 5.3, an explicit form of the solutions $y_{1,2}(x, \lambda)$, and the formula

$$\det'(-D^2 + \omega D) = -\lim_{\lambda \to 0} \frac{\det(-D^2 + \omega D - \lambda I)}{\lambda}. \qquad \square$$

Remark. Since the operator $-D^2 + \omega D$ on $[0, T]$ with periodic boundary conditions has simple eigenvalues $\lambda_n(a) = \left(\dfrac{2\pi n}{T}\right)^2 + i\omega\left(\dfrac{2\pi n}{T}\right)$, $n = -\infty, \ldots, \infty$, we can also repeat the heuristic computation in Section 4.2:

$$\frac{\det'(-D^2 + \omega D)}{\det'(-D^2)} = \prod_{n=1}^{\infty} \left(1 + \frac{\omega^2 T^2}{4\pi^2 n^2}\right) = \frac{2}{\omega T} \sinh \frac{\omega T}{2}.$$

A similar result holds for the matrix-valued Sturm-Liouville operator with periodic boundary conditions. Namely, let $U(x) = \{u_{ij}(x)\}_{i,j=1}^{n}$ be a C^1-function on $[0, T]$ which takes values in real, symmetric $n \times n$ matrices, and consider

$$A = -D^2 I_n + U(x).$$

The differential operator $\boldsymbol{A}$ with periodic boundary conditions is self-adjoint on the Hilbert space $L^2([0,T],\mathbb{C}^n)$ of $\mathbb{C}^n$-valued functions, and has a pure point spectrum accumulating to ∞. Its regularized determinant $\det'\boldsymbol{A}$ and characteristic determinant $\det(\boldsymbol{A}-\lambda\boldsymbol{I})$, where $\boldsymbol{I}$ is the identity operator in $L^2([0,T],\mathbb{C}^n)$, are defined as in the $n=1$ case. Let $\boldsymbol{Y}_1(x,\lambda)$ and $\boldsymbol{Y}_2(x,\lambda)$ be the solutions of the differential equation

$$-\boldsymbol{Y}'' + U(x)\boldsymbol{Y} = \lambda\boldsymbol{Y}$$

satisfying, respectively, the initial conditions

$$\boldsymbol{Y}_1(0,\lambda) = I_n, \quad \boldsymbol{Y}_1'(0,\lambda) = 0 \quad \text{and} \quad \boldsymbol{Y}_2(0,\lambda) = 0, \quad \boldsymbol{Y}_2'(0,\lambda) = I_n.$$

The monodromy matrix $\boldsymbol{T}(\lambda)$ is defined as the following $2n \times 2n$ block matrix:

$$\boldsymbol{T}(\lambda) = \begin{pmatrix} \boldsymbol{Y}_1(T,\lambda) & \boldsymbol{Y}_2(T,\lambda) \\ \boldsymbol{Y}_1'(T,\lambda) & \boldsymbol{Y}_2'(T,\lambda) \end{pmatrix},$$

and is a matrix-valued entire function. The analog of Theorem 5.2 is the following statement.

Proposition 5.2. *The characteristic determinant is given by the formula*

$$\det(\boldsymbol{A}-\lambda\boldsymbol{I}) = (-1)^n \det_{2n}(\boldsymbol{T}(\lambda) - I_{2n}),$$

where $\det_{2n}$ is the determinant of a $2n \times 2n$ matrix, and

$$\frac{\det(\boldsymbol{A}-\lambda\boldsymbol{I})}{\det'\boldsymbol{A}} = (-\lambda)^\delta \prod_{\lambda_n \neq 0}\left(1 - \frac{\lambda}{\lambda_n}\right),$$

where δ is the multiplicity of the eigenvalue $\lambda = 0$.

Problem 5.2. Prove Theorem 5.3.

Problem 5.3. Derive Corollary 5.4.

5.3. First order differential operators. Here we continue to assume that $u(x) \in C^1([0,T],\mathbb{R})$, and consider the first order differential operator

$$A = D + u(x)$$

on the interval $[0,T]$ with periodic boundary conditions $y(0) = y(T)$. The equation

$$y' + u(x)y = \lambda y$$

has an explicit solution $y(x) = Ce^{\lambda x - \int_0^x u(\tau)d\tau}$, which is periodic if and only if $\lambda = \lambda_n$, where

$$\lambda_n = u_0 + \frac{2\pi i n}{T}, \quad n \in \mathbb{Z}, \quad \text{and} \quad u_0 = \frac{1}{T}\int_0^T u(x)dx.$$

Thus the spectrum of the operator A coincides with the spectrum of the operator $A_0 = D + u_0$.

Proposition 5.3. *For $u_0 > 0$,*

$$\det(D + u(x)) = 1 - e^{-u_0 T},$$

and $\det' D = T$ *for* $u_0 = 0$.

Proof. The zeta-function of the operator A with $u_0 > 0$ is given by the series

$$\zeta_A(s) = \sum_{n=-\infty}^{\infty} \frac{1}{\lambda_n^s},$$

where $\lambda_n^{-s} = e^{-s \log \lambda_n}$ with the principal branch of the logarithm. This series is absolutely convergent for $\operatorname{Re} s > 1$. Introducing the Hurwitz zeta-function

$$\zeta(s, a) = \sum_{n=0}^{\infty} \frac{1}{(n+a)^s},$$

where $\operatorname{Re} a > 0$ and $\operatorname{Re} s > 1$, we can rewrite $\zeta_A(s)$ as

$$\zeta_A(s) = \left(\frac{2\pi}{T} \right)^{-s} \left(e^{-\frac{\pi i s}{2}} \zeta(s, a) + e^{\frac{\pi i s}{2}} \zeta(s, \bar{a}) \right) + \frac{1}{u_0^s}, \quad a = 1 - \frac{u_0 T}{2\pi} i.$$

It is well known that the Hurwitz zeta-function admits a meromorphic continuation to the whole s-plane with single simple pole at $s = 1$ with residue 1, and

$$\zeta(0, a) = \frac{1}{2} - a, \qquad \frac{\partial \zeta}{\partial s}(0, a) = \log \Gamma(a) - \frac{1}{2} \log 2\pi.$$

Using the classical formula

$$\Gamma(1 + z)\Gamma(1 - z) = \frac{\pi z}{\sin \pi z},$$

we obtain

$$\frac{d\zeta_A}{ds}(0) = \log |\Gamma(a)|^2 - \log u_0 T + \frac{u_0 T}{2}$$

$$= -\log(e^{\frac{u_0 T}{2}} - e^{-\frac{u_0 T}{2}}) + \frac{u_0 T}{2},$$

so that $\det(D + u(x)) = 1 - e^{-u_0 T}$. Finally, $\zeta_D(s) = \lim_{u_0 \to 0}(\zeta_A(s) - u_0^{-s})$, and we get

$$\det' D = \lim_{u_0 \to 0} \frac{\det'(D + u_0)}{u_0} = T. \qquad \square$$

Remark. For $u_0 < 0$ one should use the branch of the logarithm with the cut along the positive semi-axis, and the above arguments give

$$\det(D + u(x)) = 1 - e^{u_0 T}.$$

Remark. One can also consider the operator $A = D + u(x)$ on the interval $[0, T]$ with anti-periodic boundary conditions $y(0) = -y(T)$. The corresponding eigenvalues are

$$\lambda_n = u_0 + \frac{\pi i (2n + 1)}{T}, \quad n \in \mathbb{Z},$$

and the passage from periodic to anti-periodic boundary conditions amounts to replacing u_0 by $u_0 + \frac{\pi i}{T}$. It follows from Proposition 5.3 that for $u_0 > 0$,

$$\det(D + u(x)) = 1 + e^{-u_0 T}.$$

Remark. Proposition 5.3 is very useful for calculating Gaussian path integrals in the holomorphic representation, discussed in Section 2.4. As an example, consider the harmonic oscillator with the Wick symbol $H(\bar{a}, a) = \omega(\bar{a}a + \frac{1}{2}\hbar)$. Formula (2.16) expresses the trace $\operatorname{Tr} e^{-\frac{1}{\hbar}TH}$ as a path integral in the holomorphic representation. On the other hand, using the explicit form of the eigenvalues $E_n = \omega\hbar(n + \frac{1}{2})$, we immediately get

$$\operatorname{Tr} e^{-\frac{1}{\hbar}TH} = \sum_{n=0}^{\infty} e^{-\frac{1}{\hbar}E_n T} = \frac{e^{-\frac{1}{2}\omega T}}{1 - e^{-\omega T}} = \frac{1}{2\sinh\frac{\omega T}{2}}.$$

Comparing with (2.16), we obtain that

$$(5.22) \qquad \int_{\left\{\substack{\bar{a}(0)=\bar{a}(T) \\ a(0)=a(T)}\right\}} e^{-\frac{1}{\hbar}\int_0^T (\bar{a}\dot{a} + \omega\bar{a}a)dt} \mathscr{D}\bar{a}\mathscr{D}a = \frac{1}{\det(D + \omega)},$$

which should be considered as a special analog of the finite-dimensional Gaussian integration in the complex domain — formula (3.3).

Problem 5.4. Give a direct proof of formula (5.22).

6. Semi-classical asymptotics – II

Here we consider the semi-classical asymptotics — the asymptotics of the propagator[8] $K_\hbar(\boldsymbol{q}', t'; \boldsymbol{q}, t)$ as $\hbar \to 0$. We compare the heuristic method, based on the Feynman path integral representation (1.26), with the rigorous analysis, based on the short-wave asymptotics, derived in Section 6.1 of Chapter 3.

---

[8]Here the dependence on the Planck constant $\hbar$ is introduced explicitly.

6.1. Using the Feynman path integral. Start with the Lagrangian $L(q, \dot{q}) = \frac{1}{2}m\dot{q}^2 - V(q)$ for a classical particle with one degree of freedom. The propagator $K_\hbar(q', t'; q, t)$ is given by the Feynman path integral (1.20), and we will formally apply the stationary phase method to investigate its behavior as $\hbar \to 0$. As in Section 4, we assume that there is a unique classical trajectory $q_{\mathrm{cl}}(\tau)$ connecting points q and q' at times t and t', set $q(\tau) = q_{\mathrm{cl}}(\tau) + y(\tau)$, and consider the expansion (4.1), i.e.,

$$S(q) = S_{\mathrm{cl}} + \tfrac{1}{2}m \int_t^{t'} (\dot{y}^2 - u(\tau)y^2)d\tau + O(y^3),$$

where $u(\tau) = \frac{1}{m}V''(q_{\mathrm{cl}}(\tau))$ and

$$S_{\mathrm{cl}} = \int_t^{t'} (\tfrac{1}{2}m\dot{q}_{\mathrm{cl}}^2 - V(q_{\mathrm{cl}}))d\tau = S(q', t'; q, t).$$

According to the stationary phase method (see Section 2.3 in Chapter 2), the leading contribution to the Feynman integral (1.20) as $\hbar \to 0$ comes from a critical point of the action functional — the classical trajectory $q_{\mathrm{cl}}(\tau)$. Thus we obtain as $\hbar \to 0$,

$$K_\hbar(q', t'; q, t) \simeq e^{\frac{i}{\hbar}S_{\mathrm{cl}}} \int_{\left\{\begin{smallmatrix} y(t')=0 \\ y(t)=0 \end{smallmatrix}\right\}} e^{\frac{im}{2\hbar}\int_t^{t'} (\dot{y}^2 - u(\tau)y^2)d\tau} \mathscr{D}y$$

$$(6.1) \qquad = \sqrt{\frac{m}{\pi i \hbar \det A}} \exp\left\{\frac{i}{\hbar}S(q', t'; q, t)\right\}.$$

Here A is the corresponding Jacobi operator — a second order differential operator $-D^2 - u(\tau)$ on the interval $[t, t']$ with Dirichlet boundary conditions. (We are assuming that the potential $V(q)$ is sufficiently smooth so that $u(\tau) \in C^1([t, t'])$.) Formula (6.1) is a remarkably simple expression which shows a deep relation between semi-classical asymptotics of a quantum mechanical propagator and classical motion.

Remark. When $u(\tau) = \dfrac{1}{m}V''(q_{\mathrm{cl}}(\tau))$, the regularized determinant of the differential operator $A = -D^2 - u(\tau)$ on the interval $[t, t']$ with Dirichlet boundary conditions can be expressed entirely in terms of the classical trajectory $q_{\mathrm{cl}}(\tau)$. Namely, differentiating Newton's equation

$$(6.2) \qquad m\ddot{q} = -V'(q),$$

with respect to τ, we find that the function $y(\tau) = \dot{q}_{\mathrm{cl}}(\tau)$ satisfies the differential equation $Ay = 0$. When $y(t) = \dot{q}_{\mathrm{cl}}(t) = 0$, the function[9]

$$y_1(\tau) - \frac{y(\tau)}{\dot{y}(t)}$$

---

[9]Here we assume that $\dot{y}(t) \neq 0$, so that $V'(q) \neq 0$.

satisfies initial condition (5.6) (where the interval $[0, T]$ is replaced by the interval $[t, t']$). According to Theorem 5.1, we have in this case,

$$(6.3) \qquad \det A = 2y_1(t') = -2m \frac{\dot{q}_{\mathrm{cl}}(t')}{V'(q)}.$$

In order to find the solution $y_1(\tau)$ of the differential equation $Ay = 0$ for the case $y(t) \neq 0$, observe that the Wronskian $\dot{y}_1 y - \dot{y} y_1$ of its two solutions is constant on $[t, t']$. Using (5.6) we get

$$\dot{y}_1 y - \dot{y} y_1 = y(t),$$

and solving this differential equation we obtain

$$y_1(\tau) = y(\tau) y(t) \int_t^\tau \frac{ds}{y^2(s)}.$$

Thus we get the formula

$$(6.4) \qquad \det A = 2y(t)y(t') \int_t^{t'} \frac{d\tau}{y^2(\tau)}, \quad y(\tau) = \dot{q}_{\mathrm{cl}}(\tau),$$

which expresses the fluctuating factor in the semi-classical asymptotics of the propagator in terms of the classical motion.

Similarly, for the case of n degrees of freedom, when

$$L = \tfrac{1}{2} m \dot{\boldsymbol{q}}^2 - V(\boldsymbol{q}),$$

and $V(\boldsymbol{q}) \in C^3(\mathbb{R}^n, \mathbb{R})$, we obtain

$$(6.5) \qquad K_\hbar(\boldsymbol{q}', t'; \boldsymbol{q}, t) \simeq \left(\frac{m}{\pi i \hbar} \right)^{\frac{n}{2}} \frac{1}{\sqrt{\det \boldsymbol{A}}} \, e^{\frac{i}{\hbar} S(\boldsymbol{q}', t'; \boldsymbol{q}, t)}$$

as $\hbar \to 0$, where $\boldsymbol{A} = -D^2 - U(\tau)$ and

$$U(\tau) = \frac{\partial^2 V}{\partial \boldsymbol{q}^2}(\boldsymbol{q}_{\mathrm{cl}}(\tau)) = \left\{ \frac{\partial^2 V}{\partial q_i \partial q_j}(\boldsymbol{q}_{\mathrm{cl}}(\tau)) \right\}_{i,j=1}^n.$$

6.2. Rigorous derivation. Let $K_\hbar(q, q', t)$ be the fundamental solution of the Schrödinger equation — the solution of the Cauchy problem (1.3) and (1.5). Since $K_\hbar(q', t'; q, t) = K_\hbar(q', q, T)$, where $T = t' - t$, we need to find the asymptotics of a fundamental solution $K_\hbar(q, q', T)$ as $\hbar \to 0$. The solution of this problem can be divided into two parts.

1. Find the short-wave asymptotics — asymptotics as $\hbar \to 0$ of the solution $\psi_\hbar(q, T)$ of the Cauchy problem for the Schrödinger equation

$$i\hbar \frac{\partial \psi}{\partial t} = -\frac{\hbar^2}{2m} \frac{\partial^2 \psi}{\partial q^2} + V(q)\psi$$

   with the initial condition

$$\psi_\hbar(q, t)|_{t=0} = \varphi(q) e^{\frac{i}{\hbar} s(q)},$$

where $s(q), \varphi(q) \in C^\infty(\mathbb{R}, \mathbb{R})$, and the amplitude $\varphi(q)$ has compact support.

2. Using the representation

$$\delta(q - q_0) = \frac{\varphi(q)}{2\pi\hbar} \int_{-\infty}^{\infty} e^{\frac{i}{\hbar}\xi(q-q_0)} d\xi,$$

where φ has compact support and satisfies $\varphi(q_0) = 1$, express $K_\hbar(Q, q_0, T)$ as an integral over $d\xi$ of solutions of the Schrödinger equation with initial amplitude $\varphi(q)$ and initial phase $s(q, \xi) = \xi(q - q_0)$. Use asymptotics from part **1** to evaluate the resulting integral by the stationary phase method as $\hbar \to 0$.

The problem in the first part was solved in Section 6.1 in Chapter 3, so we proceed to the second part. For fixed ξ, let $\psi_\hbar(q, t; \xi)$ be the solution of the Schrödinger equation with the initial amplitude $\varphi(q)$ and the initial phase $s(q) = \xi(q - q_0)$. By the superposition principle,

$$(6.6) \qquad K_\hbar(Q, q_0, T) = \frac{1}{2\pi\hbar} \int_{-\infty}^{\infty} \psi_\hbar(Q, T; \xi) d\xi.$$

Let $\gamma(t; q, \xi)$ be the classical trajectory — the solution of Newton's equation (6.2) with the initial conditions

$$(6.7) \qquad \gamma(0; q, \xi) = q \quad \text{and} \quad \dot{\gamma}(0; q, \xi) = \frac{\xi}{m}.$$

As in Section 2.3 of Chapter 1, here we assume that the mapping $q \mapsto Q = \gamma(T; q, \xi)$ is a diffeomorphism, and denote by $q = q(\xi, Q)$ the corresponding inverse function,

$$(6.8) \qquad \gamma(T; q(\xi, Q), \xi) = Q.$$

The asymptotics of $\psi_\hbar(Q, T; \xi)$ as $\hbar \to 0$ is given by formula (6.13) in Section 6.1 of Chapter 3, where $p(q) = \dfrac{\partial s}{\partial q} = \xi$, so that the characteristic ending at Q has the initial momentum ξ and the initial coordinate $q = q(\xi, Q)$. Substituting this expression into (6.6), we obtain

$$K_\hbar(Q, q_0; T) \simeq \frac{1}{2\pi\hbar} \int_{-\infty}^{\infty} \varphi(q(\xi, Q)) \left| \frac{\partial \gamma}{\partial q}(T; q(\xi, Q), \xi) \right|^{-\frac{1}{2}}$$

$$\times \exp\left\{ \frac{i}{\hbar}(S(Q, q(\xi, Q); T) + \xi(q(\xi, Q) - q_0)) \right\} d\xi.$$

To apply the stationary phase method to this integral, we need to find the critical points of the function $S(Q, q(\xi, Q); T) + \xi(q(\xi, Q) - q_0)$. Using the formula

$$\frac{\partial S}{\partial q}(Q, q; T) = -p = -\xi,$$

which follows from Proposition 2.2 in Section 2.3 of Chapter 1, we obtain

$$\frac{\partial}{\partial \xi}\left(S(Q, q(\xi, Q); T) + \xi(q(\xi, Q) - q_0)\right)$$

$$= -\xi \frac{\partial q}{\partial \xi}(\xi, Q) + q(\xi, Q) - q_0 + \xi \frac{\partial q}{\partial \xi}(\xi, Q) = q(\xi, Q) - q_0,$$

so that a single critical point ξ_0 is determined from the equation $q(\xi_0, Q) = q_0$. The prefactor in the stationary phase method is given by

$$\varphi(q_0)\left|\frac{\partial \gamma}{\partial q}(T; q_0, \xi_0)\frac{\partial q}{\partial \xi}(\xi_0, Q)\right|^{-\frac{1}{2}},$$

which can be simplified using $\varphi(q_0) = 1$ and the equation

$$\frac{\partial \gamma}{\partial \xi} + \frac{\partial \gamma}{\partial q}\frac{\partial q}{\partial \xi} = 0,$$

which follows from (6.8). Thus the final expression for the semi-classical asymptotics is

$$(6.9) \qquad K_\hbar(Q, q_0; T) \simeq \frac{1}{\sqrt{2\pi i \hbar}}\left|\frac{\partial \gamma}{\partial \xi}(T; q_0, \xi_0)\right|^{-\frac{1}{2}} e^{\frac{i}{\hbar}S(Q, q_0; T)}.$$

It is remarkable that formula (6.9), after the identification $Q = q'$ and $q_0 = q$, coincides with formula (6.1)! Indeed, they have exactly the same exponential factors, and equality of the corresponding prefactors follows from the following result.

Lemma 6.1. *Let $\gamma(t; q, \xi)$ be the classical trajectory with the initial conditions (6.7). Then*

$$(6.10) \qquad \frac{\partial \gamma}{\partial \xi}(T; q, \xi) = \frac{1}{2m}\det A,$$

where A is the differential operator $-D^2 - u(t)$ on the interval $[0, T]$ with Dirichlet boundary conditions, and $u(t) = \dfrac{V''(\gamma(t, q, \xi))}{m}$.

Proof. Differentiating equation (6.2) with respect to ξ, we obtain that $y(t) = \dfrac{\partial \gamma}{\partial \xi}(t; q, \xi)$ satisfies the differential equation

$$m\ddot{y} = -V''(\gamma(t, q, \xi))y,$$

so that $Ay = 0$. Differentiating initial conditions (6.7) with respect to ξ, we get

$$y(0) = 0 \quad \text{and} \quad \dot{y} = \frac{1}{m},$$

so that by Theorem 5.1, $\det A = 2my(T)$. $\qquad\square$

Remark. One can get another remarkable formula for the prefactor in representation (6.9). Namely, differentiating $\dfrac{\partial S}{\partial q}(Q, q; T) = -\xi$ with respect to Q we get

$$\frac{\partial^2 S}{\partial q \partial Q} = -\frac{\partial \xi}{\partial Q},$$

where $Q = \gamma(T; q, \xi)$. Thus (6.9) can be rewritten entirely in terms of the classical action

$$(6.11) \qquad K_\hbar(Q, q_0; T) \simeq \frac{1}{\sqrt{2\pi i\hbar}} \left| \frac{\partial^2 S}{\partial q \partial Q}(Q, q_0; T) \right|^{\frac{1}{2}} e^{\frac{i}{\hbar} S(Q, q_0; T)}.$$

Remark. When assumptions in Section 2.3 of Chapter 1 are not satisfied, there are several characteristics connecting points q_0 and Q, and the situation becomes more complicated. In this case we have

$$K_\hbar(Q, q_0, T) \simeq \sum_j \frac{1}{\sqrt{2\pi i\hbar}} \left| \frac{\partial \gamma}{\partial \xi}(T; q_0, \xi_j) \right|^{-\frac{1}{2}} e^{\frac{i}{\hbar} S(Q, q_0, \xi_j; T) - \frac{\pi i}{2} \mu_j},$$

where ξ_j is the initial momentum, and μ_j is the Morse index of the characteristic $\gamma(t; q_0, \xi_j)$ (see Section 6.1 in Chapter 3).

The case of n degrees of freedom is considered similarly. Using short-wave asymptotics (6.14) in Section 6.1 of Chapter 3, we obtain as $\hbar \to 0$,

$$(6.12) \qquad K_\hbar(\boldsymbol{Q}, \boldsymbol{q}_0, T) \simeq (2\pi i\hbar)^{-\frac{n}{2}} \left| \det \left(\frac{\partial \boldsymbol{Q}}{\partial \boldsymbol{\xi}}(\boldsymbol{q}_0, \boldsymbol{\xi}_0) \right) \right|^{-\frac{1}{2}} e^{\frac{i}{\hbar} S(\boldsymbol{Q}, \boldsymbol{q}_0; T)},$$

where $\boldsymbol{Q} = \gamma(t, \boldsymbol{q}_0, \boldsymbol{\xi}_0)$. Using the equation

$$\frac{\partial \boldsymbol{\xi}}{\partial \boldsymbol{Q}} = -\frac{\partial^2 S}{\partial \boldsymbol{q} \partial \boldsymbol{Q}} = - \left\{ \frac{\partial^2 S}{\partial q_i \partial Q_j} \right\}_{i,j=1}^n,$$

(6.12) can be rewritten as

$$K_\hbar(\boldsymbol{Q}, \boldsymbol{q}_0; T) \simeq (2\pi i\hbar)^{-\frac{n}{2}} \left| \det \left(\frac{\partial^2 S}{\partial \boldsymbol{q} \partial \boldsymbol{Q}}(\boldsymbol{Q}, \boldsymbol{q}_0; T) \right) \right|^{\frac{1}{2}} e^{\frac{i}{\hbar} S(\boldsymbol{Q}, \boldsymbol{q}_0; T)}.$$

Here $\det \dfrac{\partial^2 S}{\partial \boldsymbol{q} \partial \boldsymbol{Q}}$ is known as the *van Vleck determinant*.

Problem 6.1. Justify all computations in this section.

7. Notes and references

The path integral approach to quantum mechanics was developed by Feynman in his 1942 Princeton thesis, and was published in [**Fey48**]. In addition to the classic text [**FH65**] — the best introduction to Feynman path integrals in configuration space, written from a physics perspective — we also refer the reader to a modern textbook [**DR01**]. The Feynman path integral in the phase space was introduced by Feynman [**Fey51**] in 1951 and by Tobocman [**Tob56**] in 1956. Nowadays it is a very useful method in quantum field theory (see, e.g., Faddeev's Les Houches lectures [**Fad76**] and the monograph [**FS91**]). Our derivation of the Feynman path integral in Section 1 is standard and follows monograph [**RS75**], which also contains a complete proof of the Kato-Lie-Trotter formula. As we mentioned in Section 1, this approach establishes only the L^2-convergence of finite-dimensional approximations like (1.15) and (1.23) to the propagator. We refer the reader to the paper [**Fuj80**] for the proof of the convergence in other topologies on functional spaces, and to the monograph [**AHK76**] for the rigorous definition of Feynman path integrals as infinite-dimensional Fresnel integrals.

In Section 2 we follow the outline in [**Ber71a**] and [**Fad76, FS91**], and rigorously derive formulas (2.1) and (2.6) for the pq and qp-symbols of the evolution operator by using explicit relations between the symbols and the propagator. However, corresponding formulas (2.9) and (2.13) for the Weyl and Wick symbols of the evolution operator were derived only heuristically. As was already emphasized in [**Ber71a**], in this case one needs to justify formulas (2.8) and (2.11). This non-trivial problem was only recently solved in [**Dyn98**] for a large class of symbols, and we refer to this paper for further details and references. However, following [**Fad76, FS91**], we carefully treat boundary conditions for the Wick symbol, deriving the correct expression (2.14) (as opposed to the formula in [**Ber71a**]). Material in Section 3 is standard, and in our exposition we made a special emphasis on the details of computation related to the Morse index. The formula for the propagator of the harmonic oscillator in Proposition 3.1 is called the *Feynman-Souriau formula* in [**DR01**]. The relation between formula (3.6) for the propagator of the harmonic oscillator and the Mehler identity for Hermite-Tchebyscheff polynomials, was stated in [**FH65**]. For the answer to Problem 3.2 we refer the reader to [**Ber71a**] (after correcting the typos), and for the elegant heuristic solution of Problem 3.3 — to [**FS91**].

Besides having a conceptual meaning, the path integral formalism is a very convenient computational tool, since it allows us to use, albeit at a heuristic level, such methods of finite-dimensional integration as change of variables, integration by parts, and stationary phase approximation. There is a vast literature devoted to the applications of Feynman path integration in quantum physics. We only mention a perturbative expansion of the propagator using Feynman diagrams [**FH65**], which is nowadays the main computational method in quantum mechanics and quantum field theory (see also lectures [**Kaz99**] for the rigorous treatment of a finite-dimensional example). For more applications see [**DR01**], and references therein.

The idea of computing Gaussian path integrals by evaluating separately the classical contribution and the fluctuation factor goes back to [**FH65**], and Problems 4.1 and 4.4 are taken from this source. In Section 4 we emphasized the role of the zeta-function regularized determinants of differential operators, by considering the simplest example of the operator $A = -D^2$. We refer the reader to the textbook [**Apo76**] for the basic properties of the Riemann zeta-function. Our proof of Lemma 4.1 can be considered as a simplified one-dimensional analog of the derivation of the first Kronecker limit formula given in [**Lan87**]. For the case of Laplace operators on compact Riemannian manifolds, the definition (4.5) of a regularized determinant $\det' A$ (under the name *analytic torsion*) was given in [**RS71**], and the operator zeta-function $\zeta_A(s)$ was introduced in [**MP49**]. A similar notion of a perturbation determinant goes back to M.G. Krein [**Kre62**]; according to Problem 2.6 in Section 2.2 of Chapter 3, the transition coefficient $a(\sqrt{\lambda})$ is the perturbation determinant of $H - \lambda I$, where H is a one-dimensional Schrödinger operator. We also note that regularized determinants of differential operators have been extensively used in quantum field theory. The corresponding definition was given by V.A. Fock in 1937 and J. Schwinger in 1951, and is similar to (4.5). It is currently known [**IZ80**] as the *Fock-Schwinger proper time method.*

In general, the proof of meromorphic continuation of the zeta-function $\zeta_A(s)$ of an elliptic operator A, and its regularity at $s = 0$, uses the theory of complex powers A^{-s}, developed in [**See67**] (or the short-time asymptotics $t \to 0$ of $\operatorname{Tr} e^{-At}$ — the trace of a heat kernel of A — when A is non-negative [**Gil95**]). In Section 5 we are using an elementary approach to the short-time asymptotics of the heat kernel, which is based on the large n asymptotics of the eigenvalues of the corresponding boundary value Sturm-Liouville problems. The monograph [**LS91**] contains all the facts used in Sections 5.1 and 5.2. The coefficients $a_{-\frac{1}{2}}$ and a_0 in formula (5.4), and in their analogs for the periodic case, are called Seeley coefficients. Our proof of the key relation (5.13) and its analog for the periodic case uses Wronskian identities (5.11), and goes back to the papers [**Fad57, BF60**] on the trace identities for one-dimensional Schrödinger operators (see also Problem 2.6 in Section 2.2 of Chapter 3). The theorem that "the operator trace is equal to the matrix trace", used in Sections 5.1 and 5.2, as well as the properties of the Fredholm determinant, can be found in the classic monograph [**GK69**]. Problem 5.1 — the Gelfand-Levitan trace identity — is taken from the paper [**GL53**] (see also [**Dik58**] for its generalization). Properties of the Hurwitz zeta-function, used in Section 5.3, are proved in [**Apo76**]. For the general approach to characteristic determinants of the n-th order differential operators on the interval with matrix coefficients and Dirichlet or periodic boundary conditions, we refer to [**BFK91, BFK95**].

Our exposition in Section 6 follows [**GS77**], with simplifications arising from considering the case of one degree of freedom. Lemma 6.1 establishes the equivalence between the heuristic approach in Section 6.1 using path integrals, and the rigorous approach in Section 6.2 using short-wave asymptotics. We refer to [**GS77**] and [**MF81**] for the details. Also, see [**Foc78**] for the relation between the van Vleck determinant and canonical transformation in classical mechanics, and [**DR01**] for the examples.

Chapter 6

Integration in Functional Spaces

In the previous chapter we studied the propagator $K(\boldsymbol{q}', t; \boldsymbol{q}, 0)$ — the integral kernel of the evolution operator $U(t) = e^{-\frac{i}{\hbar} tH}$ — by using its representation by the Feynman path integral. Here we replace the physical time t by the Euclidean time $-it$, and study the integral kernel of the semigroup $e^{-\frac{1}{\hbar} tH}$ for $t > 0$ by using its representation by the Wiener integral.

1. Gaussian measures

Here we consider the simplest example of Gaussian measures, which are naturally defined for finite-dimensional and infinite-dimensional vector spaces, and are used in many areas of analysis and probability theory. The basic result of the Gaussian integration — the Wick theorem — is the main tool of the perturbation expansion in quantum mechanics and quantum field theory, conveniently expressed through Feynman diagrams.

1.1. Finite-dimensional case. Let A be a positive-definite, real symmetric $n \times n$ matrix. A basic formula of Gaussian integration is

$$(1.1) \qquad \int_{\mathbb{R}^n} e^{-\frac{1}{2}(A\boldsymbol{q}, \boldsymbol{q})} d^n \boldsymbol{q} = \sqrt{\frac{(2\pi)^n}{\det A}}$$

(see Lemma 3.1 in Section 3.1 of Chapter 5). The corresponding *Gaussian measure* associated with the matrix A is a probability measure μ_A on $\mathbb{R}^n$, defined by

$$(1.2) \qquad d\mu_A(\boldsymbol{q}) = \sqrt{\frac{\det A}{(2\pi)^n}}\, e^{-\frac{1}{2}(A\boldsymbol{q},\boldsymbol{q})} d^n\boldsymbol{q}.$$

The measure μ_A is a mean-zero probability measure on $\mathbb{R}^n$ with the co-variance $G = A^{-1}$. When $A = I_n$ — the $n \times n$ identity matrix — the corresponding measure is denoted by μ_n.

As follows from Lemma 3.1 in Section 3.1 of Chapter 5,

$$(1.3) \qquad \int_{\mathbb{R}^n} e^{(\boldsymbol{p},\boldsymbol{q})} d\mu_A(\boldsymbol{q}) = e^{\frac{1}{2}(G\boldsymbol{p},\boldsymbol{p})},$$

and by analytic continuation,

$$(1.4) \qquad \int_{\mathbb{R}^n} e^{i(\boldsymbol{p},\boldsymbol{q})} d\mu_A(\boldsymbol{q}) = \lim_{R\to\infty} \int_{\|\boldsymbol{q}\|\leq R} e^{i(\boldsymbol{p},\boldsymbol{q})} d\mu_A(\boldsymbol{q}) = e^{-\frac{1}{2}(G\boldsymbol{p},\boldsymbol{p})}.$$

The function $(2\pi)^{-\frac{n}{2}} e^{-\frac{1}{2}\langle G\boldsymbol{p},\boldsymbol{p}\rangle}$ is a Fourier transform of the measure μ_A.

Theorem 1.1 (Wick theorem).

$$\int_{\mathbb{R}^n} (\boldsymbol{v}_1,\boldsymbol{q})\ldots(\boldsymbol{v}_N,\boldsymbol{q}) d\mu_A(\boldsymbol{q}) = \begin{cases} 0, & N \text{ is odd,} \\ \sum (G\boldsymbol{v}_{i_1},\boldsymbol{v}_{i_2})\ldots(G\boldsymbol{v}_{i_{N-1}},\boldsymbol{v}_{i_N}), & N \text{ is even,} \end{cases}$$

where the sum goes over all possible pairings $(i_1,i_2),\ldots,(i_{N-1},i_N)$ — *all partitions in pairs of the set* $\{1,2,\ldots,N\}$.

Proof. Applying to (1.3) the directional derivative along the vector $\boldsymbol{v} \in \mathbb{R}^n$ — the differential operator

$$\partial_{\boldsymbol{v}} = \boldsymbol{v}\frac{\partial}{\partial\boldsymbol{p}} = \sum_{k=1}^n v_k \frac{\partial}{\partial p_k}$$

(differentiation under the integral sign is clearly legitimate), we obtain

$$(1.5) \qquad \int_{\mathbb{R}^n} (\boldsymbol{v},\boldsymbol{q}) e^{(\boldsymbol{p},\boldsymbol{q})} d\mu_A(\boldsymbol{q}) = (G\boldsymbol{v},\boldsymbol{p}) e^{\frac{1}{2}(G\boldsymbol{p},\boldsymbol{p})}.$$

Setting here $\boldsymbol{p} = 0$ we get

$$\int_{\mathbb{R}^n} (\boldsymbol{v},\boldsymbol{q}) d\mu_A(\boldsymbol{q}) = 0,$$

while applying to (1.5) another $\partial_{\boldsymbol{v}'}$ and setting $\boldsymbol{p} = 0$ afterwards, we obtain

$$(1.6) \qquad \int_{\mathbb{R}^n} (v,\boldsymbol{q})(v',\boldsymbol{q}) d\mu_A(\boldsymbol{q}) = (G\boldsymbol{v},\boldsymbol{v}').$$

To get the general result, we differentiate (1.3) N times along the vectors $\boldsymbol{v}_1,\ldots,\boldsymbol{v}_N$ and set $\boldsymbol{p} = 0$. $\qquad\square$

Problem 1.1. Let $v_i, w_j \in \mathbb{R}^n$ be such that $(Av_i, w_j) = 0$, $i, j = 1, \ldots, N$, and let F and H be bounded measurable functions on $\mathbb{R}^N$. Show that the functions $f(q) = F((v_1, q), \ldots, (v_N, q))$ and $h(q) = H((w_1, q), \ldots, (w_N, q))$ satisfy

$$\int_{\mathbb{R}^n} f(q)h(q)d\mu_A(q) = \int_{\mathbb{R}^n} f(q)d\mu_A(q) \int_{\mathbb{R}^n} h(q)d\mu_A(q).$$

1.2. Infinite-dimensional case. Let $\mathscr{V} = \mathbb{R}^\infty$ be the Cartesian product of countably many copies of $\mathbb{R}$ equipped with the Tychonoff topology, and let

$$\mathscr{H} = \ell^2(\mathbb{R}) = \left\{ x = \{x_i\}_{i=1}^\infty \in \mathscr{V} : \|x\|^2 = \sum_{i=1}^\infty x_i^2 < \infty \right\}$$

be the real Hilbert space with the scalar product $(x, y) = \sum_{i=1}^\infty x_i y_i$. In particular, the Hilbert space $\mathscr{H}$ contains all elements of finite support: the elements $x \in \mathscr{V}$ such that $x_i = 0$ for sufficiently large i.

The Gaussian measure μ on $\mathscr{V}$ is defined by the direct product of the Gaussian measures μ_1,

$$\mu = \mu_\infty = \mu_1 \times \mu_1 \times \cdots \times \mu_1 \times \cdots .$$

More precisely, the measure μ is defined as follows. Let $\mathscr{C}$ be the set of cylindrical subsets of $\mathscr{V}$: $C \in \mathscr{C}$ if $C = p_n^{-1}(E_1 \times \cdots \times E_n)$ for some n, where $p_n : \mathscr{V} \to \mathbb{R}^n$ is the projection on the Cartesian product of the first n factors, and $E_1, \ldots, E_n$ are Borel subsets of $\mathbb{R}$. Then we set

$$\mu(C) = \mu_1(E_1) \ldots \mu_1(E_n),$$

and extend μ to the whole σ-algebra generated by $\mathscr{C}$ by using the Kolmogoroff extension theorem. In particular, if $F(x) = f(x_1, \ldots, x_n)$, where f is a bounded measurable function on $\mathbb{R}^n$, then

$$(1.7) \qquad \int_{\mathscr{V}} F d\mu = \int_{\mathbb{R}^n} f d\mu_n.$$

Equivalently, the Gaussian measure μ is characterized by the following property.

Lemma 1.1. *The measure μ is a unique probability measure on $\mathscr{V}$ such that for all $v \in \mathscr{V}$ with finite support,*

$$\int_{\mathscr{V}} e^{i(v,x)} d\mu(x) = e^{-\frac{1}{2}\|v\|^2}.$$

Proof. The proof immediately follows from (1.4), since measures μ_n are uniquely determined by their Fourier transforms. $\qquad \square$

Remark. The Gaussian measure μ can be heuristically represented by

$$d\mu = (2\pi)^{-\infty} e^{-\frac{1}{2}\|x\|^2} \prod_{i=1}^\infty dx_i.$$

Here the "divergent to 0" product $(2\pi)^{-\infty}e^{-\frac{1}{2}\|x\|^2}$" compensates the "divergent to ∞" product $\prod_{i=1}^{\infty} dx_i$.

Now for $\alpha = \{\alpha_i\} \in \mathscr{V}$ let

$$\mathscr{H}_\alpha = \left\{ x \in \mathscr{V} : \sum_{i=1}^{\infty} \alpha_i^2 x_i^2 < \infty \right\}.$$

The following result is a version of Kolmogoroff's celebrated 0–1 law in probability theory.

Proposition 1.1.

$$\mu(\mathscr{H}_\alpha) = \begin{cases} 0 & \text{if } \alpha \notin \mathscr{H}, \\ 1 & \text{if } \alpha \in \mathscr{H}. \end{cases}$$

In particular, $\mu(\mathscr{H}) = 0$.

Proof. Let χ_α be the characteristic function of the set $\mathscr{H}_\alpha \subset \mathscr{V}$,

$$\chi_\alpha(x) = \lim_{\varepsilon \to 0} \lim_{n \to \infty} \exp \left\{ -\varepsilon^2 \sum_{i=1}^{n} \alpha_i^2 x_i^2 \right\}.$$

Twice applying the dominated convergence theorem, we get

$$\mu(\mathscr{H}_\alpha) = \int_{\mathscr{V}} \chi_\alpha d\mu = \lim_{\varepsilon \to 0} \lim_{n \to \infty} \int_{\mathbb{R}^n} \exp \left\{ -\varepsilon^2 \sum_{i=1}^{n} \alpha_i^2 x_i^2 \right\} d\mu_n(x)$$

$$= \lim_{\varepsilon \to 0} \lim_{n \to \infty} \prod_{i=1}^{n} (1 + \varepsilon^2 \alpha_i^2)^{-1/2},$$

and the product $\prod_{i=1}^{\infty}(1 + \varepsilon^2 \alpha_i^2)$ is convergent if and only if $\alpha \in \mathscr{H}$. $\qquad\square$

Remark. For $v \in \mathscr{H}$ let $v^{(n)} = (v_1, \ldots, v_n, 0, 0, \ldots)$. It follows from (1.6) that

$$\int_{\mathscr{V}} (v^{(n)}, x)^2 d\mu(x) = \|v^{(n)}\|^2,$$

so that the sequence of functions $F_n(x) = (v^{(n)}, x)$, $x \in \mathscr{V}$, is a Cauchy sequence in $L^2(\mathscr{V}, d\mu)$, and it converges in L^2 to the function $F(x)$. Abusing notation, we write $F(x) = (v, x) \in L^2(\mathscr{V}, d\mu)$. Thus though $\mu(\mathscr{H}) = 0$, Lemma 1.1 and, consequently, Wick's theorem, hold for $v \in \mathscr{H}$.

Problem 1.2. Prove that there is no probability measure μ on $\mathscr{H}$ such that $\mu(C) = \mu_n(E)$ for every cylindrical subset $C = p_n^{-1}(E)$ of $\mathscr{H}$, where $p_n : \mathscr{H} \to \mathbb{R}^n$ is the natural projection, and E is a Borel subset of $\mathbb{R}^n$. However, show that there exists a finitely-additive, non-negative function ν on cylindrical subsets of $\mathscr{H}$ satisfying this property.

Problem 1.3. Show that the formula

$$\nu(\{x \in \mathscr{V} : x_1 \in I_1, \ldots, x_n \in I_n\}) = \prod_{k=1}^{n} \mu_1(kI_k),$$

where $kI_k = (k\alpha_k, k\beta_k)$ for $I_k = (\alpha_k, \beta_k) \subseteq \mathbb{R}$, defines a probability measure ν on $\mathscr{V}$ such that $\nu(\mathscr{H}) = 1$.

2. Wiener measure and Wiener integral

2.1. Definition of the Wiener measure.

Here we define the probability measure on the space $\mathscr{C} = C([0, \infty), \mathbb{R}^n; 0)$ of parametrized continuous paths in $\mathbb{R}^n$ starting at the origin, called the *Wiener measure*. It is related to the Brownian motion: the diffusion process in $\mathbb{R}^n$ with diffusion coefficient $D > 0$, which is described by the probability density

$$(2.1) \qquad P(\boldsymbol{q}', \boldsymbol{q}; t) = (4\pi Dt)^{-\frac{n}{2}} e^{-\frac{(q - q')^2}{4Dt}}$$

that a particle with initial certainty of being at point $\boldsymbol{q} \in \mathbb{R}^n$ is diffused in time t to point $\boldsymbol{q}' \in \mathbb{R}^n$.

It will be convenient to compactify $\mathbb{R}^n$ by adding a point at infinity, $\widehat{\mathbb{R}}^n = \mathbb{R}^n \cup \{\infty\} \simeq S^n$. Let

$$\Omega = \prod_{0 \leq t < \infty} \widehat{\mathbb{R}}^n$$

be the Cartesian product of copies of $\widehat{\mathbb{R}}^n$ parametrized by $\mathbb{R}_{\geq 0}$. Equipped with the Tychonoff topology, Ω is a compact topological space — the space of all parametrized paths in $\widehat{\mathbb{R}}^n$. For every partition $\boldsymbol{t}_m = \{0 \leq t_1 \leq \cdots \leq t_m\}$ and every $F \in C(\underbrace{\widehat{\mathbb{R}}^n \times \cdots \times \widehat{\mathbb{R}}^n}_{m})$, define $\varphi \in C(\Omega)$ by

$$\varphi(\boldsymbol{\gamma}) = F(\boldsymbol{\gamma}(t_1), \ldots, \boldsymbol{\gamma}(t_m)) \quad \text{for all} \quad \boldsymbol{\gamma} \in \Omega.$$

We denote by $C_{\mathrm{fin}}(\Omega)$ the subspace of $C(\Omega)$ spanned by the functions φ for all partitions $\boldsymbol{t}_m$ and for all continuous functions F. Define a linear functional l on $C_{\mathrm{fin}}(\Omega)$ by the following formula:

$$(2.2) \qquad l(\varphi) = \int_{\mathbb{R}^n} \cdots \int_{\mathbb{R}^n} F(\boldsymbol{q}_1, \ldots, \boldsymbol{q}_m) P(\boldsymbol{q}_m, \boldsymbol{q}_{m-1}; t_m - t_{m-1}) \cdots$$
$$\cdots P(\boldsymbol{q}_1, 0; t_1) d^n \boldsymbol{q}_1 \ldots d^n \boldsymbol{q}_m.$$

It follows from the semi-group property

$$\int_{\mathbb{R}^n} P(\boldsymbol{q}', \boldsymbol{q}_1; t' - t_1) P(\boldsymbol{q}_1, \boldsymbol{q}; t_1 - t) d^n \boldsymbol{q}_1 = P(\boldsymbol{q}', \boldsymbol{q}; t' - t)$$

(also called Kolmogoroff's equation in probability theory) that the functional l is well defined. The functional l is positive: $l(\varphi) \geq 0$ for $\varphi \geq 0$, satisfies $l(1) = 1$, and

$$|l(\varphi)| \leq \|\varphi\|_\infty = \sup_{\gamma \in \Omega} |\varphi(\gamma)|.$$

The subspace $C_{\text{fin}}(\Omega)$ separates points in Ω and $1 \in C_{\text{fin}}(\Omega)$, so that by the Stone-Weierstrass theorem $C_{\text{fin}}(\Omega)$ is dense in $C(\Omega)$. Now the functional l has a unique extension to a continuous positive linear functional on $C(\Omega)$ with norm 1, and by the Riesz-Markoff theorem, there exists a unique regular Borel measure μ_W on Ω with $\mu_W(\Omega) = 1$ such that

$$l(\varphi) = \int_\Omega \varphi \, d\mu_W.$$

The measure μ_W is called the Wiener measure. The integral over the Wiener measure is called the *Wiener integral.*

Remark. The Riesz-Markoff theorem provides a natural way of defining measures in various problems of functional analysis. In general, it ensures the existence of a Baire measure — a measure defined on the σ-algebra of Baire sets. However, for compact spaces a Baire measure has a unique extension to a regular Borel measure — a measure defined on the σ-algebra generated by all open subsets. A Borel measure μ is regular, if for every Borel set $E \subseteq \Omega$,

$$\mu(E) = \begin{cases} \inf \mu(U), & E \subseteq U, \ U \text{ is open}, \\ \sup \mu(K), & K \subseteq E, \ K \text{ is compact and Borel.} \end{cases}$$

The space Ω is "so large" that its σ-algebras of Baire and Borel sets are different.

Proposition 2.1. *The Wiener measure μ_W is supported on continuous paths starting at the origin, i.e., $\mu_W(\mathscr{C}) = 1$.*

Replacing the probability density $P(\boldsymbol{q}_1, 0; t_1)$ by $P(\boldsymbol{q}_1, \boldsymbol{q}_0; t_1)$ in definition (2.2), for a fixed $\boldsymbol{q}_0 \in \mathbb{R}^n$ we obtain a Wiener measure $\mu_{\boldsymbol{q}_0}$, which is supported on the space $\mathscr{C}_{\boldsymbol{q}_0} = C([0, \infty), \mathbb{R}^n; \boldsymbol{q}_0)$ of continuous paths in $\mathbb{R}^n$ that start at $\boldsymbol{q}_0$.

Remark. The support of the Wiener measure μ_W can be characterized more precisely as follows. For $0 < \alpha \leq 1$ let Ω_α be the subspace of Ω of Hölder continuous paths of order α:

$$\Omega_\alpha = \left\{ \gamma \in \Omega : \sup_{t, t' \geq 0} \frac{\|\gamma(t) - \gamma(t')\|}{|t - t'|^\alpha} < \infty \right\}.$$

Then

$$\mu_W(\Omega_\alpha) = \begin{cases} 1, & \text{if } 0 < \alpha < \frac{1}{2}, \\ 0, & \text{if } \frac{1}{2} \le \alpha \le 1. \end{cases}$$

Remark. It seems natural to define Wiener measure by the following construction. Set for simplicity $n = 1$, and for every partition $\boldsymbol{t}_m = \{0 \le t_1 \le \cdots \le t_m\}$ and intervals $(\alpha_1, \beta_1), \ldots, (\alpha_m, \beta_m)$, define the measure of a cylindrical set

$$C_{\boldsymbol{t}} = \{\gamma \in \Omega : \alpha_1 < \gamma(t_1) < \beta_1, \ldots, \alpha_m < \gamma(t_m) < \beta_m\}$$

by the formula

$$(2.3) \quad \mu(C_{\boldsymbol{t}}) = \int_{\alpha_1}^{\beta_1} \cdots \int_{\alpha_m}^{\beta_m} P(q_m, q_{m-1}; t_m - t_{m-1}) \ldots P(q_1, 0; t_1) dq_1 \ldots dq_m.$$

By the Kolmogoroff extension theorem and the semi-group property, μ extends to a measure on the σ-algebra generated by the cylindrical subsets of Ω, which we continue to denote by μ. However, the set $\mathscr{C}$ of continuous paths starting at 0 turns out to be non-measurable! Specifically, one can show that

$$\mu_*(\mathscr{C}) = 0 \quad \text{and} \quad \mu^*(\mathscr{C}) = 1,$$

where $\mu_*(E)$ and $\mu^*(E)$ denote, respectively, the inner and outer measures of the subset $E \subseteq \Omega$. To remedy this situation, one should from the beginning define cylindrical sets $C_{\boldsymbol{t}}$ as consisting of continuous paths only, and define $\mu(C_{\boldsymbol{t}})$ by the same formula as above. Then the measure μ extends to a σ-algebra generated by cylindrical subsets of $\mathscr{C}$, and coincides with the Wiener measure μ_W.

Remark. The same formula (2.3) can be used to define the Wiener measure on the space $C([0, \pi], \mathbb{R}; 0)$ of continuous functions on the interval $[0, \pi]$ which vanish at $t = 0$; in this case $\boldsymbol{t}_m$ is a partition of $[0, \pi]$. This is the classical definition of the Wiener measure given by Wiener himself.

The following result will be used for representing the integral kernel of the one-parameter semigroup $e^{-\frac{1}{\hbar}tH}$ for $t > 0$ by the Wiener integral.

Proposition 2.2. *Let the real-valued function $V \in C(\mathbb{R}^n)$ be bounded below. Then for every $t \ge 0$ the function $\mathcal{F}_t : \mathscr{C} \to \mathbb{R}$, defined by*

$$\mathcal{F}_t(\boldsymbol{\gamma}) = e^{-\int_0^t V(\boldsymbol{\gamma}(\tau))d\tau},$$

is integrable with respect to the Wiener measure, and

$$\int_{\mathscr{C}} \mathcal{F}_t \, d\mu_W = \lim_{N \to \infty} \int_{\mathbb{R}^n} \cdots \int_{\mathbb{R}^n} \exp\left\{-\sum_{k=1}^{N} V(\boldsymbol{q}_k)\Delta t\right\} P(\boldsymbol{q}_N, \boldsymbol{q}_{N-1}; \Delta t) \cdots$$

$$\cdots P(\boldsymbol{q}_1, 0; \Delta t) d^n \boldsymbol{q}_1 \ldots d^n \boldsymbol{q}_N, \quad \Delta t = \frac{t}{N}.$$

Proof. For $\gamma \in \mathscr{C}$,

$$\int_0^t V(\gamma(\tau))d\tau = \lim_{N \to \infty} \sum_{k=1}^N V(\gamma(t_k))\Delta t,$$

where $t_k = k\Delta t$. Since by definition every function $\sum_{k=1}^N V(\gamma(t_k))\Delta t$ is measurable with respect to the Wiener measure on $\mathscr{C}$, the function $\mathcal{F}_t$ is measurable as a point-wise limit of a sequence of measurable functions. The function $\mathcal{F}_t$ is bounded and, therefore, is integrable on $\mathscr{C}$ with respect to the Wiener measure. Finally, by the dominated convergence theorem,

$$\int_{\mathscr{C}} \mathcal{F}_t \, d\mu_W = \lim_{N \to \infty} \int_{\mathscr{C}} \exp\left\{ -\sum_{k=1}^N V(\gamma(t_k))\Delta t \right\} d\mu_W(\gamma),$$

and the result follows from (2.2). $\qquad\qquad\qquad\qquad\qquad\qquad\qquad\qquad\square$

Remark. Note the limit in Proposition 2.2 exists because the function $\mathcal{F}_t$ is integrable, and not the other way around. This is similar to an elementary calculus argument that the limit

$$\lim_{n \to \infty} \left(1 + \frac{1}{2} + \cdots + \frac{1}{n} - \log n \right)$$

exists because the integral

$$\int_0^1 \left(\frac{1}{x} - \left[\frac{1}{x} \right] \right) dx$$

is convergent. Here $[x]$ stands for the largest integer not greater than x.

Problem 2.1. Prove all statements about the support of Wiener measure.

Problem 2.2. Construct the Wiener measure by defining it on the cylindrical subsets of $\mathscr{C}$ by formula (2.3).

Problem 2.3. Prove that the Wiener measure μ_W on $C([0, \pi], \mathbb{R}; 0)$ for $D = \frac{1}{4}$ actually coincides with the measure ν defined in Problem 1.3. (*Hint:* Show that the mapping $\gamma(t) \mapsto \gamma(t) - \gamma(0)$ establishes the isomorphism between the space of $\hat{C}([0, \pi], \mathbb{R})$ of functions orthogonal to 1 and $C([0, \pi], \mathbb{R}; 0)$, and use the Fourier sine coefficients for the embedding $\hat{C}([0, \pi], \mathbb{R}) \hookrightarrow \mathscr{V}$.)

2.2. Conditional Wiener measure and Feynman-Kac formula. Let

$$\Omega_{q,q'} = \left\{ \gamma \in \prod_{t \leq \tau \leq t'} \widehat{\mathbb{R}}^n : \gamma(t) = q, \gamma(t') = q' \right\}$$

be the space of all parametrized paths in $\widehat{\mathbb{R}}^n$ which start at $q \in \mathbb{R}^n$ at time t and end at $q' \in \mathbb{R}^n$ at t', and let $\mathscr{C}_{q,q'}$ be the corresponding subspace of continuous paths. The conditional Wiener measure $\mu_{q,q'}$ on $\Omega_{q,q'}$ is defined similarly. We replace a positive linear functional l on $C(\Omega)$ by a positive linear functional $l_{\boldsymbol{q},\boldsymbol{q}'}$ on $C(\Omega_{\boldsymbol{q},\boldsymbol{q}'})$, which for $\varphi \in C_{\mathrm{fin}}(\Omega_{\boldsymbol{q},\boldsymbol{q}'})$ is defined by

$$l_{\boldsymbol{q},\boldsymbol{q}'}(\varphi) = \int_{\mathbb{R}^n} \cdots \int_{\mathbb{R}^n} F(\boldsymbol{q}_1,\ldots,\boldsymbol{q}_m) P(\boldsymbol{q}',\boldsymbol{q}_m; t'-t_m) \ldots$$
$$\ldots P(\boldsymbol{q}_1,\boldsymbol{q}; t_1-t) d^n\boldsymbol{q}_1 \ldots d^n\boldsymbol{q}_m,$$

where $t \leq t_1 \leq \cdots \leq t_m \leq t'$ and $\varphi(\boldsymbol{\gamma}) = F(\boldsymbol{\gamma}(t_1),\ldots,\boldsymbol{\gamma}(t_m))$. Then

$$l_{\boldsymbol{q},\boldsymbol{q}'}(\varphi) = \int_{\Omega_{\boldsymbol{q},\boldsymbol{q}'}} \varphi \, d\mu_{\boldsymbol{q},\boldsymbol{q}'}.$$

As in the case of the Wiener measure μ_W, the conditional Wiener measure $\mu_{\boldsymbol{q},\boldsymbol{q}'}$ is supported on continuous paths and

$$\mu_{\boldsymbol{q},\boldsymbol{q}'}(\mathscr{C}_{\boldsymbol{q},\boldsymbol{q}'}) = P(\boldsymbol{q}',\boldsymbol{q}; t'-t).$$

Let

$$H = H_0 + V = \frac{\boldsymbol{P}^2}{2m} + V(\boldsymbol{Q})$$

be the Schrödinger operator on $L^2(\mathbb{R}^n, d^n\boldsymbol{q})$ with continuous, real-valued and bounded below potential $V(\boldsymbol{q})$. Denote by $L_\hbar(\boldsymbol{q}',t'; \boldsymbol{q},t)$, $t' > t$, the heat kernel — the integral kernel of the diffusion operator $e^{-\frac{t'-t}{\hbar}H}$. Here is the main result of this section.

Theorem 2.1 (Feynman-Kac formula).

$$L_\hbar(\boldsymbol{q}',t'; \boldsymbol{q},t) = \int_{\mathscr{C}_{\boldsymbol{q},\boldsymbol{q}'}} e^{-\frac{1}{\hbar}\int_t^{t'} V(\boldsymbol{\gamma}(\tau))d\tau} d\mu_{\boldsymbol{q},\boldsymbol{q}'}(\boldsymbol{\gamma}),$$

where $\mu_{\boldsymbol{q},\boldsymbol{q}'}$ is the conditional Wiener measure.

Proof. We have by the Lie-Kato-Trotter product formula

$$e^{-\frac{1}{\hbar}TH} = \lim_{N \to \infty} (e^{-\frac{\Delta t}{\hbar}H_0} e^{-\frac{\Delta t}{\hbar}V})^N, \quad \Delta t = \frac{T}{N},$$

where $T = t' - t$. Let $L_\hbar^{(N)}(\boldsymbol{q}',t'; \boldsymbol{q},t)$ be the integral kernel of the operator $(e^{-\frac{\Delta t}{\hbar}H_0} e^{-\frac{\Delta t}{\hbar}V})^N$. Computing it as in Section 1.3 of Chapter 5, and using the definition of the conditional Wiener measure we obtain

$$L_\hbar^{(N)}(\boldsymbol{q}',t'; \boldsymbol{q},t) = \int_{\mathscr{C}_{\boldsymbol{q},\boldsymbol{q}'}} \exp\left\{ -\frac{1}{\hbar} \sum_{k=1}^N v(\boldsymbol{\gamma}(t_k))\Delta t \right\} d\mu_{\boldsymbol{q},\boldsymbol{q}'}(\boldsymbol{\gamma}).$$

Evaluating the limit $N \to \infty$ by the dominated convergence theorem completes the proof. $\qquad\square$

Using the conditional Wiener measure it is easy to define a measure μ_W^{loop} on the space

$$\mathcal{L} = \left\{ \gamma \in \prod_{t \leq \tau \leq t'} \widehat{\mathbb{R}}^n : \gamma(t) = \gamma(t') \right\}$$

of free parametrized loops in $\widehat{\mathbb{R}}^n$. Namely, the space $\mathcal{L}$ is a disjoint union over $\boldsymbol{q} \in \widehat{\mathbb{R}}^n$ of the spaces $\Omega_{\boldsymbol{q},\boldsymbol{q}}$ of the parametrized loops based at $\boldsymbol{q} \in \widehat{\mathbb{R}}^n$. By definition, a function $\varphi : \mathcal{L} \to \mathbb{C}$ is called integrable, if its restrictions $\varphi|_{\Omega_{\boldsymbol{q},\boldsymbol{q}}}$ are measurable with respect to $\mu_{\boldsymbol{q},\boldsymbol{q}}$ for all $\boldsymbol{q} \in \mathbb{R}^n$, and if the function $\int_{\Omega_{\boldsymbol{q},\boldsymbol{q}}} \varphi \, d\mu_{\boldsymbol{q},\boldsymbol{q}} : \mathbb{R}^n \to \mathbb{C}$ is Lebesgue integrable on $\mathbb{R}^n$. Then

$$(2.4) \qquad \int_{\mathcal{L}} \varphi \, d\mu_W^{\text{loop}} = \int_{\mathbb{R}^n} \int_{\Omega_{\boldsymbol{q},\boldsymbol{q}}} \varphi \, d\mu_{\boldsymbol{q},\boldsymbol{q}} \, d^n \boldsymbol{q}.$$

Suppose that the Schrödinger operator $H = H_0 + V$ with continuous, real-valued, and bounded below potential $V(\boldsymbol{q})$ has a pure point spectrum, and that $e^{-\frac{T}{\hbar}H}$ is of trace class[1]. The following result is a useful corollary of the Feynman-Kac formula.

Corollary 2.2.

$$\operatorname{Tr} e^{-\frac{T}{\hbar}H} = \int_{\mathcal{L}} e^{-\frac{1}{\hbar}\int_0^T V(\gamma(t))dt} d\mu_W^{\text{loop}}(\gamma).$$

Proof. The heat kernel $L_\hbar(\boldsymbol{q}', T; \boldsymbol{q}, 0)$ — the integral kernel of the trace class operator $e^{-\frac{T}{\hbar}H}$ — is a continuous function of $\boldsymbol{q}$ and $\boldsymbol{q}'$, so that by the theorem[2] that "the operator trace equals the matrix trace", used in Sections 5.1–5.2 of Chapter 5, we get

$$\operatorname{Tr} e^{-\frac{T}{\hbar}H} = \int_{\mathbb{R}^n} L_\hbar(\boldsymbol{q}, T; \boldsymbol{q}, 0) d^n \boldsymbol{q}.$$

The result now follows from the Feynman-Kac formula and equation (2.4).

$\square$

Remark. The Wiener measure on the loop space

$$\mathcal{L} = \{\gamma \in C([0,T], \mathbb{R}) : \gamma(0) = \gamma(T)\}$$

in terms of the Fourier coefficients

$$\gamma(t) = \sum_{n=-\infty}^{\infty} c_n e^{\frac{2\pi i n t}{T}}, \quad \bar{c}_n = c_{-n},$$

can be expressed as follows:

$$(2.5) \qquad d\mu_W^{\text{loop}} = \sqrt{\frac{m}{2\pi\hbar T}} \, dc_0 \prod_{n=1}^{\infty} \left(\frac{m}{\pi\hbar T}\right)(2\pi n)^2 e^{-\frac{m}{\hbar T}(2\pi n)^2|c_n|^2} d^2 c_n.$$

---

[1] This is the case when $V(\boldsymbol{q}) \to \infty$ as $\boldsymbol{q} \to \infty$ "fast enough", i.e., $V(q) \geq C\|\boldsymbol{q}\|^2$.
[2] See Section 7 of Chapter 5 for the reference.

In particular,

$$(2.6) \qquad \int_{\mathcal{L}} f(\gamma) d\mu_W^{\mathrm{loop}}(\gamma) = \sqrt{\frac{m}{2\pi\hbar T}} \int_{-\infty}^{\infty} \left(\int_{\mathcal{L}_0} f(x,\gamma_0)\, d\mu_W^{\mathrm{loop}}(\gamma_0) \right) dx.$$

Here $\mathcal{L}_0$ is the space of loops γ_0 with $c_0 = 0$, and $\mathcal{L} = \mathbb{R} \times \mathcal{L}_0$ according to the decomposition $\gamma = x + \gamma_0$, where $\int_0^T \gamma_0(t)dt = 0$ so that $x = c_0(\gamma)$.

Problem 2.4. Deduce from the Feynman-Kac formula that for every $\psi \in L^2(\mathbb{R}^n)$,

$$(e^{-\frac{1}{\hbar}T(H_0+V)}\psi)(\boldsymbol{q}) = \int_{\mathscr{C}_q} \psi(\boldsymbol{\gamma}(t))e^{-\frac{1}{\hbar}\int_0^T V(\boldsymbol{\gamma}(t))dt} d\mu_q(\boldsymbol{\gamma}).$$

Problem 2.5. Prove formula (2.5) and deduce (2.6) from it. (*Hint:* Use Problem 2.3.)

2.3. Relation between Wiener and Feynman integrals. It is very instructive to compare Feynman and Wiener integrals. Heuristically, the conditional Wiener measure can be written as

$$(2.7) \qquad \mathscr{D}\boldsymbol{q}_W = e^{-\frac{m}{2\hbar}\int_t^{t'} \dot{\boldsymbol{q}}^2 d\tau} \mathscr{D}_\hbar \boldsymbol{q},$$

where

$$(2.8) \qquad \mathscr{D}_\hbar \boldsymbol{q} = \lim_{N\to\infty} \left(\frac{m}{2\pi\hbar\Delta t}\right)^{\frac{N}{2}} \prod_{k=1}^{N-1} d^n \boldsymbol{q}_k.$$

Of course, neither the "measure" $\mathscr{D}_\hbar \boldsymbol{q}$ exists: the infinite product (2.8) is divergent, nor the trajectory $\boldsymbol{\gamma} = \boldsymbol{q}(\tau)$ is, in general, differentiable: the integral $\int_t^{t'} \dot{\boldsymbol{q}}^2 d\tau$ is divergent. However, due to the presence of a negative sign in the exponential, we have the indeterminate form "infinity over infinity", and the resulting expression (2.7) can be given a precise meaning as the Wiener measure. The corresponding "measure" for the Feynman path integral is obtained by replacing $\hbar$ by $i\hbar$ in (2.7)–(2.8), and the exponential factor no longer compensates for the divergence of $\mathscr{D}_{i\hbar}\boldsymbol{q}$, since it is a complex number of modulus 1 for a differentiable trajectory, and has no meaning for a non-differentiable one.

However, as we have seen in Sections 1.3–1.4 of Chapter 5, the Feynman path integral allows us to represent the propagator of a quantum particle in a profound way, which shows a deep relation between quantum and classical mechanics. Namely,

$$K_\hbar(\boldsymbol{q}',t';\boldsymbol{q},t) = \int_{\left\{\begin{smallmatrix} \boldsymbol{q}(t')=\boldsymbol{q}' \\ \boldsymbol{q}(t)=\boldsymbol{q} \end{smallmatrix}\right\}} e^{\frac{i}{\hbar}S(\boldsymbol{\gamma})} \mathscr{D}_{i\hbar}\boldsymbol{q},$$

where $S(\boldsymbol{\gamma}) = \int_t^{t'} L(\boldsymbol{q},\dot{\boldsymbol{q}})d\tau$ is the action functional evaluated on a trajectory $\boldsymbol{\gamma} = \boldsymbol{q}(\tau)$, and $L(\boldsymbol{q},\dot{\boldsymbol{q}}) = \frac{1}{2}m\dot{\boldsymbol{q}}^2 - V(\boldsymbol{q})$ is the corresponding Lagrangian function. One also can formally rewrite the Wiener integral for the integral kernel $L_\hbar(\boldsymbol{q}', t'; \boldsymbol{q}, t)$ — the Feynman-Kac formula — in a similar form,

$$L_\hbar(\boldsymbol{q}', t'; \boldsymbol{q}, t) = \int_{\left\{\begin{smallmatrix} \boldsymbol{q}(t')=\boldsymbol{q}' \\ \boldsymbol{q}(t)=\boldsymbol{q} \end{smallmatrix}\right\}} e^{-\frac{1}{\hbar} \int_t^{t'} E(\boldsymbol{q}, \dot{\boldsymbol{q}}) d\tau} \mathscr{D}_\hbar \boldsymbol{q},$$

where $E(\boldsymbol{q}, \dot{\boldsymbol{q}}) = \frac{1}{2} m \dot{\boldsymbol{q}}^2 + V(\boldsymbol{q})$ is the energy function. However, in this representation we no longer see the presence of the corresponding action functional, and connection with the classical mechanics is lost.

The precise relation between Wiener and Feynman integrals for a quantum particle is the following. As in the case of the Feynman-Kac formula, suppose that the real-valued potential $V(\boldsymbol{q}) \in C(\mathbb{R}^n)$ is bounded below. Using the Lie-Kato-Trotter formula, it is easy to show that the heat kernel $L_\hbar(\boldsymbol{q}', t'; \boldsymbol{q}, t)$, defined for $\hbar > 0$, admits an analytic continuation into the half-plane $\operatorname{Re} \hbar > 0$. Then

$$(2.9) \qquad K_\hbar(\boldsymbol{q}', t'; \boldsymbol{q}, t) = \lim_{\varepsilon \to 0^+} L_{i\hbar+\varepsilon}(\boldsymbol{q}', t'; \boldsymbol{q}, t).$$

Remark. Let f be a smooth bounded function on $\mathbb{R}$ such that $f'(x) = O(|x|^{-1})$ as $|x| \to \infty$. The Gaussian integral $\int_{-\infty}^{\infty} f(x) e^{-x^2} dx$ is absolutely convergent, whereas the integral $\int_{-\infty}^{\infty} f(x) e^{ix^2} dx$ is only conditionally convergent. The formula

$$\int_{-\infty}^{\infty} f(x) e^{ix^2} dx = \lim_{\varepsilon \to 0^+} \int_{-\infty}^{\infty} f(x) e^{(i-\varepsilon)x^2} dx$$

interprets it as the limit $\varepsilon \to 0$ of an integral with respect to the complex-valued Gaussian measure $e^{(i-\varepsilon)x^2} dx$. It is tempting to extend this interpretation to Wiener integrals, and for a complex diffusion coefficient D with $\operatorname{Re} D > 0$ to define a complex-valued Wiener measure by the same formula (2.2). However, a theorem of Cameron states that the linear functional l, defined by an analog of (2.2) for $\operatorname{Re} D > 0$, is no longer bounded on $C_{\text{fin}}(\Omega)$, so that this approach does not work.

Problem 2.6. Let l be the functional on $C_{\text{fin}}(\Omega)$ defined by (2.2) where $P(\boldsymbol{q}', \boldsymbol{q}, t)$ is given by (2.1) with $\operatorname{Re} D \geq 0$.

(i) Prove that when $\operatorname{Re} D = 0$,

$$\sup_{t \geq 0, F \in C(\widehat{\mathbb{R}})} \{ |l(\varphi)| : \|\varphi\|_\infty = 1, \ \varphi(\gamma) = F(\gamma(t)) \} = \infty.$$

(ii) Prove that when $\operatorname{Re} D > 0$,

$$\sup_{\boldsymbol{t}_m, F \in C(\widehat{\mathbb{R}^m})} \{ |l(\varphi)| : \|\varphi\|_\infty = 1, \ \varphi(\gamma) = F(\gamma(t_1), \dots, \gamma(t_m)) \} = \left(\frac{|D|}{\operatorname{Re} D} \right)^m.$$

3. Gaussian Wiener integrals

In Section 4 of Chapter 5 we evaluated Gaussian Feynman integrals in terms of the regularized determinants of corresponding differential operators. Here we consider the same problem for Gaussian Wiener integrals.

3.1. Dirichlet boundary conditions. Let

$$A = -D^2 + u(t), \quad D = \frac{d}{dt},$$

where $u \in C^1([0,T])$, be the Sturm-Liouville operator on the interval $[0,T]$ with Dirichlet boundary conditions. The following result is fundamental.

Theorem 3.1. *Suppose that $u(t) \geq 0$. Then*

$$\int_{\mathscr{C}_{0,0}} e^{-\frac{m}{2\hbar}\int_0^T u(t)y^2(t)dt} d\mu_{0,0}(y) = \sqrt{\frac{m}{\pi\hbar \det A}}.$$

Proof. Using Proposition 2.2 and finite-dimensional Gaussian integration formula (1.1), we get

$$\int_{\mathscr{C}_{0,0}} e^{-\frac{m}{2\hbar}\int_0^T u(t)y^2(t)dt} d\mu_{0,0}(y) = \lim_{n\to\infty} \left(\frac{m}{2\pi\hbar\Delta t}\right)^{\frac{n}{2}} \int \cdots \int_{\mathbb{R}^{n-1}}$$

$$\exp\left\{-\frac{m}{2\hbar\Delta t}\sum_{k=0}^{n-1}\left((y_{k+1}-y_k)^2 + u(t_k)(\Delta t)^2 y_k^2\right)\right\}\prod_{k=1}^{n-1} dy_k$$

$$= \lim_{n\to\infty}\sqrt{\frac{m}{2\pi\hbar\Delta t\,\det A_{n-1}}}$$

(cf. the proof of Proposition 3.1 in Section 3.2 of Chapter 5). Here $y_0 = y_n = 0$, $t_k = k\Delta t$, $\Delta t = \dfrac{T}{n}$, and

$$A_{n-1} = \begin{pmatrix} a_1 & -1 & 0 & \cdots & 0 & 0 \\ -1 & a_2 & -1 & \cdots & 0 & 0 \\ 0 & -1 & a_3 & \cdots & 0 & 0 \\ \vdots & \vdots & \vdots & \ddots & \vdots & \vdots \\ 0 & 0 & 0 & \cdots & a_{n-2} & -1 \\ 0 & 0 & 0 & \cdots & -1 & a_{n-1} \end{pmatrix},$$

where $a_k = 2 + u(t_k)(\Delta t)^2$, $k = 1,\dots,n-1$. Let $y_k^{(n)}$ be Δt times the principal minor of order $k-1$ of the matrix A_{n-1} corresponding to its upper-left corner, so that $y_n^{(n)} = \Delta t \det A_{n-1}$. The sequence $y_k^{(n)}$ satisfies the initial conditions

$$y_2^{(n)} = 2\Delta t + O((\Delta t)^3), \quad y_3^{(n)} = 3\Delta t + O((\Delta t)^3),$$

and expanding the order k determinant $y_{k+1}^{(n)}$ with respect to the last row, we obtain the recurrence relation

$$y_{k+1}^{(n)} + y_{k-1}^{(n)} - 2y_k^{(n)} = u(t_k)(\Delta t)^2 y_k^{(n)}, \quad k = 3, \ldots, n-1.$$

Since

$$\lim_{k,n\to\infty} u(t_k) = u(t) \quad \text{when} \quad \lim_{k,n\to\infty} t_k = t,$$

it follows from the method of finite-differences for solving initial value problem for ordinary differential equations that

$$\lim_{k,n\to\infty} y_k^{(n)} = y(t),$$

where $y(t)$ is the solution of the differential equation

$$-y'' + u(t)y = 0,$$

with the initial conditions

$$y(0) = \lim_{n\to\infty} y_2^{(n)} = 0, \quad y'(0) = \lim_{n\to\infty} \frac{y_3^{(n)} - y_2^{(n)}}{\Delta t} = 1.$$

Using Theorem 5.1 in Section 5.1 of Chapter 5, we finally obtain

$$\lim_{n\to\infty} \Delta t \det A_{n-1} = \lim_{n\to\infty} y_n^{(n)} = y(T) = \frac{1}{2}\det' A. \qquad \square$$

Remark. The same computation also works for Feynman path integrals, and gives a rigorous proof of the formula

$$\int_{\left\{\begin{smallmatrix} y(T)=0 \\ y(0)=0 \end{smallmatrix}\right\}} e^{\frac{im}{2\hbar} \int_0^T (\dot{y}^2 - u(t)y^2)dt} \mathscr{D}y = \sqrt{\frac{m}{\pi i\hbar \det A}},$$

where $A = -D^2 - u(t)$ is a positive-definite operator (see Section 6.1 of Chapter 5). In particular, the coefficient $c_{m,\hbar}$ from Section 4 of Chapter 5 is indeed equal to $\sqrt{\frac{m}{\pi i\hbar}}$.

3.2. Periodic boundary conditions.

Let $A = -D^2 + u(t)$, where $u \in C^1([0,T])$, be the Sturm-Liouville operator on the interval $[0,T]$ with periodic boundary conditions. Here we prove the following analog of Theorem 3.1.

Theorem 3.2. *Suppose that $u(t) > 0$. Then*

$$\int_{\mathcal{L}} e^{-\frac{m}{2\hbar} \int_0^T u(t)y^2(t)dt} d\mu_W^{\text{loop}}(y) = \frac{1}{\sqrt{\det A}},$$

where $\mathcal{L}$ is the space of free loops in $\mathbb{R}$ parametrized by the interval $[0,T]$.

Proof. As in the proof of Theorem 3.1, we have

$$\int_{\mathcal{L}} e^{-\frac{m}{2\hbar}\int_0^T u(t)y^2(t)dt} d\mu_W^{\text{loop}}(y) = \lim_{n\to\infty} \left(\frac{m}{2\pi\hbar\Delta t}\right)^{\frac{n}{2}} \underbrace{\int \cdots \int}_{\mathbb{R}^n}$$

$$\exp\left\{-\frac{m}{2\hbar\Delta t}\sum_{k=0}^{n-1}\left((y_{k+1}-y_k)^2 + u(t_k)(\Delta t)^2 y_k^2\right)\right\}\prod_{k=1}^{n} dy_k$$

$$= \lim_{n\to\infty}\frac{1}{\sqrt{\det A_n}}.$$

Here $y_0 = y_n$ and A_n is the following $n\times n$ matrix:

$$A_n = \begin{pmatrix} a_0 & -1 & 0 & \cdots & 0 & -1 \\ -1 & a_1 & -1 & \cdots & 0 & 0 \\ 0 & -1 & a_2 & \cdots & 0 & 0 \\ \vdots & \vdots & \vdots & \ddots & \vdots & \vdots \\ 0 & 0 & 0 & \cdots & a_{n-2} & -1 \\ -1 & 0 & 0 & \cdots & -1 & a_{n-1} \end{pmatrix},$$

where $a_k = 2 + u(t_k)(\Delta t)^2$, $k = 0, 1, \ldots, n-1$. We compute $\det A_n$ by the following elegant argument. First, note that the real λ is an eigenvalue of A_n if only if the difference equation

$$(3.1)\qquad -(y_{k+1} + y_{k-1} - 2y_k) + u(t_k)(\Delta t)^2 y_k = \lambda y_k, \quad k = 0, \ldots, n-1,$$

with initial conditions y_{-1} and y_0 has a "periodic solution" — a solution $\{y_k\}_{k=1}^n$ satisfying $y_{n-1} = y_{-1}$ and $y_n = y_0$. For given λ denote by $v_k^{(1)}(\lambda)$ and $v_k^{(2)}(\lambda)$ solutions of (3.1) with the corresponding initial conditions $v_{-1}^{(1)}(\lambda) = 1$, $v_0^{(1)}(\lambda) = 0$ and $v_{-1}^{(2)}(\lambda) = 0$, $v_0^{(2)}(\lambda) = 1$, and put

$$T_n(\lambda) = \begin{pmatrix} v_{n-1}^{(1)}(\lambda) & v_{n-1}^{(2)}(\lambda) \\ v_n^{(1)}(\lambda) & v_n^{(2)}(\lambda) \end{pmatrix}.$$

It is easy to show that the discrete analog of the Wronskian $v_{k-1}^{(1)}(\lambda)v_k^{(2)}(\lambda) - v_k^{(1)}(\lambda)v_{k-1}^{(2)}(\lambda)$ does not depend on k, so that $\det T_n(\lambda) = 1$. Since every solution y_k of the initial value problem for (3.1) is a linear combination of the solutions $v_k^{(1)}(\lambda)$ and $v_k^{(2)}(\lambda)$, we have

$$\begin{pmatrix} y_{n-1} \\ y_n \end{pmatrix} = T_n(\lambda) \begin{pmatrix} y_{-1} \\ y_0 \end{pmatrix}.$$

From here we conclude that λ is an eigenvalue of the matrix A_n if and only if $\det(T_n(\lambda) - I_2) = 0$, and the multiplicity of λ is the multiplicity of a root of this algebraic equation. Since

$$v_{n-1}^{(1)}(\lambda) = O(\lambda^{n-1}), \quad v_n^{(2)}(\lambda) = (-\lambda)^n + O(\lambda^{n-1}) \quad \text{as} \quad \lambda \to \infty,$$

we obtain

$$\det(A_n - \lambda I_n) = -\det(T_n(\lambda) - I_2) = v_{n-1}^{(1)}(\lambda) + v_n^{(2)}(\lambda) - 2.$$

It remains to compute $\lim_{n\to\infty} \det(A_n - \lambda I_n)$. Denote by $y_k^{(1)}(\lambda)$ and $y_k^{(2)}(\lambda)$, correspondingly, two solutions of the difference equation (3.1) with the initial conditions $y_{-1}^{(1)}(\lambda) = y_0^{(1)}(\lambda) = 1$ and $y_{-1}^{(2)}(\lambda) = 0$, $y_0^{(2)}(\lambda) = \Delta t$. We have

$$v_k^{(1)}(\lambda) = y_k^{(1)}(\lambda) - \frac{1}{\Delta t}\, y_k^{(2)}(\lambda) \quad \text{and} \quad v_k^{(2)}(\lambda) = \frac{1}{\Delta t}\, y_k^{(2)}(\lambda),$$

so that

$$(3.2) \qquad \det(A_n - \lambda I_n) = y_{n-1}^{(1)}(\lambda) + \frac{y_n^{(2)}(\lambda) - y_{n-1}^{(2)}(\lambda)}{\Delta t} - 2.$$

Now it follows from the method of finite differences that

$$(3.3) \qquad \lim_{k,n\to\infty} y_k^{(1)}(\lambda) = y_1(t, \lambda), \qquad \lim_{k,n\to\infty} y_k^{(2)}(\lambda) = y_2(t, \lambda)$$

as $\lim_{k,n\to\infty} t_k = t$, where $y_{1,2}(t, \lambda)$ are two solutions of the differential equation

$$-y'' + u(t)y = \lambda y$$

with the initial conditions $y_1(0, \lambda) = 1$, $y_1'(0, \lambda) = 0$ and $y_2(0, \lambda) = 0$, $y_2'(0, \lambda) = 1$. Using (3.2)–(3.3) and Theorem 5.2 in Section 5.2 of Chapter 5, we finally obtain

$$\lim_{n\to\infty} \det(A_n - \lambda I_n) = y_1(T, \lambda) + y_2'(T, \lambda) - 2 = \det(A - \lambda I). \qquad \square$$

Example 3.1. Corollary 2.2 and Theorem 3.2 can be used to compute $\operatorname{Tr} e^{-\frac{1}{\hbar}TH}$ for the harmonic oscillator $H = \dfrac{1}{2m}(P^2 + m^2\omega^2 Q^2)$. We have

$$\operatorname{Tr} e^{-\frac{1}{\hbar}TH} = \frac{1}{\sqrt{\det A_{i\omega}}},$$

where $A_{i\omega} = -D^2 + \omega^2$, and by formula (5.20) in Section 5.2 of Chapter 5,

$$(3.4) \qquad\qquad \operatorname{Tr} e^{-\frac{1}{\hbar}TH} = \frac{1}{2\sinh\frac{\omega T}{2}}.$$

Of course, since the eigenvalues are $E_n = \hbar\omega(n + \frac{1}{2})$, we can get the same result by using geometric series

$$\operatorname{Tr} e^{-\frac{1}{\hbar}TH} = \sum_{n=0}^{\infty} e^{-\frac{1}{\hbar}TE_n} = e^{-\frac{\omega T}{2}} \sum_{n=0}^{\infty} e^{-\omega Tn} = \frac{1}{2\sinh\frac{\omega T}{2}}.$$

Similar results hold for the general second order differential operator $A = -D^2 + v(t)D + u(t)$ on the interval $[0, T]$ with periodic boundary conditions.

Theorem 3.3. *Suppose that* $\det A > 0$, *where* $A = -D^2 + v(t)D + u(t)$. *Then*

$$\int_{\mathcal{L}} e^{-\frac{m}{2\hbar} \int_0^T (v(t)\dot{y}(t)y(t)+u(t)y^2(t))dt} d\mu_W^{\mathrm{loop}}(y) = \frac{1}{\sqrt{\det A}}.$$

Example 3.2. Let $A = -D^2 + \omega D$. It follows from Theorem 3.3 and (2.6) that for $\varepsilon > 0$

$$\frac{1}{\sqrt{\det(A + \varepsilon I)}} = \int_{\mathcal{L}} e^{-\frac{m}{2\hbar} \int_0^T (\omega \dot{y}(t)y(t)+\varepsilon y^2(t))dt} d\mu_W^{\mathrm{loop}}(y)$$

$$= \sqrt{\frac{m}{2\pi\hbar T}} \int_{-\infty}^{\infty} e^{-\frac{mT}{2\hbar}\varepsilon x^2} dx \int_{\mathcal{L}_0} e^{-\frac{m}{2\hbar} \int_0^T (\omega \dot{y}_0(t)y_0(t)+\varepsilon y_0^2(t))dt} d\mu_W^{\mathrm{loop}}(y_0)$$

$$= \frac{1}{T\sqrt{\varepsilon}} \int_{\mathcal{L}_0} e^{-\frac{m}{2\hbar} \int_0^T (\omega \dot{y}_0(t)y_0(t)+\varepsilon y_0^2(t))dt} d\mu_W^{\mathrm{loop}}(y_0),$$

where we have used the decomposition $y(t) = x + y_0(t)$, $\int_0^T y_0(t)dt = 0$. Here $\mathcal{L}_0$ is the subset of the free loop space $\mathcal{L}$ which consists of loops with zero constant term in the Fourier series expansion. Using Corollary 5.4 in Section 5.2 of Chapter 5, we obtain

$$\int_{\mathcal{L}_0} e^{-\frac{m}{2\hbar} \int_0^T \omega \dot{y}_0(t)y_0(t)dt} d\mu_W^{\mathrm{loop}}(y_0) = \lim_{\varepsilon \to 0} \frac{T\sqrt{\varepsilon}}{\sqrt{\det(A + \varepsilon I)}} = \frac{T}{\sqrt{\det' A}}$$

$$(3.5) \qquad\qquad\qquad\qquad = \left(\frac{\frac{\omega T}{2}}{\sinh \frac{\omega T}{2}} \right)^{\frac{1}{2}}.$$

We will use this result in Section 2.2 of Chapter 8.

Problem 3.1. Derive formula (3.4) using (2.5).

Problem 3.2. Prove Theorem 3.3.

4. Notes and references

There is a vast literature on Wiener integration theory and Brownian motion, and here we present, in a succinct form, only the very basic facts. Material in Section 1 is standard, and our exposition follows the exercise section in [**Rab95**]. For necessary facts from probability theory, including Kolmogoroff's extension theorem and an introduction to the stochastic processes, we refer to the classic treatise [**Loè77, Loè78**] and recent text [**Kho07**]. In particular, the statement $\mu(\mathcal{H}) = 0$ in Proposition 1.1 follows from the strong law of large numbers.

Th elegant construction of the Wiener measure in Section 2 belongs to E. Nelson [**Nel64**], and our exposition, including the proof of the Feynman-Kac formula, follows [**RS75**]. We refer the reader to Ito-McKean's classic monograph [**IM74**] for the construction of the Wiener measure from the probability theory point of view; along this way Problems 2.1 and 2.2 get solutions. The classic book [**Kac59**] and lectures [**Kac80**] by M. Kac are another excellent source of information on Wiener integration and its applications in different areas of mathematics. The relation between Wiener and Feynman path integrals in Section 2.3 belongs to E. Nelson [**Nel64**]. Problem 2.6 is taken from R.H. Cameron's paper [**Cam63**] (see also [**RS75**]); this result shows that there is no complex-valued analog of Wiener measure associated with the complex diffusion coefficient with $\operatorname{Re} D > 0$ (as opposed to the statement made in [**GY56**]).

Theorems proved in Section 3 serve as a rigorous foundation for our discussion of the Gaussian Feynman path integrals in Section 4 of Chapter 5. Our proof of Theorem 3.1 in Section 3.1, which uses the method of finite differences, follows the outline in [**GY56**] which attributed it to [**Mon52**]. The proof of Theorem 3.2 in Section 3.2 seems to be new.

The Euclidean quantum mechanics, obtained by replacing the physical time t by the Euclidean (or imaginary) time $-it$, is ultimately related to the theory of stochastic processes. It can be formulated by a set of axioms of the one-dimensional Euclidean quantum field theory, and we refer the interested reader to [**Str05**] for the detailed discussion.

Chapter 7

Fermion Systems

1. Canonical anticommutation relations

1.1. Motivation. In Sections 2.6 and 2.7 of Chapter 2 we have shown that the Hilbert space $\mathscr{H} \simeq L^2(\mathbb{R}, dq)$ of a one-dimensional quantum particle can be described in terms of the creation and annihilation operators. Namely, the operators[1]

$$a^* = \frac{1}{\sqrt{2\hbar}}\,(Q - iP) \quad \text{and} \quad a = \frac{1}{\sqrt{2\hbar}}\,(Q + iP)$$

satisfy the canonical commutation relation

$$[a, a^*] = I$$

on $W^{2,2}(\mathbb{R}) \cap \hat{W}^{2,2}(\mathbb{R})$, and the vectors

$$\psi_k = \frac{(a^*)^k}{\sqrt{k!}}\,\psi_0, \quad k = 0, 1, 2, \ldots,$$

where $\psi_0(q) = (\pi\hbar)^{-\frac{1}{4}} e^{-\frac{1}{2\hbar}q^2} \in \mathscr{H}$ satisfies $a\psi_0 = 0$, form an orthonormal basis for $\mathscr{H}$. The corresponding operator $N = a^*a$ is self-adjoint and has an integer spectrum,

$$N\psi_k = k\psi_k, \quad k = 0, 1, 2, \ldots.$$

Similarly, for several degrees of freedom $\mathscr{H} \simeq L^2(\mathbb{R}^n, d^n\boldsymbol{q})$, the creation and annihilation operators are given by

$$(1.1) \quad a_k^* = \frac{1}{\sqrt{2\hbar}}\,(Q_k - iP_k) \quad \text{and} \quad a_k = \frac{1}{\sqrt{2\hbar}}\,(Q_k + iP_k)\,, \quad k = 1, \ldots, n,$$

---

[1] Here in comparison with Section 2.6 of Chapter 2 we put $\omega = 1$.

and satisfy canonical commutation relations

$$(1.2) \qquad [a_k, a_l] = [a_k^*, a_l^*] = 0 \quad \text{and} \quad [a_k, a_l^*] = \delta_{kl} I, \quad k, l = 1, \dots, n.$$

The ground state, the vector $\psi_0(\boldsymbol{q}) = (\pi\hbar)^{-\frac{n}{4}} e^{-\frac{1}{2\hbar} \boldsymbol{q}^2} \in \mathscr{H}$, has the property

$$a_k \psi_0 = 0, \quad k = 1, \dots, n,$$

and the vectors

$$\psi_{k_1,\dots,k_n} = \frac{(a_1^*)^{k_1} \dots (a_n^*)^{k_n}}{\sqrt{k_1! \dots k_n!}} \psi_0, \quad k_1, \dots, k_n = 0, 1, 2, \dots,$$

form an orthonormal basis for $\mathscr{H}$. The operator $N = \sum_{k=1}^{n} a_k^* a_k$ is self-adjoint and has an integer spectrum:

$$N \psi_{k_1,\dots,k_n} = (k_1 + \dots + k_n) \psi_{k_1,\dots,k_n},$$

and the Hilbert space $\mathscr{H}$ decomposes into the direct sum of invariant subspaces

$$(1.3) \qquad \mathscr{H} = \bigoplus_{k=0}^{\infty} \mathscr{H}_k$$

— the eigenspaces for the operator N.

However spin operators, introduced in Chapter 4, satisfy algebraic relations of different type. Namely, consider the operators $\sigma_{\pm} = \frac{2}{\hbar} S_{\pm} = \frac{2}{\hbar}(S_1 \pm i S_2)$, where S_1 and S_2 are spin operators of a quantum particle of spin $\frac{1}{2}$ (see Section 1.1 of Chapter 4). Using the explicit representation of spin operators by Pauli matrices, we get

$$\sigma_- = \begin{pmatrix} 0 & 0 \\ 1 & 0 \end{pmatrix}, \quad \sigma_+ = \begin{pmatrix} 0 & 1 \\ 0 & 0 \end{pmatrix}.$$

The operators $\sigma_{\pm}$ are nilpotent, $\sigma_{\pm}^2 = 0$, and satisfy the *anticommutation relation*

$$\sigma_+ \sigma_- + \sigma_- \sigma_+ = I_2,$$

where I_2 is the identity operator in $\mathbb{C}^2$. Introducing the notion of an anticommutator of two operators,

$$[A, B]_+ = AB + BA,$$

we see that the operators $a = \sigma_-$ and $a^* = \sigma_+$ satisfy *canonical anticommutation relations*

$$[a, a]_+ = [a^*, a^*]_+ = 0 \quad \text{and} \quad [a, a^*]_+ = I_2.$$

The vector $e_0 = \begin{pmatrix} 0 \\ 1 \end{pmatrix}$ has the property $a e_0 = 0$, and together with the vector $a^* e_0 = \begin{pmatrix} 1 \\ 0 \end{pmatrix}$ they form an orthonormal basis of $\mathbb{C}^2$. The matrix

$$N = a^* a = \tfrac{1}{2}(\sigma_3 + I_2) = \begin{pmatrix} 1 & 0 \\ 0 & 0 \end{pmatrix}$$

has eigenvectors e_0 and $a^* e_0$ with eigenvalues 0 and 1. Thus in complete analogy with the previous discussion, we say that a and a^* are *fermion creation* and *annihilation operators* for the case of one degree of freedom. The Hilbert space of a Fermi particle is $\mathscr{H} = \mathbb{C}^2$, and the vector e_0 is the ground state.

It is straightforward to generalize this construction to the case of several degrees of freedom. Namely, canonical anticommutation relations have the form

$$(1.4) \quad [a_k, a_l]_+ = [a_k^*, a_l^*]_+ = 0 \quad \text{and} \quad [a_k, a_l^*]_+ = \delta_{kl} I, \quad k, l = 1, \ldots, n,$$

where I is the identity operator, and creation operators a_j^* are adjoint to annihilation operators a_j in the fermion Hilbert space $\mathscr{H}_F$. Canonical anticommutation relations (1.4) can be realized in the Hilbert space $\mathscr{H}_F = (\mathbb{C}^2)^{\otimes n} = \mathbb{C}^{2^n}$ as follows:

$$(1.5) \qquad a_k = \underbrace{\sigma_3 \otimes \cdots \otimes \sigma_3}_{k-1} \otimes\, a \otimes I_2 \otimes \cdots \otimes I_2,$$

$$(1.6) \qquad a_k^* = \underbrace{\sigma_3 \otimes \cdots \otimes \sigma_3}_{k-1} \otimes\, a^* \otimes I_2 \otimes \cdots \otimes I_2,$$

$k = 1, \ldots, n$. The ground state, the vector $\psi_0 = e_0 \otimes \cdots \otimes e_0 \in \mathscr{H}_F$, satisfies

$$(1.7) \qquad a_k \psi_0 = 0, \quad k = 1, \ldots, n,$$

and the vectors

$$(1.8) \qquad \psi_{k_1, \ldots, k_n} = (a_1^*)^{k_1} \ldots (a_n^*)^{k_n} \psi_0, \quad k_1, \ldots, k_n = 0, 1,$$

form an orthonormal basis for $\mathscr{H}_F$. The operator

$$N = \sum_{k=1}^{n} a_k^* a_k$$

is self-adjoint and has an integer spectrum:

$$N \psi_{k_1, \ldots, k_n} = (k_1 + \cdots + k_n) \psi_{k_1, \ldots, k_n},$$

and the Hilbert space $\mathscr{H}_F$ decomposes into the direct sum of invariant subspaces

$$(1.9) \qquad \mathscr{H}_F = \bigoplus_{k=0}^{n} \mathscr{H}_k$$

— the eigenspaces of N.

Remark. The fermion Hilbert space $\mathscr{H}_F$ is isomorphic to the spin part of the Hilbert space of n quantum particles of spin $\frac{1}{2}$, discussed in Section 3.1 of Chapter 4. The corresponding fermion creation and annihilation operators can also be used for describing spin degrees of freedom. The fundamental importance of the fermion and boson Hilbert spaces — $\mathscr{H}_F = (\mathbb{C}^2)^{\otimes n}$ and

$\mathscr{H}_B = L^2(\mathbb{R}^n, d^n\boldsymbol{q})$ — manifests itself in quantum field theory, which formally corresponds to the case $n = \infty$, and describes quantum systems with infinitely many degrees of freedom.

Lemma 1.1. *Realization* (1.5)–(1.6) *of canonical anticommutation relations in the fermion Hilbert space $\mathscr{H}_F$ is irreducible: every operator in $\mathscr{H}_F$, which commutes with all creation and annihilation operators a_k^* and a_k, is a multiple of the identity operator.*

Proof. It is sufficient to show that if $\{0\} \neq V \subseteq \mathscr{H}_F$ is an invariant subspace for all operators a_k and a_k^*, then $V = \mathscr{H}_F$. Indeed, every non-zero $\psi \in V$ can be written in the form

$$\psi = \sum_{k_1,\ldots,k_n=0,1} c_{k_1,\ldots,k_n} \psi_{k_1,\ldots,k_n},$$

and let $c_{k_1,\ldots,k_n} \psi_{k_1,\ldots,k_n}$ be any of its non-zero components with maximal degree $k_1 + \cdots + k_n$. Using canonical anticommutation relations and (1.7), we obtain

$$a_n^{k_n} \ldots a_1^{k_1} \psi = c\psi_0, \quad c = c_{k_1,\ldots,k_n} \neq 0,$$

so that $\psi_0 \in V$. Applying creation operators to ψ_0, we get $V = \mathscr{H}_F$. $\qquad\square$

Remark. The realization of canonical anticommutation relations in the fermion Hilbert space $\mathscr{H}_F$ is analogous to the representation by the occupation numbers of canonical commutation relations, discussed in Section 2.7 of Chapter 2, and is called representation by the occupation numbers for fermions. It should be emphasized that the algebraic structure of the former relations allows their realization in a finite-dimensional Hilbert space, while the algebraic structure of the latter relations warrants the infinite-dimensional Hilbert space.

Analogously to (1.1), coordinate and momentum operators for fermions are defined by

$$(1.10) \qquad Q_k = \sqrt{\tfrac{\hbar}{2}}(a_k + a_k^*), \quad P_k = -i\sqrt{\tfrac{\hbar}{2}}(a_k - a_k^*), \quad k = 1,\ldots,n.$$

As follows from (1.4), they satisfy the following anticommutation relations:

$$(1.11) \quad [Q_k, Q_l]_+ = [P_k, P_l]_+ = \hbar\delta_{kl}I \quad \text{and} \quad [P_k, Q_l]_+ = 0, \quad k, l = 1,\ldots,n$$

— a fermion analog of Heisenberg commutations relations, introduced in Section 2.1 of Chapter 2. The following result is a fermion analog of the Stone-von Neumann theorem from Section 3.1 of Chapter 2.

Theorem 1.1. *Every irreducible finite-dimensional representation of canonical anticommutation relations is unitarily equivalent to the representation by occupation numbers in the fermion Hilbert space $\mathscr{H}_F$.*

Proof. Let V be the Hilbert space which realizes the irreducible representation of canonical anticommutation relations (1.4). First of all, there is $\varphi_0 \in V$, $\|\varphi_0\| = 1$, such that

$$a_1\varphi_0 = \cdots = a_n\varphi_0 = 0.$$

Indeed, choose any non-zero $\varphi \in V$; if $a_1\varphi \neq 0$, replace it by the vector $a_1\varphi$, which obviously satisfies $a_1(a_1\varphi_1) = 0$. If $a_2(a_1\varphi) \neq 0$, replace it by $a_2a_1\varphi$, which is annihilated by a_1 and a_2, etc. In finitely many steps we arrive at a non-zero vector $\tilde{\varphi}$ annihilated by the operators $a_1, \ldots, a_n$, and $\varphi_0 = \tilde{\varphi}/\|\tilde{\varphi}\|$. Now consider the subspace V_0 of V, spanned by the vectors

$$\varphi_{k_1,\ldots,k_n} = (a_1^*)^{k_1} \ldots (a_n^*)^{k_n} \varphi_0, \quad k_1, \ldots, k_n = 0, 1.$$

It follows from (1.4) that V_0 is an invariant subspace for all operators a_k and a_k^*, so that $V_0 = V$. Since operators a_k^* and a_k are adjoint with respect to the inner product in V, it is easy to see, again using canonical anticommutation relations (1.4), that vectors $\varphi_{k_1,\ldots,k_n}$ form an orthonormal basis for V. The mapping $V \ni \varphi_{k_1,\ldots,k_n} \mapsto \psi_{k_1,\ldots,k_n} \in \mathscr{H}_F$ establishes the Hilbert space isomorphism $V \simeq \mathscr{H}_F$. $\qquad\square$

Problem 1.1. Express the spin operators $S_j^{(k)}$ of the k-th particle in the system of n spin $\frac{1}{2}$ particles (see Section 3.1 of Chapter 4) in terms of fermion creation and annihilation operators in $\mathscr{H}_F$.

1.2. Clifford algebras. We have seen in Section 2.1 of Chapter 2 that the Heisenberg Lie algebra is the fundamental mathematical structure associated with canonical commutation relations. Similarly, the fundamental mathematical structure associated with the canonical anticommutation relations is *Clifford algebra.*

Let V be a finite-dimensional vector space over the field k of characteristic zero, and let $Q : V \rightarrow k$ be a symmetric non-degenerate quadratic form on V, i.e., $Q(v) = \Phi(v, v)$, $v \in V$, where $\Phi : V \otimes_k V \rightarrow k$ is a symmetric non-degenerate bilinear form. The pair (V, Q) is called quadratic vector space.

Definition. A Clifford algebra $C(V, Q) = C(V)$ associated with a quadratic vector space (V, Q) is a k-algebra generated by the vector space V with relations

$$v^2 = Q(v) \cdot 1, \quad v \in V.$$

Equivalently, Clifford algebra is defined as a quotient algebra

$$C(V) = T(V)/J,$$

where J is a two-sided ideal in the tensor algebra $T(V)$ of V, generated by the elements $u \otimes v + v \otimes u - 2\Phi(u, v) \cdot 1$ for all $u, v \in V$, and 1 is the unit in $T(V)$. In terms of a basis $\{e_i\}_{i=1}^n$ of V, the Clifford algebra $C(V)$ is a k-algebra with the generators $e_1, \ldots, e_n$, satisfying the relations

$$[e_i, e_j]_+ = e_i e_j + e_j e_i = 2\Phi(e_i, e_j) \cdot 1, \quad i, j = 1, \ldots, n.$$

When $k = \mathbb{C}$ (or any algebraically closed field of characteristic zero), there always exists an orthonormal basis for V — a basis $\{e_i\}_{i=1}^n$ such that $\Phi(e_i, e_k) = \delta_{ik}$. In this case for every dimension n there is one (up to an isomorphism) Clifford algebra C_n with generators $e_1, \ldots, e_n$ and relations

$$e_i e_j + e_j e_i = 2\delta_{ij} \cdot 1, \quad i, k = 1, \ldots, n.$$

Remark. If $k = \mathbb{R}$, there exist non-negative integers $p + q = n$ and an isomorphism $V \simeq \mathbb{R}^n$ such that

$$Q(\boldsymbol{x}) = x_1^2 + \cdots + x_p^2 - x_{p+1}^2 - \cdots - x_n^2, \quad \boldsymbol{x} \in \mathbb{R}^n.$$

This classifies Clifford algebras over $\mathbb{R}$.

Definition. A left module S for a Clifford algebra $C(V)$ is a finite-dimensional k-vector space S with the linear map $\rho : C(V) \otimes S \to S$ such that

$$\rho(ab \otimes s) = \rho(a \otimes \rho(b \otimes s))$$

for all $a, b \in C(V)$ and $s \in S$.

The fermion Hilbert space $\mathscr{H}_F$, introduced in the previous section, is an irreducible C_{2n}-module. Indeed, it follows from canonical anticommutation relations (1.4) that self-adjoint operators

$$(1.12) \qquad\qquad \gamma_{2k-1} = a_k + a_k^*,$$

$$(1.13) \qquad\qquad \gamma_{2k} = -i(a_k - a_k^*), \quad k = 1, \ldots, n,$$

satisfy the relations

$$(1.14) \qquad\qquad \gamma_\mu \gamma_\nu + \gamma_\nu \gamma_\mu = 2\delta_{\mu\nu} I, \quad \mu, \nu = 1, \ldots, 2n,$$

where I is the identity operator in $\mathscr{H}_F$. We define the action of the Clifford algebra C_{2n} on $\mathscr{H}_F$ by setting

$$\rho(1) = I \quad \text{and} \quad \rho(e_\mu) = \gamma_\mu, \quad \mu = 1, \ldots, 2n,$$

and extending it to a $\mathbb{C}$-algebra homomorphism $\rho : C_{2n} \to \operatorname{End}(\mathscr{H}_F)$. Relations (1.14) show that the map ρ admits such an extension.

Proposition 1.1. *The homomorphism $\rho : C_{2n} \to \operatorname{End}(\mathscr{H}_F)$ is a $\mathbb{C}$-algebra isomorphism.*

Proof. It follows from Lemma 1.1 that the representation ρ is irreducible: every operator in $\mathscr{H}_F$ which commutes with all elements of the $\mathbb{C}$-algebra $\rho(C_{2n})$ is a multiple of the identity operator. Then by Wedderburn's theorem $\rho(C_{2n}) = \operatorname{End}(\mathscr{H}_F)$, and since $\dim C_{2n} = 2^{2n} = \dim \operatorname{End}(\mathscr{H}_F)$, the map ρ is an isomorphism. $\qquad\square$

Remark. The structure of a Clifford algebra with an odd number of generators is different. Thus the mapping $\rho(e_k) = \sigma_k$, where σ_k, $k = 1, 2, 3$, are Pauli matrices (see Section 1 of Chapter 4), defines an irreducible representation of C_3 in $\mathscr{H}_F = \mathbb{C}^2$. However, in this case $C_3 \simeq \mathrm{End}(\mathbb{C}^2) \otimes \mathbb{C}[\varepsilon]$, where $\varepsilon = i e_1 e_2 e_3$ and satisfies $\varepsilon^2 = 1$.

We define the *chirality operator* by $\Gamma = e^{\pi i N}$, where $N = \sum_{j=1}^{n} a_j^* a_j$. Since the operator N has an integral spectrum, $\Gamma^2 = I$. Moreover, we have

$$(1.15) \qquad [\Gamma, \gamma_\mu]_+ = 0, \quad \mu = 1, \ldots, 2n.$$

Indeed, as follows from (1.4),

$$N a_j^* = a_j^*(N + I) \quad \text{and} \quad N a_j = a_j(N - I),$$

so that

$$e^{\pi i N} a_j^* = a_j^* e^{\pi i (N+I)} = -a_j^* e^{\pi i N} \quad \text{and} \quad e^{\pi i N} a_j = a_j e^{\pi i (N-I)} = -a_j e^{\pi i N}.$$

Thus Γ anticommutes with all a_j, a_j^*, and hence with all γ_μ. Since $\Gamma^2 = I$, the operators

$$P_\pm = \frac{1}{2}(I \pm \Gamma)$$

are orthogonal projection operators and we have a decomposition

$$\mathscr{H}_F = \mathscr{H}_F^+ \oplus \mathscr{H}_F^-$$

into the subspaces of *positive* and *negative chirality spinors*. It follows from (1.15) that

$$\gamma_\mu(\mathscr{H}_F^+) = \mathscr{H}_F^-, \quad \mu = 1, \ldots, 2n.$$

Also, since $e^{\pi i a_j^* a_j} = I - 2a_j^* a_j = -i\gamma_{2j-1}\gamma_{2j}$, we have

$$\Gamma = (-i)^n \gamma_1 \ldots \gamma_{2n}.$$

Remark. When $n = 2$, 4×4 matrices $\gamma_1, \gamma_2, \gamma_3, \gamma_4$ are celebrated Dirac gamma matrices (for the Euclidean metric on $\mathbb{R}^4$), and $\Gamma = \gamma_5$.

Problem 1.2. Show that the definition of a Clifford algebra $C(V)$ is compatible with the field change: if $k \subset K$ is a field extension and $V_K = K \otimes_k V_k$, then

$$C(V_K) = K \otimes_k C(V_k).$$

Problem 1.3. Let $\mathbb{C} \cdot 1 = C^0 \subset C^1 \subset \cdots \subset C^n = C(V)$ be the natural filtration of a Clifford algebra $C(V)$, where C^r is spanned by the elements $v_1 \ldots v_s$, $s \leq r$. Let

$$C^{\mathrm{gr}}(V) = \bigoplus_{k=1}^{n} C^k / C^{k-1}$$

be the associated graded algebra. Show that the skew-symmetrizer map

$$v_1 \wedge \cdots \wedge v_r \mapsto \frac{1}{r!} \sum_{\sigma \in \mathrm{Sym}_r} (-1)^{\varepsilon(\sigma)} v_{\sigma(1)} \ldots v_{\sigma(r)}$$

establishes a $\mathbb{Z}$-graded algebra isomorphism $\Lambda^{\bullet}(V) \simeq C^{\mathrm{gr}}(V)$, where $\Lambda^{\bullet}(V)$ is the exterior algebra of V.

Problem 1.4. Formulate and prove the analog of Proposition 1.1 for Clifford algebras with an odd number of generators.

2. Grassmann algebras

Grassmann algebras — algebras with anticommuting generators — are necessary for the semi-classical description of fermions. The corresponding mathematical definition is the following.

Definition. A Grassmann algebra with n generators is a $\mathbb{C}$-algebra Gr_n with the generators $\theta_1, \ldots, \theta_n$ satisfying the relations

$$\theta_i \theta_j + \theta_j \theta_i = 0, \quad i, j = 1, \ldots, n.$$

In particular, these relations imply that generators of a Grassmann algebra are nilpotent: $\theta_1^2 = \cdots = \theta_n^2 = 0$. Equivalently,

$$\mathrm{Gr}_n = \mathbb{C}\langle \theta_1, \ldots, \theta_n \rangle / J$$

— a quotient of a free $\mathbb{C}$-algebra $\mathbb{C}\langle \theta_1, \ldots, \theta_n \rangle$, generated by $\theta_1, \ldots, \theta_n$, by the two-sided ideal J generated by the elements $\theta_i \theta_j + \theta_j \theta_i$, $i, j = 1, \ldots, n$.

Remark. It follows from (1.11) that in the semi-classical limit $\hbar \to 0$ fermion operators P_k and Q_k, $k = 1, \ldots, 2n$, satisfy the defining relations of Grassmann algebra Gr_{2n}.

Comparison with the polynomial algebra

$$\mathbb{C}[x_1, \ldots, x_n] = \mathbb{C}\langle x_1, \ldots, x_n \rangle / I$$

— a quotient of a free $\mathbb{C}$-algebra $\mathbb{C}\langle x_1, \ldots, x_n \rangle$ by the two-sided ideal generated by the elements $x_i x_j - x_j x_i$, $i, j = 1, \ldots, n$ — shows that the Grassmann algebra Gr_n can also be considered as a *polynomial algebra in anticommuting variables* $\theta_1, \ldots, \theta_n$. In what follows we will always use Roman letters for commuting variables and Greek letters for anticommuting variables, so that

$$\mathrm{Gr}_n = \mathbb{C}[\theta_1, \ldots, \theta_n].$$

Needless to say, the polynomial algebra $\mathbb{C}[x_1, \ldots, x_n]$ is isomorphic to the symmetric algebra of the vector space spanned by $x_1, \ldots, x_n$, and the Grassmann algebra $\mathbb{C}[\theta_1, \ldots, \theta_n]$ is isomorphic to the exterior algebra $\Lambda^{\bullet} V$ of the vector space $V = \mathbb{C}\theta_1 \oplus \cdots \oplus \mathbb{C}\theta_n$ with the basis $\theta_1, \ldots, \theta_n$.

The Grassmann algebra Gr_n is a complex vector space of dimension 2^n and is $\mathbb{Z}$-graded: it admits a decomposition

$$(2.1) \qquad \mathrm{Gr}_n = \bigoplus_{k=0}^{n} \mathrm{Gr}_n^k$$

into homogeneous components Gr_n^k of degree k and dimension $\binom{n}{k}$, $k = 0, \ldots, n$, where $\mathrm{Gr}_n^0 = \mathbb{C}\cdot 1$. Namely, denote by $|\cdot|$ the degree of homogeneous elements in the Grassmann algebra, $|\alpha| = k$ for $\alpha \in \mathrm{Gr}_n^k$. Then multiplication in Gr_n satisfies $\mathrm{Gr}_n^k \cdot \mathrm{Gr}_n^l \subset \mathrm{Gr}_n^{k+l}$, where $\mathrm{Gr}_n^{k+l} = 0$ if $k + l > n$, and is graded-commutative:

$$(2.2) \qquad \alpha\beta = (-1)^{|\alpha||\beta|}\beta\alpha$$

for homogenous elements $\alpha, \beta \in \mathrm{Gr}_n$. The elements of the Grassmann algebra Gr_n of even degree are called *even elements*, and those of odd degree — *odd elements*.

2.1. Realization of canonical anticommutation relations.

The Grassmann algebra provides us with the explicit representation of canonical anticommutation relations by multiplication and differentiation operators, which is analogous to the holomorphic representation for canonical commutation relations (see Section 2.7 of Chapter 2). Namely, let

$$\partial_i = \frac{\partial}{\partial\theta_i} : \mathrm{Gr}_n \to \mathrm{Gr}_n$$

be left partial differentiation operators, defined on homogeneous monomials $\theta_{i_1} \ldots \theta_{i_k}$ by

$$\frac{\partial}{\partial\theta_i}\theta_{i_1} \ldots \theta_{i_k} = \sum_{l=1}^{k}(-1)^{l-1}\delta_{ii_l}\theta_{i_1} \ldots \check{\theta}_{i_l} \ldots \theta_{i_k},$$

where $\check{\theta}_{i_l}$ denotes the omission of the factor θ_{i_l}. The differentiation operators are of degree -1 and satisfy the graded Leibniz rule,

$$\frac{\partial}{\partial\theta_i}(\alpha\beta) = \frac{\partial\alpha}{\partial\theta_i}\beta + (-1)^{|\alpha|}\alpha\frac{\partial\beta}{\partial\theta_i}.$$

Remark. One can also introduce right partial differentiation operators by

$$(\theta_{i_1} \ldots \theta_{i_k})\frac{\partial}{\partial\theta_i} = \sum_{l=1}^{k}(-1)^{k-l}\delta_{ii_l}\theta_{i_1} \ldots \check{\theta}_{i_l} \ldots \theta_{i_k},$$

which satisfy the following graded Leibniz rule:

$$(\alpha\beta)\frac{\partial}{\partial\theta_i} = \alpha\left(\beta\frac{\partial}{\partial\theta_i}\right) + (-1)^{|\beta|}\left(\alpha\frac{\partial}{\partial\theta_i}\right)\beta.$$

To distinguish between the left and right partial derivatives of $f \in \mathrm{Gr}_n$, we will denote them, respectively, by $\dfrac{\partial}{\partial\theta_i}f$ and $f\dfrac{\partial}{\partial\theta_i}$.

As a complex vector space, Grassmann algebra Gr_n carries a standard inner product defined by the property that homogeneous monomials $\theta_{i_1} \ldots \theta_{i_k}$, for all $1 \leq i_1 < \cdots < i_k \leq n$, form an orthonormal basis,

$$(2.3) \qquad (\theta_{i_1} \ldots \theta_{i_k}, \theta_{j_1} \ldots \theta_{j_l}) = \delta_{kl} \delta_{i_1 j_1} \ldots \delta_{i_k j_k}.$$

By checking on homogeneous monomials, it is elementary to verify that

$$(\partial_i f, g) = (f, \hat{\theta}_i g), \quad f, g \in \mathrm{Gr}_n,$$

where $\hat{\theta}_i$ are left-multiplication by θ_i operators in Gr_n, so that $\hat{\theta}_i = \partial_i^*$. It is also easy to verify that the operators $\hat{\theta}_i$ and ∂_i satisfy the anticommutation relations

$$[\hat{\theta}_i, \hat{\theta}_j]_+ = [\partial_i, \partial_j]_+ = 0 \quad \text{and} \quad [\hat{\theta}_i, \partial_j]_+ = \delta_{ij} I, \quad i, j = 1, \ldots, n,$$

where I is an identity operator in Gr_n. Thus we have the following result.

Proposition 2.1. *The assignment*

$$\mathscr{H}_F \ni \psi_{k_1, \ldots, k_n} \mapsto \theta_1^{k_1} \ldots \theta_n^{k_n} \in \mathrm{Gr}_n$$

establishes an isomorphism $\mathscr{H}_F \simeq \mathrm{Gr}_n$ between the fermion Hilbert space of n identical particles, and the vector space of the Grassmann algebra with n generators. It preserves decompositions (1.9) and (2.1) and has the property that

$$a_i^* \mapsto \hat{\theta}_i \quad \text{and} \quad a_i \mapsto \frac{\partial}{\partial \theta_i}, \quad i = 1, \ldots, n.$$

Using Proposition 2.1, it is also very easy to verify that the representation of canonical anticommutation relations in the fermion Hilbert space $\mathscr{H}_F$ is irreducible. Indeed, suppose that $B \in \mathrm{End}(\mathrm{Gr}_n)$ commutes with all operators $\hat{\theta}_i$ and ∂_i. Then

$$\partial_i(B(1)) = B(\partial_i(1)) = 0, \quad i = 1, \ldots, n.$$

The only solution of the equations $\partial_1 f = \cdots = \partial_n f = 0$ is $f = c \cdot 1$, so that $B(1) = c \cdot 1$. Since B commutes with all creation operators $\hat{\theta}_i$, we obtain $B = cI$.

Remark. We will show in Section 2.3 that by using the notion of Berezin integral, the inner product (2.3) in Gr_n can be written in a form (2.11), which is similar to the definition of the inner product in the holomorphic representation, given by formula (2.52) in Section 2.7 of Chapter 2.

2.2. Differential forms. An algebra of differential forms in anticommuting variables $\theta_1, \ldots, \theta_n$ is a $\mathbb{C}$-algebra $\Omega_n^\bullet$ with odd generators $\theta_1, \ldots, \theta_n$ and even generators $d\theta_1, \ldots, d\theta_n$ satisfying relations

$$\theta_i \cdot d\theta_j = d\theta_j \cdot \theta_i, \quad i,j = 1, \ldots, n.$$

Equivalently, $\Omega_n^\bullet$ is a symmetric tensor product over $\mathbb{C}$ of Grassmann algebra Gr_n and polynomial algebra $\mathbb{C}[d\theta_1, \ldots, d\theta_n]$, and every $\omega \in \Omega_n^\bullet$ can be uniquely written as

$$(2.4) \qquad \omega = \sum_{\boldsymbol{k}=0}^{\infty} f_{\boldsymbol{k}} (d\theta_1)^{k_1} \cdots (d\theta_n)^{k_n}, \quad f_{\boldsymbol{k}} \in \mathrm{Gr}_n,$$

where $\boldsymbol{k} = (k_1, \ldots, k_n)$ is a multi-index and $f_{\boldsymbol{k}} = 0$ for all $\boldsymbol{k}$, except finitely many. By definition, the degree $|\omega_{\boldsymbol{k}}|$ of a homogeneous component $\omega_{\boldsymbol{k}} = f_{\boldsymbol{k}} (d\theta_1)^{k_1} \cdots (d\theta_n)^{k_n} \in \Omega_n^\bullet$ is $|f_{\boldsymbol{k}}|$, the degree of $f_{\boldsymbol{k}} \in \mathrm{Gr}_n$.

Remark. It is instructive to compare the algebra $\Omega_n^\bullet$ with the algebra of polynomial differential forms on $\mathbb{C}^n$. On the one hand, $\Omega_n^\bullet$ is an infinite-dimensional algebra in commuting variables $d\theta_1, \ldots, d\theta_n$ with coefficients in a finite-dimensional algebra Grassmann algebra $\mathbb{C}[\theta_1, \ldots, \theta_n]$. On the other hand, the algebra of differential forms in commuting variables $x_1, \ldots, x_n$ is a finite-dimensional algebra in anticommuting variables $dx_1, \ldots, dx_n$ with coefficients in an infinite-dimensional polynomial algebra $\mathbb{C}[x_1, \ldots, x_n]$.

The analog of the exterior (de Rham) differential on the algebra $\Omega_n^\bullet$ is the mapping $d : \Omega_n^\bullet \to \Omega_n^\bullet$, defined by

$$d\omega = \sum_{\boldsymbol{k}=0}^{\infty} \sum_{i=1}^{n} \frac{\partial f_{\boldsymbol{k}}}{\partial \theta_i} d\theta_i (d\theta_1)^{k_1} \cdots (d\theta_n)^{k_n}, \quad f_{\boldsymbol{k}} \in \mathrm{Gr}_n,$$

where ω is given by (2.4). It can also be written as $d = \sum_{i=1}^{n} d\theta_i \partial_i$, where it is understood that $\partial_i(d\theta_j) = 0$ for all $i,j = 1, \ldots, n$.

Lemma 2.1. *The exterior differential d on $\Omega_n^\bullet$ satisfies the graded Leibniz rule,*

$$d(\omega_1 \omega_2) = d\omega_1\, \omega_2 + (-1)^{|\omega_1|} \omega_1 d\omega_2,$$

for homogeneous ω_1, and is nilpotent, $d^2 = 0$.

Proof. The graded Leibniz rule for d follows from the corresponding property of partial differentiation operators ∂_i. The property $d^2 = 0$ follows from the commutativity of "differentials" $d\theta_i$ and anticommutativity of the partial derivatives ∂_i. $\qquad\square$

The next result is an analog of the Poincaré lemma for differential forms in anticommuting variables.

Lemma 2.2. *Suppose that $\omega \in \Omega_n^\bullet$ is closed, $d\omega = 0$. Then ω is exact: there is $\eta \in \Omega_n^\bullet$ such that $\omega = d\eta$.*

Definition. A 2-form $\omega \in \Omega_n^\bullet$,

$$(2.5) \qquad \omega = \tfrac{1}{2} \sum_{i,j=1}^{n} \omega^{ij} \, d\theta_i d\theta_j, \quad \omega^{ij} = \omega^{ji} \in \mathrm{Gr}_n,$$

is called a *symplectic form on a Grassmann algebra* if it is closed, $d\omega = 0$, and is non-degenerate: the $n \times n$ symmetric matrix $\{\omega^{ij}\}_{i,j=1}^{n}$ is invertible in Gr_n.

In particular, the 2-form ω with constant coefficients $\omega^{ij} \in \mathbb{C}$ is always closed, and ω is symplectic if and only if the matrix $\{\omega^{ij}\}_{i,j=1}^{n}$ is non-degenerate. Since every quadratic form over $\mathbb{C}$ can be written as a sum of squares, we can always assume that generators $\theta_1, \dots, \theta_n$ are chosen such that the symplectic form ω with constant coefficients has a canonical form

$$(2.6) \qquad \omega = \tfrac{1}{2} \sum_{i=1}^{n} d\theta_i d\theta_i.$$

With every symplectic form (2.5) there is an associated *Poisson bracket on Grassmann algebra*, defined by

$$(2.7) \qquad \{f, g\} = \sum_{i,j=1}^{n} \left(f \frac{\partial}{\partial \theta_i} \right) \omega_{ij} \left(\frac{\partial}{\partial \theta_j} g \right), \quad f, g \in \mathrm{Gr}_n,$$

where $\{\omega_{ij}\}_{i,j=1}^{n}$ is the inverse matrix to $\{\omega^{ij}\}_{i,j=1}^{n}$, and we are using notation for the left and right partial differentiation operators, introduced in the previous section. The following result — an analog of Theorem 2.9 in Section 2.4 of Chapter 1 — is fundamental for formulating Hamiltonian mechanics for the systems with anticommuting variables, which we will discuss in Chapter 8.

Proposition 2.2. *Suppose that all coefficients ω^{ij} of a closed symplectic form ω are even. Then the Poisson bracket map*

$$\{\,,\,\} : \mathrm{Gr}_n \times \mathrm{Gr}_n \to \mathrm{Gr}_n$$

satisfies the following properties.

    (i) *(Graded skew-symmetry)*

$$\{f, g\} = -(-1)^{|f||g|} \{g, f\}.$$

    (ii) *(Graded Leibniz rule)*

$$\{fg, h\} = f\{g, h\} + (-1)^{|f||g|} g\{f, h\}.$$

(iii) *(Graded Jacobi identity)*

$$\{f,\{g,h\}\} + (-1)^{|f|(|g|+|h|)}\{g,\{h,f\}\} + (-1)^{|h|(|f|+|g|)}\{h,\{f,g\}\} = 0$$

for all $f, g, h \in \mathrm{Gr}_n$.

Proof. Part (i) follows from (2.2), the property

$$\frac{\partial}{\partial\theta_i}f = -(-)^{|f|}f\frac{\partial}{\partial\theta_i},$$

and the condition that coefficients ω^{ij} are even. The graded Leibniz rule follows from the corresponding property of right partial differentiation operators. The graded Jacobi identity for the Poisson bracket

$$\{f,g\} = \sum_{i=1}^{n}\left(f\frac{\partial}{\partial\theta_i}\right)\left(\frac{\partial}{\partial\theta_i}g\right),$$

which corresponds to the canonical symplectic form (2.6), can be verified by a direct computation. The proof of a general case is left to the reader. $\square$

Problem 2.1. Complete the proof of Lemma 2.1.

Problem 2.2. Prove Lemma 2.2.

Problem 2.3. Complete the proof of Proposition 2.2.

2.3. Berezin integral. There is a principal difference between differential forms in commuting and in anticommuting variables. The former can be differentiated and integrated, and the differential and integral are related by the Stokes' formula. However, the latter can only be differentiated. Still, there is an analog of the integration over anticommuting variables.

Definition. The integral on a Grassmann algebra Gr_n with an ordered set of generators $\theta_1, \ldots, \theta_n$ *(Berezin integral)* is a linear functional $B : \mathrm{Gr}_n \to \mathbb{C}$, defined by

$$B(f) = f^{12\ldots n},$$

where

$$f = \sum_{k=0}^{n}\sum_{1\le i_1 < \cdots < i_k \le n} f^{i_1\ldots i_k}\theta_{i_1}\ldots\theta_{i_k} \in \mathrm{Gr}_n.$$

It is traditional to write the Berezin integral in the form

$$B(f) = \int f(\boldsymbol{\theta})d\theta_1\ldots d\theta_n,$$

where $\boldsymbol{\theta} = (\theta_1,\ldots,\theta_n)$, as if $f = f(\theta_1,\ldots,\theta_n)$ was actually a "function of anticommuting variables". It follows from the definition of partial differentiation operators that

$$\int f(\boldsymbol{\theta})d\theta_1\ldots d\theta_n = \frac{\partial}{\partial\theta_n}\cdots\frac{\partial}{\partial\theta_1}f,$$

which implies

$$\int \frac{\partial}{\partial \theta_i} f(\boldsymbol{\theta}) d\theta_1 \dots d\theta_n = 0.$$

This leads to the following integration by parts formula for the Berezin integral:

$$\int f(\boldsymbol{\theta})\Big(\frac{\partial}{\partial \theta_i} g\Big)(\boldsymbol{\theta}) d\theta_1 \dots d\theta_n = \int \Big(f \frac{\partial}{\partial \theta_i}\Big)(\boldsymbol{\theta}) g(\boldsymbol{\theta})\, d\theta_1 \dots d\theta_n$$

for homogeneous $f, g \in \mathrm{Gr}_n$.

Remark. The Berezin integral is not an integral in the sense of integration theory. It is defined as a linear functional on a Grassmann algebra Gr_n and it depends on the ordering of the generators $\theta_1, \dots, \theta_n$ of Gr_n, which is symbolized by $d\theta_1 \dots d\theta_n$. For each $\sigma \in \mathrm{Sym}_n$,

$$\int f(\boldsymbol{\theta}) d\theta_1 \dots d\theta_n = (-1)^{\varepsilon(\sigma)} \int f(\boldsymbol{\theta}) d\theta_{\sigma(1)} \dots d\theta_{\sigma(n)},$$

where $\varepsilon(\sigma)$ is the parity of a permutation σ.

Remark. Using the embeddings $\mathrm{Gr}_k \subset \mathrm{Gr}_n$ for $k \leq n$, physicists usually define the Berezin integral as a "repeated integral" starting from the following "one-dimensional integrals":

$$\int d\theta_i = 0, \quad \int \theta_i d\theta_i = 1, \quad i = 1, \dots, n.$$

Lemma 2.3 (Change of variables for Berezin integral). *Let $\theta_1, \dots, \theta_n$ and $\tilde{\theta}_1, \dots, \tilde{\theta}_n$ be two sets of generators of the Grassmann algebra Gr_n, related by $\theta_i = \sum_{j=1}^n a_{ij} \tilde{\theta}_j$, where the $n \times n$ matrix $A = \{a_{ij}\}_{i,j=1}^n$ is non-degenerate. Then*

$$\int f(\boldsymbol{\theta}) d\theta_1 \dots d\theta_n = \frac{1}{\det A} \int \tilde{f}(\tilde{\boldsymbol{\theta}}) d\tilde{\theta}_1 \dots d\tilde{\theta}_n,$$

where $\tilde{f}(\tilde{\boldsymbol{\theta}}) = f(\boldsymbol{\theta}) = f(\sum_{j=1}^n a_{1j} \tilde{\theta}_j, \dots, \sum_{j=1}^n a_{nj} \tilde{\theta}_j)$.

Proof. By multi-linear algebra, $\tilde{f}^{12\dots n} = f^{12\dots n} \det A$. $\qquad\square$

Remark. According to the lemma, the "density" $d\theta_1 \dots d\theta_n$ has the transformation law

$$(2.8) \qquad d\theta_1 \dots d\theta_n = \frac{1}{\det A}\, d\tilde{\theta}_1 \dots d\tilde{\theta}_n$$

under the change of variables $\theta_i = \sum_{j=1}^n a_{ij} \tilde{\theta}_j$. This differs from the usual change of variables formula for the Lebesgue integral

$$(2.9) \qquad \int_{\mathbb{R}^n} f(x_1, \dots, x_n) dx_1 \dots dx_n = |\det A| \int_{\mathbb{R}^n} \tilde{f}(y_1, \dots, y_n) dy_1 \dots dy_n,$$

or $dx_1 \ldots dx_n = |\det A| dy_1 \ldots dy_n$, where $x_i = \sum_{j=1}^{n} a_{ij} y_j$. Of course, the Berezin integral is rather a multiple derivative than an integral with respect to a measure, which explains this profound difference.

Let $A = \{a_{ij}\}_{i,j=1}^{n}$ be an $n \times n$ skew-symmetric matrix. For even $n = 2m$ its *Pfaffian* $\mathrm{Pf}(A)$ is defined by

$$\mathrm{Pf}(A) = \frac{1}{m! 2^m} \sum_{\sigma \in \mathrm{Sym}_n} (-1)^{\varepsilon(\sigma)} a_{\sigma(1)\sigma(2)} \cdots a_{\sigma(n-1)\sigma(n)},$$

where $\varepsilon(\sigma)$ is the parity of a permutation σ. By definition, $\mathrm{Pf}(A) = 0$ when n is odd.

Proposition 2.3 (Gaussian integration for anticommuting variables). *Let $A = \{a_{ij}\}_{i,j=1}^{n}$ be an $n \times n$ skew-symmetric matrix. Then*

(i)

$$\int \exp\left\{ \tfrac{1}{2} \sum_{i,j=1}^{n} a_{ij} \theta_i \theta_j \right\} d\theta_1 \ldots d\theta_n = \mathrm{Pf}(A).$$

(ii) *For any non-degenerate $n \times n$ matrix C*

$$\mathrm{Pf}(CAC^t) = \mathrm{Pf}(A) \det C.$$

(iii)

$$\mathrm{Pf}(A)^2 = \det A.$$

Proof. Part (i) obviously holds for odd n since the integrand is an even element of Gr_n. By the definition of the Pfaffian, we get for $n = 2m$,

$$\frac{1}{m! 2^m} \left(\sum_{i,j=1}^{n} a_{ij} \theta_i \theta_j \right)^m = \mathrm{Pf}(A)\, \theta_1 \ldots \theta_n,$$

and expanding the exponent into power series, we obtain

$$\int \exp\left\{ \tfrac{1}{2} \sum_{i,j=1}^{n} a_{ij} \theta_i \theta_j \right\} d\theta_1 \ldots d\theta_n = \mathrm{Pf}(A) \int \theta_1 \ldots \theta_n\, d\theta_1 \ldots d\theta_n = \mathrm{Pf}(A).$$

Part (ii) follows from part (i) and Lemma 2.3. Part (iii) is a classical result, which can be proved by the Berezin integral as follows. Suppose first that A is real-valued. There exists an orthogonal matrix C with determinant 1 such that

$$CAC^{-1} = \begin{pmatrix} 0 & \lambda_1 & \cdots & 0 & 0 \\ -\lambda_1 & 0 & \cdots & 0 & 0 \\ \vdots & \vdots & \ddots & \vdots & \vdots \\ 0 & 0 & \cdots & 0 & \lambda_m \\ 0 & 0 & \cdots & -\lambda_m & 0 \end{pmatrix}$$

is a block-diagonal matrix. Using part (ii) with this matrix C, we obtain

$$\mathrm{Pf}(A) = \int e^{\lambda_1 \tilde{\theta}_1 \tilde{\theta}_2 + \cdots + \lambda_m \tilde{\theta}_{2m-1} \tilde{\theta}_{2m}} d\tilde{\theta}_1 \ldots d\tilde{\theta}_{2m} = \lambda_1 \ldots \lambda_m,$$

so that $\mathrm{Pf}(A)^2 = \det A$. This relation holds for complex-valued A, since both sides are polynomials in variables a_{ij}, $1 \leq i < j \leq n$, which coincide for real a_{ij}. $\qquad\square$

For a Grassmann algebra $\mathrm{Gr}_{2n} = \mathbb{C}[\theta_1, \ldots, \theta_n, \bar{\theta}_1, \ldots, \bar{\theta}_n]$ with $2n$ generators denote by $\int d\boldsymbol{\theta} d\bar{\boldsymbol{\theta}} = \int d\theta_1 d\bar{\theta}_1 \ldots d\theta_n d\bar{\theta}_n$ the corresponding Berezin integral,

$$\int f(\boldsymbol{\theta}, \bar{\boldsymbol{\theta}}) d\boldsymbol{\theta} d\bar{\boldsymbol{\theta}} = \frac{\partial}{\partial \bar{\theta}_n} \frac{\partial}{\partial \theta_n} \cdots \frac{\partial}{\partial \bar{\theta}_1} \frac{\partial}{\partial \theta_1} f, \quad f \in \mathbb{C}[\theta_1, \ldots, \theta_n, \bar{\theta}_1, \ldots, \bar{\theta}_n].$$

Lemma 2.4. *For any* $n \times n$ *matrix* $A = \{a_{ij}\}_{i,j=1}^n$,

$$\int \exp\left\{ \sum_{i,j=1}^n a_{ij} \theta_i \bar{\theta}_j \right\} d\boldsymbol{\theta} d\bar{\boldsymbol{\theta}} = \det A.$$

Proof. It follows from the definition of a matrix determinant that

$$(2.10) \qquad \frac{1}{n!}\left(\sum_{i,j=1}^n a_{ij} \theta_i \bar{\theta}_j \right)^n = \det A \, \theta_1 \bar{\theta}_1 \ldots \theta_n \bar{\theta}_n. \qquad\square$$

Definition. An involution on a Grassmann algebra Gr_n over $\mathbb{C}$ is a complex anti-linear mapping $\mathrm{Gr}_n \ni f \mapsto f^* \in \mathrm{Gr}_n$ satisfying $(f^*)^* = f$ and $(fg)^* = g^* f^*$ for all $f, g \in \mathrm{Gr}_n$.

The Grassmann algebra $\mathbb{C}[\theta_1, \ldots, \theta_n, \bar{\theta}_1, \ldots, \bar{\theta}_n]$ has a natural involution defined on generators by $(\theta_1)^* = \bar{\theta}_1, (\bar{\theta}_1)^* = \theta_1, \ldots, (\theta_n)^* = \bar{\theta}_n, (\bar{\theta}_n)^* = \theta_n$. In particular, for

$$f(\boldsymbol{\theta}) = \sum_{k=0}^n \sum_{1 \leq i_1 < \cdots < i_k \leq n} f^{i_1 \ldots i_k} \theta_{i_1} \ldots \theta_{i_k} \in \mathrm{Gr}_n \subset \mathrm{Gr}_{2n}$$

we have

$$f(\boldsymbol{\theta})^* = \overline{f(\boldsymbol{\theta})} = \sum_{k=0}^n \sum_{1 \leq i_1 < \cdots < i_k \leq n} \overline{f^{i_1 \ldots i_k}} \, \bar{\theta}_{i_k} \ldots \bar{\theta}_{i_1} \in \mathrm{Gr}_{2n}.$$

The next lemma expresses the inner product on the Grassmann algebra Gr_n, introduced in Section 2.1, in terms of the Berezin integral.

Lemma 2.5. *The standard inner product (2.3) on the Grassmann algebra* $\mathrm{Gr}_n = \mathbb{C}[\theta_1, \ldots, \theta_n]$ *is given by the following Berezin integral over the Grassmann algebra* $\mathrm{Gr}_{2n} = \mathbb{C}[\theta_1, \ldots, \theta_n, \bar{\theta}_1, \ldots, \bar{\theta}_n]$:

$$(2.11) \qquad (f_1, f_2) = \int f_1(\boldsymbol{\theta})\overline{f_2(\boldsymbol{\theta})}e^{-\bar{\boldsymbol{\theta}}\boldsymbol{\theta}}d\boldsymbol{\theta}d\bar{\boldsymbol{\theta}},$$

where $\bar{\boldsymbol{\theta}}\boldsymbol{\theta} = \bar{\theta}_1\theta_1 + \cdots + \bar{\theta}_n\theta_n$.

Proof. Put $f_1(\boldsymbol{\theta}) = \theta_{i_1} \ldots \theta_{i_k}$ and $f_2(\boldsymbol{\theta}) = \theta_{j_1} \ldots \theta_{j_l}$. It is clear that the integral (2.11) is 0, unless $k = l$ and $i_1 = j_1, \ldots, i_k = j_k$, in which case we have

$$(\theta_{i_1} \ldots \theta_{i_k}, \theta_{i_1} \ldots \theta_{i_k}) = \int \theta_{i_1} \ldots \theta_{i_k}\bar{\theta}_{i_k} \ldots \bar{\theta}_{i_1} e^{-(\bar{\theta}_1\theta_1 + \cdots + \bar{\theta}_n\theta_n)}d\boldsymbol{\theta}d\bar{\boldsymbol{\theta}}$$

$$= \int \theta_1 \ldots \theta_n\bar{\theta}_n \ldots \bar{\theta}_1 \, d\boldsymbol{\theta}d\bar{\boldsymbol{\theta}}$$

$$= \int \theta_1\bar{\theta}_1 \ldots \theta_n\bar{\theta}_n \, d\boldsymbol{\theta}d\bar{\boldsymbol{\theta}} = 1. \qquad \square$$

The following result was already stated in Section 2.1. Lemma 2.5 allows us to prove it in a way that is reminiscent of a holomorphic representation (see Section 2.7 of Chapter 2).

Corollary 2.1. *The operators* ∂_i *and* $\hat{\theta}_i$, $i = 1, \ldots, n$, *are adjoint with respect to the inner product on* Gr_n.

Proof. Using Lemma 2.5, formulas $\partial_i\overline{f(\boldsymbol{\theta})} = 0$, $\partial_i e^{-\bar{\boldsymbol{\theta}}\boldsymbol{\theta}} = \bar{\theta}_i e^{-\bar{\boldsymbol{\theta}}\boldsymbol{\theta}}$, and the integration by parts formula, we obtain

$$(\partial_i f_1, f_2) = \int \partial_i f_1(\boldsymbol{\theta})\overline{f_2(\boldsymbol{\theta})}e^{-\bar{\boldsymbol{\theta}}\boldsymbol{\theta}}d\boldsymbol{\theta}d\bar{\boldsymbol{\theta}}$$

$$= -(-1)^{|f_1|+|f_2|} \int f_1(\boldsymbol{\theta})\overline{f_2(\boldsymbol{\theta})}\,\bar{\theta}_i e^{-\bar{\boldsymbol{\theta}}\boldsymbol{\theta}}d\boldsymbol{\theta}d\bar{\boldsymbol{\theta}}$$

$$= -(-1)^{|f_1|+|f_2|} \int f_1(\boldsymbol{\theta})\overline{\hat{\theta}_i f_2(\boldsymbol{\theta})}e^{-\bar{\boldsymbol{\theta}}\boldsymbol{\theta}}d\boldsymbol{\theta}d\bar{\boldsymbol{\theta}}$$

$$= (f_1, \hat{\theta}_i f_2),$$

since the last integral and $(f_1, \hat{\theta}_i f_2)$ are both 0 unless $|f_1| + |f_2|$ is odd. $\square$

Problem 2.4. Evaluate the Berezin integral

$$\int \exp\left\{\tfrac{1}{2}\sum_{i,j=1}^{n} a_{ij}\theta_i\theta_j + \sum_{k=1}^{n} \eta_k\theta_k\right\}d\theta_1 \ldots d\theta_n,$$

where $\eta_1, \ldots, \eta_n$ are Grassmann variables.

Problem 2.5. Prove formula (2.10).

Problem 2.6. Prove that

$$\int f(\boldsymbol{\theta})e^{-\bar{\boldsymbol{\theta}}\boldsymbol{\theta}}\,d\boldsymbol{\theta}\,d\bar{\boldsymbol{\theta}} = f(0)$$

— the constant term of expansion of $f(\boldsymbol{\theta})$ into the sum of monomials in $\mathbb{C}[\theta_1,\ldots,\theta_n]$.

3. Graded linear algebra

3.1. Graded vector spaces and superalgebras.

The notions of graded vector spaces and superalgebras, introduced in this section, allow us to consider commuting and anticommuting variables on the same footing.

Definition. A $\mathbb{Z}/2\mathbb{Z}$ *graded vector space* (*graded vector space* for brevity, or *super vector space*) over $\mathbb{C}$ is a vector space W with a decomposition

$$W = W^0 \oplus W^1$$

into even and odd subspaces. The elements in $W^0 \cup W^1 \setminus \{0\}$ are called homogeneous and the parity is a map $|\cdot| : W^0 \cup W^1 \setminus \{0\} \to \{0,1\}$ such that $|w| = 0$ for $w \in W^0$ and $|w| = 1$ for $w \in W^1$.

We reserve the notation V for ordinary (even) vector spaces, denoting graded vector spaces by W. If W is finite-dimensional, we define *graded dimension* as a pair $(\dim W^0, \dim W^1)$, usually denoted by $n_0|n_1$, where $n_i = \dim W^i$, $i = 0,1$. When $W^0 = \mathbb{C}^p$ and $W^1 = \mathbb{C}^q$, the corresponding graded vector space W is denoted by $\mathbb{C}^{p|q}$. A fermion Hilbert space $\mathscr{H}_F$ is a graded vector space with the even and odd subspaces given by decomposition (1.9):

$$\mathscr{H}_F^0 = \bigoplus_{k \text{ even}} \mathscr{H}_k, \quad \mathscr{H}_F^1 = \bigoplus_{k \text{ odd}} \mathscr{H}_k.$$

The graded dimension of $\mathscr{H}_F$ is $2^{n-1}|2^{n-1}$.

Direct sums and tensor products of graded vector spaces are defined in the same way as for ordinary vector spaces. For the direct sums the homogeneous subspaces are defined by

$$(W_1 \oplus W_2)^k = W_1^k \oplus W_2^k, \quad k = 0,1,$$

and for the tensor products they are defined as

$$(W_1 \otimes W_2)^k = \bigoplus_{i+j=k} W_1^i \otimes W_2^j, \quad k = 0,1.$$

The difference between ordinary and graded vector spaces becomes transparent in the definition of the corresponding tensor categories. Namely, the associativity morphism

$$c_{W_1 W_2 W_3} : W_1 \otimes (W_2 \otimes W_3) \to (W_1 \otimes W_2) \otimes W_3$$

for graded vector spaces is defined by the same formula

$$c_{W_1 W_2 W_3}(w_1 \otimes (w_2 \otimes w_3)) = (w_1 \otimes w_2) \otimes w_3$$

as in the case of ordinary vector spaces, whereas the commutativity morphism

$$\sigma_{W_1 W_2} : W_1 \otimes W_2 \to W_2 \otimes W_1$$

is defined for the homogeneous elements by

$$\sigma_{W_1 W_2}(w_1 \otimes w_2) = (-1)^{|w_1||w_2|} w_2 \otimes w_1.$$

The tensor algebra $T(W)$ of a graded vector space W is defined using the associativity morphism. However the exterior algebra $\Lambda^\bullet W$ and symmetric algebra $\mathrm{Sym}(W)$ of W are defined as quotient algebras of $T(W)$ by using the commutativity morphism. Namely,

$$\mathrm{Sym}(W) = T(W)/I,$$

where I is the two-sided ideal in $T(W)$ generated by $w_1 \otimes w_2 - \sigma(w_2 \otimes w_1)$, $w_1, w_2 \in W$, and

$$\Lambda^\bullet(W) = T(W)/J,$$

where J is the two-sided ideal in $T(W)$ generated by $w_1 \otimes w_2 + \sigma(w_2 \otimes w_1)$, $w_1, w_2 \in W$. Here $\sigma = \sigma_{W,W}$ is the commutativity morphism.

Definition. Let $W = W^0 \oplus W^1$ be a graded vector space. A *parity-reversed* vector space ΠW is a graded vector space with $(\Pi W)^0 = W^1$ and $(\Pi W)^1 = W^0$.

It immediately follows from the definitions that for even vector space V

$$(3.1) \qquad \mathrm{Sym}(\Pi V) = \Lambda^\bullet(V) \quad \text{and} \quad \Lambda^\bullet(\Pi V) = \mathrm{Sym}(V).$$

Definition. A *superalgebra* over $\mathbb{C}$ is a graded vector space $A = A^0 \oplus A^1$ with a $\mathbb{C}$-algebra structure such that $1 \in A^0$ and

$$A^0 \cdot A^0 \subset A^0, \quad A^0 \cdot A^1 \subset A^1, \quad A^1 \cdot A^1 \subset A^0.$$

A superalgebra A is a *commutative superalgebra* if

$$a \cdot b = (-1)^{|a||b|} b \cdot a$$

for homogeneous elements $a, b \in A$.

An even superalgebra A is just an ordinary $\mathbb{C}$-algebra.

Definition. A left (super) module for a superalgebra A is a graded vector space M with the linear map

$$A \otimes M \ni a \otimes m \mapsto a \cdot m \in M$$

such that $|a \cdot m| = (|a| + |m|) \bmod 2$ for homogeneous $a \in A$ and $m \in M$, and $ab \cdot m = a \cdot (b \cdot m)$ for all $a, b \in A$ and $m \in M$.

Problem 3.1. Let $\sigma \in \mathrm{Sym}_n$. Show that the isomorphism

$$W_1 \otimes \cdots \otimes W_n \simeq W_{\sigma^{-1}(1)} \otimes \cdots \otimes W_{\sigma^{-1}(n)},$$

induced by the commutativity morphism of graded vector spaces, is well defined — does not depend on the representation of σ as a product of transpositions in Sym_n.

Problem 3.2. Show that for a graded vector space W the algebras $\mathrm{Sym}(W)$ and $\Lambda^\bullet(W)$ are superalgebras.

3.2. Examples of superalgebras.

Example 3.1 (Tensor algebra). The tensor algebra

$$T(V) = \bigoplus_{k=0}^{\infty} V^{\otimes k}, \quad V^0 = \mathbb{C} \cdot 1,$$

of an even vector space V is a superalgebra. The multiplication is given by the tensor product, and even and odd subspaces — by

$$T(V)^0 = \bigoplus_{k \text{ even}} V^{\otimes k}, \quad T(V)^1 = \bigoplus_{k \text{ odd}} V^{\otimes k}.$$

Example 3.2 (Symmetric algebra). The symmetric algebra $\mathrm{Sym}(V)$ of an even vector space V is a commutative algebra. A choice of a basis $x_1, \ldots, x_n$ of V establishes the isomorphism $\mathrm{Sym}(V) \simeq \mathbb{C}[x_1, \ldots, x_n]$ — a polynomial algebra in commuting variables $x_1, \ldots, x_n$.

Example 3.3 (Exterior algebra). The exterior algebra $\Lambda^\bullet V$ of an even vector space V is a commutative superalgebra, with multiplication given by the wedge product; the even and odd subspaces are images of the corresponding subspaces of $T(V)$ under the surjective mapping

$$T(V) \to \Lambda^\bullet V = T(V)/J,$$

where J is the two-sided ideal in $T(V)$, generated by the elements $u \otimes v + v \otimes u$, $u, v \in V$. According to (3.1), $\Lambda^\bullet(V) = \mathrm{Sym}(\Pi V)$. The choice of a basis $\theta_1, \ldots, \theta_n$ of the odd vector space ΠV establishes the isomorphism

$$\Lambda^\bullet(V) \simeq \mathbb{C}[\theta_1, \ldots, \theta_n].$$

Here $\mathbb{C}[\theta_1, \ldots, \theta_n]$ is the Grassmann algebra — a polynomial algebra in anticommuting variables $\theta_1, \ldots, \theta_n$.

Example 3.4 (Algebra of differential forms). Let M be an n-dimensional manifold. The graded algebra $\mathcal{A}^\bullet(M)$ of smooth differential forms on M is a commutative superalgebra.

Example 3.5 (Clifford algebra). The Clifford algebra $C(V)$ of a quadratic vector space (V, Q) is a superalgebra, with the multiplication and grading descending from the tensor algebra $T(V)$ under the surjective mapping

$$T(V) \to \Lambda^\bullet V = T(V)/J,$$

where now J is the two-sided ideal in $T(V)$, generated by the elements $u \otimes v + v \otimes u - 2\Phi(u,v) \cdot 1$, $u, v \in V$. The natural map $V \hookrightarrow C(V)$ is injective, and V is identified with its image in $C(V)$. The elements of V are odd in $C(V)$. The fermion Hilbert space $\mathscr{H}_F$ is a left supermodule for C_{2n}, and the $\mathbb{C}$-algebra isomorphism $\rho : C_{2n} \to \operatorname{End}(\mathscr{H}_F)$ (see Proposition 1.1) is an isomorphism of superalgebras.

Example 3.6 (Graded matrix algebra). Let W be a graded vector space. The vector space $\operatorname{End}(W)$ of all endomorphisms of W is a graded vector space: the even subspace consists of all endomorphisms which preserve the grading in W, and the odd subspace consists of those which reverse the grading. The vector space $\operatorname{End}(W)$ is a superalgebra with the product given by the composition of endomorphisms, and W is a module for $\operatorname{End}(W)$. When $W = \mathbb{C}^{p|q}$, the superalgebra $\operatorname{End}(W)$ is usually denoted by $\operatorname{Mat}(p|q)$. Its elements can be conveniently represented by 2×2 block matrices

$$A = \begin{pmatrix} A_{11} & A_{12} \\ A_{21} & A_{22} \end{pmatrix},$$

where A_{11}, A_{12}, A_{21}, and A_{22} are, respectively, matrices of orders $p \times p$, $p \times q$, $q \times p$, and $q \times q$. The even and odd elements of $\operatorname{Mat}(p|q)$ are, respectively, block-diagonal and anti-diagonal matrices $\left(\begin{smallmatrix} A_{11} & 0 \\ 0 & A_{22} \end{smallmatrix}\right)$ and $\left(\begin{smallmatrix} 0 & A_{12} \\ A_{21} & 0 \end{smallmatrix}\right)$.

Example 3.7 (Lie superalgebra). A graded vector space $\mathfrak{g}$ is called a *Lie superalgebra* (or, *super Lie algebra*) if it carries a Lie superbracket — a linear mapping $[\ ,\ ] : \mathfrak{g} \otimes \mathfrak{g} \to \mathfrak{g}$ satisfying the following properties.

(i) (Super skew-symmetry)

$$[x,y] = -(-1)^{|x||y|}[y,x]$$

for homogeneous $x, y \in \mathfrak{g}$.

(ii) (Super Jacobi identity)

$$[x,[y,z]] + (-1)^{|x|(|y|+|z|)}[y,[z,x]] + (-1)^{|z|(|x|+|y|)}[z,[x,y]] = 0$$

for homogeneous $x, y, z \in \mathfrak{g}$.

According to Proposition 2.2, a Grassmann algebra Gr_n with a Poisson bracket (2.7) is a Lie superalgebra.

Corresponding to classical simple Lie algebras there are associated Lie superalgebras.

Problem 3.3. Verify all the statements in this section.

Problem 3.4. Show that a superalgebra A carries a Lie superalgebra structure with a Lie superbracket defined by

$$[a,b] = ab - (-1)^{|a||b|}ba$$

for homogeneous $a, b \in A$.

3.3. Supertrace and Berezinian. Let $\mathcal{A} = \mathcal{A}^0 \oplus \mathcal{A}^1$ be a commutative superalgebra. We have the following general notion of a graded matrix algebra.

Definition. A graded matrix algebra with coefficients in $\mathcal{A}$ is a superalgebra $\mathrm{Mat}_{\mathcal{A}}(p|q)$ of 2×2 block-matrices

$$A = \begin{pmatrix} A_{11} & A_{12} \\ A_{21} & A_{22} \end{pmatrix}$$

where A_{11}, A_{12}, A_{21}, and A_{22} are, respectively, matrices of orders $p \times p$, $p \times q$, $q \times p$, and $q \times q$ with elements in $\mathcal{A}$. The element $A \in \mathrm{Mat}_{\mathcal{A}}(p|q)$ is even if corresponding matrices A_{11} and A_{22} consist of even elements of $\mathcal{A}$, and matrices A_{12} and A_{21} consist of odd elements of $\mathcal{A}$. The element $A \in \mathrm{Mat}_{\mathcal{A}}(p|q)$ is odd if matrices A_{11} and A_{22} consist of odd elements of $\mathcal{A}$, and A_{12} and A_{21} consist of even elements of $\mathcal{A}$. The graded vector space $\mathrm{Mat}_{\mathcal{A}}(p|q)$ is a superalgebra with the product given by the matrix multiplication.

The algebra $\mathrm{Mat}(p|q)$ in Example 3.6 corresponds to the case $\mathcal{A} = \mathbb{C}$; another interesting example is when $\mathcal{A}$ is a Grassmann algebra.

It is quite remarkable that such basic notions of linear algebra as trace and determinant admit non-trivial generalizations for graded matrix algebras.

Definition. A *supertrace* on a graded matrix algebra is a linear mapping $\mathrm{Tr}_s : \mathrm{Mat}_{\mathcal{A}}(p|q) \to \mathcal{A}$, defined by

$$\mathrm{Tr}_s A = \mathrm{Tr}\, A_{11} - (-1)^{|A|}\, \mathrm{Tr}\, A_{22}, \quad A = \begin{pmatrix} A_{11} & A_{12} \\ A_{21} & A_{22} \end{pmatrix} \in \mathrm{Mat}_{\mathcal{A}}(p|q),$$

where Tr is the ordinary matrix trace.

Proposition 3.1 (The cyclic property of the supertrace).

$$\mathrm{Tr}_s AB = (-1)^{|A||B|}\, \mathrm{Tr}_s BA, \quad A, B \in \mathrm{Mat}_{\mathcal{A}}(p|q).$$

Proof. We check only the case when both 2×2 block-matrices A and B are even elements of $\mathrm{Mat}_{\mathcal{A}}(p|q)$. Other cases are treated similarly and are left to the reader. We have, using the cyclic property of the trace,

$$\begin{aligned} \mathrm{Tr}_s AB &= \mathrm{Tr}(A_{11}B_{11} + A_{12}B_{21}) - \mathrm{Tr}(A_{21}B_{12} + A_{22}B_{22}) \\ &= \mathrm{Tr}(B_{11}A_{11} - B_{21}A_{12}) - \mathrm{Tr}(-B_{12}A_{21} + B_{22}A_{22}) = \mathrm{Tr}_s BA. \quad \square \end{aligned}$$

The superalgebra $\mathrm{End}(\mathscr{H}_F)$ corresponds to the case $\mathcal{A} = \mathbb{C}$ and is isomorphic to $\mathrm{Mat}(2^{n-1}|2^{n-1})$. The supertace on $\mathrm{End}(\mathscr{H}_F)$ is given by

$$(3.2) \qquad \mathrm{Tr}_s A = \mathrm{Tr}\, A_{11} - \mathrm{Tr}\, A_{22} = \mathrm{Tr}\, A\Gamma, \quad A \in \mathrm{End}(\mathscr{H}_F),$$

where Γ is the chirality operator (see Section 1.2). It has the following 2×2 block-matrix form:

$$\Gamma = \begin{pmatrix} I & 0 \\ 0 & -I \end{pmatrix},$$

where I stands for the identity operator in $\mathscr{H}_F^0$ and $\mathscr{H}_F^1$.

It is a non-trivial problem to define a *superdeterminant* — a natural analog of the determinant for graded matrix algebras, which is multiplicative and generalizes the rule $\det e^A = e^{\operatorname{Tr} A}$. The corresponding notion, which is defined only for invertible $A \in \operatorname{Mat}_{\mathcal{A}}(p|q)$, was introduced by F.A. Berezin. It is now commonly called *Berezinian* and is denoted by $\operatorname{Ber}(A)$.

Consider first the case of even diagonal $A = \begin{pmatrix} A_{11} & 0 \\ 0 & A_{22} \end{pmatrix}$. The relation

$$\operatorname{Ber}(e^A) = e^{\operatorname{Tr}_s A}$$

defines the Berezinian by

$$\operatorname{Ber}(A) = \det A_{11} \det A_{22}^{-1}.$$

Thus in this case the even $q \times q$ matrix A_{22} is necessarily invertible, i.e., $\det A_{22}$ is an invertible element of the commutative algebra $\mathcal{A}^0$, so that the inverse matrix A_{22}^{-1} exists. Now consider the general even $A \in \operatorname{Mat}_{\mathcal{A}}(p|q)$, and suppose that A_{22} is invertible. It is easy to verify the following analog of the Gauss decomposition:

$$\begin{pmatrix} A_{11} & A_{12} \\ A_{21} & A_{22} \end{pmatrix} = \begin{pmatrix} I_p & A_{12}A_{22}^{-1} \\ 0 & I_q \end{pmatrix} \begin{pmatrix} A_{11} - A_{12}A_{22}^{-1}A_{21} & 0 \\ 0 & A_{22} \end{pmatrix} \begin{pmatrix} I_p & 0 \\ A_{22}^{-1}A_{21} & I_q \end{pmatrix},$$

where I_p and I_q are, respectively $p \times p$ and $q \times q$ identity matrices. This justifies the following definition.

Definition. Let $A \in \operatorname{Mat}_{\mathcal{A}}(p|q)$ be even and such that the corresponding even $q \times q$ matrix A_{22} is invertible. Then the Berezinian (superdeterminant) of A is given by

$$\operatorname{Ber}(A) = \det(A_{11} - A_{12}A_{22}^{-1}A_{21}) \det A_{22}^{-1}.$$

Assuming that the $p \times p$ matrix A_{11} is invertible, we get the decomposition

$$\begin{pmatrix} A_{11} & A_{12} \\ A_{21} & A_{22} \end{pmatrix} = \begin{pmatrix} I_p & 0 \\ A_{21}A_{11}^{-1} & I_q \end{pmatrix} \begin{pmatrix} A_{11} & 0 \\ 0 & A_{22} - A_{21}A_{11}^{-1}A_{12} \end{pmatrix} \begin{pmatrix} I_p & A_{11}^{-1}A_{12} \\ 0 & I_q \end{pmatrix},$$

which suggests that also

$$\operatorname{Ber}(A) = \det A_{11} \det(A_{22} - A_{21}A_{11}^{-1}A_{12})^{-1}.$$

It is a fundamental fact that for even invertible $A \in \operatorname{Mat}_{\mathcal{A}}(p|q)$ these two formulas for the Berezinian coincide.

Theorem 3.1. *Let*

$$A = \begin{pmatrix} A_{11} & A_{12} \\ A_{21} & A_{22} \end{pmatrix}$$

be an even element of $\mathrm{Mat}_{\mathcal{A}}(p|q)$. *Then*

    (i) *A is invertible if and only if the matrices A_{11} and A_{22} are invertible.*

    (ii) *For invertible A,* $\mathrm{Ber}(A)$ *is an invertible element of $\mathcal{A}^0$ satisfying*

$$\mathrm{Ber}(A) = \det(A_{11} - A_{12}A_{22}^{-1}A_{21})\det A_{22}^{-1} = \det A_{11}\det(A_{22} - A_{21}A_{11}^{-1}A_{12})^{-1}$$

    and

$$\mathrm{Ber}(A^{-1}) = \mathrm{Ber}(A)^{-1}.$$

    (iii) *If even $A, B \in \mathrm{Mat}_{\mathcal{A}}(p|q)$ are invertible, then*

$$\mathrm{Ber}(AB) = \mathrm{Ber}(A)\mathrm{Ber}(B).$$

Problem 3.5. Complete the proof of Proposition 3.1.

Problem 3.6. Prove Theorem 3.1.

4. Path integrals for anticommuting variables

4.1. Wick and matrix symbols. Here we describe the calculus of Wick and matrix symbols of operators in the fermion Hilbert space $\mathscr{H}_F$, which is analogous to the treatment of Wick and matrix symbols in the holomorphic representation in Section 2.7 of Chapter 2. As well as in the bosonic case, it is a tradition to work in anti-holomorphic representation by using $\mathscr{H}_F = \mathbb{C}[\bar{\theta}_1, \dots, \bar{\theta}_n]$ as the fermion Hilbert space with the annihilation and creation operators

$$(4.1) \qquad\qquad a_k = \hat{\bar{\theta}}_k \quad \text{and} \quad a_k^* = \frac{\partial}{\partial \bar{\theta}_k}, \quad k = 1, \dots, n.$$

The inner product (2.11) takes the form

$$(4.2) \qquad\qquad (f_1, f_2) = \int f_1(\bar{\boldsymbol{\theta}})\overline{f_2(\bar{\boldsymbol{\theta}})}e^{-\boldsymbol{\theta}\bar{\boldsymbol{\theta}}}d\bar{\boldsymbol{\theta}}d\boldsymbol{\theta},$$

and the monomials

$$f_I(\bar{\boldsymbol{\theta}}) = \bar{\theta}_{i_1}\dots\bar{\theta}_{i_k}, \quad 1 \le i_1 < \dots < i_k \le n,$$

parametrized by subsets $I = \{i_1, \dots, i_k\} \subseteq \{1, \dots, n\}$, form an orthonormal basis in $\mathscr{H}_F$.

Definition. A *matrix symbol* of an operator $A : \mathscr{H}_F \to \mathscr{H}_F$ is an element $\tilde{A}(\bar{\boldsymbol{\theta}}, \boldsymbol{\theta}) \in \mathbb{C}[\bar{\theta}_1, \ldots, \bar{\theta}_n, \theta_1, \ldots, \theta_n]$, defined by

$$\tilde{A}(\bar{\boldsymbol{\theta}}, \boldsymbol{\theta}) = \sum_{I,J} (Af_J, f_I) f_I(\boldsymbol{\theta}) \overline{f_J(\bar{\boldsymbol{\theta}})}$$

$$= \sum_{I,J} (Af_J, f_I)\, \bar{\theta}_{i_1} \ldots \bar{\theta}_{i_k} \theta_{j_l} \ldots \theta_{j_1}.$$

Here summation goes over all subsets $I = \{i_1, \ldots, i_k\}$ and $J = \{j_1, \ldots, j_l\}$ of the set $\{1, \ldots, n\}$, and as in Section 2.3, we denote by $\overline{f(\bar{\boldsymbol{\theta}})}$ a natural involution on the Grassmann algebra $\mathbb{C}[\bar{\theta}_1, \ldots, \bar{\theta}_n, \theta_1, \ldots, \theta_n]$,

$$\overline{f_J(\bar{\boldsymbol{\theta}})} = \theta_{j_l} \ldots \theta_{j_1} \quad \text{for} \quad f_J(\bar{\boldsymbol{\theta}}) = \bar{\theta}_{j_1} \ldots \bar{\theta}_{j_l}.$$

According to Proposition 1.1, $C_{2n} \simeq \mathrm{End}(\mathscr{H}_F)$, so that every operator $A : \mathscr{H}_F \to \mathscr{H}_F$ can be uniquely represented in a *Wick normal form* as follows:

$$A = \sum_{I,J} K_{IJ}\, a_{i_1}^* \ldots a_{i_k}^* a_{j_1} \ldots a_{j_l}.$$

Definition. A *Wick symbol* of an operator $A : \mathscr{H}_F \to \mathscr{H}_F$ is an element $A(\bar{\boldsymbol{\theta}}, \boldsymbol{\theta}) \in \mathbb{C}[\bar{\theta}_1, \ldots, \bar{\theta}_n, \theta_1, \ldots, \theta_n]$, defined by

$$A(\bar{\boldsymbol{\theta}}, \boldsymbol{\theta}) = \sum_{I,J} K_{IJ}\, \bar{\theta}_{i_1} \ldots \bar{\theta}_{i_k} \theta_{j_1} \ldots \theta_{j_l}.$$

Remark. The definition of matrix and Wick symbols in the fermion case repeats verbatim the corresponding definition for the bosonic case in Section 2.7 of Chapter 2. Note, however, that in the fermion case the product $\theta_{j_l} \ldots \theta_{j_1}$ in the definition of the matrix symbol has reverse ordering, as is required by the inner product (4.2) in $\mathbb{C}[\bar{\theta}_1, \ldots, \bar{\theta}_n]$.

To matrix and Wick symbols $\tilde{A}(\bar{\boldsymbol{\theta}}, \boldsymbol{\theta})$ and $A(\bar{\boldsymbol{\theta}}, \boldsymbol{\theta})$ of an operator A one canonically associates elements $\tilde{A}(\bar{\boldsymbol{\theta}}, \boldsymbol{\alpha})$, $A(\bar{\boldsymbol{\theta}}, \boldsymbol{\alpha})$ and $\tilde{A}(\bar{\boldsymbol{\alpha}}, \boldsymbol{\theta})$, $A(\bar{\boldsymbol{\alpha}}, \boldsymbol{\theta})$ in the larger Grassmann algebra

$$\mathbb{C}[\boldsymbol{\alpha}, \bar{\boldsymbol{\alpha}}, \boldsymbol{\theta}, \bar{\boldsymbol{\theta}}] = \mathbb{C}[\alpha_1, \ldots, \alpha_n, \bar{\alpha}_1, \ldots, \bar{\alpha}_n, \theta_1, \ldots, \theta_n \bar{\theta}_1, \ldots, \bar{\theta}_n],$$

by replacing, correspondingly, θ_i by α_i and $\bar{\theta}_i$ by $\bar{\alpha}_i$. The *incomplete Berezin integral* $\int d\boldsymbol{\alpha} d\bar{\boldsymbol{\alpha}}$ on $\mathbb{C}[\boldsymbol{\alpha}, \bar{\boldsymbol{\alpha}}, \boldsymbol{\theta}, \bar{\boldsymbol{\theta}}]$ is defined by

$$\int f d\boldsymbol{\alpha} d\bar{\boldsymbol{\alpha}} = \frac{\partial}{\partial \bar{\alpha}_n} \frac{\partial}{\partial \alpha_n} \cdots \frac{\partial}{\partial \bar{\alpha}_1} \frac{\partial}{\partial \alpha_1} f, \quad f \in \mathbb{C}[\boldsymbol{\alpha}, \bar{\boldsymbol{\alpha}}, \boldsymbol{\theta}, \bar{\boldsymbol{\theta}}],$$

and has the property

$$(4.3) \qquad \int h(\boldsymbol{\theta}, \bar{\boldsymbol{\theta}}) g(\boldsymbol{\alpha}, \bar{\boldsymbol{\alpha}}) d\boldsymbol{\alpha} d\bar{\boldsymbol{\alpha}} = h(\boldsymbol{\theta}, \bar{\boldsymbol{\theta}}) \int g(\boldsymbol{\alpha}, \bar{\boldsymbol{\alpha}}) d\boldsymbol{\alpha} d\bar{\boldsymbol{\alpha}}.$$

We will also use the incomplete Berezin integral $\int d\bar{\theta}d\theta$, defined by

$$\int f d\bar{\theta}d\theta = \frac{\partial}{\partial \theta_n}\frac{\partial}{\partial \bar{\theta}_n}\cdots\frac{\partial}{\partial \theta_1}\frac{\partial}{\partial \bar{\theta}_1}f, \quad f \in \mathbb{C}[\alpha, \bar{\alpha}, \theta, \bar{\theta}].$$

As follows from the proof of Corollary 2.1,

$$(4.4) \qquad \int (a_k^* f_1)\overline{f_2}e^{-\theta\bar{\theta}}d\bar{\theta}d\theta = -(-1)^{|f|_1+|f_2|}\int f_1\overline{(a_k f_2)}e^{-\theta\bar{\theta}}d\bar{\theta}d\theta,$$

where operators a_k^* and a_k are given by (4.1), and $f_1, f_2 \in \mathbb{C}[\alpha, \bar{\alpha}, \theta, \bar{\theta}]$.

The next result shows that the matrix symbol of an operator A in $\mathscr{H}_F$, which is just a $2^n \times 2^n$ matrix, can be considered as an integral kernel in anticommuting variables!

Lemma 4.1. *Let* $\tilde{A}(\bar{\theta}, \theta)$ *be the matrix symbol of an operator* A *in* $\mathscr{H}_F$. *Then for every* $f(\bar{\theta}) \in \mathscr{H}_F$,

$$(4.5) \qquad\qquad (Af)(\bar{\theta}) = \int \tilde{A}(\bar{\theta}, \alpha)f(\bar{\alpha})e^{-\bar{\alpha}\alpha}d\alpha d\bar{\alpha}.$$

Proof. It is sufficient to verify (4.5) for $f = f_K$, where $K = \{k_1, \ldots, k_m\} \subseteq \{1, \ldots, n\}$. Using (4.3) and Lemma 2.5, we get

$$\int \tilde{A}(\bar{\theta}, \alpha)f_K(\bar{\alpha})e^{-\bar{\alpha}\alpha}d\alpha d\bar{\alpha} = \sum_{I,J}(Af_J, f_I)f_I(\bar{\theta})\int \overline{f_J(\bar{\alpha})}f_K(\bar{\alpha})e^{-\bar{\alpha}\alpha}d\alpha d\bar{\alpha}$$

$$= \sum_I (Af_K, f_I)f_I(\bar{\theta}) = (Af_K)(\bar{\theta}). \qquad \square$$

Next, we introduce Grassmann analogs of the coherent states (see Section 2.7 of Chapter 2). Put

$$\Phi_{\boldsymbol{\alpha}}(\bar{\theta}) = e^{\bar{\theta}\alpha} = \sum_I f_I(\bar{\theta})\overline{f_I(\bar{\alpha})} \quad \text{and} \quad \tilde{\Phi}_{\boldsymbol{\alpha}}(\theta) = e^{-\theta\alpha} = \sum_I \overline{f_I(\bar{\alpha})}f_I(\bar{\theta}).$$

As in the bosonic case, elements $\Phi_{\boldsymbol{\alpha}}, \tilde{\Phi}_{\boldsymbol{\alpha}} \in \mathbb{C}[\alpha, \bar{\alpha}, \theta, \bar{\theta}]$ satisfy

$$(4.6) \qquad a_k\Phi_{\boldsymbol{\alpha}} = \alpha_k\Phi_{\boldsymbol{\alpha}} \quad \text{and} \quad a_k\tilde{\Phi}_{\boldsymbol{\alpha}} = -\alpha_k\tilde{\Phi}_{\boldsymbol{\alpha}}, \quad k = 1, \ldots, n.$$

It is very easy to express the matrix symbol of an operator in terms of the coherent states.

Lemma 4.2. *Let* $\tilde{A}(\bar{\alpha}, \alpha)$ *be the matrix symbol of an operator* A *in* $\mathscr{H}_F$. *Then*

$$\tilde{A}(\bar{\alpha}, \alpha) = (A\Phi_{\boldsymbol{\alpha}}, \tilde{\Phi}_{\boldsymbol{\alpha}}).$$

Proof. The proof is a straightforward computation (cf. the proof of Lemma 2.4 in Section 2.7 of Chapter 2):

$$\tilde{A}(\bar{\alpha}, \alpha) = \sum_{I,J} \left(Af_J, f_I\right) f_I(\bar{\alpha})\overline{f_J(\bar{\alpha})}$$

$$= \sum_{I,J} \left(Af_J\overline{f_J(\bar{\alpha})}, \overline{f_I(\bar{\alpha})}f_I\right) = (A\Phi_\alpha, \tilde{\Phi}_\alpha). \qquad \square$$

The next result is an exact analog of Lemma 2.4 in Section 2.7 of Chapter 2.

Lemma 4.3. *The matrix and Wick symbols of an operator A in $\mathscr{H}_F$ are related by*

$$\tilde{A}(\bar{\alpha}, \alpha) = e^{\bar{\alpha}\alpha} A(\bar{\alpha}, \alpha).$$

Moreover, $\tilde{A}(\bar{\alpha}, \theta) = e^{\alpha\theta} A(\bar{\alpha}, \theta)$ and $\tilde{A}(\bar{\theta}, \alpha) = e^{\bar{\theta}\alpha} A(\bar{\theta}, \alpha)$.

Proof. As follows from the definition of coherent states,

$$(\Phi_\alpha, \tilde{\Phi}_\alpha) = e^{\bar{\alpha}\alpha}.$$

Now representing the operator A in a Wick normal form and using properties (4.4) and (4.6), we obtain

$$(A\Phi_\alpha, \tilde{\Phi}_\alpha) = \sum_{I,J} K_{IJ}(a^*_{i_1} \dots a^*_{i_k} a_{j_1} \dots a_{j_l}\Phi_\alpha, \tilde{\Phi}_\alpha)$$

$$= \sum_{I,J}(-1)^{kl+k} K_{IJ}(a_{j_1} \dots a_{j_l}\Phi_\alpha, a_{i_k} \dots a_{i_1}\tilde{\Phi}_\alpha)$$

$$= \sum_{I,J}(-1)^{kl} K_{IJ}(\alpha_{j_1} \dots \alpha_{j_l}\Phi_\alpha, \alpha_{i_k} \dots \alpha_{i_1}\tilde{\Phi}_\alpha)$$

$$= \sum_{I,J}(-1)^{kl} K_{IJ}\alpha_{j_1} \dots \alpha_{j_l}\bar{\alpha}_{i_1} \dots \bar{\alpha}_{i_k}(\Phi_\alpha, \tilde{\Phi}_\alpha)$$

$$= A(\bar{\alpha}, \alpha)(\Phi_\alpha, \tilde{\Phi}_\alpha) = e^{\bar{\alpha}\alpha} A(\bar{\alpha}, \alpha).$$

The same computation, after replacing Φ_α by Φ_θ, or $\tilde{\Phi}_\alpha$ by $\tilde{\Phi}_\theta$, proves the remaining two formulas. $\qquad \square$

Thus we have shown that $2^n \times 2^n$ matrices — operators in the fermion Hilbert space $\mathscr{H}_F$ — can be considered as integral operators in anticommuting variables with the integral kernels given by matrix or Wick symbols. The next result is an exact analog of Theorem 2.2 in Section 2.7 of Chapter 2, and establishes the calculus of symbols for operators in $\mathscr{H}_F$.

Theorem 4.1. *Let A_1 and A_2 be operators in $\mathscr{H}_F$ with matrix symbols $\tilde{A}_1(\bar{\theta}, \theta)$ and $\tilde{A}_2(\bar{\theta}, \theta)$ and Wick symbols $A_1(\bar{\theta}, \theta)$ and $A_2(\bar{\theta}, \theta)$. Then the following formulas hold.*

(i) *The matrix and Wick symbols of the operator* $A = A_1 A_2$ *are given by*

$$\tilde{A}(\bar{\boldsymbol{\theta}}, \boldsymbol{\theta}) = \int \tilde{A}_1(\bar{\boldsymbol{\theta}}, \boldsymbol{\alpha})\tilde{A}_2(\bar{\boldsymbol{\alpha}}, \boldsymbol{\theta})e^{-\bar{\boldsymbol{\alpha}}\boldsymbol{\alpha}}d\boldsymbol{\alpha}d\bar{\boldsymbol{\alpha}},$$

$$A(\bar{\boldsymbol{\theta}}, \boldsymbol{\theta}) = \int A_1(\bar{\boldsymbol{\theta}}, \boldsymbol{\alpha})A_2(\bar{\boldsymbol{\alpha}}, \boldsymbol{\theta})e^{-(\bar{\boldsymbol{\theta}}-\bar{\boldsymbol{\alpha}})(\boldsymbol{\theta}-\boldsymbol{\alpha})}d\boldsymbol{\alpha}d\bar{\boldsymbol{\alpha}}.$$

(ii) *The trace and supertrace of an operator* A *in* $\mathscr{H}_F$ *are given by*

$$\operatorname{Tr} A = \int \tilde{A}(\bar{\boldsymbol{\theta}}, \boldsymbol{\theta})e^{-\boldsymbol{\theta}\bar{\boldsymbol{\theta}}}d\bar{\boldsymbol{\theta}}d\boldsymbol{\theta} = \int A(\bar{\boldsymbol{\theta}}, \boldsymbol{\theta})e^{-2\boldsymbol{\theta}\bar{\boldsymbol{\theta}}}d\bar{\boldsymbol{\theta}}d\boldsymbol{\theta},$$

$$\operatorname{Tr}_s A = \int \tilde{A}(\bar{\boldsymbol{\theta}}, \boldsymbol{\theta})e^{-\bar{\boldsymbol{\theta}}\boldsymbol{\theta}}d\boldsymbol{\theta}d\bar{\boldsymbol{\theta}} = \int A(\bar{\boldsymbol{\theta}}, \boldsymbol{\theta})d\boldsymbol{\theta}d\bar{\boldsymbol{\theta}}.$$

Proof. Part (i) for matrix symbols is proved by the following straightforward computation:

$$\int \tilde{A}_1(\bar{\boldsymbol{\theta}}, \boldsymbol{\alpha})\tilde{A}_2(\bar{\boldsymbol{\alpha}}, \boldsymbol{\theta})e^{-\bar{\boldsymbol{\alpha}}\boldsymbol{\alpha}}d\boldsymbol{\alpha}d\bar{\boldsymbol{\alpha}}$$

$$= \sum_{I,J}\sum_{K,L}(A_1 f_J, f_I)(A_2 f_L, f_K)\int f_I(\bar{\boldsymbol{\theta}})\overline{f_J(\bar{\boldsymbol{\alpha}})}f_K(\bar{\boldsymbol{\alpha}})\overline{f_L(\bar{\boldsymbol{\theta}})}e^{-\bar{\boldsymbol{\alpha}}\boldsymbol{\alpha}}d\boldsymbol{\alpha}d\bar{\boldsymbol{\alpha}}$$

$$= \sum_{I,J,L}(A_1 f_J, f_I)(A_2 f_L, f_J)f_I(\bar{\boldsymbol{\theta}})\overline{f_L(\bar{\boldsymbol{\theta}})}$$

$$= \sum_{I,J,L}(A_2 f_L, f_J)(f_J, A_1^* f_I)f_I(\bar{\boldsymbol{\theta}})\overline{f_L(\bar{\boldsymbol{\theta}})}$$

$$= \sum_{I,L}(A_2 f_L, A_1^* f_I)f_I(\bar{\boldsymbol{\theta}})\overline{f_L(\bar{\boldsymbol{\theta}})} = \sum_{I,L}(A_1 A_2 f_L, f_I)f_I(\bar{\boldsymbol{\theta}})\overline{f_L(\bar{\boldsymbol{\theta}})}$$

$$= \tilde{A}(\bar{\boldsymbol{\theta}}, \boldsymbol{\theta}).$$

The corresponding formula for the Wick symbols now follows from Lemma 4.3. The proof of part (ii) is also straightforward. We have

$$\int \tilde{A}(\bar{\boldsymbol{\theta}}, \boldsymbol{\theta})e^{-\boldsymbol{\theta}\bar{\boldsymbol{\theta}}}d\bar{\boldsymbol{\theta}}d\boldsymbol{\theta} = \sum_{I,J}(A f_J, f_I)\int f_I(\bar{\boldsymbol{\theta}})\overline{f_J(\bar{\boldsymbol{\theta}})}e^{-\boldsymbol{\theta}\bar{\boldsymbol{\theta}}}d\bar{\boldsymbol{\theta}}d\boldsymbol{\theta}$$

$$= \sum_I (A f_I, f_I) = \operatorname{Tr} A.$$

Similarly,

$$\int \tilde{A}(\bar{\boldsymbol{\theta}}, \boldsymbol{\theta})e^{-\bar{\boldsymbol{\theta}}\boldsymbol{\theta}}d\boldsymbol{\theta}d\bar{\boldsymbol{\theta}} = \sum_{I,J}(A f_J, f_I)\int f_I(\bar{\boldsymbol{\theta}})\overline{f_J(\bar{\boldsymbol{\theta}})}e^{-\bar{\boldsymbol{\theta}}\boldsymbol{\theta}}d\boldsymbol{\theta}d\bar{\boldsymbol{\theta}}$$

$$= \sum_I (-1)^{|I|}(A f_I, f_I) = \operatorname{Tr}_s A,$$

where $|I|$ denotes the cardinality of the subset $I \subseteq \{1, \ldots, n\}$. Corresponding formulas for the Wick symbol follow from Lemma 4.3. $\qquad\square$

Problem 4.1. Verify directly all results in this section for the simplest case of one degree of freedom, when $\mathscr{H}_F = \mathbb{C}^2$.

Problem 4.2. Show that the Wick symbol of the product $A = A_l \ldots A_1$ is given by

$$A(\bar{\boldsymbol\theta}, \boldsymbol\theta) = \int \cdots \int A_l(\bar{\boldsymbol\theta}, \boldsymbol\alpha_{l-1}) \ldots A_1(\bar{\boldsymbol\alpha}_1, \boldsymbol\theta) \exp\left\{ \sum_{k=1}^{l-1} \bar{\boldsymbol\alpha}_k(\boldsymbol\alpha_{k-1} - \boldsymbol\alpha_k) + \bar{\boldsymbol\theta}(\boldsymbol\alpha_{l-1} - \boldsymbol\theta) \right\}$$
$$d\boldsymbol\alpha_1 d\bar{\boldsymbol\alpha}_1 \ldots d\boldsymbol\alpha_{l-1} d\bar{\boldsymbol\alpha}_{l-1}$$

where $\boldsymbol\alpha_0 = \boldsymbol\theta$ and $A_k(\bar{\boldsymbol\theta}, \boldsymbol\theta)$ are the Wick symbols of the operators A_k.

Problem 4.3. Prove that the Wick symbol $\Gamma(\bar{\boldsymbol\theta}, \boldsymbol\theta)$ of the chirality operator Γ is $e^{-2\bar{\boldsymbol\theta}\boldsymbol\theta}$.

4.2. Path integral for the evolution operator.

Let H be a Hamiltonian of a system of n fermions — an operator in $\mathscr{H}_F$ with the Wick symbol $H(\bar{\boldsymbol\theta}, \boldsymbol\theta)$. Here we express the Wick symbol $U(\bar{\boldsymbol\theta}, \boldsymbol\theta; T)$ of the evolution operator $U(T) = e^{-iTH}$ by using the path integral over Grassmann variables. Our exposition will be parallel to that in Section 2.4 of Chapter 5, with obvious simplification due to the fact that fermion Hilbert space $\mathscr{H}_F$ is finite-dimensional. Namely, the following elementary result replaces assumption (2.11), made in Section 2.4 of Chapter 5.

Lemma 4.4. *Let $\widetilde{U}(\Delta t)$ be the operator with the Wick symbol $e^{-iH(\bar{\boldsymbol\theta}, \boldsymbol\theta)\Delta t}$. Then*

$$U(T) = \lim_{N \to \infty} \widetilde{U}(\Delta t)^N, \quad \text{where} \quad \Delta t = \frac{T}{N}.$$

Proof. The Wick symbol of the operator $R(\Delta t) = I - i\hat{H}\Delta t - \widetilde{U}(\Delta t)$ is a polynomial in Δt with Grassmann algebra coefficients, which starts with the term $(\Delta t)^2$. It is easy to see that $\|R(\Delta t)\| \leq c(\Delta t)^2$ for some $c > 0$, and

$$U(T) = \lim_{N \to \infty} (I - iH\Delta t)^N = \lim_{N \to \infty} (U(\Delta t) + R(\Delta t))^N = \lim_{N \to \infty} \widetilde{U}(\Delta t)^N. \quad\square$$

Using the formula for the composition of Wick symbols (see Theorem 4.1 and Problem 4.2), we can represent the Wick symbol $U_N(\bar{\boldsymbol\theta}, \boldsymbol\theta; T)$ of the operator $\widetilde{U}(\Delta t)^N$ as an $(N - 1)$-fold Berezin integral. Namely, consider the anticommuting variables $\boldsymbol\alpha_k = \{\alpha_k^1, \ldots, \alpha_k^n\}$, $\bar{\boldsymbol\alpha}_k = \{\bar{\alpha}_k^1, \ldots, \bar{\alpha}_k^n\}$, $k = 1, \ldots, N-1$ — generators of the Grassmann algebra with involution — and denote $\bar{\alpha}_k \alpha_k = \sum_{l=1}^{n} \bar{\alpha}_k^l \alpha_k^l$, etc. Then

$$U_N(\bar{\boldsymbol{\theta}}, \boldsymbol{\theta}; T) = \int \cdots \int \exp\left\{ \sum_{k=1}^{N} (\bar{\boldsymbol{\alpha}}_k(\boldsymbol{\alpha}_{k-1} - \boldsymbol{\alpha}_k) + \bar{\boldsymbol{\theta}}(\boldsymbol{\alpha}_N - \boldsymbol{\theta}) \right.$$
$$\left. -iH(\bar{\boldsymbol{\alpha}}_k, \boldsymbol{\alpha}_{k-1})\Delta t) \right\} \prod_{k=1}^{N-1} d\boldsymbol{\alpha}_k d\bar{\boldsymbol{\alpha}}_k,$$

where $\boldsymbol{\alpha}_0 = \boldsymbol{\theta}$, and we put $\bar{\boldsymbol{\alpha}}_N = \bar{\boldsymbol{\theta}}$. It follows from Lemma 4.4 that

$$(4.7) \qquad U(\bar{\boldsymbol{\theta}}, \boldsymbol{\theta}; T) = \lim_{N \to \infty} U_N(\bar{\boldsymbol{\theta}}, \boldsymbol{\theta}; T) = \lim_{N \to \infty} \int \cdots \int$$
$$\exp\left\{ \sum_{k=1}^{N} (\bar{\boldsymbol{\alpha}}_k(\boldsymbol{\alpha}_{k-1} - \boldsymbol{\alpha}_k) + \bar{\boldsymbol{\theta}}(\boldsymbol{\alpha}_N - \boldsymbol{\theta}) - iH(\bar{\boldsymbol{\alpha}}_k, \boldsymbol{\alpha}_{k-1})\Delta t) \right\} \prod_{k=1}^{N-1} d\boldsymbol{\alpha}_k d\bar{\boldsymbol{\alpha}}_k.$$

This formula looks exactly the same as the corresponding formula (2.13) for the Wick symbol of the evolution operator in Section 2.4 of Chapter 5! Accordingly, we interpret the limit $N \to \infty$ as the following *Feynman path integral for Grassmann variables* (or *Grassmann path integral*):

$$(4.8) \qquad U(\bar{\boldsymbol{\theta}}, \boldsymbol{\theta}; T) = \int_{\left\{ \substack{\bar{\boldsymbol{\alpha}}(T) = \bar{\boldsymbol{\theta}} \\ \boldsymbol{\alpha}(0) = \boldsymbol{\theta}} \right\}} e^{i \int_0^T (i\bar{\boldsymbol{\alpha}}\dot{\boldsymbol{\alpha}} - H(\bar{\boldsymbol{\alpha}}, \boldsymbol{\alpha}))dt + \bar{\boldsymbol{\theta}}(\boldsymbol{\alpha}(T) - \boldsymbol{\theta})} \mathscr{D}\boldsymbol{\alpha} \mathscr{D}\bar{\boldsymbol{\alpha}}.$$

Here the "integration" goes over all functions $\boldsymbol{\alpha}(t), \bar{\boldsymbol{\alpha}}(t)$ with anticommuting values[2] on the interval $[0, T]$, satisfying boundary conditions $\boldsymbol{\alpha}(0) = \boldsymbol{\theta}$, $\bar{\boldsymbol{\alpha}}(T) = \bar{\boldsymbol{\theta}}$, and

$$\mathscr{D}\boldsymbol{\alpha} \mathscr{D}\bar{\boldsymbol{\alpha}} = \prod_{0 \le t \le T} d\boldsymbol{\alpha}(t) d\bar{\boldsymbol{\alpha}}(t) = \lim_{N \to \infty} \prod_{k=1}^{N-1} d\boldsymbol{\alpha}_k d\bar{\boldsymbol{\alpha}}_k.$$

As in Section 2.4 of Chapter 5, for $0 < t < T$ variables $\bar{\boldsymbol{\alpha}}(t)$ are conjugated to $\boldsymbol{\alpha}(t)$ with respect to the Grassmann algebra involution, while $\bar{\boldsymbol{\alpha}}(0)$ and $\boldsymbol{\alpha}(T)$ — also variables of integration — are not conjugated to the boundary values $\boldsymbol{\alpha}(0) = \boldsymbol{\theta}$, $\bar{\boldsymbol{\alpha}}(T) = \bar{\boldsymbol{\theta}}$.

Remark. It should be emphasized that the only rigorous meaning of the Grassmann path integral (4.8) is the limit of multiple Berezin integrals in (4.7). However, as we have seen already in Chapter 5, it is very useful to pretend that the path integral has an independent definition, and formally work with it as if it was an actual integral.

---

[2]For every $0 \le t \le T$ there is a copy of the independent Grassmann algebra with generators $\alpha^1(t), \ldots, \alpha^n(t), \bar{\alpha}^1(t), \ldots, \bar{\alpha}^n(t)$.

Using Theorem 4.1, it is easy to express the supertrace of the evolution operator $U(T)$ — an operator in a finite-dimensional Hilbert space $\mathscr{H}_F$ — as a Grassmann path integral. We have

$$\mathrm{Tr}_s\, e^{-iTH} = \lim_{N\to\infty} \int U_N(\bar{\boldsymbol{\theta}},\boldsymbol{\theta};T)d\boldsymbol{\theta} d\bar{\boldsymbol{\theta}} = \lim_{N\to\infty} \int\cdots\int \exp\Big\{\bar{\boldsymbol{\theta}}(\boldsymbol{\alpha}(T)-\boldsymbol{\theta})$$

$$+\sum_{k=1}^{N}(\bar{\boldsymbol{\alpha}}_k(\boldsymbol{\alpha}_{k-1}-\boldsymbol{\alpha}_k)-iH(\bar{\boldsymbol{\alpha}}_k,\boldsymbol{\alpha}_{k-1})\Delta t)\Big\}\prod_{k=1}^{N-1} d\boldsymbol{\alpha}_k d\bar{\boldsymbol{\alpha}}_k d\boldsymbol{\theta} d\bar{\boldsymbol{\theta}}$$

$$= \int_{\left\{\begin{smallmatrix}\bar{\boldsymbol{\alpha}}(0)=\bar{\boldsymbol{\alpha}}(T)\\\boldsymbol{\alpha}(0)=\boldsymbol{\alpha}(T)\end{smallmatrix}\right\}} e^{i\int_0^T(i\bar{\boldsymbol{\alpha}}\dot{\boldsymbol{\alpha}}-H(\bar{\boldsymbol{\alpha}},\boldsymbol{\alpha}))dt}\mathscr{D}\boldsymbol{\alpha}\mathscr{D}\bar{\boldsymbol{\alpha}}$$

— a Grassmann path integral with periodic boundary conditions. Here periodic boundary conditions emerge from $\boldsymbol{\alpha}_0 = \boldsymbol{\theta}$ and $\bar{\boldsymbol{\alpha}}_N = \bar{\boldsymbol{\theta}}$ in the same way as in Section 2.4 of Chapter 5. Namely, since $\bar{\boldsymbol{\theta}}$ and $\boldsymbol{\theta}$ are now variables of integration, the identity (2.17) in Section 2.4 of Chapter 5, which is valid in the case of anticommuting variables as well, ensures that $\boldsymbol{\alpha}_N = \boldsymbol{\theta}$ and $\bar{\boldsymbol{\alpha}}_0 = \bar{\boldsymbol{\theta}}$. Denoting by Λ the corresponding "Grassmann loop space" — the space of all functions with anticommuting values $\boldsymbol{\alpha}(t)$ and $\bar{\boldsymbol{\alpha}}(t)$ conjugated with respect to the Grassmann algebra involution and satisfying periodic boundary conditions $\boldsymbol{\alpha}(0) = \boldsymbol{\alpha}(T)$ and $\bar{\boldsymbol{\alpha}}(0) = \bar{\boldsymbol{\alpha}}(T)$ — we can rewrite the previous formula as

$$\mathrm{Tr}_s\, e^{-iTH} = \int_{\Lambda} e^{i\int_0^T(i\bar{\boldsymbol{\alpha}}\dot{\boldsymbol{\alpha}}-H(\bar{\boldsymbol{\alpha}},\boldsymbol{\alpha}))dt}\mathscr{D}\boldsymbol{\alpha}\mathscr{D}\bar{\boldsymbol{\alpha}}.$$

Replacing the physical time t by the Euclidean time $-it$ and T by $-iT$, we get the Grassmann integral representation for the Wick symbol of the operator $U(-iT) = e^{-TH}$,

$$U(\bar{\boldsymbol{\theta}},\boldsymbol{\theta};-iT) = \int_{\left\{\begin{smallmatrix}\bar{\boldsymbol{\alpha}}(T)=\bar{\boldsymbol{\theta}}\\\boldsymbol{\alpha}(0)=\boldsymbol{\theta}\end{smallmatrix}\right\}} e^{-\int_0^T(\bar{\boldsymbol{\alpha}}\dot{\boldsymbol{\alpha}}+H(\bar{\boldsymbol{\alpha}},\boldsymbol{\alpha}))dt}\mathscr{D}\boldsymbol{\alpha}\mathscr{D}\bar{\boldsymbol{\alpha}},$$

and for the supertrace,

$$(4.9)\qquad \mathrm{Tr}_s\, e^{-TH} = \int_{\Lambda} e^{-\int_0^T(\bar{\boldsymbol{\alpha}}\dot{\boldsymbol{\alpha}}+H(\bar{\boldsymbol{\alpha}},\boldsymbol{\alpha}))dt}\mathscr{D}\boldsymbol{\alpha}\mathscr{D}\bar{\boldsymbol{\alpha}}.$$

Problem 4.4. Express the matrix symbol of the evolution operator as the Grassmann path integral.

Problem 4.5. Show that

$$\mathrm{Tr}\, e^{-TH} = \int_{\left\{\begin{smallmatrix}\bar{\boldsymbol{\alpha}}(0)=-\bar{\boldsymbol{\alpha}}(T)\\\boldsymbol{\alpha}(0)=-\boldsymbol{\alpha}(T)\end{smallmatrix}\right\}} e^{-\int_0^T(\bar{\boldsymbol{\alpha}}\dot{\boldsymbol{\alpha}}+H(\bar{\boldsymbol{\alpha}},\boldsymbol{\alpha}))dt}\mathscr{D}\boldsymbol{\alpha}\mathscr{D}\bar{\boldsymbol{\alpha}}$$

— the Grassmann path integral with anti-periodic boundary conditions.

4.3. Gaussian path integrals over Grassmann variables. For simplicity, here we consider only the case $n = 1$. As in Section 5.3 of Chapter 5, for $u(t) \in C^1([0, T], \mathbb{R})$ we put

$$u_0 = \frac{1}{T} \int_0^T u(t)dt \quad \text{and} \quad D = \frac{d}{dt},$$

and consider on the interval $[0, T]$ the first-order differential operator $D + u(t)$ with periodic boundary conditions. The following result evaluates the simplest Gaussian path integral for Grassmann variables.

Theorem 4.2. *We have*

$$\int_\Lambda e^{-\int_0^T (\bar{\alpha}\dot{\alpha}+u(t)\bar{\alpha}\alpha)dt} \mathscr{D}\alpha\mathscr{D}\bar{\alpha} = \det(D + u(t)) = 1 - e^{-u_0 T}.$$

Proof. Using Lemma 2.4, we get

$$\int_\Lambda e^{-\int_0^T (\bar{\alpha}\dot{\alpha}+u(t)\bar{\alpha}\alpha)dt} \mathscr{D}\alpha\mathscr{D}\bar{\alpha}$$

$$= \lim_{N\to\infty} \int \cdots \int e^{\sum_{k=1}^N \left(\bar{\alpha}_k(\alpha_{k-1}-\alpha_k)-u(t_k)\bar{\alpha}_k\alpha_{k-1}\Delta t\right)} \prod_{k=1}^N d\alpha_k d\bar{\alpha}_k$$

$$= \lim_{N\to\infty} \det A_N.$$

Here $\alpha_0 = \alpha_N, \bar{\alpha}_0 = \bar{\alpha}_N, t_k = k\Delta t$, and A_N is the following $N \times N$ matrix:

$$A_N = \begin{pmatrix} 1 & 0 & 0 & \cdots & 0 & b_N \\ b_1 & 1 & 0 & \cdots & 0 & 0 \\ 0 & b_2 & 1 & \cdots & 0 & 0 \\ \vdots & \vdots & \vdots & \ddots & \vdots & \vdots \\ 0 & 0 & 0 & \cdots & 1 & 0 \\ 0 & 0 & 0 & \cdots & b_{N-1} & 1 \end{pmatrix},$$

where $b_k = -1 + u(t_k)\Delta t$. We have

$$\det A_N = 1 - (-1)^N \prod_{k=1}^N b_k = 1 - \prod_{k=1}^N (1 - u(t_k)\Delta t),$$

so that

$$\lim \det A_N = 1 - e^{-\int_0^T u(t)dt} = 1 - e^{-u_0 T}.$$

Using Proposition 5.3 in Section 5.3 of Chapter 5 completes the proof. $\quad\square$

Example 4.1 (The fermion harmonic oscillator). The fermion analog of the harmonic oscillator is the Hamiltonian

$$H = \tfrac{1}{2}\omega(a^*a - aa^*) = \omega(a^*a - \tfrac{1}{2}I) = \omega(N - \tfrac{1}{2}I),$$

where a^* and a are creation and annihilation operators in the one fermion Hilbert space $\mathscr{H}_F = \mathbb{C}^2$ (see Section 1.1). The Wick symbol H is $H(\bar{\alpha}, \alpha) = \omega(\bar{\alpha}\alpha - \frac{1}{2})$. Now using (4.9) and Theorem 4.2, we obtain

$$\mathrm{Tr}_s\, e^{-TH} = \int_\Lambda e^{-\int_0^T (\bar{\alpha}\dot{\alpha} + H(\bar{\alpha},\alpha))dt}\, \mathscr{D}\alpha\mathscr{D}\bar{\alpha}$$

$$= e^{\frac{\omega T}{2}} \int_\Lambda e^{-\int_0^T (\bar{\alpha}\dot{\alpha} + \omega\bar{\alpha}\alpha)dt}\, \mathscr{D}\alpha\mathscr{D}\bar{\alpha} = e^{\frac{\omega T}{2}}(1 - e^{-\omega T}) = 2\sinh\frac{\omega T}{2}.$$

Of course, the same result can be obtained directly since e^{-TH} is just a 2×2 matrix with eigenvalues $e^{\frac{\omega T}{2}}$ and $e^{-\frac{\omega T}{2}}$ which correspond, respectively, to the eigenspaces $\mathscr{H}_F^0$ and $\mathscr{H}_F^1$. Thus

$$\mathrm{Tr}_s\, e^{-TH} = e^{\frac{\omega T}{2}} - e^{-\frac{\omega T}{2}} = 2\sinh\frac{\omega T}{2}.$$

Remark. We have the following analog of formula (2.6) of Chapter 6 for path integrals over Grassmann variables:

$$(4.10) \qquad \int_\Lambda f(\alpha, \bar{\alpha})\mathscr{D}\alpha\mathscr{D}\bar{\alpha} = \int \left(\int_{\Lambda_0} f(\beta, \theta; \bar{\beta}, \bar{\theta})\mathscr{D}\beta\mathscr{D}\bar{\beta} \right) d\theta d\bar{\theta},$$

where

$$\alpha(t) = \theta + \beta(t) \quad \text{and} \quad \int_0^T \beta(t)dt = 0,$$

and Λ_0 is the space of Grassmann loops $\beta(t), \bar{\beta}(t)$ with zero constant term.

Example 4.2. Using formula (4.10) and Theorem 4.2, we get

$$\det(D + \omega) = \int_\Lambda e^{-\int_0^T (\bar{\alpha}\dot{\alpha} + \omega\bar{\alpha}\alpha)dt}\, \mathscr{D}\alpha\mathscr{D}\bar{\alpha}$$

$$= \int e^{-\omega T\bar{\theta}\theta} d\theta d\bar{\theta} \int_{\Lambda_0} e^{-\int_0^T (\bar{\beta}\dot{\beta} + \omega\bar{\beta}\beta)dt}\, \mathscr{D}\beta\mathscr{D}\bar{\beta} = \omega T \int_{\Lambda_0} e^{-\int_0^T (\bar{\beta}\dot{\beta} + \omega\bar{\beta}\beta)dt}\, \mathscr{D}\beta\mathscr{D}\bar{\beta},$$

so that

$$\int_{\Lambda_0} e^{-\int_0^T \bar{\beta}\dot{\beta}dt}\, \mathscr{D}\beta\mathscr{D}\bar{\beta} = \lim_{\omega \to 0} \frac{\det(D + \omega)}{\omega T} = \lim_{\omega \to 0} \frac{1 - e^{-\omega T}}{\omega T} = 1.$$

Equivalently,

$$(4.11) \qquad \int_{\Lambda_0} e^{-\int_0^T \bar{\beta}\dot{\beta}dt}\, \mathscr{D}\beta\mathscr{D}\bar{\beta} = \frac{1}{T}\det{}' D,$$

where $\det' D = T$ (see Proposition 5.3 in Section 5.3 of Chapter 5). Introducing

$$\theta_1(t) = \frac{1}{\sqrt{2}}(\beta(t) + \bar{\beta}(t)), \quad \theta_2(t) = \frac{1}{i\sqrt{2}}(\beta(t) - \bar{\beta}(t)),$$

so that $\bar{\theta}_j(t) = \theta_j(t)$, $j = 1, 2$, we can rewrite (4.11) as

$$\int_{\Lambda_0} e^{-\frac{1}{2}\int_0^T (\theta_1\dot{\theta}_1 + \theta_2\dot{\theta}_2)dt}\, \mathscr{D}\theta_1\mathscr{D}\theta_2 = \frac{1}{T}\det{}' D = 1.$$

Since $\mathrm{Pf}'(D) = \sqrt{\det' D} = \sqrt{T}$, we have

$$(4.12) \qquad \int_{\Lambda_0(\mathbb{R})} e^{-\frac{1}{2}\int_0^T \theta\dot\theta\,dt}\,\mathscr{D}\theta = \frac{1}{\sqrt{T}}\,\mathrm{Pf}'(D) = 1.$$

Here the domain of integration $\Lambda_0(\mathbb{R})$ consists of all functions $\theta(t)$ with values in the Grassmann algebra over $\mathbb{R}$, which have zero constant term and satisfy periodic boundary conditions $\theta(0) = \theta(T)$. We will use formulas (4.10) and (4.12) in Section 2.2 of Chapter 8.

Problem 4.6. Show that

$$\int_{\left\{\begin{smallmatrix}\bar\alpha(0)=-\bar\alpha(T)\\ \alpha(0)=-\alpha(T)\end{smallmatrix}\right\}} e^{-\int_0^T (\bar\alpha\dot\alpha + u(t)\bar\alpha\alpha)\,dt}\,\mathscr{D}\alpha\mathscr{D}\bar\alpha = 1 + e^{-u_0 T}$$

— the regularized determinant of the operator $D + u(t)$ on $[0, T]$ with anti-periodic boundary conditions.

Problem 4.7. For the fermion harmonic oscillator verify that $\mathrm{Tr}\, e^{-TH} = 2\cosh\frac{\omega T}{2}$ directly, and also by using results of Problems 4.5 and 4.6.

Problem 4.8. Prove formula (4.10). (*Hint:* Follow the proof of Theorem 4.2, and use the decomposition $\alpha_k = \theta + \beta_k$, where $\beta_0 = \beta_N$ and $\sum_{k=1}^{N}\beta_k = 0$.)

Problem 4.9. Give a direct proof of formula (4.12).

5. Notes and references

Canonical anticommutation relations, discussed in Section 1, were introduced by P. Jordan and E. Wigner in 1928 [**JW28**]. We refer the reader to F.A. Berezin's classical monograph [**Ber66**] for a comprehensive mathematical treatment of canonical commutation and anticommutation relations. Though [**Ber66**] is mainly devoted to quantum systems with infinitely many degrees of freedom, which are studied by quantum field theory, it also discusses the simpler case of finitely many degrees of freedom. A self-contained mathematical introduction to Clifford algebras, their representations, and other topics can be found in [**Var04**] and references therein. The fundamental idea that systems with Grassmann variables appear as a semi-classical limit of fermions was formulated by I.L. Martin in 1959 [**Mar59b, Mar59a**]. Independently, F.A. Berezin in [**Ber61**] introduced Grassmann variables for rigorous mathematical description of the second quantization for fermion systems by using generating functionals for vectors and operators. Material in Section 2 — differential and integral calculus on the Grassmann algebra — belongs to F.A. Berezin, and our exposition follows [**Ber66**] and [**Ber87**]. The superalgebra — graded linear algebra — which we very briefly describe in Section 3, was also discovered and developed by F.A. Berezin [**Ber87**]. The references [**Man97**], [**Fre99**], and [**DM99**] provide the reader with a more abstract mathematical treatment, while lecture notes [**Var04**] supplement the category theory approach of [**DM99**] with motivation from physics. We refer to [**Ber61, Ber66**] for the general discussion of the path integral over Grassmann variables, and of the matrix and Wick symbols of operators in the fermion Hilbert space. In Section 4 we follow the elegant exposition in [**FS91**]; as in Section 2.4 of Chapter 5, we carefully treat boundary conditions for Grassmann path integrals for the Wick symbols (as opposed to the formulas [**Ber71a**]).

Chapter 8

Supersymmetry

1. Supermanifolds

The coordinate vector space $V = \mathbb{R}^n$ has a natural structure of a smooth manifold, which to every open subset $U \subseteq \mathbb{R}^n$ assigns a commutative $\mathbb{R}$-algebra $C^\infty(U)$ of all smooth functions on U. The assignment

$$U \mapsto C^\infty(U)$$

for all open $U \subseteq \mathbb{R}^n$ defines a sheaf of commutative $\mathbb{R}$-algebras (commutative rings) on a topological space $\mathbb{R}^n$, and turns it into a *ringed space*. Every smooth n-dimensional manifold M is a ringed space — a topological space with a sheaf of commutative rings, which is locally isomorphic to the ringed space $\mathbb{R}^n$. It is easy to see that this definition is equivalent to the standard one, defined by gluing coordinate charts. The notion of a *supermanifold* generalizes this idea by using local models associated with graded vector spaces.

Namely, let $W = \mathbb{R}^{p|q}$ be the coordinate graded vector space of dimension $p|q$ over $\mathbb{R}$. The following definition formalizes the intuitive idea that odd coordinates on W are anticommuting.

Definition. A supermanifold $\mathbb{R}^{p|q}$ is a topological space $\mathbb{R}^p$ with a sheaf of commutative $\mathbb{R}$-superalgebras (supercommutative rings) over $\mathbb{R}$, called the structure sheaf, defined by the assignment

$$U \mapsto C^\infty(U)[\theta^1, \ldots, \theta^q]$$

for all open $U \subseteq \mathbb{R}^p$, where $C^\infty(U)[\theta^1, \ldots, \theta^q]$ is a Grassmann algebra with generators $\theta^1, \ldots, \theta^q$ over the commutative ring $C^\infty(U)$.

Remark. Elements of $C^\infty(U)[\theta^1,\ldots,\theta^q]$ have the form

$$f = \sum_I f_I \theta^I, \quad \theta^I = \theta^{i_1}\ldots\theta^{i_k} \quad \text{and} \quad f_I \in C^\infty(U)$$

for $I = \{i_1,\ldots,i_k\} \subset \{1,\ldots,q\}$, and are called *functions on a supermanifold* $\mathbb{R}^{p|q}$ over U. Thus $\mathbb{R}^{p|q}$ is a coordinate space with *even coordinates* $\boldsymbol{x} = (x^1,\ldots,x^p)$ and *odd coordinates* $\boldsymbol{\theta} = (\theta^1,\ldots,\theta^q)$, and we will write $f \in C^\infty(U)[\theta^1,\ldots,\theta^q]$ as $f(\boldsymbol{x},\boldsymbol{\theta})$.

Definition. A supermanifold of dimension $p|q$ is a pair $(X,\mathcal{O}_X)$ — a topological space X with a sheaf $\mathcal{O}_X$ of supercommutative rings over $\mathbb{R}$, called the structure sheaf, which is locally isomorphic to $\mathbb{R}^{p|q}$.

Supermanifolds form a category: a morphism between supermanifolds $(X,\mathcal{O}_X)$ and $(Y,\mathcal{O}_Y)$ is a continuous map $\varphi : X \to Y$, together with a sheaf map $\varphi^* : \mathcal{O}_Y \to \mathcal{O}_X$ over φ, a collection of homomorphisms of supercommutative rings over $\mathbb{R}$,

$$\varphi^*_V : \mathcal{O}_Y(V) \to \mathcal{O}_X(\varphi^{-1}(V)), \quad V \subseteq Y \quad \text{open},$$

which commute with restriction maps of the sheaves.

To every vector bundle E of rank q over an ordinary p-dimensional manifold M there is an associated supermanifold ΠE of dimension $p|q$, defined by reversing parity in the fibers of E. Namely, for every open $U \subseteq M$ let $C^\infty(U,E)$ be the space of all smooth sections of E over U. It is a free $C^\infty(U)$-module generated by q sections $\theta^1,\ldots,\theta^q$. Consider them as generators of the Grassmann algebra (this explains using Greek letters), and define the structure sheaf of ΠE by the assignment $U \mapsto C^\infty(U)[\theta^1,\ldots,\theta^q]$. It can be shown that every supermanifold is isomorphic (non-canonically) to ΠE for some vector bundle E over M.

Remark. Supermanifolds have many more morphisms than vector bundles, since one is allowed to mix even and odd variables. Thus for the supermanifold $\mathbb{R}^{1|2}$ with even coordinate x and odd coordinates θ_1,θ_2 the mapping

$$\varphi(x) = x + \theta^1\theta^2, \quad \varphi(\theta^i) = \theta^i, \quad i = 1,2,$$

is an isomorphism of $\mathbb{R}^{1|2}$, but it is not induced by an isomorphism of a trivial rank 2 vector bundle over $\mathbb{R}$.

Using these definitions, one can develop differential geometry of supermanifolds and the corresponding theory of integration, which is quite similar to that for ordinary manifolds (see Section 7 for the references). Here we will only consider the simplest example of supermanifolds ΠTM, where M is an ordinary manifold. It is quite remarkable that the basic notions of differential geometry on a manifold M can be formulated in terms of the supermanifold ΠTM.

Namely, let $\mathcal{A}^{\bullet}(M)$ be a commutative superalgebra of smooth differential forms on the n-dimensional manifold M, and let $C^{\infty}(\Pi TM)$ be a commutative superalgebra of global sections of the structure sheaf of ΠTM — the superalgebra of functions on a supermanifold ΠTM. The isomorphism

$$\mathcal{A}^{\bullet}(M) \simeq C^{\infty}(\Pi TM)$$

allows us to interpret differential forms on M as functions on ΠTM. Indeed, to every $\omega_p \in \mathcal{A}^p(M)$, given in local coordinates on $U \subset M$ by

$$\omega_p = \sum_{1 \leq i_1 < \cdots < i_p \leq n} a_{i_1 \ldots i_p}(\boldsymbol{x}) dx^{i_1} \wedge \cdots \wedge dx^{i_p},$$

we assign

$$\omega_p(\boldsymbol{x}, \boldsymbol{\theta}) = \sum_{1 \leq i_1 < \cdots < i_p \leq n} a_{i_1 \ldots i_p}(\boldsymbol{x})\theta^{i_1} \ldots \theta^{i_p} \in C^{\infty}(U)[\theta^1, \ldots, \theta^n].$$

It trivially follows from the definition of differential form and of ΠTM that under a change of coordinates $a_{i_1 \ldots i_p}(\boldsymbol{x})$ transform like coefficients of differential forms, and $\theta^1, \ldots, \theta^n$ transform like components of tangent vectors, so that $\omega_p(\boldsymbol{x}, \boldsymbol{\theta})$ is a well-defined function on the supermanifold ΠTM. Correspondingly, the de Rham differential d gives rise to an odd vector field δ on ΠTM,

$$\delta = \sum_{k=1}^{n} \theta^k \frac{\partial}{\partial x^k},$$

with the property that $\delta^2 = 0$. Finally, the supermanifold ΠTM carries a canonical volume form $d\boldsymbol{x}d\boldsymbol{\theta} = dx^1 \ldots dx^n d\theta^1 \ldots d\theta^n$, which is well defined due to the opposite change of variables formulas for the ordinary and Berezin's integrals (see Section 2.3 of Chapter 7). The integration of a top differential form over M reduces to the integration of a corresponding function over ΠTM with respect to the canonical volume form,

$$\int_M \omega_n = \int_{\Pi TM} \omega_n(\boldsymbol{x}, \boldsymbol{\theta})d\boldsymbol{x}d\boldsymbol{\theta}.$$

Remark. Sometimes in the mathematics literature, in order to emphasize the opposite change of variables formulas for the ordinary and Berezin's integrals, the volume form on the supermanifold ΠTM is denoted by $d\boldsymbol{x}d\boldsymbol{\theta}^{-1}$.

Problem 1.1. Prove all statements in this section.

2. Equivariant cohomology and localization

2.1. Finite-dimensional case. Let M be a compact orientable n-dimensional manifold with the circle action — an action of abelian group $\mathrm{U}(1) =$

S^1 — and let $V \in \mathrm{Vect}(M)$ be the vector field corresponding to this action,

$$V(f)(x) = \frac{d}{dt}\bigg|_{t=0} f(e^{it} \cdot x), \quad x \in M.$$

Consider the linear operator

$$D = d - i_V : \mathcal{A}^\bullet(M) \to \mathcal{A}^\bullet(M),$$

where i_V is the inner product operator with V. Using the identification $\mathcal{A}^\bullet(M) \simeq C^\infty(\Pi TM)$, in local coordinates $\boldsymbol{x} = (x^1, \ldots, x^n)$ on $U \subset M$ we can represent D by

$$(2.1) \qquad D = \sum_{\mu=1}^{n} \left(\theta^\mu \frac{\partial}{\partial x^\mu} - v^\mu \frac{\partial}{\partial \theta^\mu} \right), \quad \text{where} \quad V = \sum_{\mu=1}^{n} v^\mu \frac{\partial}{\partial x^\mu}.$$

It follows from the Cartan formula (see Chapter 1), or directly from (2.1), that

$$D^2 = -\mathcal{L}_V,$$

where $\mathcal{L}_V$ is a Lie derivative. Thus D is a differential on the subcomplex $\mathcal{A}^\bullet(M)^{S^1}$ of S^1-invariant differential forms on M. The cohomology of this complex is an *equivariant cohomology* in the Cartan formulation. Since d and i_V have, respectively, degrees 1 and -1, equivariantly closed forms on M necessarily have several components. Namely, the equation $D\alpha = 0$ for $\alpha = \sum_{p=0}^{n} \alpha_p \in \mathcal{A}^\bullet(M)$ is equivalent to the following system of equations:

$$d\alpha_p = i_V \alpha_{p+2}, \quad p = 0, \ldots, n-2, \quad \text{and} \quad d\alpha_{n-1} = 0.$$

Let M_V be a zero locus of the vector field V — the set of fixed points of the circle action. We have the following fundamental property of equivariantly closed differential forms.

Lemma 2.1. *An equivariantly closed differential form an a compact manifold M is equivariantly exact on $M \setminus M_V$.*

Proof. Let $\alpha \in \mathcal{A}^\bullet(M)$ be such that $D\alpha = 0$. We want to find a differential form λ on $M \setminus M_V$ such that $\alpha = D\lambda$ on $M \setminus M_V$. Suppose there is a form ξ on $M \setminus M_V$, with components of odd degrees, satisfying $D\xi = 1$. Setting $\lambda = \xi \wedge \alpha$ we get

$$D\lambda = D\xi \wedge \alpha - \xi \wedge D\alpha = \alpha.$$

To construct such a form ξ, choose an S^1-invariant Riemannian metric g on M, and denote by $\beta \in \mathcal{A}^1(M)$ the 1-form on M dual to the vector field V with respect to the Riemannian metric, $\beta = \langle V, \cdot \rangle$. Since the metric g is S^1-invariant, $\mathcal{L}_V g = 0$ and, therefore, $\mathcal{L}_V \beta = 0$. We have

$$D\beta = K + \Omega, \quad \text{where} \quad K = -\|V\|^2 \quad \text{and} \quad \Omega = d\beta.$$

Since $K \in \mathcal{A}^0(M)$ does not vanish on $M \setminus M_V$,

$$\xi = \beta(D\beta)^{-1} = \frac{\beta}{K}(1 + K^{-1}\Omega)^{-1} = \frac{\beta}{K}\sum_{i=0}^{l}(-1)^i \frac{\Omega^i}{K^i}, \quad l = \left[\frac{n}{2}\right],$$

is a well-defined form on $M \setminus M_V$ satisfying $D\xi = 1$. $\qquad\square$

Corollary 2.1. *The top component of an equivariantly closed form on M is exact on $M \setminus M_V$.*

Remark. In local coordinates $\boldsymbol{x} = (x^1, \dots, x^n)$ on M the Riemannian metric g has the form[1] $ds^2 = g_{\mu\nu}dx^\mu dx^\nu$, and equation $\mathcal{L}_V g = 0$ — the condition that V is a Killing vector field with respect to the metric g — can be written as

$$(2.2) \qquad g_{\mu\lambda}\nabla_\nu v^\lambda + g_{\nu\lambda}\nabla_\mu v^\lambda = 0.$$

Here ∇_μ is a covariant derivative with respect to the vector field $\dfrac{\partial}{\partial x^\mu}$. For the 1-form $\beta = g_{\mu\nu}v^\nu dx^\mu$ we also have

$$(2.3) \qquad D\beta = -g_{\mu\nu}v^\mu v^\nu + \omega_{\mu\nu}dx^\mu \wedge dx^\nu, \quad \text{where} \quad \omega_{\mu\nu} = g_{\nu\lambda}\nabla_\mu v^\lambda.$$

By (2.2), the $n \times n$ matrix $\omega(\boldsymbol{x}) = \{\omega_{\mu\nu}(\boldsymbol{x})\}_{\mu,\nu=1}^n$ is skew-symmetric. For even $n = 2l$, orientation on M allows us to define $\mathrm{Pf}(\omega(\boldsymbol{x}))$ as the product $\lambda_1 \dots \lambda_l$ of above-diagonal elements in the canonical form[2] of a skew-symmetric matrix $\omega(\boldsymbol{x})$ (see Section 2.3 of Chapter 7). Correspondingly, $\mathrm{Pf}(\omega)(\boldsymbol{x})(\det g(\boldsymbol{x}))^{-\frac{1}{2}}$ does not depend on a choice of local coordinates, and defines a function on M. For every $x \in M$ there is also a linear mapping $L_x : T_x M \to T_x M$, defined by $L_x W = (\nabla_W V)(x)$, where ∇_W is a covariant derivative with respect to $W \in T_x M$. Since $\det L_x \det g(x) = \mathrm{Pf}(\omega(x))^2$, we define

$$(2.4) \qquad \sqrt{\det L_x} = \frac{\mathrm{Pf}(\omega(x))}{\sqrt{\det g(x)}}.$$

For $x \in M_V$ the mapping L_x does not depend on the choice of an S^1-invariant metric g, and is given by the matrix $\left\{\dfrac{\partial v^\mu}{\partial x^\nu}(\boldsymbol{x})\right\}_{\mu,\nu=1}^n$.

According to Lemma 2.1, the integral $\int_M \alpha$ of an equivariantly closed differential form α localizes on the zero locus M_V. A quantitative result is given by the following statement.

Proposition 2.1. *Suppose that $D\alpha = 0$ on M. Then for every S^1-invariant $\beta \in \mathcal{A}^1(M)$ the integral $\int_M \alpha e^{tD\beta}$ does not depend on t.*

---

[1] We are assuming the summation over repeated indices.
[2] Obtained by using orthogonal transformation with determinant 1.

Proof. Denote this integral by $Z(t)$. We have

$$\frac{dZ}{dt} = \int_M \alpha D\beta e^{tD\beta} = \int_M D(\beta \alpha e^{tD\beta}) + \int_M \beta D(\alpha e^{tD\beta}) = 0,$$

where the first integral is zero by Stokes' theorem, and the second integral is zero by the conditions $D\alpha = 0$ and $D^2\beta = 0$. $\qquad\square$

This result is usually used to evaluate $Z(0) = \int_M \alpha$ by computing the asymptotics of $Z(t)$ as $t \to \infty$. For a suitably chosen 1-form β the leading contribution is given by the integral over a tubular neighborhood of the zero locus of the component $K = (D\beta)_0 \in \mathcal{A}^0(M)$, which can be computed in the closed form as $t \to \infty$. The simplest case of a localization phenomenon is when the circle action on M has only isolated fixed points, i.e., when M_V is a finite set and L_x is non-degenerate for $x \in M_V$. In particular, in this case n is even.

Theorem 2.2 (Berline-Vergne localization theorem). *Let M be a compact oriented $2l$-dimensional manifold with a circle action which has only isolated fixed points. Then for every equivariantly closed form $\alpha \in \mathcal{A}^\bullet(M)$,*

$$\int_M \alpha = (2\pi)^l \sum_{x \in M_V} \frac{\alpha_0(x)}{\sqrt{\det L_x}}.$$

Proof. Let β be the 1-form introduced in the proof of Lemma 2.1, so that in local coordinates $\Omega = d\beta = \omega_{\mu\nu} dx^\mu \wedge dx^\nu$. Using Proposition 2.1 and the identification $\mathcal{A}^\bullet(M) \simeq C^\infty(\Pi TM)$, we get

$$\int_M \alpha = \lim_{t \to \infty} \int_{\Pi TM} \alpha(\boldsymbol{x}, \boldsymbol{\theta}) e^{-t g_{\mu\nu}(\boldsymbol{x}) v^\mu(\boldsymbol{x}) v^\nu(\boldsymbol{x}) + t \omega_{\mu\nu}(\boldsymbol{x}) \theta^\mu \theta^\nu} d\boldsymbol{x} d\boldsymbol{\theta}.$$

We have in the distributional sense,

$$(2.5) \qquad \lim_{t \to \infty} \left(\frac{t}{\pi}\right)^l \sqrt{\det g(\boldsymbol{x})} \, e^{-t g_{\mu\nu}(\boldsymbol{x}) v^\mu(\boldsymbol{x}) v^\nu(\boldsymbol{x})} = \delta(V(\boldsymbol{x}))$$

— the main formula of the method of steepest descent (cf. the method of stationary phase in Section 2.3 of Chapter 2) — and

$$(2.6) \qquad \lim_{t \to \infty} (2t)^{-l} \frac{1}{\mathrm{Pf}(\omega(\boldsymbol{x}))} e^{t \omega_{\mu\nu}(\boldsymbol{x}) \theta^\mu \theta^\nu} = \delta(\boldsymbol{\theta})$$

— the main formula of Gaussian integration over Grassmann variables (see Proposition 2.3 in Section 2.3 of Chapter 7). Since $\alpha_0(\boldsymbol{x}) = \alpha(\boldsymbol{x}, 0)$, we obtain the result by using formulas (2.5)–(2.6), (2.4), and

$$\delta(V(\boldsymbol{x})) = \sum_{x_i \in M_V} \frac{1}{|\det L_{x_i}|} \delta(\boldsymbol{x} - \boldsymbol{x}_i). \qquad\square$$

Problem 2.1. Prove formulas (2.2)–(2.4).

Problem 2.2. Prove formulas (2.5)–(2.6) and complete the proof of Theorem 2.2.

2.2. Infinite-dimensional case. An important class of infinite-dimensional manifolds is given by the loop spaces. Let M be a compact orientable n-dimensional manifold and let

$$\mathcal{L}(M) = C^{\infty}(S^1, M),$$

where $S^1 = \mathbb{R}/\mathbb{Z}$, be its free loop space. The loop space $\mathcal{L}(M)$ is an infinite-dimensional Fréchet manifold, and the tangent space $T_{\gamma}\mathcal{L}(M)$ to $\mathcal{L}(M)$ at a point $\gamma \in \mathcal{L}(M)$ is

$$T_{\gamma}\mathcal{L}(M) = \Gamma(S^1, \gamma^*(TM))$$

— the vector space of smooth vector fields $V = \{v(t) \in T_{\gamma(t)}M, \, 0 \leq t \leq 1\}$ along the loop γ in M.

The loop space $\mathcal{L}(M)$ has a canonical circle action given by the action of S^1 on itself by rotations. The vector field corresponding to this action is $\dot{\gamma}$ — the velocity vector field $\dot{\gamma}(t)$ along the loop γ in M. The fixed points of the action are constant loops, so that $\mathcal{L}(M)_{\dot{\gamma}} = M$.

The vector field $\dot{\gamma}$ on $\mathcal{L}(M)$ — a generator of the circle action on $\mathcal{L}(M)$ — is analogous to the vector field V on M, considered in the previous section. As in the finite-dimensional case, denote by $D = d - i_{\dot{\gamma}}$ the equivariant differential on the complex $\mathcal{A}^{\bullet}(\mathcal{L}(M))^{S^1}$ of equivariant differential forms on $\mathcal{L}M$. Using the identification $\mathcal{A}^{\bullet}(\mathcal{L}(M)) \simeq C^{\infty}(\Pi T\mathcal{L}(M))$, we have the following infinite-dimensional analog of representation (2.1):

$$(2.7) \quad DF = \int_0^1 \left(\theta^{\mu}(t) \frac{\delta F}{\delta x^{\mu}(t)} - \dot{x}^{\mu}(t) \frac{\delta F}{\delta \theta^{\mu}(t)} \right) dt, \quad F \in C^{\infty}(\Pi T\mathcal{L}(M)).$$

Here we are using functional derivatives (as in calculus of variations), local coordinates $\boldsymbol{x} = (x^1, \dots, x^n)$ on M, and associated standard coordinates in $T_{\gamma(t)}M$ and $\Pi T_{\gamma(t)}M$, so that $\dot{\boldsymbol{\gamma}}(t) = (\dot{x}^1(t), \dots, \dot{x}^n(t)) \in T_{\gamma(t)}M$ and $\boldsymbol{\theta}(t) = (\theta^1(t), \dots, \theta^n(t)) \in \Pi T_{\gamma(t)}M$. As in the finite-dimensional case,

$$(2.8) \qquad D^2 = -\int_0^1 \left(\dot{x}^{\mu}(t) \frac{\delta}{\delta x^{\mu}(t)} + \dot{\theta}^{\mu}(t) \frac{\delta}{\delta \theta^{\mu}(t)} \right) dt = -\mathcal{L}_{\dot{\gamma}}.$$

To define the infinite-dimensional analog of a 1-form β, we choose a Riemannian metric $g_{\mu\nu}dx^{\mu}dx^{\nu}$ on M and consider an S^1-invariant Riemannian metric on $\mathcal{L}(M)$, obtained by the integration of g along a loop[3],

$$\langle V_1, V_2 \rangle_{\gamma} = \int_0^1 \langle v_1(t), v_2(t) \rangle dt, \quad V_1, V_2 \in T_{\gamma}\mathcal{L}(M).$$

---

[3]Here and in what follows $\langle v_1(t), v_2(t) \rangle$ always stands for the inner product in $T_{\gamma(t)}M$ given by the Riemannian metric g.

The 1-form β on $\mathcal{L}(M)$ is dual to the vector field $\dot{\gamma}$ with respect to the Riemannian metric on $\mathcal{L}(M)$, and is given by the following function on $\Pi T\mathcal{L}(M)$:

$$(2.9) \qquad \beta(\gamma,\theta) = \int_0^1 \langle \dot{\gamma}(t), \boldsymbol{\theta}(t)\rangle dt = \int_0^1 g_{\mu\nu}(\gamma(t))\dot{x}^\mu(t)\theta^\nu(t)dt.$$

Lemma 2.2. *The 1-form β on $\mathcal{L}(M)$ is S^1 invariant, and the function $D\beta = -2S$ on $\Pi T\mathcal{L}(M)$ is given by*

$$(2.10) \qquad S(\gamma,\theta) = \frac{1}{2}\int_0^1 \left(\|\dot{\gamma}(t)\|^2 + \langle \boldsymbol{\theta}(t), \nabla_{\dot{\gamma}(t)}\boldsymbol{\theta}(t)\rangle\right)dt,$$

where $\nabla_{\dot{\gamma}}$ is a covariant derivative along γ. The first term in $-2S$ is an analog of the function K in the finite-dimensional case, and the second term is an analog of the 2-form Ω.

Proof. The 1-form β is S^1-invariant by definition. We have

$$D\beta = d\beta - i_{\dot{\gamma}}\beta,$$

where the second term, a function $-i_{\dot{\gamma}}\beta$ on $\Pi T\mathcal{L}(M)$, clearly gives the first term in (2.10). Now using that in the distributional sense

$$\frac{\delta}{\delta x^\rho(s)}x^\mu(t) = \delta^{\rho\mu}\delta(t-s)$$

and

$$\frac{\delta}{\delta x^\rho(s)}\dot{x}^\mu(t) = \delta^{\rho\mu}\frac{d}{dt}\delta(t-s) = -\delta^{\rho\mu}\frac{d}{ds}\delta(t-s),$$

we compute the term $d\beta$ as follows:

$$d\beta(\gamma,\theta) = \int_0^1\int_0^1 \theta^\rho(s)\frac{\delta}{\delta x^\rho(s)}\left(g_{\mu\nu}(\boldsymbol{x}(t))\dot{x}^\mu(t)\right)\theta^\nu(t)dsdt$$

$$= \int_0^1\int_0^1 \theta^\rho(s)\left(\frac{\partial g_{\mu\nu}}{\partial x^\rho}\dot{x}^\mu(t)\delta(t-s) - \delta^{\rho\mu}g_{\mu\nu}\frac{d}{ds}\delta(t-s)\right)\theta^\nu(t)dsdt$$

$$= \int_0^1\left(g_{\mu\nu}(\boldsymbol{x}(t))\dot{\theta}^\mu(t)\theta^\nu(t) + \frac{\partial g_{\mu\nu}}{\partial x^\rho}(\boldsymbol{x}(t))\dot{x}^\mu(t)\theta^\rho(t)\theta^\nu(t)\right)dt.$$

It follows from the definition of Christoffel's symbols (see Example 1.7 in Section 1.3 of Chapter 1) and anticommutativity of θ^μ that

$$(2.11) \qquad \frac{\partial g_{\mu\nu}}{\partial x^\rho}\dot{x}^\mu\theta^\rho\theta^\nu = g_{\mu\nu}\Gamma^\nu_{\lambda\sigma}\dot{x}^\lambda\theta^\mu\theta^\sigma,$$

and we obtain

$$d\beta(\gamma,\theta) = \int_0^1 g_{\mu\nu}(\boldsymbol{x}(t))\left(\dot{\theta}^\mu(t) + \Gamma^\mu_{\sigma\lambda}(\boldsymbol{x}(t))\dot{x}^\sigma(t)\theta^\lambda(t)\right)\theta^\nu(t)dt$$

$$= \int_0^1 \langle \nabla_{\dot{\gamma}(t)}\boldsymbol{\theta}(t), \boldsymbol{\theta}(t)\rangle dt. \qquad \square$$

As in Chapter 6, one can develop the integration theory on $\mathcal{L}(M)$ with respect to the Wiener measure associated with a Riemannian metric on M by using the corresponding heat kernel[4]. One can also integrate differential forms over $\mathcal{L}(M)$, provided that $\mathcal{L}(M)$ is orientable. As in the finite-dimensional case, the latter is defined by the condition that the structure group of an $\mathcal{L}(SO(n))$-bundle $\mathcal{L}(FM) \to \mathcal{L}(M)$, where $FM \to M$ is the frame bundle, reduces to the connected component of the identity. It follows from the transgression homomorphism

$$H^2(M, \mathbb{Z}_2) \to H^1(\mathcal{L}(M), \mathbb{Z}_2)$$

that the image of the second Stiefel-Whitney class of M is the obstruction to orientability of $\mathcal{L}(M)$. In particular, if M is a spin manifold, then $\mathcal{L}(M)$ is orientable, and if M is simply connected, then this is also a necessary condition.

In what follows we will not attempt to develop the integration theory on $\mathcal{L}(M)$, and will formulate and "prove" the infinite-dimensional version of the general Berline-Vergne localization theorem at a heuristic level only. The key observation, by Lemma 2.2, is that the differential form e^{-S} on $\mathcal{L}(M)$ is equivariantly closed, therefore the functional integral $\int_{\mathcal{L}(M)} e^{-S}$ localizes — reduces to the finite-dimensional integral over $\mathcal{L}(M)_{\dot{\gamma}} = M$, the zero locus of the vector field $\dot{\gamma}$.

To give the quantitative formulation of this result, recall the notion of the $\hat{A}$-genus. Namely, the $\hat{A}$-genus of a Riemannian manifold (M, g) is a differential form $\hat{A}(M)$, defined as follows. Let $R \in \mathcal{A}^2(M, \mathfrak{so}(TM))$ be the Riemannian curvature of M, and let e_μ be a local orthonormal frame of TM. Put

$$\mathrm{R}_{\mu\nu} = (Re_\mu, e_\nu) \in \mathcal{A}^2(M),$$

and let

$$j(\mathrm{R}) = \det\left(\frac{\sinh \mathrm{R}/2}{\mathrm{R}/2}\right)$$

be an even degree differential form on M. It does not depend on the choice of a local orthonormal frame of TM, and the $\hat{A}$-genus of the manifold M is defined by[5]

$$\hat{A}(M) = j(\mathrm{R})^{-\frac{1}{2}} = \det\left(\frac{\mathrm{R}/2}{\sinh \mathrm{R}/2}\right)^{\frac{1}{2}} \in \mathcal{A}^\bullet(M).$$

It is a closed differential form, whose cohomology class does not depend on the choice of a Riemannian metric on M.

---

[4]In Chapter 6 we considered the case of $M = \mathbb{R}^n$ with the standard Euclidean metric.
[5]This is the geometer's definition of the $\hat{A}$-genus; in topology one replaces R by $(2\pi i)^{-1}\mathrm{R}$.

Remark. In local coordinates on M,

$$\mathrm{R}_{\mu\nu} = \frac{1}{2} R_{\mu\nu\rho\sigma} dx^\rho \wedge dx^\sigma,$$

where $R_{\mu\nu\rho\sigma}$ is the Riemann curvature tensor.

Theorem 2.3. *Let M be a compact orientable manifold such that the free loop space $\mathcal{L}(M)$ is orientable. Then*

$$\int_{\mathcal{L}(M)} e^{-S} = (2\pi i)^{-\frac{n}{2}} \int_M \hat{A}(M).$$

Proof. The proof is at a heuristic level of rigor. The integral

$$\int_{\mathcal{L}(M)} e^{-S} = \int_{\Pi T \mathcal{L}(M)} e^{-S(\gamma,\theta)} \mathscr{D}\boldsymbol{x} \mathscr{D}\boldsymbol{\theta}$$

localizes at the zero locus $\mathcal{L}(M)_{\dot{\gamma}} = M$, so it is sufficient to integrate over a small tubular neighborhood of $\Pi T M$ in $\Pi T \mathcal{L}(M)$. For this aim, consider Riemann normal coordinates of the normal bundle $\mathcal{N}(\Pi T M)$ to $\Pi T M$ at a point $(x_0, \theta_0) \in \Pi T_{x_0} M$, given by $\boldsymbol{y}(t) = (y^1(t), \ldots, y^n(t)) \in T_{x_0} M$ and $\boldsymbol{\eta}(t) = (\eta^1(t), \ldots, \eta^n(t)) \in \Pi T_{x_0} M$, satisfying

$$\int_0^1 y^\mu(t) dt = 0 \quad \text{and} \quad \int_0^1 \eta^\mu(t) dt = 0, \quad \mu = 1, \ldots, n.$$

Coordinates $(\varepsilon\boldsymbol{y}(t), \varepsilon\boldsymbol{\eta}(t))$, where real ε is sufficiently small, correspond to a point $(\exp_{x_0}(\varepsilon\boldsymbol{y}(t)), P_{(\varepsilon,t)}(\theta_0 + \boldsymbol{\eta}(t)))$, in the normal bundle $\mathcal{N}(\Pi T M)$, where $\exp_{x_0}$ is the exponential map, and $P_{(\varepsilon,t)}$ is a parallel transport in $\Pi T M$ from x_0 along the path $\exp_{x_0}(s\boldsymbol{y}(t))$ as s changes from 0 to ε.

Since we are integrating over an arbitrarily small tubular neighborhood of $\Pi T M$ in $\Pi T \mathcal{L}(M)$, it is sufficient to expand $S(\gamma, \theta)$ up to terms of second order in local coordinates $(\boldsymbol{y}(t), \boldsymbol{\eta}(t))$. Since in Riemann normal coordinates around x_0 on M

$$g_{\mu\nu}(\boldsymbol{x}) = \delta_{\mu\nu} + O(\|\boldsymbol{x}\|^2) \quad \text{and} \quad R_{\mu\nu\rho\sigma}(x_0) = \frac{\partial \Gamma^\rho_{\nu\sigma}}{\partial x^\mu} - \frac{\partial \Gamma^\rho_{\mu\sigma}}{\partial x^\nu},$$

we readily obtain that at $(x_0, \theta_0) \in \Pi T_{x_0} M$

$$S(\gamma, \theta) = S_0(x_0, \theta_0; \boldsymbol{y}, \boldsymbol{\eta}) + \text{higher order terms in } y^\mu(t), \eta^\mu(t),$$

where

$$S_0(x_0, \theta_0; \boldsymbol{y}, \boldsymbol{\eta})$$

$$(2.12) \qquad = \frac{1}{2} \int_0^1 \left(\dot{y}^\mu(t)\dot{y}^\mu(t) - \tilde{\mathrm{R}}_{\mu\nu}(x_0)\dot{y}^\mu(t)y^\nu(t) - \dot{\eta}^\mu(t)\eta^\mu(t) \right) dt,$$

and

$$\tilde{\mathrm{R}}_{\mu\nu}(x_0) = \frac{1}{2} R_{\mu\nu\rho\sigma}(x_0)\theta_0^\rho \theta_0^\sigma.$$

Now using property (2.6) in Section 2.2 of Chapter 6 (where $m = \hbar = 1$ and $T = 1$), formula (4.10) in Section 4.3 of Chapter 7, and performing Gaussian integration, we get

$$\int_{\mathcal{L}(M)} e^{-S} = (2\pi)^{-\frac{n}{2}} \int_{\Pi T M} \left(\int_{\mathcal{N}(\Pi T M)} e^{-S_0(x_0,\theta_0;\boldsymbol{y},\boldsymbol{\eta})} \mathscr{D}\boldsymbol{y}\,\mathscr{D}\boldsymbol{\eta} \right) d\boldsymbol{x}_0 d\boldsymbol{\theta}_0$$

$$= (2\pi)^{-\frac{n}{2}} \int_{\Pi T M} \det'(-D^2\delta_{\mu\nu} - \tilde{\mathrm{R}}_{\mu\nu}(x_0,\theta_0)D)^{-\frac{1}{2}} d\boldsymbol{x}_0 d\boldsymbol{\theta}_0,$$

where we have also used formula (4.12) in Section 4.3 of Chapter 7. Since $\tilde{\mathrm{R}} = \{\tilde{\mathrm{R}}_{\mu\nu}\}_{\mu,\nu=1}^{n}$ is a skew-symmetric matrix with even matrix elements, there is an orthogonal matrix C of determinant 1 such that

$$C\tilde{\mathrm{R}}C^{-1} = \begin{pmatrix} 0 & r_1 & \cdots & 0 & 0 \\ -r_1 & 0 & \cdots & 0 & 0 \\ \vdots & \vdots & \ddots & \vdots & \vdots \\ 0 & 0 & \cdots & 0 & r_m \\ 0 & 0 & \cdots & -r_m & 0 \end{pmatrix}, \quad m = \frac{n}{2}.$$

It is easy to see that second order matrix differential operators

$$-D^2 I_2 + \begin{pmatrix} 0 & -r \\ r & 0 \end{pmatrix} D \quad \text{and} \quad -D^2 I_2 + \begin{pmatrix} -ir & 0 \\ 0 & ir \end{pmatrix} D,$$

where I_2 is the 2×2 identity matrix, have the same spectrum. Using formula (3.5) in Section 3.2 of Chapter 6 with $\omega = \pm ir_k$, we get

$$\det'(-D^2\delta_{\mu\nu} - \tilde{\mathrm{R}}_{\mu\nu}D) = \prod_{k=1}^{m} \det'\left(-D^2 I_2 + \begin{pmatrix} 0 & -r_k \\ r_k & 0 \end{pmatrix} D \right)$$

$$= \prod_{k=1}^{m} \det'(-D^2 - ir_k D)\det'(-D^2 + ir_k D)$$

$$= \prod_{k=1}^{m} \left(\frac{\sin r_k/2}{r_k/2} \right)^2 = \det\left(\frac{\sin \tilde{\mathrm{R}}/2}{\tilde{\mathrm{R}}/2} \right)$$

$$= j(i\tilde{\mathrm{R}}).$$

Therefore, using the identification $C^\infty(\Pi T M) \simeq \mathcal{A}^\bullet(M)$ we finally obtain

$$\int_{\mathcal{L}(M)} e^{-S} = (2\pi)^{-\frac{n}{2}} \int_{\Pi T M} j(i\tilde{\mathrm{R}}(x,\theta))^{-\frac{1}{2}} d\boldsymbol{x} d\boldsymbol{\theta}$$

$$= (2\pi i)^{-\frac{n}{2}} \int_M \hat{A}(M). \qquad \square$$

Remark. Using formulas (2.6) in Section 2.2 of Chapter 6 and (4.10) in Section 4.3 of Chapter 7, we see that the same result holds for the space

$\mathcal{L}_I(M)$ of loops on M parametrized by the interval $I = [0, T]$,

$$\int_{\Pi T \mathcal{L}_I(M)} e^{-\frac{1}{2} \int_0^T (\|\dot{\gamma}(t)\|^2 + \langle \boldsymbol{\theta}(t), \nabla_{\dot{\gamma}(t)} \boldsymbol{\theta}(t) \rangle) dt} \mathscr{D}\boldsymbol{x} \mathscr{D}\boldsymbol{\theta} = (2\pi i)^{-\frac{n}{2}} \int_M \hat{A}(M).$$

Problem 2.3. Derive formula (2.8).

Problem 2.4. Derive formulas (2.11) and (2.12).

Problem 2.5. Justify the arguments in the "proof" of Theorem 2.3. (*Hint:* See the references in Section 7.)

3. Classical mechanics on supermanifolds

As was mentioned in Chapter 4, spin is a pure quantum notion which has no classical analog, and the same applies to fermion systems, considered in Chapter 7. However, one can formally consider particles with anticommuting coordinates and formulate classical mechanics on supermanifolds. Though such systems have no physical interpretation[6], their formal quantization yields fermion systems. This allows us to interpret particles with odd degrees of freedom as semi-classical limit of fermions.

3.1. Functions with anticommuting values. Any smooth map $f : M \to N$ of smooth manifolds M and N gives rise to a Fréchet algebra homomorphism

$$f^* : C^\infty(N) \to C^\infty(M),$$

where $f^*(\varphi) = \varphi \circ f$ for $\varphi \in C^\infty(N)$. Conversely, any Fréchet algebra homomorphism $F : C^\infty(N) \to C^\infty(M)$ is of this form for some smooth map $f : M \to N$.

By definition, a map (morphism) between supermanifolds X and Y is a homomorphism of superalgebras

$$F : C^\infty(Y) \to C^\infty(X),$$

where $C^\infty(X)$ and $C^\infty(Y)$ are commutative superalgebras of global sections of corresponding structure sheaves $\mathcal{O}_X$ and $\mathcal{O}_Y$ (see Section 1). We will denote by $\mathrm{Map}(X, Y)$ the space of all maps between supermanifolds X and Y. The simplest cases are the following.

1. $X = \mathbb{R}^{0|1}$ — odd one-dimensional supermanifold, and $Y = M$ — an ordinary (even) manifold.

2. $X = \mathbb{R}$ (or S^1 and $I = [t_0, t_1]$) — even one-dimensional manifold, and $Y = \mathbb{R}^{0|n}$ — odd n-dimensional coordinate vector space.

---

[6]Classical mechanics, which describes physical phenomena at the macroscopic level, necessarily uses commuting coordinates.

In the first case every homomorphism of superalgebras $F : C^\infty(M) \to \mathbb{R}[\theta]$ has the form
$$F(\varphi) = A(\varphi) + B(\varphi)\theta, \quad \varphi \in C^\infty(M),$$
where $A : C^\infty(M) \to \mathbb{R}$ is an $\mathbb{R}$-algebra homomorphism, and $B : C^\infty(M) \to C^\infty(M)$ is a derivation of the commutative algebra $C^\infty(M)$. By the Gelfand-Naimark theorem, there exists $x \in M$ such that $A(\varphi) = \varphi(x)$, and $B(\varphi) = d\varphi(v)$ where $v \in T_x M$. Thus every map $f : \mathbb{R}^{0|1} \to M$ is given by a point $(x, v) \in TM$.

However, in the second case we have a homomorphism of superalgebras
$$F : \mathbb{R}[\theta^1, \dots, \theta^n] \to C^\infty(\mathbb{R}),$$
and since the commutative algebra $C^\infty(\mathbb{R})$ contains no nilpotents,
$$F(\theta^1) = \cdots = F(\theta^n) = 0.$$
This result is clearly unsatisfactory, since every map $\mathbb{R} \to \mathbb{R}^{0|n}$ is trivial: $\theta^k(t) = 0$, $k = 1, \dots, n$. To remedy this situation, one should consider the analog of Grothendieck's functor of points from algebraic geometry. Namely, introduce an auxiliary Grassmann algebra $W = \mathbb{R}[\eta^1, \dots, \eta^N]$ and consider the product[7] $W \times \mathbb{R}$ with the superalgebra of functions $C^\infty(\mathbb{R})[\eta^1, \dots, \eta^N]$. The "$W$-points" of $\mathrm{Map}(\mathbb{R}, \mathbb{R}^{0|n})$ are defined by homomorphisms of commutative superalgebras
$$F : \mathbb{R}[\theta^1, \dots, \theta^n] \to \mathbb{C}^\infty(\mathbb{R})[\eta^1, \dots, \eta^N].$$

In this case it is customary to write

(3.1) $$\theta^k(t) = F(\theta^k) = a_1^k(t)\eta^1 + \cdots + a_N^k(t)\eta^N, \quad k = 1, \dots, n,$$

where real-valued $a_1^k(t), \dots, a_N^k(t) \in C^\infty(\mathbb{R})$. It is convenient to think of $\theta^k(t)$ as smooth functions of variable t with anticommuting values, and define

(3.2) $$\dot{\theta}^k(t) = \dot{a}_1^k(t)\eta^1 + \cdots + \dot{a}_N^k(t)\eta^N,$$

where the dot stands for the derivative. To emphasize that Grassmann variables $\theta^k(t)$ are "real-valued", we will always assume that W is a set of fixed points in a Grassmann algebra with involution over $\mathbb{C}$, so that
$$\bar{\theta}^k(t) = \theta^k(t), \quad k = 1, \dots, n.$$
In the same way one defines W-points of $\mathrm{Map}(S^1, \mathbb{R}^{0|n})$ and $\mathrm{Map}(I, \mathbb{R}^{0|n})$.

Remark. This definition of the space of maps $\mathrm{Map}(I, \mathbb{R}^{0|n})$ (and also of the spaces $\mathrm{Map}(\mathbb{R}, \mathbb{R}^{0|n})$ and $\mathrm{Map}(S^1, \mathbb{R}^{0|n})$) is motivated by the definition used by physicists, who introduce an infinite-dimensional auxiliary Grassmann algebra W, so that for every t the values $\theta^k(t)$ are "independent real Grassmann variables".

---

[7] Geometrically this corresponds to the family of spaces parametrized by W.

Remark. One can also define functions on the space $\mathrm{Map}(I, \mathbb{R}^{0|n})$ — the functionals of functions with anticommuting values $\theta^k(t)$. Thus for the simplest example of the quadratic functional

$$(3.3) \qquad S(\theta) = \frac{i}{2} \int_{t_0}^{t_1} \theta(t)\dot{\theta}(t)dt,$$

where we put $n = 1$ and $\theta^1(t) = \theta(t)$, using (3.1)–(3.2) we obtain

$$S(\theta) = \frac{i}{2} \int_{t_0}^{t_1} \big(a_k(t)\dot{a}_l(t) - \dot{a}_k(t)a_l(t)\big)dt\, \eta^k \eta^l \in \mathbb{R}[\eta^1, \ldots, \eta^N].$$

Here the factor $i = \sqrt{-1}$ ensures that $i\eta^k\eta^l$ are real Grassmann elements, i.e., $\overline{i\eta^k\eta^l} = -i\eta^l\eta^k = i\eta^k\eta^l$. In general, the space of "functions" on $\mathrm{Map}(I, \mathbb{R}^{0|n})$ is $\Lambda^\bullet(\mathrm{Map}(I, \mathbb{R}^n)^*) \otimes W$, where $\mathrm{Map}(I, \mathbb{R}^n)^*$ is the (topological) dual to the vector space $\mathrm{Map}(I, \mathbb{R}^n)$, and W is a Grassmann algebra with infinitely many generators.

Vector fields with anticommuting values are naturally associated with the path space of a smooth manifold M. Namely, let

$$P_I(M) = \{\gamma : I \to M\},$$

where $I = [t_0, t_1]$, be the space of smooth parametrized paths in M. Consider the supermanifold ΠTM (see Section 1), and for every $\gamma \in P_I(M)$ let $\gamma^*(\Pi TM)$ be the pullback bundle over I. By definition, a vector field with anticommuting values $\theta(t)$ is a section of $\gamma^*(\Pi TM)$ over I. In local coordinates $x = (x^1, \ldots, x^n)$,

$$\boldsymbol{\theta}(t) = \theta^\mu(t)\frac{\partial}{\partial x^\mu} \in \Pi T_{\gamma(t)}M.$$

Thus

$$\Pi TP_I(M) = \{(\gamma, \theta) : \gamma \in P_I(M),\ \theta \in \Gamma(I, \gamma^*(\Pi TM))\}.$$

Problem 3.1. Show that $S \in \Lambda^\bullet(\mathrm{Map}(I, \mathbb{R})^*) \otimes W$, where S is the quadratic functional given by (3.3).

3.2. Classical systems. Here we consider the basic examples of classical systems on supermanifolds.

Example 3.1 (Free classical particle of spin $\frac{1}{2}$). The configuration space is the supermanifold $\Pi T\mathbb{R}^3 \simeq \mathbb{R}^{3|3}$ — the tangent bundle to $\mathbb{R}^3$ with reversed parity of the fibers, with even and odd coordinates $x = (x^1, x^2, x^3)$ and $\boldsymbol{\theta} = (\theta^1, \theta^2, \theta^3)$. The action functional $S : \Pi TP_I(\mathbb{R}^3) \to W$, where W is some auxiliary Grassmann algebra, is defined by

$$S(\boldsymbol{x}(t), \boldsymbol{\theta}(t)) = \int_{t_0}^{t_1} L(\boldsymbol{x}(t), \dot{\boldsymbol{x}}(t), \boldsymbol{\theta}(t))dt$$

$$(3.4) \qquad = \frac{1}{2}\int_{t_0}^{t_1}(m\dot{\boldsymbol{x}}^2 + i\boldsymbol{\theta}\dot{\boldsymbol{\theta}})dt, \quad (\boldsymbol{x}(t), \boldsymbol{\theta}(t)) \in \Pi T P_I(\mathbb{R}^3).$$

Here

$$L(\boldsymbol{x}(t), \dot{\boldsymbol{x}}(t), \boldsymbol{\theta}(t)) = \frac{1}{2}(m\dot{\boldsymbol{x}}^2(t) + i\boldsymbol{\theta}(t)\dot{\boldsymbol{\theta}}(t))$$

is the Lagrangian function, which is real-valued since

$$\overline{i\boldsymbol{\theta}(t)\dot{\boldsymbol{\theta}}(t)} = -i\dot{\boldsymbol{\theta}}(t)\boldsymbol{\theta}(t) = i\boldsymbol{\theta}(t)\dot{\boldsymbol{\theta}}(t).$$

As in Section 1.2 of Chapter 1, we readily derive classical equations of motion as the following Euler-Lagrange equations:

$$\ddot{\boldsymbol{x}}(t) = 0 \quad \text{and} \quad \dot{\boldsymbol{\theta}}(t) = 0.$$

Canonically conjugated momenta[8] are defined by

$$p^k = \frac{\partial L}{\partial \dot{x}^k} = m\dot{x}^k \quad \text{and} \quad \pi^k = \frac{\partial}{\partial \dot{\theta}^k}L = -\frac{i}{2}\theta^k, \quad k = 1, 2, 3,$$

and the Hamiltonian function H is given by the Legendre transformation,

$$H = \boldsymbol{p}\dot{\boldsymbol{x}} + \dot{\boldsymbol{\theta}}\boldsymbol{\pi} - L = \frac{\boldsymbol{p}^2}{2m}.$$

In agreement with classical equations of motion $\dot{\boldsymbol{\theta}}(t) = 0$, the Hamiltonian H does not depend on the odd variables. The phase space is a supermanifold $\mathbb{R}^{6|3}$ with real coordinates $\boldsymbol{p}, \boldsymbol{x}, \boldsymbol{\theta}$ and the symplectic form

$$\omega = d\boldsymbol{p} \wedge d\boldsymbol{x} - \frac{i}{2}d\boldsymbol{\theta}d\boldsymbol{\theta}.$$

In accordance with Section 2.2 of Chapter 7, Poisson brackets between the odd variables are given by

$$(3.5) \qquad \{\theta^k, \theta^l\} = i\delta^{kl}, \quad k, l = 1, 2, 3.$$

Poisson brackets satisfy the following involution property:

$$(3.6) \qquad \overline{\{f_1, f_2\}} = (-1)^{|f_1||f_2|}\{\bar{f}_1, \bar{f}_2\}, \quad f_1, f_2 \in \mathbb{C}[\theta^1, \theta^2, \theta^3].$$

We will see in Section 5 that quantization of this system describes the free quantum particle of spin $\frac{1}{2}$.

---

[8]Here the indices are raised by the standard Euclidean metric on $\mathbb{R}^3$.

Example 3.2 (Classical spin $\frac{1}{2}$ particle in constant magnetic field). This system is described by the configuration space $\mathbb{R}^{3|3}$ with real coordinates $\boldsymbol{x}, \boldsymbol{\theta}$ and Lagrangian function

$$L = \frac{1}{2}(m\dot{\boldsymbol{x}}^2(t) + i\boldsymbol{\theta}(t)\dot{\boldsymbol{\theta}}(t) - i(\boldsymbol{B} \times \boldsymbol{\theta})\boldsymbol{\theta}).$$

The corresponding phase space and symplectic form are the same as in the previous example, and the Hamiltonian function is

$$(3.7) \qquad\qquad H = \frac{\boldsymbol{p}^2}{2m} + \frac{i}{2}(\boldsymbol{B} \times \boldsymbol{\theta})\boldsymbol{\theta}.$$

We will see in Section 5 that quantization of the Hamiltonian (3.7) yields the Pauli Hamiltonian.

Example 3.3 (Free particle on $\mathbb{R}^{n|n}$). Generalizing Example 3.1, consider the configuration space $\Pi T\mathbb{R}^n \simeq \mathbb{R}^{n|n}$ — the tangent bundle to $\mathbb{R}^n$ with reversed parity of the fibers, with even and odd real coordinates $\boldsymbol{x} = (x^1, \dots, x^n)$ and $\boldsymbol{\theta} = (\theta^1, \dots, \theta^n)$. The Lagrangian function

$$(3.8) \qquad L(\boldsymbol{x}(t), \dot{\boldsymbol{x}}(t), \boldsymbol{\theta}(t)) = \frac{1}{2}(m\dot{x}^\mu\dot{x}^\mu + i\theta^\mu\dot{\theta}^\mu)$$

yields the same Euler-Lagrange equations as in Example 3.1,

$$\ddot{\boldsymbol{x}} = 0 \quad \text{and} \quad \dot{\boldsymbol{\theta}} = 0.$$

The corresponding phase space is a supermanifold $\mathbb{R}^{2n|n}$ with real coordinates $\boldsymbol{p}, \boldsymbol{x}, \boldsymbol{\theta}$ and the symplectic form

$$\omega = d\boldsymbol{p} \wedge d\boldsymbol{x} - \frac{i}{2}d\boldsymbol{\theta}d\boldsymbol{\theta},$$

and the Hamiltonian function is

$$H = \frac{\boldsymbol{p}^2}{2m}.$$

Example 3.4 (Free particle on ΠTM). Let (M, g) be a Riemannian manifold with Riemannian metric $g_{\mu\nu}dx^\mu dx^\nu$. Similar to the construction in Section 2.2, we consider the phase space ΠTM with the Lagrangian function

$$(3.9) \qquad L = \frac{1}{2}(\|\dot{\boldsymbol{x}}\|^2 + i\langle\boldsymbol{\theta}, \nabla_{\dot{\boldsymbol{x}}}\boldsymbol{\theta}\rangle),$$

where $\nabla_{\dot{\boldsymbol{x}}}$ is a covariant derivative along the path $\boldsymbol{x}(t)$ (which is also customarily denoted by D/Dt), and $\boldsymbol{\theta}(t) \in \Pi T_x M$. The corresponding action functional is defined by

$$S = \frac{1}{2}\int_{t_0}^{t_1}(\|\dot{\boldsymbol{x}}\|^2 + i\langle\boldsymbol{\theta}, \nabla_{\dot{\boldsymbol{x}}}\boldsymbol{\theta}\rangle)dt,$$

or in local coordinates $\boldsymbol{x} = (x^1, \dots, x^n)$ on M,

$$S = \frac{1}{2} \int_{t_0}^{t_1} g_{\mu\nu}(\boldsymbol{x}(t)) \left(\dot{x}^\mu \dot{x}^\nu + \theta^\mu \frac{D\theta^\nu}{Dt} \right) dt$$

$$= \frac{1}{2} \int_{t_0}^{t_1} g_{\mu\nu}(\boldsymbol{x}(t)) \left(\dot{x}^\mu \dot{x}^\nu + \theta^\mu \dot{\theta}^\nu + \Gamma^\nu_{\lambda\rho}(\boldsymbol{x}(t)) \dot{x}^\lambda \theta^\mu \theta^\rho \right) dt.$$

We will see in the next section that this system describes a supersymmetric particle on a Riemannian manifold.

Problem 3.2. Develop Lagrangian and Hamiltonian formalisms for classical mechanics on supermanifolds.

Problem 3.3. Prove formula (3.6).

Problem 3.4. Show that Euler-Lagrange equations for the system with Lagrangian (3.9) are

$$g_{\lambda\mu} \frac{D\dot{x}^\mu}{Dt} = \frac{1}{2} R_{\lambda\rho\mu\nu} \dot{x}^\rho \theta^\mu \theta^\nu \quad \text{and} \quad \frac{D\theta^\mu}{Dt} = 0$$

— equations for a spinning particle in a gravitational field.

Problem 3.5. Describe the phase space and find the Hamiltonian function for Example 3.4.

4. Supersymmetry

The Lagrangian system considered in Example 3.3 in the previous section possesses remarkable symmetries.

4.1. Total angular momentum. First of all, the Lagrangian (3.8) is obviously invariant with respect to the action of the orthogonal group $G = \mathrm{SO}(n)$

(4.1) $$L(g \cdot \dot{\boldsymbol{x}}, g \cdot \boldsymbol{\theta}) = L(\dot{\boldsymbol{x}}, \boldsymbol{\theta}), \quad g \in G.$$

The corresponding conserved quantity — the Noether charge $\boldsymbol{J} \in \mathfrak{g}^*$ — the dual space to the Lie algebra $\mathfrak{g} = \mathfrak{so}(n)$, can be obtained as follows (cf. the proof of Theorem 1.3 in Section 1.4 of Chapter 1).

Consider an infinitesimal change of even and odd coordinates

$$\boldsymbol{x}(t) \mapsto \tilde{\boldsymbol{x}}(t) = \boldsymbol{x}(t) + \delta\boldsymbol{x}(t), \quad \boldsymbol{\theta}(t) \mapsto \tilde{\boldsymbol{\theta}}(t) = \boldsymbol{\theta}(t) + \delta\boldsymbol{\theta}(t),$$

and using $\dot{\boldsymbol{\theta}}\,\delta\boldsymbol{\theta} = -\delta\boldsymbol{\theta}\,\dot{\boldsymbol{\theta}}$, compute $\delta L = L(\dot{\boldsymbol{x}} + \delta\dot{\boldsymbol{x}}, \boldsymbol{\theta} + \delta\boldsymbol{\theta}) - L(\dot{\boldsymbol{x}}, \boldsymbol{\theta})$ up to the second order terms in $\delta\boldsymbol{x}$ and $\delta\boldsymbol{\theta}$,

$$\delta L = m\dot{\boldsymbol{x}}\,\delta\dot{\boldsymbol{x}} + \frac{i}{2}\left(\delta\boldsymbol{\theta}\,\dot{\boldsymbol{\theta}} + \boldsymbol{\theta}\,\delta\dot{\boldsymbol{\theta}} \right)$$

$$= -m\ddot{\boldsymbol{x}}\,\delta\boldsymbol{x} + m\frac{d}{dt}(\dot{\boldsymbol{x}}\,\delta\boldsymbol{x}) + \frac{i}{2}\left(\delta\boldsymbol{\theta}\,\dot{\boldsymbol{\theta}} - \dot{\boldsymbol{\theta}}\,\delta\boldsymbol{\theta} \right) + \frac{i}{2}\frac{d}{dt}(\boldsymbol{\theta}\,\delta\boldsymbol{\theta})$$

$$= -m\ddot{\boldsymbol{x}}\,\delta\boldsymbol{x} + m\frac{d}{dt}(\dot{\boldsymbol{x}}\,\delta\boldsymbol{x}) + i\delta\boldsymbol{\theta}\,\dot{\boldsymbol{\theta}} + \frac{i}{2}\frac{d}{dt}(\boldsymbol{\theta}\,\delta\boldsymbol{\theta}).$$

Thus on the solutions of Euler-Lagrange equations $\ddot{\boldsymbol{x}} = 0$, $\dot{\boldsymbol{\theta}} = 0$ we have

$$\delta L = \frac{d}{dt}\left(m\dot{\boldsymbol{x}}\,\delta\boldsymbol{x} + \frac{i}{2}\boldsymbol{\theta}\,\delta\boldsymbol{\theta}\right).$$

Now using (4.1) with $g = e^{\varepsilon\boldsymbol{u}}$, where $\boldsymbol{u} = \{u^{\mu}_{\nu}\}^{n}_{\mu,\nu=1}$ is a skew-symmetric $n\times n$ matrix, we see that $\delta L = 0$ for $\delta x^{\mu} = \varepsilon u^{\mu}_{\nu}x^{\nu}$ and $\delta\theta^{\mu} = \varepsilon u^{\mu}_{\nu}\theta^{\nu}$. Choosing the standard basis in the space of skew-symmetric $n \times n$ matrices, we obtain that components

$$(4.2) \qquad\qquad J^{\mu\nu} = m(x^{\mu}\dot{x}^{\nu} - x^{\nu}\dot{x}^{\mu}) - i\theta^{\mu}\theta^{\nu}$$

are integrals of motion:

$$\frac{d}{dt}J^{\mu\nu} = 0$$

on solutions of the Euler-Lagrange equations. In particular, for $n = 3$ we get

$$J^{1} := J^{23} = M^{1} - i\theta^{2}\theta^{3},$$
$$J^{2} := J^{31} = M^{2} - i\theta^{3}\theta^{1},$$
$$J^{3} := J^{12} = M^{3} - i\theta^{1}\theta^{2},$$

where M^{1}, M^{2}, M^{3} are components of the angular momentum $\boldsymbol{M}$ of a particle in $\mathbb{R}^{3}$ (see Section 1.4 in Chapter 1). We will see in the next section that after quantization the vector $\boldsymbol{J} = (J^{1}, J^{2}, J^{3})$ becomes the total angular momentum operator of a quantum particle of spin $\frac{1}{2}$ in $\mathbb{R}^{3}$.

Remark. In fact, the Lagrangian (3.8) is invariant under the action of $G\times G$ on $\mathbb{R}^{n|n}$, so that both the angular momentum $m(x^{\mu}\dot{x}^{\nu} - x^{\nu}\dot{x}^{\mu})$ in $\mathbb{R}^{n}$ and the "Grassmann angular momentum" $-i\theta^{\mu}\theta^{\nu}$ in $\mathbb{R}^{0|n}$ are conserved.

4.2. Supersymmetry transformation. It is quite remarkable that in addition to the symmetries discussed in the previous section, the Lagrangian (3.8) is also invariant under special transformations on $\mathbb{R}^{n|n}$ that mix even and odd coordinates. Namely, let $(\boldsymbol{\gamma}(t), \boldsymbol{\theta}(t)) \in \Pi TP_{I}(\mathbb{R}^{n})$, where $\boldsymbol{\theta}(t) \in \Pi T_{\gamma}P_{I}(\mathbb{R}^{n})$ is a section over $I = [t_{0}, t_{1}]$ of the pull-back by γ of the tangent bundle of $T\mathbb{R}^{n}$ with reverse parity of the fibers,

$$\boldsymbol{\theta}(t) = \theta^{\mu}(t)\frac{\partial}{\partial x^{\mu}} \in \Pi T_{\gamma(t)}M.$$

Consider the following infinitesimal change of coordinates which mixes even and odd variables:

$$(4.3) \qquad \boldsymbol{x}(t) \mapsto \boldsymbol{x}(t) + \delta_{\varepsilon}\boldsymbol{x}(t) \quad \text{and} \quad \boldsymbol{\theta}(t) \mapsto \boldsymbol{\theta}(t) + \delta_{\varepsilon}\boldsymbol{\theta}(t),$$

where

$$(4.4) \quad \delta_{\varepsilon}\boldsymbol{x}(t) = i\varepsilon\boldsymbol{\theta}(t) \in T_{\boldsymbol{x}(t)}\mathbb{R}^{n} \quad \text{and} \quad \delta_{\varepsilon}\boldsymbol{\theta}(t) = -m\varepsilon\dot{\boldsymbol{x}}(t) \in \Pi T_{\boldsymbol{x}(t)}\mathbb{R}^{n},$$

and ε is an odd real element. Then for

$$\delta_{\varepsilon}L = L(\boldsymbol{x} + \delta_{\varepsilon}\boldsymbol{x}, \boldsymbol{\theta} + \delta_{\varepsilon}\boldsymbol{\theta}) - L(\boldsymbol{x}, \boldsymbol{\theta})$$

we obtain

$$\delta_\varepsilon L = m\dot{\boldsymbol{x}}\,\delta_\varepsilon\dot{\boldsymbol{x}} + \frac{i}{2}(\delta_\varepsilon\boldsymbol{\theta}\,\dot{\boldsymbol{\theta}} + \boldsymbol{\theta}\,\delta_\varepsilon\dot{\boldsymbol{\theta}})$$

$$= im\dot{\boldsymbol{x}}\,\varepsilon\dot{\boldsymbol{\theta}} - \frac{im}{2}(\varepsilon\dot{\boldsymbol{x}}\,\dot{\boldsymbol{\theta}} + \boldsymbol{\theta}\,\varepsilon\ddot{\boldsymbol{x}})$$

$$= \frac{im}{2}(\dot{\boldsymbol{x}}\,\varepsilon\dot{\boldsymbol{\theta}} + \varepsilon\boldsymbol{\theta}\,\ddot{\boldsymbol{x}})$$

$$= \frac{im\varepsilon}{2}\frac{d}{dt}(\boldsymbol{\theta}\,\dot{\boldsymbol{x}}).$$

Thus for periodic boundary conditions the action functional

$$S(\gamma,\theta) = \int_{t_0}^{t_1} L(\boldsymbol{\gamma}(t),\boldsymbol{\theta}(t))dt$$

is invariant under the change of coordinates (4.3),

$$\delta_\varepsilon S(\gamma,\theta) = 0 \quad \text{for all periodic } (\gamma,\theta) \in \Pi T P_I(\mathbb{R}^n).$$

The infinitesimal transformation (4.3) is called the *supersymmetry transformation*.

Remark. We emphasize that invariance of the action under the supersymmetry transformations holds for all $(\boldsymbol{\gamma}(t),\boldsymbol{\theta}(t))$ with periodic boundary conditions, and not only on equations of motion!

Introducing the quantity

$$Q = im\boldsymbol{\theta}\dot{\boldsymbol{x}} = i\boldsymbol{\theta}\boldsymbol{p} = i\theta^\mu p_\mu,$$

called the *generator of supersymmetry* (or the *supercharge*), we can summarize the above computation as

$$(4.5) \qquad \delta_\varepsilon L = \frac{\varepsilon}{2}\frac{dQ}{dt}.$$

Another remarkable property is that the Lagrangian L can be recovered from the supercharge Q. Namely, the simple computation

$$\delta_\varepsilon Q = m(\delta_\varepsilon\boldsymbol{\theta}\,\dot{\boldsymbol{x}} + \boldsymbol{\theta}\,\delta_\varepsilon\dot{\boldsymbol{x}}) = m\varepsilon(-m\dot{\boldsymbol{x}}^2 - i\boldsymbol{\theta}\,\dot{\boldsymbol{\theta}})$$

gives

$$(4.6) \qquad -2mi\varepsilon L = \delta_\varepsilon Q.$$

Using (3.5) we obtain

$$(4.7) \qquad \{Q,Q\} = -i\boldsymbol{p}^2 = -2miH,$$

which shows that the Hamiltonian H can also be recovered from the supercharge Q.

Remark. Geometrically, supersymmetry transformation is just an equivariant differential on the space $\mathcal{L}_I(\mathbb{R}^n)$ of free loops on $\mathbb{R}^n$ parametrized by the interval $I = [t_0, t_1]$, considered in Section 2.2 (for the interval $[0, 1]$). Namely, set $m = 1$ and consider the Wick rotation $t \mapsto -it$ to the Euclidean time, so that the sypersymmetry transformation (4.3)–(4.4) becomes

$$(4.8) \qquad \delta_\varepsilon \boldsymbol{x}(t) = i\varepsilon\boldsymbol{\theta}(t), \quad \delta_\varepsilon\boldsymbol{\theta}(t) = -i\varepsilon\dot{\boldsymbol{x}}(t).$$

Now we immediately see that formulas (4.8) can be written as

$$\delta_\varepsilon\boldsymbol{x}(t) = i\varepsilon D\boldsymbol{x}(t) \quad \text{and} \quad \delta_\varepsilon\boldsymbol{\theta}(t) = i\varepsilon D\boldsymbol{\theta}(t),$$

where D is the equivariant differential (2.7),

$$D = \int_{t_0}^{t_1} \left(\theta^\mu(t)\frac{\delta}{\delta x^\mu(t)} - \dot{x}^\mu(t)\frac{\delta}{\delta\theta^\mu(t)} \right)dt,$$

which satisfies $D^2 = -\mathcal{L}_{\dot\gamma}$. The corresponding Euclidean supercharge Q coincides with the function β on $\Pi T\mathcal{L}(\mathbb{R}^n)$, given by formula (2.9), so that

$$Q = \int_{t_0}^{t_1} \theta^\mu(t)\dot{x}^\mu(t)dt,$$

where we are using the standard Euclidean metric on $\mathbb{R}^n$. It follows from Lemma 2.2 that

$$DQ = -\int_{t_0}^{t_1} (\dot{\boldsymbol{x}}(t)\dot{\boldsymbol{x}}(t) + \boldsymbol{\theta}(t)\dot{\boldsymbol{\theta}}(t))dt = -2S(\boldsymbol{x}(t), \boldsymbol{\theta}(t)),$$

where now S stands for the Euclidean action. The invariance of the action S under the supersymmetry transformation is the property that S is equivariantly closed, $DS = 0$.

Problem 4.1. Verify that

$$[\delta_{\varepsilon_1}, \delta_{\varepsilon_2}] = 2i\varepsilon_1\varepsilon_2\frac{d}{dt},$$

where $\varepsilon_1, \varepsilon_2$ are odd variables, and deduce from it formula (4.7).

4.3. Supersymmetric particle on a Riemannian manifold.

The classical system with Lagrangian (3.9), considered in Example 3.4 in Section 3.2, describes a supersymmetric particle on a Riemannian manifold (M, g). Namely, for $(\gamma, \theta) \in \Pi T P_I(M)$ define the supersymmetry transformation $\boldsymbol{x}(t) \mapsto \boldsymbol{x}(t) + \delta_\varepsilon\boldsymbol{x}(t)$ and $\boldsymbol{\theta}(t) \mapsto \boldsymbol{\theta}(t) + \delta_\varepsilon\boldsymbol{\theta}(t)$, in local coordinantes $\boldsymbol{x}(t)$ and $\boldsymbol{\theta}(t)$, by the same formula (4.4),

$$(4.9) \qquad \delta_\varepsilon\boldsymbol{x}(t) = i\varepsilon\boldsymbol{\theta}(t) \in \Pi T_{\gamma(t)}M, \quad \delta_\varepsilon\boldsymbol{\theta}(t) = -m\varepsilon\dot{\boldsymbol{x}}(t) \in \Pi T_{\gamma(t)}M,$$

where ε is an odd real element.

Lemma 4.1. *Supersymmetry transformation (4.9) does not depend on the choice of local coordinates.*

Proof. Let $\tilde{\boldsymbol{x}} = (\tilde{x}^1, \ldots, \tilde{x}^n)$ be another local coordinates on M, $\tilde{x}^\mu = f^\mu(\boldsymbol{x})$, $\mu = 1, \ldots, n$. Along the path $\gamma = \boldsymbol{x}(t)$,

$$\dot{\tilde{x}}^\mu(t) = \frac{\partial f^\mu}{\partial x^\nu}(\boldsymbol{x}(t))\dot{x}^\nu(t) \quad \text{and} \quad \boldsymbol{\theta}(t) = \tilde{\theta}^\mu(t)\frac{\partial}{\partial \tilde{x}^\mu},$$

where

$$\tilde{\theta}^\mu(t) = \frac{\partial f^\mu}{\partial x^\nu}(\boldsymbol{x}(t))\theta^\nu(t).$$

We have

$$\delta_\varepsilon \tilde{x}^\mu(t) = \frac{\partial f^\mu}{\partial x^\nu}(\boldsymbol{x}(t))\delta_\varepsilon x^\nu(t) = i\frac{\partial f^\mu}{\partial x^\nu}(\boldsymbol{x}(t))\varepsilon\theta^\nu(t) = i\varepsilon\tilde{\theta}^\mu(t),$$

and

$$\delta_\varepsilon \tilde{\theta}^\mu(t) = \frac{\partial f^\mu}{\partial x^\nu}(\boldsymbol{x}(t))\delta_\varepsilon \theta^\nu(t) + \frac{\partial^2 f^\mu}{\partial x^\nu \partial x^\sigma}(\boldsymbol{x}(t))\delta_\varepsilon x^\sigma(t)\theta^\nu(t)$$

$$= -m\varepsilon\frac{\partial f^\mu}{\partial x^\nu}(\boldsymbol{x}(t))\dot{x}^\nu(t) + i\varepsilon\frac{\partial^2 f^\mu}{\partial x^\nu \partial x^\sigma}(\boldsymbol{x}(t))\theta^\sigma(t)\theta^\nu(t)$$

$$= -m\varepsilon\dot{\tilde{x}}^\mu(t),$$

since $\theta^\mu(t)$ anticommute. $\qquad\square$

As in the previous section, we define the supercharge Q by

$$Q(t) = im\langle\dot{\boldsymbol{x}}(t), \boldsymbol{\theta}(t)\rangle.$$

The main result of this section is the following statement.

Proposition 4.1. *Under the supersymmetry transformation* (4.9),

$$(4.10) \qquad\qquad \delta_\varepsilon Q = -2mi\varepsilon L,$$

where

$$(4.11) \qquad\qquad L = \frac{m}{2}\|\dot{\boldsymbol{x}}\|^2 + \frac{i}{2}\langle\boldsymbol{\theta}, \nabla_{\dot\gamma}\boldsymbol{\theta}\rangle$$

is the Lagrangian of a free particle on $\Pi T M$. *Also*

$$(4.12) \qquad\qquad \delta_\varepsilon L = \frac{\varepsilon}{2}\frac{dQ}{dt}.$$

Proof. The derivation of (4.10) is a straightforward computation using formula (2.11). Formula (4.12) is proved by another computation, similar to the one used to prove (4.5) in the previous section. $\qquad\square$

Corollary 4.1. *The action functional*

$$S(\gamma, \theta) = \int_{t_0}^{t_1} L(\boldsymbol{\gamma}(t), \boldsymbol{\theta}(t))dt$$

is invariant under the supersymmetry transformations,

$$\delta_\varepsilon S(\gamma, \theta) = 0 \quad \text{for all periodic } (\gamma, \theta) \in \Pi T P_I(M).$$

Remark. As for the example of a free particle on $\mathbb{R}^{n|n}$, considered in Section 4.2, the Euclidean version of the sypersymmetry transformation (4.9) corresponds to the equivariant differential (2.7) on the space $\mathcal{L}_I(M)$ of free loops on M parametrized by the interval $I = [t_0, t_1]$. The corresponding Euclidean supercharge Q (for $m = 1$) coincides with the function β on $\Pi T \mathcal{L}_I(M)$ given by formula (2.9), and Proposition 4.1 reduces to the statements that $DQ = -2S$ and $DS = 0$, where S is the Euclidean action, proved in Lemma 2.2.

Problem 4.2. Prove all the formulas in this section.

Problem 4.3 (Superfield formalism). For $(\gamma, \theta) \in \Pi T M$ define the *superfield* by $X(t) = \boldsymbol{x}(t) + \eta \boldsymbol{\theta}(t)$, where η is an auxiliary Grassmann variable, and let $\mathcal{D} = \dfrac{\partial}{\partial \eta} - \eta \dfrac{\partial}{\partial t}$. Show that

$$\int g_{\mu\nu}(X(t))\dot{X}(t)\mathcal{D}(X)(t)dt\,d\eta = -\int_{t_0}^{t^1} (\|\dot{\boldsymbol{x}}\|^2 + \langle \boldsymbol{\theta}, \nabla_{\dot{\gamma}}\boldsymbol{\theta}\rangle)dt$$

— twice the Euclidean action of a supersymmetric particle on a Riemannian manifold M.

5. Quantum mechanics on supermanifolds

Here we describe quantum systems which correspond to classical systems in Section 3.2. According to the correspondence principle (see Section 2 of Chapter 2), for quantization of even coordinates we replace Poisson brackets $\{\,,\,\}$ by $i[\,,\,]$, where $i = \sqrt{-1}$ and $[\,,\,]$ is the commutator[9]. For quantization of odd coordinates, in accordance with Section 1 of Chapter 7, we replace corresponding Poisson brackets by $i[\,,\,]_+$, where $[\,,\,]_+$ is the anticommutator.

Example 5.1 (Quantum particle of spin $\frac{1}{2}$). The corresponding phase space is a supermanifold $\mathbb{R}^{6|3}$ with even coordinates $\boldsymbol{p} = (p_1, p_2, p_3)$ and $\boldsymbol{x} = (x_1, x_2, x_3)$, and odd coordinates[10] $\boldsymbol{\theta} = (\theta_1, \theta_2, \theta_3)$, with canonical Poisson brackets

$$\{p_\mu, x_\nu\} = \delta_{\mu\nu} \quad \text{and} \quad \{\theta_\mu, \theta_\nu\} = i\delta_{\mu\nu}, \quad \mu, \nu = 1, 2, 3.$$

Quantum operators $\boldsymbol{P} = (P_1, P_2, P_3)$ and $\boldsymbol{Q} = (Q_1, Q_2, Q_3)$, which correspond to canonical coordinates $\boldsymbol{p}$ and $\boldsymbol{x}$, satisfy Heisenberg commutation relations (see Section 2.1 of Chapter 2), while operators $\boldsymbol{\Theta} = (\Theta_1, \Theta_2, \Theta_3)$, which correspond to the anticommuting coordinates $\boldsymbol{\theta}$, satisfy the following anticommutation relations:

$$(5.1) \qquad\qquad [\Theta_\mu, \Theta_\nu]_+ = \delta_{\mu\nu} I, \quad \mu, \nu = 1, 2, 3.$$

---

[9]Here we put $\hbar = 1$.

[10]Here it is convenient to lower all indices by the standard Euclidean metric on $\mathbb{R}^3$.

Since operators $\sqrt{2}\,\Theta_\mu$ define a representation of a Clifford algebra C_3, the only irreducible realization of (5.1) is given by

$$(5.2) \qquad \Theta_\mu = \frac{1}{\sqrt{2}}\sigma_\mu, \quad \mu = 1,2,3,$$

where σ_μ are Pauli matrices (see Section 1.1 of Chapter 4). Thus the Hilbert space of the system is $\mathscr{H} = L^2(\mathbb{R}^3) \otimes \mathbb{C}^2$ — the Hilbert space of a quantum particle of spin $\frac{1}{2}$ — and the Hamiltonian operator is

$$H = \frac{\boldsymbol{P}^2}{2m}.$$

Using (5.2) and the multiplication table of Pauli matrices, we obtain the following form for quantum Noether integrals of motion (4.2):

$$\boldsymbol{J} = \boldsymbol{M} + \boldsymbol{S},$$

where $\boldsymbol{M}$ is the angular momentum operator (see Section 3.1 of Chapter 3), and $\boldsymbol{S} = \frac{1}{2}\boldsymbol{\sigma}$. Thus $\boldsymbol{J}$ is the total angular momentum operator of a quantum spin $\frac{1}{2}$ particle (see Section 1.2 of Chapter 4).

Example 5.2 (Quantum spin $\frac{1}{2}$ particle in constant magnetic field). The Hilbert space is the same as in the previous example, while the Hamiltonian operator which corresponds to (3.7) takes the form

$$H = \frac{\boldsymbol{P}^2}{2m} + \frac{i}{2}(\boldsymbol{B} \times \boldsymbol{\Theta})\boldsymbol{\Theta} = \frac{\boldsymbol{P}^2}{2m} - \frac{\boldsymbol{B}\cdot\boldsymbol{\sigma}}{2}.$$

Thus operator H is the Pauli Hamiltonian with total magnetic moment $\mu = \frac{1}{2}$ (see Section 2.1 of Chapter 4).

Example 5.3 (Supersymmetric quantum particle on $\mathbb{R}^n$). The phase space is a supermanifold $\mathbb{R}^{2n|n}$ with even coordinates $\boldsymbol{p} = (p_1, \ldots, p_n)$ and $\boldsymbol{x} = (x_1, \ldots, x_n)$, and odd coordinates $\boldsymbol{\theta} = (\theta_1, \ldots, \theta_n)$, with canonical Poisson brackets

$$\{p_\mu, x_\nu\} = \delta_{\mu\nu} \quad \text{and} \quad \{\theta_\mu, \theta_\nu\} = i\delta_{\mu\nu}, \quad \mu, \nu = 1, \ldots, n.$$

Quantum operators $\boldsymbol{P} = (P_1, \ldots, P_n)$ and $\boldsymbol{Q} = (Q_1, \ldots, Q_n)$, which correspond to canonical coordinates $\boldsymbol{p}$ and $\boldsymbol{x}$, satisfy Heisenberg commutation relations, and operators $\boldsymbol{\Theta} = (\Theta_1, \ldots, \Theta_n)$, which correspond to the anti-commuting coordinates $\boldsymbol{\theta}$, satisfy the following anticommutation relations:

$$(5.3) \qquad [\Theta_\mu, \Theta_\nu]_+ = \delta_{\mu\nu}I, \quad \mu, \nu = 1, \ldots, n.$$

The operators $\sqrt{2}\,\Theta^\mu$ define a representation of a Clifford algebra C_n. When n is even, the only irreducible realization of (5.3) is

$$\Theta_\mu = \frac{1}{\sqrt{2}}\gamma_\mu,$$

where γ_μ are operators in $\mathcal{H}_F = (\mathbb{C}^2)^{\otimes d}$, $d = \frac{n}{2}$, given by formulas (1.12)–(1.13) in Section 1.2 of Chapter 7. For odd n, operators γ act in $(\mathbb{C}^2)^{\otimes d}$, where $d = \left[\frac{n}{2}\right]$ (see Problem 1.4 in Section 1.2 of Chapter 7). In both cases, the Hilbert space of the system is $\mathcal{H} = L^2(\mathbb{R}^n) \otimes \mathcal{H}_F = L^2(\mathbb{R}^n) \otimes \mathbb{C}^{2^d}$, the Hamiltonian operator is given by

$$H = \frac{\boldsymbol{P}^2}{2m},$$

and quantum Noether integrals of motion are

$$J_{\mu\nu} = Q_\mu P_\nu - Q_\nu P_\mu - i\Theta_\mu\Theta_\nu.$$

Quantization of the supercharge gives the operator

$$(5.4) \qquad Q = i\Theta_\mu P_\mu = \frac{i}{\sqrt{2}}\,\gamma_\mu P_\mu = \frac{1}{\sqrt{2}}\,\gamma_\mu \frac{\partial}{\partial x_\mu} = \frac{1}{\sqrt{2}}\,\slashed{\partial},$$

where

$$\slashed{\partial} = \gamma_\mu \frac{\partial}{\partial x_\mu}$$

is the Dirac operator on $\mathbb{R}^n$. The Dirac operator is skew-adjoint in $\mathcal{H}$,

$$\slashed{\partial}^* = -\slashed{\partial}.$$

When n is even, decomposition $\mathcal{H}_F = \mathcal{H}_F^+ \oplus \mathcal{H}_F^-$ into the subspaces of positive and negative chirality spinors (see Section 1.2 of Chapter 7) gives decomposition

$$\mathcal{H} = \mathcal{H}_+ \oplus \mathcal{H}_-, \quad \text{where} \quad \mathcal{H}_\pm = L^2(\mathbb{R}^n) \otimes \mathcal{H}_F^\pm,$$

and using that $\gamma_\mu(\mathcal{H}_F^+) = \mathcal{H}_F^-$ (see Section 1.2 of Chapter 7), we can represent the Dirac operator in the following 2×2 block-matrix form:

$$\slashed{\partial} = \begin{pmatrix} 0 & -\slashed{\partial}_+^* \\ \slashed{\partial}_+ & 0 \end{pmatrix}.$$

Here the operator $\slashed{\partial}_+ : \mathcal{H}_+ \to \mathcal{H}_-$ is called the chiral Dirac operator. We have

$$[Q, Q]_+ = 2Q^2 = \slashed{\partial}^2 = -2mH,$$

so that $2mH$ is the Dirac Laplacian $-\slashed{\partial}^2$. In matrix form,

$$2mH = \begin{pmatrix} \slashed{\partial}_+^*\slashed{\partial}_+ & 0 \\ 0 & \slashed{\partial}_+\slashed{\partial}_+^* \end{pmatrix}.$$

Example 5.4 (Supersymmetric quantum particle on Riemannian manifold). We have seen in Example 2.4 in Section 2.4 of Chapter 2 that though for a free quantum particle on a Riemannian manifold (M, g) it is not possible to construct operators $\boldsymbol{P}$ and $\boldsymbol{Q}$ which correspond to standard local coordinates $(\boldsymbol{p}, \boldsymbol{x})$ on T^*M, the Hamiltonian operator H is well defined as the Laplace-Beltrami operator of the Riemannian metric. It is remarkable that consistent quantization of a supersymmetric particle requires M to be a *spin manifold*, and for even $n = \dim M$ the corresponding supercharge operator Q coincides with the Dirac operator!

Namely, let $\mathrm{Spin}(n)$ be the spin group: connected, simply connected Lie group which is a double cover of $\mathrm{SO}(n)$. The oriented Riemannian manifold (M, g) of dimension n is said to have a *spin structure* (and is called a spin manifold) if the bundle $\mathrm{SO}(M)$ of oriented orthonormal frames over M — principal $\mathrm{SO}(n)$-bundle — can be extended to the principal $\mathrm{Spin}(n)$-bundle $\mathrm{Spin}(M)$. This is equivalent to the condition that for some open covering $M = \bigcup_{\alpha \in A} U_\alpha$ transition functions $t_{\alpha\beta} : U_\alpha \cap U_\beta \to \mathrm{SO}(n)$ of a tangent bundle TM can be lifted to the transition functions $\tau_{\alpha\beta} : U_\alpha \cap U_\beta \to \mathrm{Spin}(n)$; $t_{\alpha\beta} = \rho(\tau_{\alpha\beta})$, where $\rho : \mathrm{Spin}(n) \to \mathrm{SO}(n)$ is the canonical projection. The manifold M is a spin manifold if and only if its second Stieffel-Whitney class $w_2 \in H^2(M, \mathbb{Z}_2)$ vanishes; in this case different spin structures are parametrized by $H^1(M, \mathbb{Z}_2) \simeq \mathrm{Hom}(\pi_1(M), \mathbb{Z}_2)$, where $\pi_1(M)$ is the fundamental group of M. For even $n = 2d$, the irreducible representation ρ of the Clifford algebra C_n in $\mathscr{H}_F \simeq \mathbb{C}^{2^d}$ (see Section 1.2 of Chapter 7) defines unitary representation R of the spin group $\mathrm{Spin}(n)$ in $\mathscr{H}_F$, which commutes with the parity operator Γ. By definition, the spinor bundle $\mathcal{S}$ on an even-dimensional spin manifold M is a Hermitian vector bundle associated with the principal $\mathrm{Spin}(n)$-bundle $\mathrm{Spin}(M)$ by the unitary representation R. In other words, $\mathcal{S}$ is a complex vector bundle with the transition functions $R(\tau_{\alpha\beta}) : U_\alpha \cap U_\beta \to \mathrm{U}(2^d)$, where $\mathrm{U}(2^d)$ is the group of unitary $2^d \times 2^d$ matrices. Decomposition of vector spaces $\mathscr{H}_F = \mathscr{H}_F^+ \oplus \mathscr{H}_F^-$ defines the decomposition

$$\mathcal{S} = \mathcal{S}_+ \oplus \mathcal{S}_-$$

of the spinor bundle $\mathcal{S}$ into the bundles $\mathcal{S}_+$ and $\mathcal{S}_-$ of positive and negative chirality spinors.

The Dirac operator $\slashed{\partial} : C^\infty(M, \mathcal{S}) \to C^\infty(M, \mathcal{S})$ of the even-dimensional spin manifold M is defined as follows. Let $\nabla^{\mathcal{S}}$ be the connection on the spinor bundle $\mathcal{S}$ induced by the Levi-Civita connection in the tangent bundle TM. Then in the coordinate chart $U \subset M$ with local coordinates $\boldsymbol{x} = (x^1, \ldots, x^n)$ we have

$$(5.5) \qquad \slashed{\partial} = \gamma^\mu(\boldsymbol{x}) \nabla^{\mathcal{S}}_\mu,$$

where $\nabla^{\mathcal{S}}_\mu$ is the covariant derivative with respect to the vector field $\dfrac{\partial}{\partial x^\mu}$ over U, and $\gamma^\mu(\boldsymbol{x})$ are endomorphisms of the spinor bundle $\mathcal{S}$ over U satisfying

$$(5.6) \qquad [\gamma^\mu(\boldsymbol{x}), \gamma^\nu(\boldsymbol{x})]_+ = 2g^{\mu\nu}(\boldsymbol{x}) I,$$

where I is the identity endomorphism. It is easy to show that there is an open covering $M = \bigcup_{\alpha \in A} U_\alpha$ such that $\gamma^\mu(\boldsymbol{x})$ exist over each U_α, and that local expressions (5.5) give rise to a globally defined operator $\not{\partial} : C^\infty(M, \mathcal{S}) \to C^\infty(M, \mathcal{S})$. Equivalently, if $\xi \in C^\infty(M, \mathcal{S})$ is given by $\xi = \{\xi_\alpha\}_{\alpha \in A}$, $\xi_\alpha : U_\alpha \to \mathscr{H}_F$, where $\xi_\alpha = R(\tau_{\alpha\beta})\xi_\beta$ on $U_\alpha \cap U_\beta$, then

$$(5.7) \qquad \not{\partial}_\alpha \xi_\alpha = R(\tau_{\alpha\beta})\not{\partial}_\beta \xi_\beta,$$

where $\not{\partial}_\alpha$ is given by (5.5) for $U = U_\alpha$. We also have

$$\not{\partial} : C^\infty(M, \mathcal{S}_\pm) \to C^\infty(M, \mathcal{S}_\mp).$$

The Hermitian metric $\| \ \|_\mathcal{S}$ in the spinor bundle $\mathcal{S}$ and the Riemannian metric g on M allow us to define the Hilbert space $\mathscr{H}$ of square integrable global sections of $\mathcal{S}$,

$$\mathscr{H} = \left\{ \xi \in \Gamma(M, \mathcal{S}) : \|\xi\|^2 = \int_M \|\xi(x)\|_\mathcal{S}^2 \, d\mu(x) < \infty \right\}.$$

Initially defined on $C^\infty(M, \mathcal{S})$, the Dirac operator extends to the skew-adjoint operator in $\mathscr{H}$, which we continue to denote by $\not{\partial}$. As in the previous example, using the Hilbert space decomposition $\mathscr{H} = \mathscr{H}_+ \oplus \mathscr{H}_-$, we can represent the Dirac operator in the following 2×2 block-matrix form:

$$\not{\partial} = \begin{pmatrix} 0 & -\not{\partial}_+^* \\ \not{\partial}_+ & 0 \end{pmatrix},$$

where $\not{\partial}_+ : \mathscr{H}_+ \to \mathscr{H}_-$ is the chiral Dirac operator on a spin manifold M.

Remark. The Dirac operator from the previous example is the Dirac operator on the spin manifold $\mathbb{R}^n$ with the standard Euclidean metric; the corresponding spinor bundle is a trivial Hermitian vector bundle $\mathbb{R}^n \times \mathscr{H}_F$.

The space of states of a quantum supersymmetric particle on a spin manifold (M, g) is the Hilbert space $\mathscr{H}$ of square-integrable sections of the spinor bundle S over M. The corresponding supercharge operator Q and Hamiltonian operator H are given by the same formulas as in the previous example,

$$Q = \frac{1}{\sqrt{2}}\not{\partial} \quad \text{and} \quad [Q, Q]_+ = 2Q^2 = \not{\partial}^2 = -2mH.$$

The operator

$$2mH = \begin{pmatrix} \not{\partial}_+^* \not{\partial}_+ & 0 \\ 0 & \not{\partial}_+ \not{\partial}_+^* \end{pmatrix}$$

is the Dirac Laplacian on a spin manifold M.

Problem 5.1. Identify the spin group $\mathrm{Spin}(n)$ as a subgroup in the group of invertible elements of the Clifford algebra C_n.

Problem 5.2. Prove relations (5.7).

6. Atiyah-Singer index formula

Let M be an even-dimensional, oriented, compact spin manifold with the Riemannian metric g. Operators $\partial_+^* \partial_+$ and $\partial_+ \partial_+^*$ are, respectively, self-adjoint on Hilbert spaces $\mathcal{H}_+$ and $\mathcal{H}_-$, and have pure point spectra consisting of non-negative eigenvalues of finite multiplicity with the only accumulation point at ∞. In particular, complex vector spaces $\ker \partial_+^* \partial_+ = \ker \partial_+$ and $\ker \partial_+ \partial_+^* = \ker \partial_+^*$ are finite-dimensional, and we define the index $\operatorname{ind} \partial_+$ of the chiral Dirac operator ∂_+ by

$$\operatorname{ind} \partial_+ = \dim \ker \partial_+ - \dim \ker \partial_+^*.$$

We have the following basic result.

Theorem 6.1 (McKean-Singer). *For every $T > 0$,*

$$\operatorname{ind} \partial_+ = \operatorname{Tr}_s e^{T\partial^2} = \operatorname{Tr} e^{-T\partial_+^* \partial_+} - \operatorname{Tr} e^{-T\partial_+ \partial_+^*}.$$

On the other hand, we have seen in Example 5.4 in the last section that $-\partial^2 = 2H$, where H is the Hamiltonian operator of free supersymmetric particle of mass $m = 1$ on the spin manifold M. Thus for every $T > 0$,

$$\operatorname{ind} \partial_+ = \operatorname{Tr}_s e^{-2TH}.$$

In Section 2.2 of Chapter 6 and in Sections 4.2 and 4.3 of Chapter 7 we developed a formalism for expressing traces and supertraces of the evolution operator in Euclidean time by path integrals. Using these results, we can, at a physical level of rigor, represent the supertrace $\operatorname{Tr}_s e^{-2TH}$ by the path integral

$$(6.1) \qquad \operatorname{Tr}_s e^{-2TH} = \int_{\Pi T \mathcal{L}_I(M)} e^{-S_{\mathrm{E}}(\gamma, \theta)} \mathscr{D}\boldsymbol{x}\, \mathscr{D}\boldsymbol{\theta}.$$

Here

$$S_{\mathrm{E}}(\gamma, \theta) = \frac{1}{2} \int_0^{2T} (\|\dot{\gamma}\|^2 + \langle \boldsymbol{\theta}(t), \nabla_{\dot{\gamma}} \boldsymbol{\theta}(t) \rangle) dt$$

is the Euclidean action of a supersymmetric particle on a Riemannian manifold M, obtained from the Lagrangian function (4.11) by replacing time t by the Euclidean time $-it$, and $\mathcal{L}_I(M)$ is the space of free loops on M parametrized by the interval $[0, 2T]$. When[11] $2T = 1$, the integral in (6.1) coincides with integral $\int_{\mathcal{L}(M)} e^{-S}$, considered in Section 2.2. Using Theorem 2.3, we obtain

$$\operatorname{ind} \partial_+ = \operatorname{Tr}_s e^{-H} = (2\pi i)^{-\frac{n}{2}} \int_M \hat{A}(M),$$

which is the celebrated Atiyah-Singer formula for the index of the Dirac operator on a spin manifold!

---

[11]According to a remark in Section 2.2, the same result holds for every $T > 0$.

Remark. Integrating over Grassmann variables in (6.1), we obtain

$$(6.2) \qquad \mathrm{Tr}_s\, e^{-H} = \int_{\mathcal{L}(M)} \mathrm{Pf}(\nabla_{\dot{\gamma}})\, d\mu_W,$$

where $d\mu_W$ is the Wiener measure on the loop space $\mathcal{L}(M)$ associated with the Riemannian metric g on M. It can be shown that when M is a spin manifold, the Pfaffian $\mathrm{Pf}(\nabla_{\dot{\gamma}})$ of a covariant derivative operator along $\gamma \in \mathcal{L}(M)$ is a well-defined function on $\mathcal{L}(M)$. Thus formula (6.2), as opposed to the local computation in the derivation of Theorem 2.3, captures global properties of the manifold M.

Problem 6.1. Prove the McKean-Singer theorem.

Problem 6.2. "Derive" formula (6.1). (*Hint:* See the references in the next section.)

7. Notes and references

The goal of Section 1, besides giving a definition of a supermanifold, was to introduce the isomorphism $\mathcal{A}^{\bullet}(M) \simeq C^{\infty}(\Pi TM)$, emphasized by E. Witten [**Wit82a, Wit82b**], and commonly used in the physics literature. For a systematic introduction to supermanifolds, we refer to the classic texts [**Kos77, Ber87**], as well as to the modern sources [**Man97, DM99, Var04**] and references therein. A detailed exposition of equivariant cohomology and localization in the finite-dimensional case can be found in the monograph [**BGV04**]; our proof of the Berline-Vergne localization theorem in Section 2.1 follows [**Sza00**]. Our presentation of the infinite-dimensional case, based on [**BT95, Sza00**], is at a physical level of rigor. Theorem 2.3 was originally formulated by E. Witten in his famous path integral derivation of the Atiyah-Singer formula for the index of the Dirac operator. Witten's original approach was lucidly presented by Atiyah [**Ati85**]. Our exposition follows [**Wit99b**], with special attention to constant factors; see also [**AG83, Alv95**], as well as [**BT95, Sza00**].

For a detailed explanation of the functor of points and a discussion of classical mechanics on supermanifolds, see [**Fre99**]; our examples of classical systems are taken from [**Alv95**]. Supersymmetry introduced in Sections 4.2–4.3 is called $N = \frac{1}{2}$ supersymmetry and is obtained as a reduction of $N = 1$ supersymmetry; see lectures [**Alv95, DF99b, Fre99, Wit99a, Wit99b**] and [**CFKS08**] for more details and references. In particular, [**Fre99, Wit99a**] describe useful superfield formalism for producing supersymmetric Lagrangians, sketched in Problem 4.3 in Section 4.3. Material in Section 5 is based on [**Alv95**]; we refer the reader to the monograph [**BGV04**] for the invariant definition of Dirac operators on spin manifolds and related topics.

It should be emphasized that our derivation of the Atiyah-Singer formula in Section 6 is purely heuristic. Rigorous justification of this approach using Ito-Malliavin stochastic calculus was made by J.-M. Bismut [**Bis84a, Bis84b, Bis85**]; it turns out that formula (6.2) in Section 6 requires an extra factor: the exponential of the integral along the path of a multiple of a scalar curvature. When $T \to 0$, this rigorous formula for the index coincides with the heuristic expression (6.2). Still, the integration of a differential form of "top degree" over the loop space in [**Bis85**] remains formal, and is a challenging open problem. Finally, we refer the interested reader to [**ASW90**] for a path integral derivation of the H. Weyl character formula, and to [**Wit99b**] for the treatment of the Dirac operator on the loop space.

Bibliography

[AM78]   R. Abraham and J.E. Marsden, *Foundations of mechanics*, Benjamin/Cummings Publishing Co. Inc., Reading, MA, 1978.

[AG93]   N.I. Akhiezer and I.M. Glazman, *Theory of linear operators in Hilbert space*, Dover Publications Inc., New York, 1993.

[AHK76]   S.A. Albeverio and R.J. Høegh-Krohn, *Mathematical theory of Feynman path integrals*, Lecture Notes in Mathematics, vol. 523, Springer-Verlag, Berlin, 1976.

[AE05]   S.T. Ali and M. Engliš, *Quantization methods: a guide for physicists and analysts*, Rev. Math. Phys. **17** (2005), no. 4, 391–490.

[AG83]   L. Alvarez-Gaumé, *Supersymmetry and the Atiyah-Singer index theorem*, Comm. Math. Phys. **90** (1983), no. 2, 161–173.

[ASW90]   O. Alvarez, I.M. Singer, and P. Windey, *Quantum mechanics and the geometry of the Weyl character formula*, Nuclear Phys. B **337** (1990), no. 2, 467–486.

[Alv95]   O. Alvarez, *Lectures on quantum mechanics and the index theorem*, Geometry and quantum field theory (Park City, UT, 1991), IAS/Park City Math. Ser., vol. 1, Amer. Math. Soc., Providence, RI, 1995, pp. 271–322.

[Apo76]   T.M. Apostol, *Introduction to analytic number theory*, Undergraduate Texts in Mathematics, Springer-Verlag, New York, 1976.

[Arn89]   V.I. Arnol'd, *Mathematical methods of classical mechanics*, Graduate Texts in Mathematics, vol. 60, Springer-Verlag, New York, 1989.

[AG90]   V.I. Arnol'd and A.B. Givental', *Symplectic geometry*, Dynamical systems IV, Encyclopaedia of Mathematical Sciences, vol. 4, Springer-Verlag, Berlin, 1990, pp. 1–136.

[AKN97]   V.I. Arnol'd, V.V. Kozlov, and A.I. Neishtadt, *Mathematical aspects of classical and celestial mechanics*, Encyclopaedia of Mathematical Sciences, vol. 3, Springer-Verlag, Berlin, 1997.

[Ati85]   M.F. Atiyah, *Circular symmetry and stationary-phase approximation*, Astérisque (1985), no. 131, 43–59, Colloquium in honor of Laurent Schwartz, Vol. 1 (Palaiseau, 1983).

[BI66a]   M. Bander and C. Itzykson, *Group theory and the hydrogen atom. Part I*, Rev. Mod. Phys. **38** (1966), 330–345.

[BI66b]   ———, *Group theory and the hydrogen atom. Part II*, Rev. Mod. Phys. **38** (1966), 346–358.

[Bar61]   V. Bargmann, *On a Hilbert space of analytic functions*, Comm. Pure Appl. Math. **3** (1961), 215–228.

[BR86]   A.O. Barut and R. Rączka, *Theory of group representations and applications*, second ed., World Scientific Publishing Co., Singapore, 1986.

[BW97]   S. Bates and A. Weinstein, *Lectures on the geometry of quantization*, Berkeley Mathematics Lecture Notes, vol. 8, Amer. Math. Soc., Providence, RI, 1997.

[BFF$^+$78a]   F. Bayen, M. Flato, C. Fronsdal, A. Lichnerowicz, and D. Sternheimer, *Deformation theory and quantization. I. Deformations of symplectic structures*, Ann. Physics **111** (1978), no. 1, 61–110.

[BFF$^+$78b]   ———, *Deformation theory and quantization. II. Physical applications*, Ann. Physics **111** (1978), no. 1, 111–151.

[Bel87]   J.S. Bell, *Speakable and unspeakable in quantum mechanics*, Collected papers on quantum philosophy, Cambridge University Press, Cambridge, 1987.

[Ber61]   F.A. Berezin, *Canonical operator transformation in representation of secondary quantization*, Soviet Physics Dokl. **6** (1961), 212–215 (in Russian).

[Ber66]   ———, *The method of second quantization*, Pure and Applied Physics, vol. 24, Academic Press, New York, 1966.

[Ber71a]   ———, *Non-Wiener path integrals*, Teoret. Mat. Fiz. **6** (1971), no. 2, 194–212 (in Russian), English translation in Theoret. and Math. Phys. **6** (1971), 141–155.

[Ber71b]   ———, *Wick and anti-Wick symbols of operators*, Mat. Sb. (N.S.) **86(128)** (1971), 578–610 (in Russian), English translation in Math. USSR Sb. **15** (1971), 577–606.

[Ber74]   ———, *Quantization*, Izv. Akad. Nauk SSSR Ser. Mat. **38** (1974), 1116–1175 (in Russian), English translation in Math. USSR-Izv. **78** (1974), 1109–1165.

[Ber87]   ———, *Introduction to superanalysis*, Mathematical Physics and Applied Mathematics, vol. 9, D. Reidel Publishing Co., Dordrecht, 1987.

[BS91]   F.A. Berezin and M.A. Shubin, *The Schrödinger equation*, Mathematics and its Applications (Soviet Series), vol. 66, Kluwer Academic Publishers Group, Dordrecht, 1991.

[BGV04]   N. Berline, E. Getzler, and M. Vergne, *Heat kernels and Dirac operators*, Corrected reprint of the 1992 original, Grundlehren Text Editions, Springer-Verlag, Berlin, 2004.

[BS87]   M.Sh. Birman and M.Z. Solomjak, *Spectral theory of selfadjoint operators in Hilbert space*, Mathematics and its Applications (Soviet Series), D. Reidel Publishing Co., Dordrecht, 1987.

[Bis84a]   J.-M. Bismut, *The Atiyah-Singer theorems: a probabilistic approach. I. The index theorem*, J. Funct. Anal. **57** (1984), no. 1, 56–99.

[Bis84b]   ———, *The Atiyah-Singer theorems: a probabilistic approach. II. The Lefschetz fixed point formulas*, J. Funct. Anal. **57** (1984), no. 3, 329–348.

[Bis85]   ———, *Index theorem and equivariant cohomology on the loop space*, Comm. Math. Phys. **98** (1985), no. 2, 213–237.

[BT95]   M. Blau and G. Thompson, *Localization and diagonalization: a review of functional integral techniques for low-dimensional gauge theories and topological field theories*, J. Math. Phys. **36** (1995), no. 5, 2192–2236.

[Bry95]   R.L. Bryant, *An introduction to Lie groups and symplectic geometry*, Geometry and quantum field theory (Park City, UT, 1991), IAS/Park City Math. Ser., vol. 1, Amer. Math. Soc., Providence, RI, 1995, pp. 5–181.

[BFK91]   D. Burghelea, L. Friedlander, and T. Kappeler, *On the determinant of elliptic differential and finite difference operators in vector bundles over S^1*, Comm. Math. Phys. **138** (1991), no. 1, 1–18.

[BFK95]   ______, *On the determinant of elliptic boundary value problems on a line segment*, Proc. Amer. Math. Soc. **123** (1995), no. 10, 3027–3038.

[BF60]   V.S. Buslaev and L.D. Faddeev, *Formulas for traces for a singular Sturm-Liouville differential operator*, Dokl. Akad. Nauk. SSSR **132** (1960), 13–16 (in Russian), English translation in Soviet Math. Dokl. **1** (1960), 451–454.

[Cam63]   R.H. Cameron, *The Ilstow and Feynman integrals*, J. Analyse Math. **10** (1962/1963), 287–361.

[Cra83]   M. Crampin, *Tangent bundle geometry for Lagrangian dynamics*, J. Phys. A **16** (1983), no. 16, 3755–3772.

[CFKS08]   H.L. Cycon, R.G. Froese, W. Kirsch, and B. Simon, *Schrödinger operators with application to quantum mechanics and global geometry*, Corrected and extended 2nd printing, Texts and Monographs in Physics, Springer-Verlag, Berlin, 2008.

[Dav76]   A.S. Davydov, *Quantum mechanics*, International Series in Natural Philosophy, vol. 1, Pergamon Press, Oxford, 1976.

[DEF+99]   P. Deligne, P. Etingof, D.S. Freed, L.C. Jeffrey, D. Kazhdan, J.W. Morgan, D.R. Morrison, and E. Witten (eds.), *Quantum fields and strings: a course for mathematicians. Vol. 1, 2*, Amer. Math. Soc., Providence, RI, 1999, Material from the Special Year on Quantum Field Theory held at the Institute for Advanced Study, Princeton, NJ, 1996–1997.

[DF99a]   P. Deligne and D.S. Freed, *Classical field theory*, Quantum fields and strings: a course for mathematicians, Vol. 1, 2 (Princeton, NJ, 1996/1997), Amer. Math. Soc., Providence, RI, 1999, pp. 137–225.

[DF99b]   ______, *Supersolutions*, Quantum fields and strings: a course for mathematicians, Vol. 1, 2 (Princeton, NJ, 1996/1997), Amer. Math. Soc., Providence, RI, 1999, pp. 227–355.

[DM99]   P. Deligne and J.W. Morgan, *Notes on supersymmetry (following J. Bernstein)*, Quantum fields and strings: a course for mathematicians, Vol. 1, 2 (Princeton, NJ, 1996/1997), Amer. Math. Soc., Providence, RI, 1999, pp. 41–97.

[Dik58]   L.A. Dikiĭ, *Trace formulas for Sturm-Liouville differential operators*, Uspekhi Mat. Nauk (N.S.) **13** (1958), no. 3(81), 111–143 (in Russian), English translation in Amer. Math. Soc. Transl. (2) **18** (1961), 81–115.

[Dir47]   P.A.M. Dirac, *The principles of quantum mechanics*, Clarendon Press, Oxford, 1947.

[DR01]   W. Dittrich and M. Reuter, *Classical and quantum dynamics. From classical paths to path integrals*, third ed., Advanced Texts in Physics, Springer-Verlag, Berlin, 2001.

[Dri83]   V.G. Drinfel′d, *Constant quasiclassical solutions of the Yang-Baxter quantum equation*, Dokl. Akad. Nauk SSSR **273** (1983), no. 3, 531–535 (in Russian), English translation in Soviet Math. Dokl. **28** (1983), 667–671.

[Dri86]   ———, *Quantum groups*, Zap. Nauchn. Sem. Leningrad. Otdel. Mat. Inst. Steklov. (LOMI) **155** (1986), 18–49 (in Russian), English translation in J. Soviet Math. **41** (1988), no. 2, 898–915.

[Dri87]   ———, *Quantum groups*, Proceedings of the International Congress of Mathematicians, Vol. 1, 2 (Berkeley, Calif., 1986) (Providence, RI), Amer. Math. Soc., 1987, pp. 798–820.

[DFN84]   B.A. Dubrovin, A.T. Fomenko, and S.P. Novikov, *Modern geometry — methods and applications. Part I*, Graduate Texts in Mathematics, vol. 93, Springer-Verlag, New York, 1984.

[DFN85]   ———, *Modern geometry — methods and applications. Part II*, Graduate Texts in Mathematics, vol. 104, Springer-Verlag, New York, 1985.

[Dyn98]   A. Dynin, *A rigorous path integral construction in any dimension*, Lett. Math. Phys. **44** (1998), no. 4, 317–329.

[Enr01]   B. Enriquez, *Quantization of Lie bialgebras and shuffle algebras of Lie algebras*, Selecta Math. (N.S.) **7** (2001), no. 3, 321–407.

[Erd56]   A. Erdélyi, *Asymptotic expansions*, Dover Publications Inc., New York, 1956.

[EK96]   P. Etingof and D. Kazhdan, *Quantization of Lie bialgebras. I*, Selecta Math. (N.S.) **2** (1996), no. 1, 1–41.

[Fad57]   L.D. Faddeev, *An expression for the trace of the difference between two singular differential operators of the Sturm-Liouville type*, Dokl. Akad. Nauk SSSR (N.S.) **115** (1957), 878–881 (in Russian).

[Fad59]   ———, *The inverse problem in the quantum theory of scattering*, Uspekhi Mat. Nauk **14** (1959), no. 4 (88), 57–119 (in Russian), English translation in J. Math. Phys. **4** (1963), 72–104.

[Fad64]   ———, *Properties of the S-matrix of the one-dimensional Schrödinger equation*, Trudy Mat. Inst. Steklov. **73** (1964), 314–336 (in Russian), English translation in Amer. Math. Soc. Transl. (2), Vol. 65: Nine papers on partial differential equations and functional analysis (1967), 139–166.

[Fad74]   ———, *The inverse problem in the quantum theory of scattering. II*, Current problems in mathematics, Vol. 3, Akad. Nauk SSSR Vsesojuz. Inst. Naučn. i Tehn. Informacii, Moscow, 1974, pp. 93–180 (in Russian), English translation in J. Soviet Math. **5** (1976), no. 3, 334–396.

[Fad76]   ———, *Course 1. Introduction to functional methods*, Méthodes en théorie des champs/Methods in field theory (École d'Été Phys. Théor., Session XXVIII, Les Houches, 1975), North-Holland, Amsterdam, 1976, pp. 1–40.

[Fad98]   ———, *A mathematician's view of the development of physics*, Les relations entre les mathématiques et la physique théorique, Inst. Hautes Études Sci., Bures, 1998, pp. 73–79.

[Fad99]   ———, *Elementary introduction to quantum field theory*, Quantum fields and strings: a course for mathematicians, Vol. 1, 2 (Princeton, NJ, 1996/1997), Amer. Math. Soc., Providence, RI, 1999, pp. 513–550.

[FY80]   L.D. Faddeev and O.A. Yakubovskiĭ, *Lectures on quantum mechanics for mathematics students*, Leningrad University Publishers, Leningrad, 1980 (in Russian).

[FS91]   L.D. Faddeev and A.A. Slavnov, *Gauge fields. Introduction to quantum theory*, Frontiers in Physics, vol. 83, Addison-Wesley Publishing Company, Redwood City, CA, 1991.

[FT07]   L.D. Faddeev and L.A. Takhtajan, *Hamiltonian methods in the theory of solitons*, Reprint of 1987 English edition, Classics in Mathematics, Springer-Verlag, New York, 2007.

[Fey48]   R.P. Feynman, *Space-time approach to non-relativistic quantum mechanics*, Rev. Modern Physics **20** (1948), 367–387.

[Fey51]   ______, *An operator calculus having applications in quantum electrodynamics*, Physical Rev. (2) **84** (1951), 108–128.

[FH65]   R.P. Feynman and A.R. Hibbs, *Quantum mechanics and path integrals*, Mc-Graw Hill, New York, 1965.

[Fla82]   M. Flato, *Deformation view of physical theories*, Czechoslovak J. Phys. **B32** (1982), 472–475.

[Foc32]   V.A. Fock, *Konfigurationsraum und zweite Quantelung*, Z. Phys. **75** (1932), no. 9-10, 622–647.

[Foc78]   ______, *Fundamentals of quantum mechanics*, Mir Publishers, Moscow, 1978.

[Fre99]   D.S. Freed, *Five lectures on supersymmetry*, Amer. Math. Soc., Providence, RI, 1999.

[Fuj80]   D. Fujiwara, *Remarks on convergence of the Feynman path integrals*, Duke Math. J. **47** (1980), no. 3, 559–600.

[FH91]   W. Fulton and J. Harris, *Representation theory*, Graduate Texts in Mathematics, vol. 129, Springer-Verlag, New York, 1991.

[GL53]   I.M. Gel′fand and B.M. Levitan, *On a simple identity for the characteristic values of a differential operator of the second order*, Dokl. Akad. Nauk SSSR (N.S.) **88** (1953), 593–596 (in Russian).

[GY56]   I.M. Gel′fand and A.M. Yaglom, *Integration in function spaces and its application to quantum physics*, Uspekhi Mat. Nauk (N.S.) **11** (1956), no. 1(67), 77–114, English translation in J. Math. Phys. **1** (1960), 48–69.

[Ger50]   Ya.L. Geronimus, *Teoriya ortogonal′nyh mnogočlenov*, Gosudarstv. Izdat. Tehn.-Teor. Lit., Moscow-Leningrad, 1950 (in Russian).

[Gil95]   P.B. Gilkey, *Invariance theory, the heat equation, and the Atiyah-Singer index theorem*, second ed., CRC Press, Boca Raton, FL, 1995.

[God69]   C. Godbillon, *Géometrie différentielle et mécanique analytique*, Hermann, Paris, 1969.

[GK69]   I.C. Gohberg and M.G. Kreĭn, *Introduction to the theory of linear nonselfadjoint operators*, Translations of Mathematical Monographs, Vol. 18, Amer. Math. Soc., Providence, RI, 1969.

[Gol80]   H. Goldstein, *Classical mechanics*, Addison Wesley, 1980.

[GW98]   R. Goodman and N.R. Wallach, *Representations and invariants of the classical groups*, Encyclopedia of Mathematics and its Applications, vol. 68, Cambridge University Press, Cambridge, 1998.

[GS77]   V. Guillemin and S. Sternberg, *Geometric asymptotics*, Mathematical Surveys, No. 14, Amer. Math. Soc., Providence, RI, 1977.

[GS03]   S.J. Gustafson and I.M. Sigal, *Mathematical concepts of quantum mechanics*, Universitext, Springer-Verlag, Berlin, 2003.

[HS96]	P.D. Hislop and I.M. Sigal, *Introduction to spectral theory: With applications to Schrödinger operators*, Applied Mathematical Sciences, vol. 113, Springer-Verlag, New York, 1996.
[IM74]	K. Itô and H.P. McKean, Jr., *Diffusion processes and their sample paths*, Second printing, corrected, Die Grundlehren der mathematischen Wissenschaften, Band 125, Springer-Verlag, Berlin, 1974.
[IZ80]	C. Itzykson and J.B. Zuber, *Quantum field theory*, International Series in Pure and Applied Physics, McGraw-Hill International Book Co., New York, 1980.
[Jim85]	M. Jimbo, *A q-difference analogue of $U(\mathfrak{g})$ and the Yang-Baxter equation*, Lett. Math. Phys. **10** (1985), no. 1, 63–69.
[JW28]	P. Jordan and E. Wigner, *Über das Paulische Äquivalenzverbot*, Z. Phys. **47** (1928), 631–658.
[Kac59]	M. Kac, *Probability and related topics in physical sciences*, Lectures in Applied Mathematics. Proceedings of the Summer Seminar, Boulder, CO, 1957, vol. 1, Interscience Publishers, London-New York, 1959.
[Kac80]	______, *Integration in function spaces and some of its applications*, Lezioni Fermiane, Accademia Nazionale dei Lincei, Pisa, 1980.
[Kaz99]	D. Kazhdan, *Introduction to QFT*, Quantum fields and strings: a course for mathematicians, Vol. 1, 2 (Princeton, NJ, 1996/1997), Amer. Math. Soc., Providence, RI, 1999, pp. 377–418.
[Kho07]	D. Khoshnevisan, *Probability*, Graduate Studies in Mathematics, vol. 80, Amer. Math. Soc., Providence, RI, 2007.
[Kir76]	A.A. Kirillov, *Elements of the theory of representations*, Grundlehren der Mathematischen Wissenschaften, Band 220, Springer-Verlag, Berlin, 1976.
[Kir90]	______, *Geometric quantization*, Dynamical systems IV, Encyclopaedia of Mathematical Sciences, vol. 4, Springer-Verlag, Berlin, 1990, pp. 137–172.
[Kir04]	______, *Lectures on the orbit method*, Graduate Studies in Mathematics, vol. 64, Amer. Math. Soc., Providence, RI, 2004.
[KSV02]	A.Yu. Kitaev, A.H. Shen, and M.N. Vyalyi, *Classical and quantum computation*, Graduate Studies in Mathematics, vol. 47, Amer. Math. Soc., Providence, RI, 2002.
[Kon03]	M. Kontsevich, *Deformation quantization of Poisson manifolds*, Lett. Math. Phys. **66** (2003), no. 3, 157–216.
[Kos77]	B. Kostant, *Graded manifolds, graded Lie theory, and prequantization*, Differential geometrical methods in mathematical physics (Proc. Sympos., Univ. Bonn, Bonn, 1975), Lecture Notes in Math., vol. 570, Springer-Verlag, Berlin, 1977.
[Kre62]	M.G. Kreĭn, *On perturbation determinants and a trace formula for unitary and self-adjoint operators*, Dokl. Akad. Nauk SSSR **144** (1962), 268–271 (in Russian), English translation in Soviet Math. Dokl. **3** (1962), 707–710.
[LL58]	L.D. Landau and E.M. Lifshitz, *Quantum mechanics: non-relativistic theory. Course of Theoretical Physics, Vol. 3*, Pergamon Press Ltd., London-Paris, 1958.
[LL76]	L.D. Landau and E.M. Lifshitz, *Mechanics. Course of theoretical physics. Vol. 1*, Pergamon Press, Oxford, 1976.
[Lan87]	S. Lang, *Elliptic functions*, second ed., Springer-Verlag, New York, 1987.

[Ler81]   J. Leray, *Lagrangian analysis and quantum mechanics*, A mathematical structure related to asymptotic expansions and the Maslov index, MIT Press, Cambridge, MA, 1981.

[LS91]   B.M. Levitan and I.S. Sargsjan, *Sturm-Liouville and Dirac operators*, Mathematics and its Applications (Soviet Series), vol. 59, Kluwer Academic Publishers Group, Dordrecht, 1991.

[LV80]   G. Lion and M. Vergne, *The Weil representation, Maslov index and theta series*, Progress in Mathematics, vol. 6, Birkhäuser, Boston, MA, 1980.

[Loè77]   M. Loève, *Probability theory. I*, fourth ed., Graduate Texts in Mathematics, vol. 45, Springer-Verlag, New York, 1977.

[Loè78]   ______, *Probability theory. II*, fourth ed., Graduate Texts in Mathematics, vol. 46, Springer-Verlag, New York, 1978.

[Mac04]   G.W. Mackey, *Mathematical foundations of quantum mechanics*, Reprint of the 1963 original, Dover Publications Inc., Mineola, NY, 2004.

[Man97]   Y.I. Manin, *Gauge field theory and complex geometry*, second ed., Grundlehren der Mathematischen Wissenschaften, vol. 289, Springer-Verlag, Berlin, 1997.

[Mar86]   V.A. Marchenko, *Sturm-Liouville operators and applications*, Operator Theory: Advances and Applications, vol. 22, Birkhäuser Verlag, Basel, 1986.

[Mar59a]   J.L. Martin, *The Feynman principle for a Fermi system*, Proc. Roy. Soc. Ser. A **251** (1959), no. 1267, 543–549.

[Mar59b]   ______, *Generalized classical dynamics, and the 'classical analogue' of a Fermi oscillator*, Proc. Roy. Soc. Ser. A **251** (1959), no. 1267, 536–542.

[MF81]   V.P. Maslov and M.V. Fedoriuk, *Semiclassical approximation in quantum mechanics*, Mathematical Physics and Applied Mathematics, vol. 7, D. Reidel Publishing Co., Dordrecht, 1981.

[Mes99]   A. Messiah, *Quantum mechanics*, Dover Publications Inc., Mineola, NY, 1999.

[MP49]   S. Minakshisundaram and Å. Pleijel, *Some properties of the eigenfunctions of the Laplace-operator on Riemannian manifolds*, Canadian J. Math. **1** (1949), 242–256.

[Mon52]   E.W. Montroll, *Markoff chains, Wiener integrals, and quantum theory*, Comm. Pure Appl. Math. **5** (1952), 415–453.

[Nel59]   E. Nelson, *Analytic vectors*, Ann. of Math. (2) **70** (1959), 572–615.

[Nel64]   ______, *Feynman integrals and the Schrödinger equation*, J. Math. Phys. **5** (1964), 332–343.

[New02]   R.G. Newton, *Scattering theory of waves and particles*, Reprint of the 1982 second edition, with list of errata prepared by the author, Dover Publications Inc., Mineola, NY, 2002.

[Olv97]   F.W.J. Olver, *Asymptotics and special functions*, Reprint of the 1974 original, AKP Classics, A K Peters Ltd., Wellesley, MA, 1997.

[PW35]   L. Pauling and E.B. Wilson, *Introduction to quantum mechanics. With applications to chemistry*, McGraw-Hill Book Company, New York and London, 1935.

[Rab95]   J.M. Rabin, *Introduction to quantum field theory for mathematicians*, Geometry and quantum field theory (Park City, UT, 1991), IAS/Park City Math. Ser., vol. 1, Amer. Math. Soc., Providence, RI, 1995, pp. 183–269.

[RS71]   D.B. Ray and I.M. Singer, *R-torsion and the Laplacian on Riemannian manifolds*, Advances in Math. **7** (1971), 145–210.

[RS80]   M. Reed and B. Simon, *Methods of modern mathematical physics. I*, Academic Press, New York, 1980.

[RS75]   ——, *Methods of modern mathematical physics. II. Fourier analysis, self-adjointness*, Academic Press, New York, 1975.

[RS79]   ——, *Methods of modern mathematical physics. III. Scattering theory*, Academic Press, New York, 1979.

[RS78]   ——, *Methods of modern mathematical physics. IV. Analysis of operators*, Academic Press, New York, 1978.

[RTF89]   N.Yu. Reshetikhin, L.A. Takhtadzhyan, and L.D. Faddeev, *Quantization of Lie groups and Lie algebras*, Algebra i Analiz **1** (1989), no. 1, 178–206 (in Russian), English translation in Leningrad Math. J. **1** (1990), 193–225.

[Ros04]   J. Rosenberg, *A selective history of the Stone-von Neumann theorem*, Operator algebras, quantization, and noncommutative geometry, Contemp. Math., vol. 365, Amer. Math. Soc., Providence, RI, 2004, pp. 331–353.

[RSS94]   G.V. Rozenblyum, M.Z. Solomyak, and M.A. Shubin, *Spectral theory of differential operators*, Partial differential equations. VII, Encyclopaedia of Mathematical Sciences, vol. 64, Springer-Verlag, Berlin, 1994.

[Rud87]   W. Rudin, *Real and complex analysis*, third ed., McGraw-Hill Book Co., New York, 1987.

[Sak94]   J.J. Sakurai, *Modern quantum mechanics*, Addison-Wesley Publishing Company, 1994.

[See67]   R.T. Seeley, *Complex powers of an elliptic operator*, Singular Integrals (Proc. Sympos. Pure Math., Chicago, IL, 1966), Amer. Math. Soc., Providence, RI, 1967, pp. 288–307.

[STS85]   M.A. Semenov-Tian-Shansky, *Dressing transformations and Poisson group actions*, Publ. Res. Inst. Math. Sci. **21** (1985), no. 6, 1237–1260.

[SW76]   D.J. Simms and N.M.J. Woodhouse, *Lectures in geometric quantization*, Lecture Notes in Physics, 53, Springer-Verlag, Berlin, 1976.

[Ste83]   S. Sternberg, *Lectures on differential geometry*, second ed., Chelsea Publishing Co., New York, 1983.

[Str05]   F. Strocchi, *An introduction to the mathematical structure of quantum mechanics. A short course for mathematicians*, Advanced Series in Mathematical Physics, vol. 27, World Sci. Publishing, London-Singapore, 2005.

[Sza00]   R.J. Szabo, *Equivariant cohomology and localization of path integrals*, Lecture Notes in Physics. New Series: Monographs, vol. 63, Springer-Verlag, Berlin, 2000.

[Sze75]   G. Szegő, *Orthogonal polynomials*, Amer. Math. Soc., Providence, RI, 1975.

[Tak90]   L. A. Takhtajan, *Lectures on quantum groups*, Introduction to quantum group and integrable massive models of quantum field theory (Nankai, 1989), Nankai Lectures Math. Phys., World Sci. Publishing, River Edge, NJ, 1990, pp. 69–197.

[Tob56]   W. Tobocman, *Transition amplitudes as sums over histories*, Nuovo Cimento (10) **3** (1956), 1213–1229.

[Var04]   V.S. Varadarajan, *Supersymmetry for mathematicians: an introduction*, Courant Lecture Notes in Mathematics, vol. 11, New York University Courant Institute of Mathematical Sciences, New York, 2004.

[Vil68]   N.Ja. Vilenkin, *Special functions and the theory of group representations*, Translations of Mathematical Monographs, Vol. 22, Amer. Math. Soc., Providence, RI, 1968.

[vN31]   J. von Neumann, *Die Eindeutigkeit der Schrödingershen Operatoren*, Mathematische Annalen **104** (1931), 570–578.

[vN96]   ______, *Mathematical foundations of quantum mechanics*, Princeton Landmarks in Mathematics, Princeton University Press, Princeton, NJ, 1996.

[Vor05]   A.A. Voronov, *Notes on universal algebra*, Graphs and patterns in mathematics and theoretical physics (Stony Brook, NY, 2001), Proc. Sympos. Pure Math., vol. 73, Amer. Math. Soc., Providence, RI, 2005, pp. 81–103.

[Wey50]   H. Weyl, *Theory of groups and quantum mechanics*, Dover Publications, New York, 1950.

[Wig59]   E.P. Wigner, *Group theory: And its application to the quantum mechanics of atomic spectra*, Pure and Applied Physics, Vol. 5, Academic Press, New York, 1959.

[Wit82a]   E. Witten, *Constraints on supersymmetry breaking*, Nuclear Phys. B **202** (1982), no. 2, 253–316.

[Wit82b]   ______, *Supersymmetry and Morse theory*, J. Differential Geom. **17** (1982), no. 4, 661–692 (1983).

[Wit99a]   ______, *Homework*, Quantum fields and strings: a course for mathematicians, Vol. 1, 2 (Princeton, NJ, 1996/1997), Amer. Math. Soc., Providence, RI, 1999, pp. 609–717.

[Wit99b]   ______, *Index of Dirac operators*, Quantum fields and strings: a course for mathematicians, Vol. 1, 2 (Princeton, NJ, 1996/1997), Amer. Math. Soc., Providence, RI, 1999, pp. 475–511.

[Woo92]   N.M.J. Woodhouse, *Geometric quantization*, second ed., Oxford Mathematical Monographs, The Clarendon Press Oxford University Press, New York, 1992.

[Yaf92]   D.R. Yafaev, *Mathematical scattering theory. General theory*, Translations of Mathematical Monographs, vol. 105, Amer. Math. Soc., Providence, RI, 1992.

[YI73]   K. Yano and S. Ishihara, *Tangent and cotangent bundles: differential geometry*, Pure and Applied Mathematics, No. 16, Marcel Dekker Inc., New York, 1973.

Index

For a complete list of titles in this series, visit the
AMS Bookstore at **www.ams.org/bookstore/**.